Genetic Engineering

AN INTRODUCTION TO GENE ANALYSIS

AND EXPLOITATION IN EUKARYOTES

S.M. Kingsman MA, PhD
Lecturer in Biochemistry
and Fellow of Trinity College
Oxford

A.J. Kingsman MA, PhD
Lecturer in Biochemistry
and Fellow of St Catherine's College
Oxford
and Associate Director of Research
British Biotechnology Ltd
Oxford

OXFORD

BLACKWELL SCIENTIFIC PUBLICATIONS

LONDON EDINBURGH BOSTON

MELBOURNE PARIS BERLIN VIENNA

First published 1988
Reprinted 1990

Set by Setrite Typesetters Ltd
Hong Kong
Printed and bound in Great Britain
at the Alden Press, Oxford

DISTRIBUTORS
Marston Book Services Ltd
PO Box 87
Oxford OX2 0DT
(*Orders*: Tel: 0865 791155
         Fax: 0865 791927
         Telex: 837515)

USA
Publishers' Business Services
PO Box 447
Brookline Village
Massachusetts 02147
(*Orders*: Tel: (617) 524–7678)

Canada
Oxford University Press
70 Wynford Drive
Don Mills
Ontario M3C 1J9
(*Orders*: Tel: (416) 441–2941)

Australia
Blackwell Scientific Publications
(Australia) Pty Ltd
54 University Street
Carlton, Victoria 3053
(*Orders*: Tel: (03) 347-0300)

British Library
Cataloguing in Publication Data

Kingsman, S.M.
  Genetic Engineering
  1. Organisms. Eukaryotic cells. Genetic
  engineering
  I. Title.   II. Kingsman, A.J.
  574.87'322
  ISBN 0–632–01519–5
  ISBN 0–632–01521–7 Pbk

Library of Congress
Cataloging in Publication Data

Kingsman, S.M.
  Genetic Engineering
  Bibliography: p.
  Includes index.
  1. Genetic engineering.   2. Eukaryotic
  cells.
  I. Kingsman, A.J.   II. Title.
  QH442. K53 1988   574.87'3282
  87–36755
  ISBN 0–632–01519–5
  ISBN 0–632–01521–7 (pbk.)

Genetic Engineering

Contents

PART II: GENE TRANSFER SYSTEMS

PART III: THE USE AND EXPLOITATION OF GENE TRANSFER TECHNOLOGY

CONTENTS

List of Abbreviations

Restriction enzyme cleavage sites

H	Hind III
R	Eco RI
Bg	Bgl II
Bc	Bcl I
Pv	Pvu II
Rv	Eco RV
B	Bam HI
S	Sal I
P	Pst I
K	Kpn I
Xb	Xba I
Sm	Sma I
Mb	Mbo I
Sau	Sau 3a
Ss	Sst I
Ac	Acc I
C	Cla I
Hp	Hpa I
Sp	Sph I
Hc	Hinc II
(s)	synthetic restriction site

Nucleosides

A	Adenosine
G	Guanosine
C	Cytidine
T	Thymidine
U	Uridine
Y	Pyrimidine nucleoside
R	Purine nucleoside
X	Any nucleoside

Single letter amino acid code

A	Alanine
C	Cysteine
D	Aspartic acid
E	Glutamic acid
F	Phenyl alanine
G	Glycine
H	Histidine
I	Isoleucine
K	Lysine
L	Leucine
M	Methionine
N	Asparagine
P	Proline
Q	Glutamine
R	Arginine
S	Serine
T	Threonine
V	Valine
W	Tryptophan
Y	Tyrosine

Miscellaneous

pA	poly adenylation site
$(A)_n$	poly A tail
5'SS	5' splice site or splice donor
3'SS	3' splice site or splice acceptor
T	TATA box
I	mRNA initiation site
En	enhancer
UAS	upstream activator sequence
P	promoter
5'UTR	5' untranslated region
3'UTR	3' untranslated region
ORI	replication origin
ARS	autonomously replicating sequence
CEN	centromere
TEL	telomere
Py	polyoma
pBR	pBR 322
$2\mu m$	yeast 2 micron circle
bp	base pairs
kbp	kilobase pairs
nt	nucleotide
kD	kilo Dalton
LTR	long terminal repeat
X-GAL	5-bromo-4-chloro-3-indolyl-β-D-galactoside

LIST OF ABBREVIATIONS

Preface

The ability to isolate individual genes from complex genomes by using recombinant DNA techniques has had a profound influence upon the way scientists probe the mysteries of biology. Whole areas of research have been transformed by recombinant DNA technology and we are in a period of the most rapid accumulation of knowledge that biology has ever experienced. In addition to providing a new research tool, recombinant DNA techniques have led to an amazing extension of the science of biotechnology. We are seeing the genes and genomes of a wide range of different organisms being manipulated for the benefit of man. The first stage in this revolution was marked by the development of techniques for cloning genes. The second stage is the development of ways of modifying those genes and returning them to living cells; this is often called genetic engineering.

There are a number of excellent texts that describe the various ways of cloning genes but we thought that it might be useful to provide a text that described the kinds of studies that are possible once a gene has been cloned. We want to show you the ways of discovering how genes work and how to make genes work for you. This is not a methods and recipe book but a concepts book. We describe the theory behind key techniques and approaches in the context of real experiments. We have restricted the book to a discussion of eukaryotic systems because this is where some of the most spectacular advances are being made both in fundamental biology and in biotechnology.

The key technique in genetic engineering is called *gene transfer*. This encompasses the variety of methods for returning cloned genes to cells. Part I contains two chapters that are gentle introductions, first to the eukaryotic genome and then to general principles of gene transfer techniques. Part II contains four chapters devoted to explaining the different methods of gene transfer and genetic analysis that are used for the major experimental systems. These are: microbial cells, cultured animal cells, animals and plants. Part III describes the various uses of cloned genes. Chapters 7, 8, 9 and 10 show how genetic engineering is being applied to problems in the areas of gene isolation, gene expression, genome organisation and replication and protein structure and function. Chapters 11 and 12 describe how genes can be exploited to increase understanding and treatment of human diseases and how they may be manipulated for industrial, agricultural and pharmaceutical applications of biotechnology. Throughout the text we have highlighted key topics when they are first mentioned. We have not aimed to provide comprehensive treatments of each area of biology and biotechnology but by illustrating the basic concepts, questions and approaches we should equip you to fill in any gaps. It is also obvious that in a such a fast-moving field the most recent observations will not be included in this book and it is possible that radically new ideas may emerge as this goes to press. In most areas, however, any changes will be embellishments of basic concepts. We have tried to use examples that either illustrate a technique particularly well or that were historically important for establishing principles. We apologise to many colleagues whose work may not have been cited. This is not a research level review.

Much of this book is based on a course that we teach to third-year biochemistry students at Oxford and it should also be useful for the other biological and medical sciences. We hope that it will encourage researchers to identify new approaches to solve their research problems. The biotechnologist should also find it useful to see the range and limitations of different eukaryotic systems that can now be manipulated. We are assuming only a basic knowledge of biochemistry, cloning techniques and a little genetics.

We are indebted to the following colleagues for their helpful comments on individual chapters: John Draper, Anne Glover, Chris Graham, Frank Grosveld, Ron Laskey, Sandy MacLeod, Keith McCullagh, Tim Mohun and Steve Oliver. Thank you to Simon Rallison and Nick Parsons at Blackwell Scientific Publications for coordinating the effort. Also a huge thank you to Jackie Hewitt. Her commitment to the project and her skill and enthusiasm at the word-processor ensured that the organisation and presentation of the first draft proceeded with maximum efficiency. Finally we welcome all constructive comments on this first edition and if you spot mistakes and omissions please let us know.

S.M.K.
A.J.K.

I   Basic Concepts

A Guide to the Eukaryotic Genome

Introduction

A full appreciation of the design of gene transfer systems and the aims of gene manipulation experiments requires a good working knowledge of the molecular genetics of eukaryotes. This chapter provides, therefore, a 'refresher course' on the organisation of the eukaryotic genome. It is necessarily superficial and can be skipped but, if you don't know the difference between pol I, pol II and pol III genes, or are not sure about SINEs, LINEs, transposons, orphons, pseudogenes, fragile sites, Z-DNA, gene superfamilies, 'snurps' (snRNPs) and 'scyrps' (scRNPs), read on.

Gene structure and expression

There are three RNA polymerases in eukaryotic cells that are responsible for transcribing distinct classes of genes. They are RNA polymerases (pol) I, II and III (or occasionally referred to as pol A, pol B and pol C). Pol I transcribes the ribosomal RNA (rRNA) genes (except 5S RNA), pol II transcribes genes coding for proteins and a class of genes coding for small nuclear RNAs (snRNA) that are largely involved in RNA processing, and pol III transcribes genes that code for a number of small RNAs including transfer RNA (tRNA), 5S RNA and 7S LRNA. Most of this book deals with the molecular biology of protein coding, i.e. pol II genes so we will begin by looking at the structure of these genes.

Pol II genes: proteins and small nuclear RNA

The general plan of a eukaryotic gene that encodes a protein is shown in Figure 1.1 and the transfer of information from a pol II gene to a functional protein is shown in Figure 1.2. It has three major components, the *5′ flanking region*, the *transcribed region* and the *3′ flanking region*. The DNA sequences which precede the mRNA start site are the *5′ flanking sequences* and the DNA sequences beyond the mRNA terminus are the *3′ flanking sequences*. The 5′ flanking region is also referred to as the *upstream region* and the 3′ flanking region as the *downstream region*. Landmarks on a gene are often referred to as upstream or downstream of each other or sometimes *infront* (upstream, 5′) or *behind* (downstream, 3′). The 5′ and 3′ flanking regions largely control transcription initiation and termination.

A GUIDE TO THE EUKARYOTIC GENOME

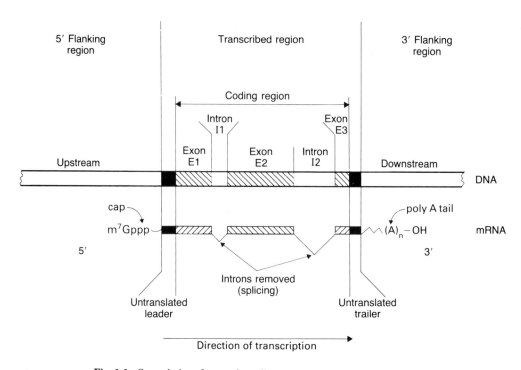

Fig. 1.1. General plan of a protein-coding gene.

The flanking regions

The 5′ flanking region contains many of the signals that control transcription. These include signals that specify the *mRNA start site*, signals that fix the maximum *rate of transcription initiation* and signals that *regulate transcription* either in response to environmental stimuli or in a developmental programme. Some of these signals are unique to individual genes but others are found more generally. Examples of this second class are the *TATA box* and the *CCAAT box* that are found in the 5′ flanking region of many genes from many different organisms and play a role in determining efficient transcription initiation. Examples of regulatory elements that are restricted to individual genes are the heat-shock response sequences that control the synthesis of heat-shock proteins and steroid hormone/receptor binding sites. Another important control region is the enhancer (Serfling *et al.*, 1985). Enhancers are sequences that generally increase the efficiency of transcription. They often function at a distance (several kb) from the mRNA start and although usually located in the 5′ flanking region they have also been found within the transcribed region and in the 3′ flanking region. Enhancers may also mediate regulatory effects by being active only under certain physiological conditions. The 5′ flanking region is often loosely referred to as the *promoter region*, where we use the term 'promoter' in a broad sense to mean any combination of signals that control

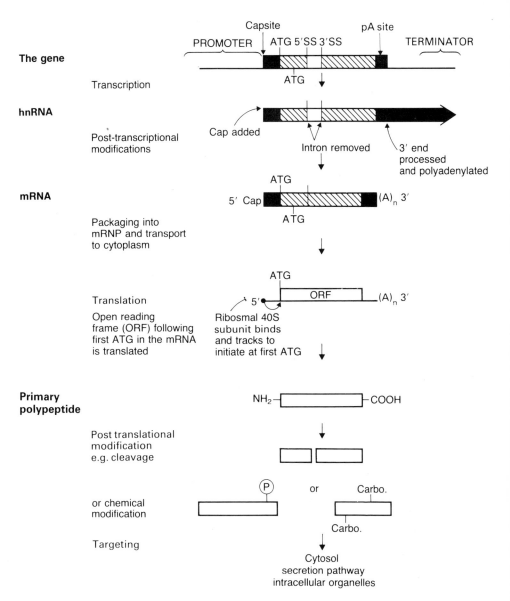

Fig. 1.2. General features of the expression of a pol II gene.

transcription. Parts of the transcribed region also contain sequences which affect transcription and these may also be referred to as promoter elements (Theisen *et al.*, 1986). A variety of different proteins interact with promoter regions to control transcription.

The 3′ flanking region contains signals for the termination of transcription and for post-transcriptional processing of the 3′ end of the mRNA. In many genes transcription terminates at specific sequences and then the 3′ end is degraded by nucleases until a second signal is reached which specifies the end of the mRNA. Most transcripts, except some of the histone mRNAs, have a tract of polyadenosine residues added to this end. This is the *poly A tail*. It is usually about 200 residues long but can be as short as 50 residues as is the case for most transcripts in *Saccharomyces cerevisiae*. The second signal is usually called the *poly A signal*, and in most genes it is associated with the sequence 5' AAUAAA in the mRNA (Birnstiel, 1984).

In some special cases the formation of secondary structures in the RNA upstream from the RNA polymerase causes the enzyme to fall off the DNA template and therefore terminate transcription. This often constitutes a control of transcription by premature termination and has been called *attenuation* (Yanofsky, 1983). It is, of course, mechanistically distinct from attenuation in prokaryotes. Transcription control signals will be discussed in Chapter 8.

The transcribed regions and messenger RNA

The synthesis of mRNA begins in the 5' flanking region some distance upstream from the initiation codon, proceeds through the coding region and terminates in the 3' flanking region. This transcribed region has three components, the *5' untranslated region (5' UTR)* or *leader sequence*, the *coding region* and the *3' untranslated region (3' UTR)* or *trailer*. The gene and its mRNA are usually given numerical coordinates but unfortunately there are two different conventions. In one, the first nucleotide of the mRNA is +1. In the second, the A of the initiating ATG (*Note*, the sequence is ATG in DNA and AUG in RNA, when discussing nucleotide coordinates we always refer to the DNA) is denoted +1. Nucleotides upstream of +1 in either case are given negative coordinates and downstream, positive coordinates.

The untranslated leader sequence has a variable length depending on the particular gene. For example, it is 35 nucleotides in the human immunoglobulin lambda light-chain message (Kelley *et al.*, 1982) whereas rat hydroxymethyl methyl glutaryl (HMG) CoA reductase RNA has a 670 nucleotide leader sequence (Reynolds *et al.*, 1984).

The 3' untranslated sequence is usually 50 to 200 nucleotides but very long trailer sequences have been found, e.g. 520 nucleotides in rat α-2 crystallin (Piatigorsky, 1984) and a mouse thymidylate synthetase mRNA has no 3' untranslated region (Jenk *et al.*, 1986). In addition, the mRNAs encoding a single protein may also be heterogeneous in the 3' UTR. For example, mouse dihydrofolate reductase mRNAs have four sizes of 3' UTR from 80 to 930 nucleotides (Setzer *et al.*, 1980). The leader sequence can be important in controlling translation and the trailer may affect both translation (Fink, 1986) and mRNA stability (Littauer and Soreg, 1982).

The coding region specifies the linear sequence of amino acids in the polypeptide. This coding information can be discontinuous. The coding region can be interrupted by one or more stretches of DNA that do not encode any part of the polypeptide. The coding sequences are called *exons* and the interrupting or intervening sequences are called *introns*. When the gene is transcribed the exons and introns are all copied into a long *primary transcript* which forms a major component of the RNA in the nucleus, the *heterogeneous nuclear RNA (hnRNA)*. The introns are removed by a complex enzymatic process and the exons are ligated together to produce the mature mRNA. The process is called *splicing* and there are signals at the exon/intron boundaries that direct the correct splicing. The signal at the 3' end of an exon is called the *splice donor site (SD)* or *5' splice site (5'SS)* and at the 5' end of the next exon is the *splice acceptor site (SA)* or *3' splice site (3' SS)*. These signals are largely conserved between different genes in a range of eukaryotes (Reed and Maniatis, 1985; Padgett *et al.*, 1986). The mature mRNA is subsequently transported to the cytoplasm where it is translated.

The number of introns in a coding region is highly variable ranging from none, as in the human interferon-alpha genes (Nagata *et al.*, 1980) to more than 50, for example in the rat α-2 procollagen gene (Wozney *et al.*, 1981). The size of the introns can also vary. The most frequent lengths are between 75 and 2000 nucleotides but introns may be very large, for example 17 kb in the rat thyroglobulin gene (Avvidimento *et al.*, 1984) at least 80 kb in several developmentally important genes, such as *Antennapoaedia*, in *Drosophila* (Garber *et al.*, 1983; Scott, 1987). The presence of introns means that a eukaryotic transcriptional unit may be rather large in relation to the size of the protein that is encoded. The human gene for clotting factor VIII has 9 kb of coding sequence spread over 186 kb of DNA (Chapter 12). Introns appear to be less common in lower eukaryotes. Fewer than 20 intron containing genes have been identified in the yeast *S. cerevisiae*.

In addition to splicing and polyadenylation there is usually a third post-transcriptional modification of the mRNA at the 5' end. A guanosine nucleotide methylated at position 7 in the base is added, via a guanylyl transferase, to the first nucleotide of the mRNA to be copied from the DNA template, creating an unusual $5'-p-p-p-5'$ linkage. This modification is called the *cap* and the transcription initiation site in the 5' flanking DNA is often called the *cap site*. The cap influences the efficiency of translation and mRNA stability (Shatkin, 1984).

Translation

The translation of eukaryotic mRNA differs in some features from the process that has been well-characterised in prokaryotes (reviewed by Kozak, 1983a; Moldave, 1985). In particular eukaryotic ribosomes cannot bind to and initiate translation at internal AUG initiation sites. The 40S subunit must interact with the free 5' end of the mRNA via the cap and the subunit appears to track along the mRNA until a suitable initiation site is found. In the majority of cases the first AUG in the message is used for initiation. When this is reached the ribosome is assembled and translation begins. This eukaryotic mRNA is *monocistronic*, i.e. only a single primary translation product is generally produced from a single mRNA. This can be contrasted with the polycistronic mRNAs of bacteria. In higher eukaryotes, there is a preferred sequence environment around the initiating AUG with respect to nucleotides −3 and +4 (where A of ATG is +1). These are usually adenine and guanine respectively (Kozak, 1984).

There are, however, some exceptions to the 'first AUG' rule, most notably in viral mRNAs, e.g. retroviruses, where downstream AUGs are used in preference or in addition to the 5′ proximal AUG. There are also some cases of chromosomal mRNAs with very long leaders which have one or more AUGs followed some way downstream by a translation termination codon, e.g. *S. cerevisiae GCN*4 (Mueller and Hinnebusch, 1986). These 5′ leaders mediate the regulation of translation (Chapter 8).

Eukaryotic mRNAs are not free nucleic acid molecules but are associated with proteins to form *messenger ribonucleoprotein particles (mRNPs)*. The proteins within these mRNPs determine whether the mRNA is sequestered in the cytoplasm in an untranslatable form as so called *repressed mRNA*, or whether it will associate with ribosomes and be translated. There are a number of example where mRNA undergoes a transition between the repressed and translated state. In *Drosophila*, for example, ribosomal protein mRNPs are selectively translated at specific stages during embryogenesis according to the demand for ribosomes (Kay and Jacobs-Lorena, 1985). The repressed state may be affected by interaction with *small cytoplasmic ribonucleoprotein particles (scRNPs)* called *prosomes* (Schmid *et al.*, 1984). Other interactions may affect the stability of the mRNA in the cytoplasm, for instance oestrogen stabilises the vitellogenin mRNA which it induces (Brock and Shapiro, 1983) and cyclic AMP stabilises RNA at specific developmental stages in *Dictyostelium* (Mangiarotti *et al.*, 1983). There may also be feedback control of translation by proteins that have been overproduced. Actin can be detected in free mRNPs in muscle cells and may inhibit translation of actin mRNA (reviewed by Spirin and Ajtkhozkin, 1985). Translation is also regulated by specific translation factors that are present in the mRNPs and that are required for binding the cap and initiating translation. Their activity can be regulated, for example, by phosphorylation. Interferon-treated cells contain high levels of a protein kinase that is only activated by double-stranded regions of mRNA. Double-stranded RNA is relatively common in viral rather than cellular mRNAs. The kinase is, therefore, activated only in viral mRNPs where it inactivates initiation factor eIF-2 by phosphorylating the α subunit. In this way the translation of viral mRNAs but not cellular mRNAs may be inhibited (De Benedetti and Baglioni, 1984).

The nucleotide sequence of the coding region can also affect the efficiency of translation, as a result of *codon usage bias*. Because of the degeneracy of the genetic code, different codons can specify the same amino acid. In many organisms, however, alternative codons are not used at random. Instead there is a bias towards a preferred set. This phenomenon is particularly pronounced for genes encoding the most abundant proteins. The abundance of different charged tRNAs often reflects this codon bias, i.e. a charged tRNA for a rarely used codon is correspondingly rare. Any mRNA that contains many rare codons may therefore be translated poorly because of a limiting tRNA concentration.

Versatility in pol II gene expression

There are a variety of mechanisms for increasing the coding potential of a pol II gene. The simplest involves differential post-translational processing.

For example, several pituitary hormones are derived from proopiomelano-cortin (POMC) which is synthesised as a single-chain precursor polypeptide. The precursor hormone is processed by proteolysis to produce several different active cores which have different hormonal activities. The post-translational cleavages are tissue-specific so that different hormones are produced in the anterior cells of the pituitary as compared to the neuro-intermediate cells (Figure 1.3) (reviewed by Douglass *et al.*, 1984; Lynch and Snyder, 1986).

Another mechanism for increasing versatility is via *complex transcription units* that produce multiple different mRNAs that translate into different polypeptides. Differential splicing produces distinct myosin light chains in chicken skeletal muscle. One pattern of splicing is seen predominantly in early embryos and there is a switch to a second pattern in adult tissue (Nabeshima *et al.*, 1984) (Figure 1.4A). In the case of the human fibronectin gene, sequence analysis suggests that differential splicing may generate up to ten different polypeptides (Kornblihtt *et al.*, 1985) and many peptide hormones are produced from complex patterns of differential RNA processing (Rosenfeld *et al.*, 1984).

The differential use of polyadenylation sites can also affect the pattern of splicing yielding polypeptides with different carboxyl sequences. For example immunoglobulin heavy chains such as IgM can be either membrane-bound or secreted depending upon sequences at the C-terminus (Figure

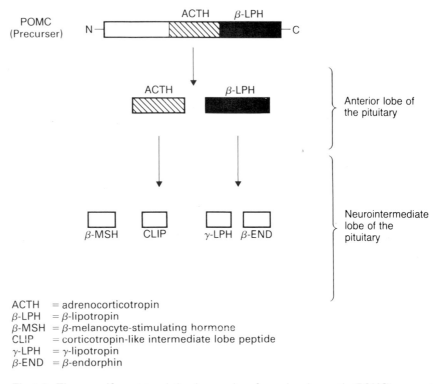

ACTH	= adrenocorticotropin
β-LPH	= β-lipotropin
β-MSH	= β-melanocyte-stimulating hormone
CLIP	= corticotropin-like intermediate lobe peptide
γ-LPH	= γ-lipotropin
β-END	= β-endorphin

Fig. 1.3. Tissue specific post-translational processing of proopiomelanocortin (POMC) (simplified from Douglass *et al.*, 1984).

A GUIDE TO THE EUKARYOTIC GENOME

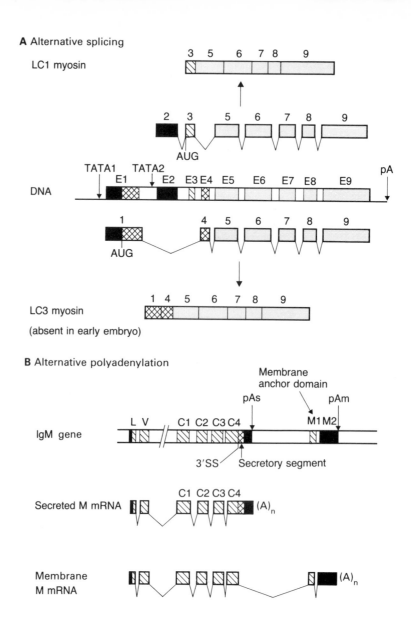

A Alternative splicing

LC1 myosin

DNA

AUG

AUG

LC3 myosin

(absent in early embryo)

B Alternative polyadenylation

Membrane anchor domain

pAs

pAm

IgM gene

3'SS   Secretory segment

Secreted M mRNA

Membrane M mRNA

1.4B). There are two polyadenylation signals in the IgM transcriptional unit. One of these, pAs, is located within an intron and the other, pAm, is at the end of the coding region. If transcription terminates at pAs then secreted IgM is produced because the C-terminus is derived from exon C4, which contains a hydrophilic secretory segment. If transcription terminates at pAm then a splice donor site within exon C4 can be used to fuse part of C4 with the exon M1 which encodes a hydrophobic membrane anchor. This produces an IgM with a new C-terminus that now anchors it in the membrane (Figure 1.4B; Early *et al.*, 1980). This membrane-bound IgM is made early in B cell development but in mature antibody-producing plasma cells secreted IgM is produced (see page 98 for a description of the B cell lineage). Tissue specific gene expression can also be achieved through

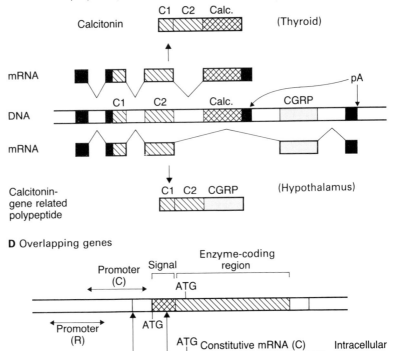

C Alternative polyadenylation and alternative splicing

D Overlapping genes

Fig. 1.4. Complex transcriptional units. The dark shading represents 5' and 3' untranslated regions: the coding exons are variously shaded by hatching, cross hatching or stippling to distinguish them in the final spliced mRNA. A, Two forms of skeletal myosin from one gene. B, Secreted and membrane-bound forms of IgM heavy chains. C, Two peptide hormones from one mRNA. D, Secreted and intracellular invertase from the yeast *SUC2* gene.

differential splicing and polyadenylation for the production of calcitonin and calcitonin gene related hormone (Figure 1.4C).

Differential mRNA initiation can also produce multiple RNA species. For example the *SUC2* gene in *S. cerevisiae* encodes both an intracellular and a secreted form of invertase. The mRNA for the secreted form is longer, encodes a signal sequence for secretion and is regulated. The mRNA for the intracellular form is initiated downstream and translation is initiated at the mature coding sequence to produce the enzyme without a secretion signal. This transcript occurs at low level and is constitutive (Figure 1.4D) (Carlson and Botstein, 1982).

Although the majority of transcriptional units are probably simple there are an increasing number of examples known where tissue-specific or developmental stage-specific expression is associated with the production of multiple mRNAs via the mechanisms outlined. The factors that determine the

A GUIDE TO THE EUKARYOTIC GENOME

choices in these transcription units are being actively investigated (Leff *et al.*, 1986; Breitbart *et al.*, 1987).

Small nuclear RNA (snRNA) genes

There is a group of genes transcribed by pol II that do not code for proteins. These are the *small nuclear RNA (snRNA) genes*, the best characterised of which are the U-snRNAs (Reddy and Busch, 1983). The U genes are multigene families containing 10−100 copies of each gene. They encode RNAs of 90−400 nucleotides. The RNAs are capped at the 5′ end but are distinguished from other pol II RNAs by the high content of uridylic acid, hence the name URNA. They also lack poly A tails. Several of the URNAs show extensive post-transcriptional modifications, particularly methylation and the formation of pseudouridine. The URNAs are abundant, about 10^5 to 10^6 molecules per cell, suggesting that their promoters are efficient, although, unlike many other pol II gene promoters they lack a TATA box (Ares *et al.*, 1985). URNAs are very stable and they are all complexed with proteins as ribonucleoprotein particles (RNPs). These snRNPs are found in specific locations. U1, U2, U4 and U5 are in the nucleoplasm, U3 is in the nucleolus and U6 in perichromatin granules. They interact with precursor hnRNA and pre-rRNA and are involved in processing these RNAs. In particular U1 may be involved in splicing the introns from mRNA and U4 has been implicated in polyadenylation. URNA genes are found in most eukaryotes and their sequences are highly conserved; for example, U1 is 95% homologous in chicken, rodents and man. Small RNAs have been found in *S. cerevisiae* but their relationship to the URNAs is unknown (Maniatis and Reed, 1987; Dreyfuss, 1986).

Pol I genes: ribosomal RNA

In eukaryotic cells there are four ribosomal RNAs, 5S, 18S, 28S and 5.8S. Three of these, 18S, 28S and 5.8S, are produced by post-transcriptional cleavage of a precursor RNA (pre-rRNA) (Figure 1.5). The rDNA transcriptional unit appears to be the only DNA recognised by RNA polymerase I (Mandal, 1984). The regions coding for 18S, 5.8S and 28S are separated by a spacer called the *internal transcribed spacer (ITS)* which is removed from the primary transcript by endonucleases during processing. The rDNA in most eukaryotes is tandemly repeated at one or more sites in the genome and each repeat is separated by a segment of non-coding DNA. This DNA is called the *non-transcribed spacer (NTS)* to distinguish it from the internal spacer regions within the pre-rRNA coding region but it is, in fact, transcribed as a result of read-through from the upstream repeat (Baker and Platt, 1986; Sollner-Webb and Tower, 1986). The number of rDNA repeats and their expression can change dramatically during oogenesis and embryogenesis to match the demand for ribosomes in particular cells. In *Xenopus* the rDNA is amplified several thousand times in oocytes to produce extrachromosomal copies which are very actively transcribed during oogenesis and then lost at meiosis. A typical rDNA repeat from *Xenopus laevis* is shown in Figure 1.5. A promoter element is located between −140 and +6

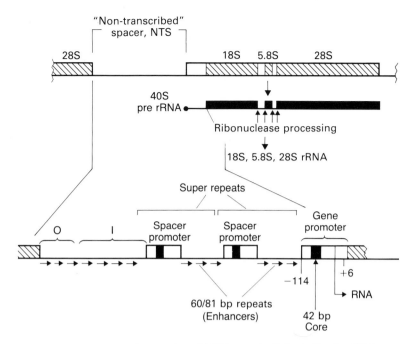

Fig. 1.5. The organization and expression of rRNA genes in *X. laevis* (Reeder, 1984).

with respect to the pre-rRNA start site at +1. The rest of the NTS is composed of an array of different repeated elements. The most striking of these is a repeat of the promoter. There may be between 2 and 7 of these repeats and they are separated by clusters of repeating enhancer elements of 60 and 81 bp that contain a 42 bp core derived from the promoter. The promoter and 60/81 bp repeats therefore comprise a super repeat element. There are two additional clusters of repeats called O and I and some non-repeated DNA. Within a group of tandemly repeated rDNA genes the NTS regions can be of variable lengths largely due to variable numbers of super repeat elements. The 42 bp core is a key component of the promoter and increasing the number of super repeats increases the frequency of transcription (reviewed by Reeder, 1984; Sollner-Webb and Tower, 1986). Despite the fact that the structure and function of rDNA transcription units are highly conserved between different eukaryotes there is very little sequence conservation in the non-transcribed spacer region except for sequences around the pre-rRNA initiation site particularly from +1 to +18. In addition the transcription factors isolated to date all seem to be species-specific (Sommerville, 1984).

Pol III genes: 5S RNA, tRNA and small cytoplasmic RNA

Pol III transcribes a number of different genes generating mostly short transcripts; these are 5S RNA, tRNA and several small cytoplasmic cellular (scRNA) and viral RNAs. All genes transcribed by pol III contain a promoter sequence within the transcribed region, called the *internal control region (ICR)*. There are two types of ICR, the 5S RNA type and the tRNA type and different protein factors are required to interact with pol III for transcription to occur at these different ICRs. The transcripts differ from pol II transcripts because they have a 5′ triphosphate rather than a cap and they usually terminate with a polyuridine tract. This reflects the fact that a stretch of T residues acts as the signal for transcription termination.

5S RNA genes

The 5S genes of *Xenopus* have been studied in some detail. Indeed this was the first system where modern gene manipulation techniques were used to investigate the control of eukaryotic gene expression. The 5S gene is present in about 20 000 copies, split into three different multigene families. The smallest multigene family, with 400 copies, is expressed at all times in oocytes and somatic cells. These are called somatic-type 5S RNA genes. The other two families are under developmental control. They are expressed in growing oocytes and silent in somatic cells, and are called the oocyte-type 5S RNA genes. This mechanism of increasing expression contrasts with that ensuring adequate 18S and 28S rRNA in oocytes, by gene amplification rather than by selective gene expression.

5S RNA genes are 120 nucleotides long and transcripts are not processed. The oocyte-type 5S and somatic-type 5S rRNAs differ by only six nucleotides and their expression is largely controlled by the ICR that is located at +41 to +87 within the coding region (Figure 1.6A) (Brown, 1982). A variety of proteins that interact with the ICR have been identified and these will be discussed in more detail in Chapter 8.

tRNA genes

The tRNAs are a family of RNAs about 80 nucleotides in length. The genes for the different tRNAs are often arranged in clusters; a 3.1 kb region of *X. laevis*, for example, contains eight tRNA genes (Figure 1.6) but each gene is transcribed separately, unlike in *E. coli* where several tRNA genes are transcribed polycistronically. The tRNA genes also contain intragenic control regions but unlike the 5S RNA genes the essential signals are split into two blocks, A and B, which have the approximate coordinates 8−19 and 52−62. This is usually called the *split promoter* (Hall *et al.*, 1982).

The ICR blocks correspond to the most conserved regions of tRNAs in eukaryotes and prokaryotes in the D and T arms of the three-dimensional tRNA molecule (Figure 1.6). Some invariant nucleotides in this region may therefore have been conserved for both promoter activity and tRNA functions although the three-dimensional structure is not essential for promoter activity.

The B block contains some sequence similarities to the −35 sequence in bacterial promoters. The distance between the two blocks may affect the efficiency of transcription and some tRNA genes contain intervening sequences between the blocks which alter the spacing. In yeast, 40 of the 400 tRNA genes have introns of 14−60 nucleotides in length producing a series of tRNA genes with different distances between blocks A and B (Guthrie and Abelson, 1982). The optimal distance appears to be 30−40 bp.

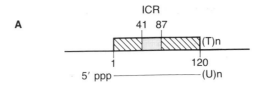

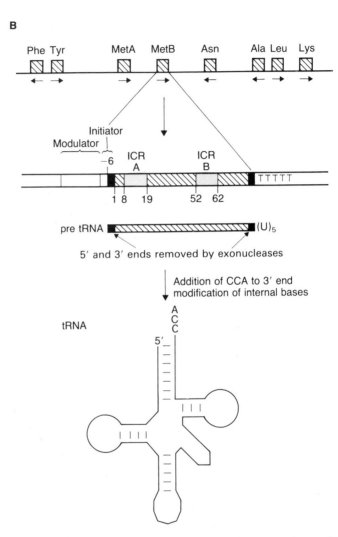

Fig. 1.6. The organization of 5S RNA and tRNA genes. A, 5S RNA. B, Cluster of tRNA genes in the chromosome.

The A block is involved in determining the RNA initiation site and both regions bind specific transcription factors (Marzooki *et al.*, 1986). Sequences upstream from the transcript start also influence the transcriptional activity and sequences have been identified that both promote and reduce transcription. These are called *modulator sequences*. The region surrounding the

A GUIDE TO THE EUKARYOTIC GENOME

transcription start between $+1$ and $+30$ also appears to be important for specific transcription initiation (Sollner-Webb, 1988).

Small cytoplasmic RNA (scRNA) genes

There are several classes of small RNA produced by pol III and some of these have essential functions in the cell. For example, the 7SL RNA genes encode an abundant cytoplasmic RNA that is incorporated into a ribonucleoprotein complex called the *signal recognition particle (SRP)*. The SRP is a crucial component of the protein secretion process (Walter and Blobel, 1982). There are four functional 7SL RNA genes in mammalian cells and the sequences are highly conserved between different eukaryotes: mammals, *Drosophila* and *Xenopus*. The 7SL RNA genes contain a split ICR which has homologies with the tRNA ICR, 5' modulater sequences are also required for maximum expression (Ullu and Weiner, 1985). The 7SL RNA itself appears to have been the precursor of the most abundant family of middle repetitive DNA sequences in primate and rodent DNA, the so-called *Alu* sequences.

A number of viruses, e.g. adenovirus, SV40 and EBV, contain genes that are transcribed by pol III. The best characterised are the adenovirus virus-associated (VA) RNAs, VA1 and VA2. The genes contain ICRs that are related to the A and B blocks in tRNA (Foulkes and Shenk, 1980) and the RNAs are found in cytoplasmic RNPs. VA1 RNA is required for the formation of a stable 48S translational pre-initiation complex and seems to preserve the activity of one of the translation initiation factors, eIF-2, possibly by preventing its phosphorylation by the kinase that is activated by dsRNA in virus-infected cells (Schneider *et al.*, 1984; Reickl *et al.*, 1984; Kitajewski *et al.*, 1986). The VA RNAs therefore promote viral translation, but they also appear to enhance translation of some cellular RNAs (see Chapter 4).

Pseudogenes and processed genes

Pseudogenes are partial copies of functional genes but are inactive and often not transcribed. Pseudogenes, of all the different classes of genes that we have discussed, have been found in most eukaryotes (Weiner *et al.*, 1986). The small pol III and pol II RNA genes often have hundreds of pseudogenes. For example, the U1 snRNA has about $50-100$ functional genes and $500-1000$ pseudogenes and the 7SL RNA gene family, in mammals, contains four active genes and several hundred pseudogenes. In the case of protein-coding genes there are usually only a few, less than twenty, pseudogenes.

Many pseudogenes are linked to the 'parental' gene and share extensive homology with the coding and flanking DNA of the 'parental' gene. The most likely mechanism for generating this type of pseudogene is by tandem duplication of a region of the chromosome containing the functional gene. One gene in the pair is then free to mutate without causing loss of production of the particular protein. In some cases this leads to an evolutionary dead end, i.e. a pseudogene (reviewed by Little, 1982). For example, the human α-globin locus contains five genes, an embryonic globin (E1), two adult globins (A1 and A2) and two pseudogenes (ψE1 and ψA1) (Figure 1.7A). ψA1 appears to be a duplication of A1. The genes are 73% homologous, they both contain introns and show the conserved 5′ control elements and significant homologies in the 5′ and 3′ flanking DNA. The ψA1 gene, however, contains a number of mutations that render it inactive. It was probably generated from the wild-type gene 45 million years ago (Proudfoot and Maniatis, 1980).

There is a second class of pseudogenes that are unlinked to the parental gene and have structures that resemble the transcript rather than the gene. These pseudogenes lack introns and often contain a run of adenosine

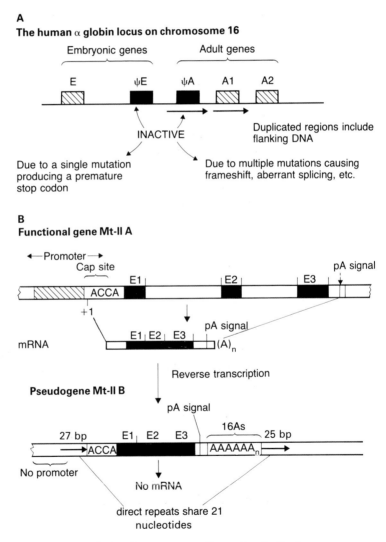

Fig. 1.7. Pseudogenes. A, Pseudogenes generated by tandem duplication.
B, Pseudogenes derived from mRNA: processed genes, e.g. human metallothionein II (Mt-II).

A GUIDE TO THE EUKARYOTIC GENOME

residues at their 3' end. The functional human metallothionein IIA (*Mt*-IIA) gene is compared with the non-functional *Mt*-IIB pseudogene in Figure 1.7B. The pseudogene appears to be an exact copy of a functional *Mt*-IIA mRNA. It begins at the authentic cap site, lacks introns and ends at the poly A site. However, the pseudogene cannot be transcribed because it lacks transcription control signals and is therefore inactive (Karin and Richards, 1982). Pseudogenes with a similar structure may be transcribed, presumably using fortuitous promoters in the flanking DNA, but they may be inactive for other reasons such as the presence of stop codons in the coding region.

Pseudogenes lacking some of the features of normal transcripts have also been found. Many snRNA pseudogenes, for example, have deletions at the 3' end and may not have a poly A tract (Bernstein *et al.*, 1983). In some cases the poly A tract contains additional sequences, particularly runs of CA or TA dinucleotides.

Many of the features of this second class of pseudogene suggest that they arose as a DNA copy of a processed RNA and they are, therefore, often referred to as *processed genes*. In addition, they are often flanked by short 5 to 30 bp directly repeated sequences and the flanking DNA is totally unrelated to that of the parental gene (Figure 1.7). Formation of processed pseudogenes is likely to involve the enzyme reverse transcriptase, which makes a DNA copy from an RNA template. Insertion of the double-stranded DNA copy into the genome would then follow. This usually involves the production of an initial staggered break which, when repaired during the integration process, generates the short direct repeats seen at the ends of processed genes (Sharp, 1983). The fact that processed genes are produced by reverse transcriptase and that they can integrate apparently at random throughout the genome has led to them being called *retroposons* (Weiner *et al.*, 1986).

A potential source of enzymes for generating retroposons in eukaryotic cells is the endogenous retroviruses and transposons which encode reverse transcriptase and integrase (Adams *et al.*, 1987a).

Genome organisation

The amount of DNA in the haploid genome of a eukaryotic cell is known as its *C-value*. C-values vary widely, for instance from 10^4 kb for the lower eukaryote *S. cerevisiae* (yeast) to 10^8 kb for some amphibia.

All higher eukaryotes contain significantly more DNA than appears to be necessary to encode their structures and functions. For example, the human haploid genome has a C-value of about 3×10^6 kb yet estimates of the number of genes in the human genome, calculated by a variety of means, suggest that there are only 4×10^4 to 10^5. If one takes 3 kb as a generous estimate for the coding sequence of an average gene then there exists at least a 10-fold excess of DNA in the genome over that required to produce a human being (reviewed by Gall, 1981).

This problem is taken further by the ranges of C-value seen within some classes of eukaryotes. The increase in complexity of the organisms that comprise a class is paralleled by an increase in the minimum C-value for that class. However, the maximum C-value within a class can be 100 times the minimum. One of the largest ranges of C-value is seen in the amphibia where the minimum is 7×10^5 and the maximum is 10^8. This phenomenon is known as the *C-value paradox*. Some genome sizes are shown in Table 1.1.

Table 1.1. The size of genomes.

	kbp
SV40	5.1
Vaccinia virus	190
E. coli	4000
S. cerevisiae	13,500
Drosophila	165,000
Man	2,900,000
South American lung fish	102,000,000

Much of the genome comprises sequences that are repeated many times. The number of repeats can vary from a few to millions. It has been convenient to classify repeated sequences broadly according to their repetition frequency. Sequences that are repeated at about $10^4 - 10^6$ copies per haploid genome are called *highly repetitive* and other repeated sequences are called *moderately repetitive*. In reality there is a spectrum of repetition frequencies. Much of the highly repetitive DNA is composed of simple sequences of 6 to 200 bp repeated often in large tandem arrays; these are known as *satellite DNA*.

Other repetitive elements are scattered throughout the genome interspersed with coding DNA. Some of these elements are short, usually less than 500 bp and they are called *short interspersed elements* or *SINEs*. Other elements are long, about 5−7 kb and these are called *long interspersed elements* or *LINEs*.

While the eukaryotic genome can be regarded generally as stable or fixed there is a certain degree of *genome fluidity*. This arises because some regions of the genome can become rearranged or amplified in response to specific stimuli or as a result of a variety of specific or non-specific recombination events. Most eukaryotic genomes also contain discrete segments of DNA called *transposons* that are capable of moving from one part of the genome to another and, as we have seen already, information from RNA can return to the genome as retroposons.

Genome organisation in man has been reviewed by Kao (1985).

Genomic components

Until recently treatments of eukaryotic genome organisation have divided the genome into coding and non-coding DNA. This distinction is becoming less clear with the analyses made possible by the advent of recombinant DNA technology. Sequences such as the repetitive transposons that were once classified as non-coding are now known to encode proteins.

Furthermore the classical division into unique, middle repetitive and

A GUIDE TO THE EUKARYOTIC GENOME

highly repetitive sequences based on reassociation kinetics is also becoming less useful as more detailed descriptions of defined segments of the genome emerge.

We will, therefore, ignore the more usual broad classifications and simply list the major components of the eukaryotic genome.

Genes

Most of the protein-coding regions are represented only once per haploid genome and are therefore often referred to as *unique* or *single-copy* sequences. Some genes, however, have evolved into *gene families* with varied numbers of member genes that are related to each other through their primary nucleotide sequence (reviewed by Long and David, 1980; Kao, 1985; Maeda and Smithies, 1986). The members of a gene family may be clustered at the same chromosomal locus in long, tandem arrays, i.e. adjacent repeats. This arrangement is often found for the rRNA and histone genes, but there are great variations between species. For example, there are several thousand rRNA genes in *Zea mays* organised as a tandem array on chromosome 6 whereas there are only 50–200 rRNA genes in man scattered over five separate chromosomes (reviewed by Mandal, 1984). Genes organised in tandem arrays are often highly conserved and this appears to be due to unequal crossing over between chromosome homologues followed by *tandem duplication* to maintain gene dosage. Although a gene may be a member of a tandem multigene array there are frequently isolated single copies of the gene, called *orphons* (Childs *et al.*, 1981) dispersed throughout the genome. In the sea urchin *Lytechinus pictus* most of the histone genes are arranged in the order H1, H4, H2B, H3, H2A and the whole cluster is repeated several hundreds of times to form a tandem array. There are however 5–20 additional copies of each of the histone genes as orphons in other regions of the genome. The functional differences, if any, between genes as orphons and in tandem arrays is not known but in the sea urchin the tandem arrays appear to be expressed predominantly in early development and orphons are expressed later (Kedes and Maxson, 1981).

Other gene families also appear to have arisen by tandem duplication but they are frequently separated by large spacer regions. For example, the Class I genes of the mouse histocompatibility locus (H-2 locus) are a multigene family that contains about 36 members distributed as 13 clusters spread over 837 kb of DNA. The largest clusters contain seven genes spread over 191 kb with 7 – 28 kb of spacer DNA between them. They encode transplantation antigens and haematopoietic differentiation antigens which have the same general structure (see Chapter 10) (Flavell et al., 1986). Further gene families comprise structurally related genes which encode functionally equivalent proteins but are expressed under different conditions. For example, in the human α-globin family on chromosome 16 one of the genes, αE, is expressed only in the embryo and two, αA1 and αA2 are expressed only in the adult. The other two members of this family are pseudogenes (Figure 1.6A).

Several apparently distinct gene families which do not share identical functions and which are not highly homologous at the nucleotide sequence level are related through subdomains of the proteins that they encode. These are called *gene superfamilies* or *supergene families*. The first of these to be described was the immunoglobulin supergene family. The immunoglobulin molecule is composed of four polypeptide chains of two types, two heavy and two light chains, held together by disulphide bridges to form a multifunctional molecule. The functions are carried out by separate *domains* of the protein. In the immunoglobulin molecule the separate functional domains are encoded by different exons, as is true of many multifunctional proteins. It is likely that complex proteins like immunoglobulins evolved by a series of duplications of ancestral exons. A number of different functional exons may have recombined to form a variety of related proteins with novel functions. The Class I and Class II histocompatibility antigens, the Thy-1 antigen and the T lymphocyte receptors all share domains with immunoglobulin and a geneological tree has been constructed that relates these

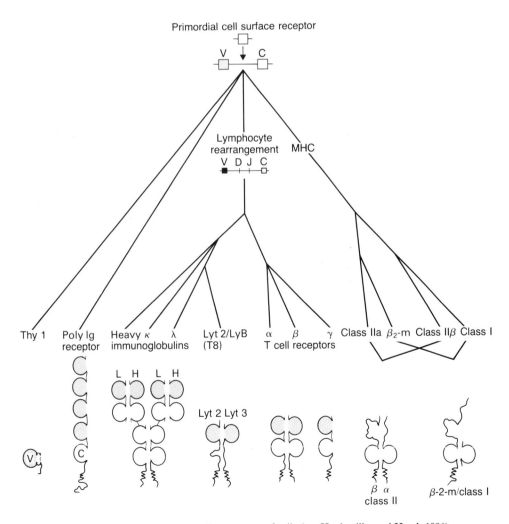

Fig. 1.8. The immunoglobulin gene super family (see Hunkapiller and Hood, 1986).

molecules (Hood *et al.*, 1985; Hunkapiller and Hood, 1986) (Figure 1.8). A number of other superfamilies have been identified. The serine proteases involved in blood coagulation and fibrinolysis appear to have evolved from modules (Patthy, 1985, Chapter 12).

In eukaryotes, unlike the situation in prokaryotes such as *E. coli*, there is no significant clustering of genes that encode functionally related proteins. Many polypeptides that form multi-subunit proteins are encoded by unlinked genes. For example the human immunoglobulin heavy-chain gene family is sited on chromosome 14, the lambda light-chain cluster on chromosome 22 and the kappa light-chain genes on chromosome 2. Human α-globin genes are on chromosome 16 and β-globin genes are on chromosome 11. Similarly the enzymes of single biochemical pathways are not clustered. In *E. coli* the enzymes galactose-4-epimerase, galactose-1-phosphate uridyl transferase and galactokinase involved in galactose metabolism are arranged as an operon at a single chromosomal location, but in man they are found on chromosomes 1, 9, 17 respectively (O'Brien, 1987). Although there is some clustering of genes in multigene families, reflecting evolution by tandem duplication, there are usually significant regions of spacer DNA between coding regions. The 'clustered' genes may not be expressed, therefore, under the same control. For example, the different members of the globin families are expressed at different developmental stages. Interestingly, in this case the order of the genes on the chromosome reflects the time at which they are expressed during development, suggesting that the clustering may have some developmental significance (Karlsson and Nienhuis, 1985). It is quite clear, however, that despite the scattering of genes there is *coordinate control* of gene expression in eukaryotes, whereby groups of genes are switched on and off in response to a common stimulus. It is likely that there are a small number of controlling genes that activate or repress many other genes sharing some common feature. This will be discussed in more detail in Chapter 8.

Genetic elements

SINEs

Each family of SINEs may have up to 10^5 copies and there may be several different families of SINEs in a particular genome. SINEs are found between and within transcriptional units both in the introns and untranslated regions but, with the exception of some inserts in rDNA genes, not in coding regions. About $1 - 5\%$ of the hnRNA is homologous to SINE families which are transcribed either as a result of being within a pol III transcriptional unit or because the SINEs contain a pol III promoter. This is of the tRNA split promoter type and therefore efficient transcription of the SINEs also depends upon chromosomal location to provide appropriate $5'$ modulater sequences (Paolella *et al.*, 1983). The best characterised SINEs are the *Alu*I family in primates and the related B1 family in rodents. A number of other SINEs are *Alu*-like in their general structure and properties.

Alu sequences are the major dispersed repeated DNA in primates. In the human genome, for example, they are repeated about $3 - 5 \times 10^5$ times and constitute about $3 - 6\%$ of the total DNA. The repeats contain a site for the restriction enzyme *Alu* I that gives them their name. A likely source of the *Alu* elements is by copying of a processed 7SL RNA (Ullu and Tschudi, 1984).

It is likely that *Alu*-like elements have been inserted into genomes over a long evolutionary time-span. Some insertions are ancient. In the β-globin cluster there is an *Alu* sequence in the same location in man and chimpanzee. Other insertions must be more recent, for example the related R.dre.I. element is only found in the rat and several chromosomal loci contain a R.dre.I. element at only one allele. On the whole, however, the pattern of *Alu*-like elements at a given locus in any one species is conserved.

Another SINE, first identified in mouse DNA, is highly conserved and is found in a wide range of eukaryotes, including pigeons, slime mould, yeast, *Drosophila* and corn. This 250 bp element called EC1 contains a region of 17 TG dinucleotides that has been found at the boundaries of tandem repeats of regions such as the β-globin locus and may act as a hot spot for gene conversion and recombination (Miesfeld *et al.*, 1981). Blocks of alternating pyrimidine/purine residues, i.e. $(CA)_n$ or $(GT)_n$, are widespread in most eukaryotic genomes and many of these blocks appear to be insertions, because they are flanked by short direct repeats. The telomeres of eukaryotic chromosomes are rich in these dinucleotides (see later) and there is some evidence that various DNA cleavage reactions occur in the vicinity of a CA dinucleotide. For example, all eukaryotic transposons end in CA; the consensus recognition sequence for Ig V-D-J joining is CACAGTG; the cleavage site for the initiation of yeast mating type conversion is CAACA; and *Tetrahymena* and yeast telomeres are nicked in the AC portion of the telomere repeats $(C_4A_2)_n$ (reviewed by Rogers, 1983).

LINEs

In primates there is a single LINE family originally called *Kpn* because of the presence of this restriction site. In rodents there is also a single family called MIF-1 or R or Bam5-R. These elements in mice and man are related and they have been renamed the L1 family. They look like retroposons, have poly A tails, are flanked by short direct repeats and many are truncated at the 5′ end, which is characteristic of premature termination by reverse transcriptase. Rearrangement of the internal sequence is however, common. The LINEs are transcribed at a low level by pol II; many of the transcripts are not polyadenylated and are confined to the nucleus but there are some transcripts associated with polysomes. Interestingly, there is a significant open reading frame (ORF) in a human element and several rodent L1 elements have been shown to contain two long ORFs (Loeb *et al.*, 1986). The distribution of mutations within the L1 sequences and the conservation of the ORFs between diverse species suggests that there has been selection for a protein-coding function (Martin *et al.*, 1984). There is no clue as to the function of the protein but it could be involved in the survival of the

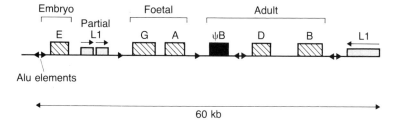

Fig. 1.9. Coding and non-coding DNA at the human β globin locus (adapted from Karlsson and Nienhuis, 1985).

element itself or be required as a part of the cell regulatory apparatus for gene expression.

In Figure 1.9 we show a region of human chromosome 11 to indicate the distribution of SINEs, LINEs, pseudogenes and genes in the β-globin region.

TRANSPOSONS AND ENDOGENOUS PROVIRUSES

Transposons can be regarded as special types of SINEs or LINEs. They are discrete, dispersed, repetitive elements that are commonly 2–7 kb long but they are highly mobile compared to the sequences that we have discussed above. SINEs and LINEs have clearly moved within genomes but this has largely occurred during an evolutionary history of millions of years. Transposons can move in the genome within a single generation of an organism. They can also move repeatedly and distribute themselves throughout the genome more or less randomly. The insertion of a transposon can produce a mutation and this is how many transposons were first identified. As early as 1948 Barbara McClintock deduced that unstable mutations in the coloration of *Zea mays* kernals were due to the insertion of mobile elements, termed controlling elements (McClintock, 1948). The ability of a transposon to move is inferred either from the production of mutations or by comparing regions of DNA from different strains of the same organism and showing that one contains a new piece of DNA. Some of the information about the best studied transposons is summarised in Figure 1.10.

Transposons have now been isolated from a number of different eukaryotes and they fall into several structural classes (reviewed by Finnegan, 1985; Doring and Starlinger, 1984; Adams *et al.*, 1987a). The best characterised are the Ty elements of *Saccharomyces* and the *copia*-like elements of *Drosophila*. These elements have the same general structure as an integrated retroviral provirus. It is likely that these transposons are either the progenitors or descendents of retroviruses and they have been called *retrotransposons* (reviewed by Adams *et al.*, 1987a).

Another transposon, the P element, has a precise and conserved structure that is apparently completely different from retrotransposons. It has short, 31 bp, identical inverted repeats flanking a 2 097 bp internal sequence that could potentially encode four polypeptides (O'Hare and Rubin, 1983). The mechanism of transposition is not known but it is controlled by the genetic

Element	Organism	General structure

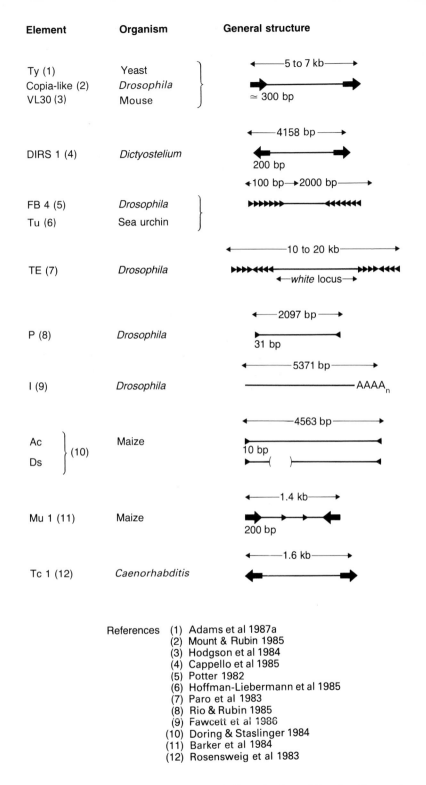

Ty (1)   Yeast
Copia-like (2) *Drosophila*
VL30 (3)   Mouse

DIRS 1 (4)   *Dictyostelium*

FB 4 (5)   *Drosophila*
Tu (6)   Sea urchin

TE (7)   *Drosophila*

P (8)   *Drosophila*

I (9)   *Drosophila*

Ac
  } (10)   Maize
Ds

Mu 1 (11)   Maize

Tc 1 (12)   *Caenorhabditis*

References (1) Adams et al 1987a
            (2) Mount & Rubin 1985
            (3) Hodgson et al 1984
            (4) Cappello et al 1985
            (5) Potter 1982
            (6) Hoffman-Liebermann et al 1985
            (7) Paro et al 1983
            (8) Rio & Rubin 1985
            (9) Fawcett et al 1986
           (10) Doring & Staslinger 1984
           (11) Barker et al 1984
           (12) Rosensweig et al 1983

Fig. 1.10. Eukaryotic transposons. The arrows indicate the size and direction of repeated sequences in each transposon. The bracket in the Ds element indicates a deletion.

A GUIDE TO THE EUKARYOTIC GENOME

background (see Chapter 5). Another *Drosophila* transposon, the I element, has significant homology with mammalian LINEs (Fawcett *et al.*, 1986).

Other transposons have a less conserved structure. The fold-back (FB) elements of *Drosophila* have inverted terminal repeats (TRs) that look like satellite DNA, i.e. they are composed of short tandem repeats separated by A-rich spacers. The TRs can be variable in length and they flank an internal sequence that is also highly variable and internally repetitious (Potter, 1982). Two FB elements can flank a large piece of coding DNA to produce a large element that can transpose several genes, for instance the TE element of *Drosophila* carries two genes, *white* and *roughest*, that can transpose to new locations at a frequency of 10^{-3} (Paro *et al.*, 1983).

Many mammalian genomes contain sequences that are clearly related to retrovirus proviruses. In many cases these can produce an infectious virus but this is suppressed in the particular host cell. In other cases the structure is defective. These are called *endogenous proviruses* because they are inherited rather than acquired by infection. They have been found in many rodents but there is no good evidence for the existence of these sequences in the human genome.

All transposons and proviruses are flanked by short direct repeats of DNA. It seems that a staggered break in the DNA is a prerequisite for the insertion of any DNA element whether it is a retroviral provirus, a transposon, a pseudogene, a SINE or a LINE. We will discuss the retrotransposons in more detail in Chapter 9, and P-elements and controlling elements in Chapters 5 and 6.

Satellite DNA

The sequences that make up satellite DNA often have an atypical base composition causing them to form a satellite band distinct from the rest of the DNA in various isopycnic centrifugation procedures, hence the general name of this class of sequences. Satellites usually consist of long tandem arrays of thousands of related sequences; in many cases the repeats are short oligonucleotides but there are examples of repeats of several thousand base pairs. The repeat units usually show sequence variation, probably reflecting the evolution of satellites by tandem duplication followed by mutation, recombination and deletion. The amount of satellite DNA varies between eukaryotes ranging between 2% and 50% of the total genome. It can even vary between closely related species, for example satellite DNA comprises 40% of the *Drosophila virilis* genome but only 18% of the *D. melanogaster* genome. Certain organisms have multiple distinguishable satellites, some of which are interrelated via core sequences. Many satellites are located around the centromeres and telomeres where they appear by microscopy as constrictions or densely staining regions called *heterochromatin* in the metaphase chromosome. They are usually not transcribed.

There are also *minisatellite* regions in eukaryotic DNA comprising only a few tandem repeats of a simple sequence that are dispersed through the genome. One human minisatellite consists of four tandem repeats of a 33 bp sequence flanked by a 9 bp duplication, suggesting integration of the minisatellite at a staggered break. The minisatellites are highly polymorphic. The number of tandem repeats at a given locus in the DNA from one individual may contain a short minisatellite while another may contain a long minisatellite. These polymorphisms can be used to make an individual specific DNA fingerprint that can be used to follow chromosomes in genetic analysis (Jeffreys *et al.*, 1985).

Centromeres and telomeres

Centromeres are the sites of interaction between the chromosome and the spindle apparatus that is essential for proper chromosome segregation at mitosis and meiosis. They are embedded in satellite DNA and are probably not transcribed. The telomeres are the very ends of the chromosomes, they are also heterochromatic and they function to ensure the faithful replication of chromosome ends. Both these DNA components of the chromosomes interact with a number of proteins and therefore probably function as nucleoprotein complexes. Their properties are discussed in detail in Chapter 9.

DNA rearrangements

Rearrangements of chromosomes can be divided into two types. The first is where a highly specific, ordered rearrangement occurs in particular cell types and/or in response to specific stimuli. In these cases rearrangement is used as a means of achieving some specific regulatory event. The second is where a rearrangement is not ordered. Many rearrangements resulting from recombination events between repetitive elements fall into this category. In this section we will briefly review one ordered rearrangement that brings different coding domains of an immunoglobulin chain together and we will discuss the role of non-ordered rearrangements in 'abnormal' cells. Other 'classical' ordered rearrangements are: antigenic variation in trypanosomes (van der Pleog, 1987) and a simple developmental switch that controls mating type in *Saccharomyces* (Klar, 1987).

Immunoglobulin gene rearrangements

The best-characterised and possibly the only rearrangement that occurs in normal mammalian cells is the specific recombination in bone marrow-derived (B) lymphocyte genomes that produces functional antibody molecules (immunoglobulins) (reviewed by Tonegawa, 1983; Wall and Kuehl, 1983; Yancopoulos and Alt, 1986). The structure of an immunoglobulin (Ig) is outlined in Figure 1.11 (and in more detail later in Figure 12.15). There are two identical heavy (H) chains and two identical light chains (L). Two types of light chains kappa (κ) and lambda (λ) are encoded by distinct gene families. Each of the chains has two major domains, a *variable* (*V*) and a *constant* (*C*) domain. The function of an immunoglobulin is to combine with foreign antigens, e.g. viral or bacterial surface proteins, and inactivate them. The immunoglobulin must interact specifically with an antigen so

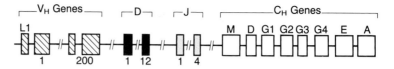

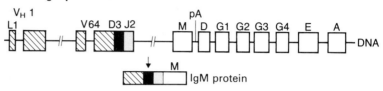

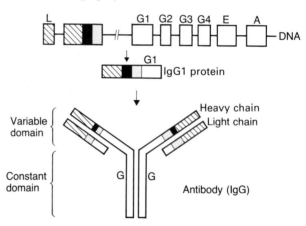

Fig. 1.11. DNA rearrangements on mouse chromosome 12: the generation of an immunoglobulin heavy chain gene (based on Tonegawa, 1983). The variable (V) exons are hatched, the diversity (D) regions are dark, the joining (J) regions are stippled and the constant regions are unshaded.

that only foreign proteins and not normal, host proteins are inactivated. To achieve this specific recognition there must be as many different antibody molecules as there are antigens. It is estimated that there are between 10^6 and 10^8 different immunoglobulins. The five different classes of immunoglobulins differ in their C regions, and are termed IgM, IgD, IgG (this has four subclasses IgG1−IgG4) IgE and IgA. The C regions of the heavy chains of these classes are called Cμ, Cγ, Cδ, Cε and Cα respectively. The C regions have the same general structure and all immunoglobulins within a class have identical C regions. All immunoglobulins, however, are different in their V region and it is this region that provides the variety of antibodies to deal with all the different antigens; this is called *antibody diversity*.

The three types of chain (H, λL and κL) are encoded by three unlinked gene families in the mouse. The λ light chain family is on chromosome 16, the κ light chain family on chromosome 6 and the heavy chain family on chromosome 12. The coding information for each of the polypeptide chains is contained in multiple gene segments that are brought together by DNA recombination and RNA splicing. The heavy-chain gene family is shown in Figure 1.11. In germ line cells and somatic cells that do not produce immunoglobulins the gene family is organised as shown in Figure 1.11A. There are 100 − 200 segments of DNA encoding V regions that fall into about ten subfamilies and five segments encoding the five different C regions of the Ig classes. Each C region is composed of multiple exons each corresponding to functional domains in the protein. For simplicity this exon structure is not shown. The V regions contain two exons, the L1 exon encoding a signal peptide. In between the V and C regions there are two additional clusters of segments. These are the diversity (D) and joining (J) regions of which there are 12 and 4 copies respectively. The expression of the immunoglobulin genes is tissue-specific, occurring primarily in B lymphocytes. The arrangement of the Ig heavy-chain in B-cell genomes is shown in Figure 1.12B. In these cells the V, D and J segments have been joined together by a DNA/DNA recombination event. In this hypothetical example V gene number 64 is joined to D segment 3 and J segment 2. Any V segment can join with any D segment which can, in turn, join with any J segment. In addition the junctions at the ends of the V and J segments can vary by several nucleotides. The variety of V, D and J segments and the flexibility of joining can create many possible amino acid sequences in the variable domain. The same sort of gene organisation and rearrangement is shown by the light chains, although they lack a D region and have the structure V to J to C. Heavy and light chains associate randomly and this produces even more variety. In addition the coding regions of the V segments display a 5% higher rate of somatic mutation than flanking DNA, generating further variability.

Joining occurs in stages during the differentiation of a precursor stem cell to the mature B lymphocyte (the B-cell lineage is shown on page 98, Figure 4.3). The D and J segments join first and then join with the proximal V segment. This is followed by a further deletion or inversion of V segments to produce the final gene (Yancopoulos *et al.*, 1984; Alt, 1986; Yancopoulos and Alt, 1986). The recombination seems to be mediated by specific sequences. In the germ line, the V and D segments are followed by the sequences 5′−CACAGTG and 5′−ACAAAAACC separated by a 23 bp non-conserved spacer. The D and J segments are preceded by the sequence 5′−GGTTTTTGT and 5′−CACTTGTG separated by 12 bp. These sequences are complementary and are referred to as the heptamer−nonamer and 12/23 bp spacer signal. They may be recognised by a specific endonuclease.

The class of antibody is determined by which C region is used. B cells always synthesise IgM and IgD initially (IgM and IgD are produced by alternative splicing of a 25 kb primary transcript) and then IgG or IgE or IgA is synthesised as a result of another DNA/DNA recombination to delete the intervening C segments. This is known as *class switching*. There are two types of antibody molecules, membrane-bound and secreted and this reflects variations in the C-terminus generated by differential poly-adenylation (Figure 1.4B).

B cells only produce antibodies with one specificity of combining site. This means that only one of the chromosome pairs is used to form the active gene, a phenomenon known as *allelic exclusion*. In addition only one of the light-chain genes, either kappa or lambda, is expressed in one cell. This is called *isotypic exclusion*. We will discuss B-cell development and the regulation of rearrangements and antibody synthesis in more detail in Chapters 8 and 9.

Unordered rearrangements in the genomes of abnormal mammalian cells

Several rearrangements in mammalian genomes are associated with pathological changes in the cells, notably the transformation of normal cells into tumour cells, known as *oncogenesis* (reviewed by Yunis, 1983). It is not known if the rearrangement is the cause or the effect of oncogenesis but in several types of tumours the same rearrangement is observed. For example in chronic myelocytic leukaemia (CML; a malignancy of white blood cells) an abnormal chromosome called the Philadelphia chromosome (Ph[1]) is present in most of the tumour cells. This is produced by a *translocation*. A region of chromosome 9 is swapped over for a region of chromosome 22. The abnormal chromosome 22 is the Ph[1] chromosome. The break point within chromosome 22 falls within a 5.8 kb region, *bcr* (breakpoint cluster region) whereas the break points in chromosome 9 are scattered over 50 kb (reviewed by Adams, 1985). Many translocations involve the immuno-globulin genes (reviewed by Leder *et al.*, 1983) and may be a function of aberrant immunoglobulin gene rearrangements. Burkitt's lymphoma is a malignancy of human B cells involving a translocation between chromosome 8 and either chromosome 14 (t8:14) or chromosome 2 (t8:2) which carry the heavy-chain and kappa light-chain genes respectively. About 17 sites in human chromosomes have been identified as common break points, and are called *fragile sites* (Le Beau and Rowley, 1984).

There is a group of normal cellular genes, *oncogenes*, whose altered expression is often associated with the tumour phenotype. In both of the translocations described an oncogene has been moved to a new location. One idea is that the new chromosomal location alters the regulation of oncogene expression so that the cell has too much or too little of the protein or the protein is made at the wrong time. Another idea is that the oncogene is mutated as a result of translocation. The inappropriate expression or the expression of an altered oncogene protein may be a key step in the formation of tumour cells and we will discuss this further in Chapter 11.

DNA amplification

The bulk of the DNA in the eukaryotic chromosome is replicated only once per cell division. There appears to be a block to re-initiation of replication. There are, however, some exceptions to this. Firstly, in many insects all of the chromosomes are repeatedly replicated without cell division in certain cells, e.g. salivary glands. This phenomenon is called *polyteny*. Secondly, several DNA viruses, e.g. SV40, are also able to circumvent the normal restriction of replication to once per cell cycle. Thirdly, while whole chromosomes are rarely re-replicated without cell division the over-replication of specific sub-sections of a chromosome is a normal occurrence in a number of cell types and developmental stages. This is called *DNA amplification* (reviewed by Stark and Wahl, 1984). The major function of localised amplification is to provide much larger quantities of a gene product than could be achieved by even very efficient transcription of a single gene. In many cases amplification is specifically regulated. A classical example of this is the amplification of ribosomal RNA during oogenesis in many species. For example in *Xenopus laevis* oocytes the rDNA is amplified 1000-fold to allow the accumulation of 10^{12} ribosomes per cell. Any of the normally repeated rDNA genes can be amplified and the new DNA is present as extrachromosomal rings. Protein-coding genes can also be amplified. The chorion genes that encode the major eggshell proteins of *Drosophila* are specifically amplified 15−60 fold in follicle cells.

It is likely that many sections of the genome undergo amplification at random but often this is only detected when selection is imposed. For example, the treatment of cells with the cytotoxic drug methotrexate (MTX) results in the appearance of a few resistant cells (MTX^R). Many of the MTX^R cells contain multiple copies of a region of DNA containing the gene encoding dihydrofolate reductase (DHFR), the enzyme that is inhibited by MTX. This DNA may be amplified 100 to 1000 times and is often present as small extrachromosomal elements called *double minute chromosomes (DM)* (Schimke, 1984). The extent of the amplified DNA can be large, about 100 kb, so that several genes may be co-amplified. Amplified DNA can also remain in the chromosome either at the original site or transposed to a new site. It tends to form a distinct structure called by cytogeneticists a *homogeneously stained region (HSR)*.

There is increasing evidence that gene amplification may be involved in oncogenesis and cellular ageing. A region of DNA containing the *myc* oncogene is amplified 10 − 200 fold in primary retinoblastomas (Lee *et al.*, 1984; Turner *et al.*, 1985) and many carcinogens such as U.V. and hydroxyuria also induce DNA amplification. The mechanisms and control of DNA amplification will be discussed in more detail in Chapter 9.

The structure of DNA, chromatin and chromosomes

DNA can assume a variety of physiochemical properties and it is never naked. In eukaryotes the DNA interacts with a variety of proteins and it is coiled and supercoiled to form the chromosomes.

Physicochemical properties of DNA

There are three major forms of DNA (Dickerson *et al.*, 1982). The right-handed helical A and B forms and left-handed Z-DNA. The most familiar B form of the DNA helix exists as two antiparallel strands wound around each other to produce a right-handed helix with a major and minor groove and 10.5 bp per turn. A-form DNA has 11 − 12 bp per turn with shallower minor and deeper major grooves. Left-handed or Z-DNA (Rich *et al.*, 1984), on the other hand, is also antiparallel and base-paired but the strands are coiled around each other in the opposite direction. This produces a molecule with 13.6 bp per turn and only one groove. The structures of Z- and B-DNA are so different that antibodies specific for Z-DNA can be produced. They are also readily distinguished by a variety of chemical reagents and specific Z-DNA-binding proteins have been isolated, e.g. from *Drosophila* cells. The formation of Z-DNA is favoured by pairs of alternating pyrimidine and purine dinucleotides, the most effective being d(CG) followed by d(CA) and d(TG). The exception is d(TA) which rarely forms Z-DNA. The B and Z forms of DNA are in equilibrium but under physiological conditions the Z form is less stable because of the proximity of the phosphate groups. It has been proposed that Z-DNA plays a role in regulating gene expression and it might also be important in determining effective pairing during recombination.

The physicochemical properties of DNA can markedly influence gene expression either because specific proteins can recognise small local differences in DNA forms or because gross topological changes either enhance or restrict the access of proteins to a transcriptional unit. The local base composition of stretches of DNA can influence conformation. For example regions that are very A+T rich are susceptible to local *unwinding* or *breathing* of the double helix, particularly in response to thermal fluctuations, because of the relatively unstable Watson−Crick hydrogen bonding between AT as opposed to GC base pairs. Many 5′ flanking regions are AT rich and this predisposition to single-strandedness may affect DNA/protein interactions required for transcription. Different nucleotide compositions can also cause minor modulations in the double helix by producing deviations in the sugar−phosphate back-bone. These are known as *twist angle* variations (Zimmerman, 1982).

The normal B helix can be further wound so that its long axis also describes a helix; this is a *super-helix*. If the helix is wound in the right-hand direction the B-helix will tend to become overwound, and in the left-hand direction it will tend to become underwound. This tendency to under or overwind is accommodated by *supercoiling*. Underwinding induces *negative supercoiling* which produces a strain on the DNA molecule called *torsional strain*. Negative supercoiling exists in all naturally occurring DNA and any factors that relieve the strain will be thermodynamically favoured. Such factors are, for example, breathing at AT-rich regions, interaction with nucleosomes (see later) and intra-strand base pairing producing cruciform structures. Because Z-DNA can accommodate more base pairs per turn, this also reduces negative supercoiling. A number of DNA-binding proteins induce *bends* and *kinks* in DNA (Wu and Crothers, 1984; Widom, 1984).

Certain cytosines are enzymatically methylated at position 5 in the base to produce 5-methyl cytosine (me^5C). This occurs most frequently in

CHAPTER 1

cytosines in d(CG) dinucleotides and this dinucleotide is distributed non-randomly in the genome, it is most abundant in the 5' flanking regions of genes. Methylation stabilises Z-DNA by increasing hydrophobic interaction and therefore could predispose the control region of genes to form Z-DNA that cannot switch to the B form. This may prevent the local unwinding that is necessary for transcription to occur (reviewed by Doerfler, 1983).

It is difficult to establish the significance of different DNA conformations in determining genome functions and there is some controversy over whether some of these structures, particularly Z-DNA, actually form *in vivo*.

Structure of chromatin

Eukaryotic DNA is always condensed or packaged with a number of different proteins to form a *nucleoprotein complex* called *chromatin* (Butler, 1983; Eissenberg *et al.*, 1985; Pederson *et al.*, 1986). There are two broad classes of proteins, the *histone* and the *non-histone chromosomal (NHC)* proteins. There are five major histones, H2A, H2B, H3, H4 and H1 (H5 in birds) and they form the structural components of chromatin. The histones are abundant proteins, about 6×10^6 molecules per cell. They are small and rich in the basic amino acids lysine and arginine. The primary sequence of histones H2A, H2B, H3 and H4 is highly conserved between different eukaryotes whereas H1 has a conserved central core but otherwise shows a divergence between species and shows several variant forms within a species. The basic unit of packaging is the *nucleosome*, which consists of a disc-shaped protein core composed of two copies of each of the histones, H2A, H2B, H3 and H4, this is the *histone octamer*.

The DNA double helix is wound around the outside of this octamer to produce a left-handed (negative) super-helix of 1.8 turns comprising 145 bp of DNA. This 145 bp of DNA plus the histone octamer is called the *nucleosome core particle*. Adjacent nucleosome core particles are separated by about 60 bp of *linker DNA* so that a stretch of chromatin looks like an array of beads on a string. The linker plus nucleosome core is the nucleosome or *nucleosome repeat*, a single nucleosome is a *mononucleosome*. On average there is 200 bp of DNA per nucleosome repeat but this can vary because of different lengths of linker DNA. The histone H1 interacts with the linker DNA with the tightest association at the DNA exit and entry points at the nucleosome core particle. This effectively stabilises two complete turns of DNA around the histone octamer. The nucleosome core particle has been crystallised and there is a detailed biophysical description of the DNA−protein interactions (Richmond *et al.*, 1984).

The NHC proteins are a vast collection of different proteins with functions ranging from the enzymes involved in DNA replication and transcription through to structural proteins. Some of the NHC proteins are as abundant as the histones and show the same distribution and these probably serve a general structural role. Some are restricted to particular regions. Other NHC proteins are moderately abundant and are found generally associated specifically with *active chromatin* which is capable of being transcribed (see later). In particular there is a group of small highly charged proteins called the *high mobility group* (HMG) proteins which associate with active chromatin. Some other NHC proteins are extremely rare and act as the specific effector molecules controlling transcription of individual genes e.g. TFIIIA (transcription factor III A), which is required for pol III-directed transcription of *Xenopus* 5S RNA genes, Spl (first identified as required for SV40 transcription) and the heat-shock transcription factor (HSTF) required for transcription of the *Drosophila* heat-shock genes. These specific transcriptional effector proteins are reviewed by Enver (1985), Dynan and Tjian (1985) and Maniatis *et al*. (1987) and will be discussed further in Chapter 8.

The factors that determine the position of the nucleosomes along a stretch of DNA are not known. It appears that for much of the chromatin the position of nucleosomes may be random (discussed in Kornberg, 1981) but there is some evidence for a more precise positioning of nucleosomes that has been called *nucleosome phasing*. This implies that a specific site fixes a nucleosome and therefore if the linker length remains constant adjacent nucleosomes will also be fixed. There are also examples of nucleosome-free DNA such as the yeast centromeres and the SV40 origin of replication (Chapter 8).

The most significant perturbation of the nucleosome structure and organisation is found in active chromatin which constitutes less than 10% of the bulk chromatin in higher eukaryotes. The rest of the chromatin is *inactive chromatin*. In different cell types different regions of the DNA will be packaged as active chromatin reflecting the differences in gene expression. There are regions of chromatin which are not transcribed in any cell type, referred to as *heterochromatin* and these are usually located in the centromeric and telomeric regions of the chromosome. In the extreme case of X-chromosome inactivation in female mammals, an entire chromosome is heterochromatic and not expressed (Martin, 1982). The nucleosome structure in heterochromatin is highly regular and Strauss and Varshavsky (1984) have shown that, in the case of the α-satellite, this is due to the binding of a specific protein which phases the nucleosomes. These workers have speculated that similar proteins may set local phasing in active chromatin.

Active and inactive chromatin

There are a number of differences between active and inactive chromatin (Spiker, 1984). An indication of these differences comes from studies such as those by Wu (1980) who showed that a region of *Drosophila* chromatin containing a heat-shock gene, *hsp70*, was hypersensitive to attack by endonucleases. This *hypersensitive region* did not form the expected protected mononucleosome fragments but appeared to be completely degraded. If only a brief digestion was used a few highly preferred target sites for endonuclease attack could be detected. These are known as *hypersensitive sites*. Nuclease hypersensitive sites have now been found in many genes. They are often in the 5' flanking region and many appear prior to the onset of transcription. Their distribution can reflect tissue-specific or developmental stage-specific transcription (Elgin, 1984). The region of chromatin displaying an altered sensitivity to endonucleases extends far beyond the transcribed portion and can be as large as 100 kb (Lawson *et al.*, 1982). The chromatin structure in this region is proposed to be more open and in the transcribed area itself there may be nucleosome-free regions, as observed in the *Drosophila hsp70* gene (Karpov *et al.*, 1984). This open structure may facilitate access to the DNA of RNA polymerase and regulatory proteins.

There may be a number of explanations for increased DNase sensitivity in addition to the absence of nucleosomes. Specific nucleotide sequences are inherently more susceptible to cleavage (Keene and Elgin, 1984) and torsional stress increases sensitivity (Villeponteau *et al.*, 1984). The formation of Z-DNA which decreases torsional stress could reduce DNase hypersensitivity.

The histones associated with active genes themselves also show various modifications. H2A becomes linked to a small protein ubiquitin to form uH2A and there is often hyperacylation, phosphorylation and poly-ADP ribosylation of histones. It is not known, however, if these modifications are a cause or consequence of transcriptional activity. The histone H1 appears to act as a general repressor of transcription of, for example, globin genes (Weintraub, 1984) and *Xenopus* 5S RNA genes (Schlissel and Brown, 1984). Clearly understanding the mechanistic basis for the difference between active and inactive chromatin will give very important information about the control of gene expression.

Higher order chromatin structure

The familiar structure of chromosomes results from higher order packaging of chromatin. The next unit of structure is a *supranucleosome organisation* which is visible as a *30 nm chromatin fibre* in E.M. sections (Felsenfeld and McGhee, 1986). The string of nucleosomes is thought to supercoil with 6−7 nucleosomes per turn. The histone H1 is essential for this higher order structure as it binds to the linker via its globular central domain and to the nucleosome via its hydrophilic carboxy and aminoterminal arms to bring them together. Binding to successive nucleosome cores is cooperative, opening up the possibility of *combinatorial gene regulation* where whole regions can be condensed and decondensed rapidly by a single interaction with one H1 molecule.

The 30 nm fibre is then organised into chromosomal domains which are loops of 20−100 kb of DNA held together, at their bases, by protein. This model is based on direct observation of *lampbrush chromosomes* in *Chironomus* oocytes and electron microscopy of chromatin after gentle cell lysis. Biochemical analysis of chromatin prepared in this way also shows a fraction of DNA that is freely accessible to nucleases and a fraction that is bound to protein. The protein is referred to as the *nuclear scaffold* or *nuclear matrix*. Attachment to the nuclear matrix appears to be mediated by specific nucleotide sequences called *matrix association regions (MAR)*. It has been suggested that these loops represent functional domains for transcription. Ciejek *et al.* (1983) have analysed 11 genomic sequences and found that only actively expressed genes were bound to the nuclear matrix. Within the 100 kb DNase I sensitive ovalbumin chromosomal domain only the transcribed

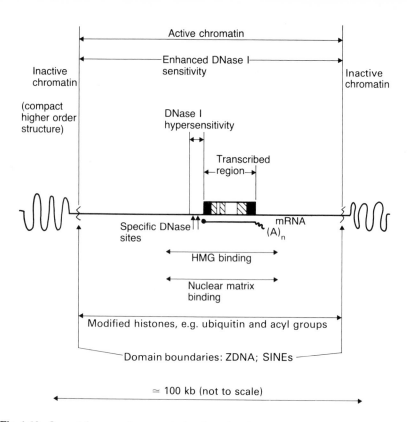

Fig. 1.12. General features of a transcriptionally active chromatin domain.

region was associated with the matrix and this association was lost when transcription was switched off by the removal of oestrogen (Strumph *et al.*, 1983). Two other important observations are that topoisomerase II is one of the scaffold proteins (Earnshaw *et al.*, 1985) and that there is one 50 bp long d(CA/GT) tract every 50—100 kb which could mark the boundaries of transcriptional domains (Hamada *et al.*, 1982). We have already seen that for many genes in eukaryotes there is no obvious clustering of similarly regulated genes, therefore the chromatin loops may simply represent the smallest region that can be transcribed.

The chromosome loops are further coiled on themselves to produce the final 7000 to 1 condensation from linear B-helix to chromosome. This final coiling can vary in compactness and this gives rise to the characteristic light-and-dark banding pattern observed in all metaphase eukaryotic chromosomes. These banding patterns are constant for a given chromosome and may reflect functional differences between the regions although there is little evidence to support this.

In Figure 1.12 we have summarised some of the features of chromosome structure as they may relate to gene expression.

Summary

The eukaryotic genome contains a large number of different types of DNA elements. There are the genes that produce the essential proteins and RNAs, there are structural components that are required for chromosome segregation, i.e. centromeres and telomeres, there are replication origins and there are sequences that function to control the processes of gene expression and chromosome replication. In addition, there are mutated genes, i.e. pseudogenes and repeated sequences that in some cases such as transposons, can frequently move to new sites in the chromosomes. DNA can be packaged with histones into a transcriptionally active or inactive structure. DNA can also rearrange both specifically and randomly and parts of chromosomes or whole chromosomes can be amplified. Gene expression can be controlled at any point from the availability of the DNA template to the stability of the final protein.

The challenges that we present in this book are (1) to discover how these DNA elements interact and function and (2) to develop ways of exploiting these elements. The key technology that is meeting these challenges is the transfer of cloned genes into living eukaryotic cells and organisms; this is genetic engineering.

2 The Aims and Principles of Gene Transfer

Introduction

The discovery of methods for transferring cloned DNA back into living cells has provided the key to understanding a vast spectrum of fundamental biological processes.

One of the best ways of establishing the functions of genes and their products is to observe what happens when the genes are mutated. This is the simple principle of classical genetic analysis. However, one of the major limitations of classical genetics is that only mutations that produce a readily detectable phenotype can be studied. There is also the problem that in diploid organisms many mutations can only be detected after appropriate breeding strategies to generate homozygotes for the mutation. This means that many mutations may be missed and experiments with animals with long breeding cycles can take months or years. It is also of course not possible to subject man to genetic analysis except by retrospective studies of family pedigrees. Another problem is that some mutations are lethal in homozygotes. Many key steps in early embryogenesis and differentiation fall into this category and so cannot be studied in depth by classical genetics. Finally the existence of a gene is really only proved by the discovery of a mutation in a gene, genetic maps are not established by studying normal genes but by locating the mutant forms *(alleles)*. Any gene for which mutants or variants have not been detected, officially does not exist.

Gene transfer techniques have revolutionised the approaches for genetic analysis. It is now possible to clone genes from any organism, often without any knowledge of the function of the gene product. This is extending the genetic map of many organisms. The genes can then be introduced into virtually any cell including human cells. All possible mutations can be studied by simply changing the cloned DNA before re-introducing it into cells. For the first time genes can also be transferred between species allowing conservation or divergence in biological processes to be established. This *heterologous gene transfer* often allows the biochemical properties of cloned genes to be established more readily because the gene may confer new properties on a cell which does not normally have the gene. Fundamental processes such as DNA replication and chromosome behaviour can be studied because the key DNA components have been artificially reconstructed. The complex processes of development and differentiation have long been suspected to be controlled by hierarchies of interacting gene products. These complex regulatory interactions can now be analysed by introducing different combinations of genes into the same cell and directly testing their interactions. Many of these experiments are done with simple cell culture systems that generate answers in weeks rather than years.

Gene transfer technology has also extended the ways that man exploits nature. Useful gene products can be produced in huge quantities by transferring genes into simple cells such as yeasts or cultured animal cells and then growing these cells in large-scale cultures. For example, human proteins such as interferons and hormones that can only be found in minute quantities in human tissues can now be manufactured by the kilogramme. New organisms can also be created by introducing new genes. For example, pest resistance genes can be introduced into food crops. This is the essence of the new science of *biotechnology*.

Gene transfer is also producing great advances in medicine. Gene-transfer techniques have allowed genes that are involved in cancer to be isolated and then characterised. Precise models for specific human diseases such as cancer, diabetes, muscular dystrophies and growth defects are being established by introducing new genes and gene combinations into mice. There is also the possibility that human diseases that result from defects in single genes might be treated by transferring normal genes into some of the cells of these patients. This is *gene therapy*.

As with all areas of science there are some concepts and terminologies that are fundamental to many aspects of gene transfer. We will now briefly introduce these basic ideas which will be repeated and extended in later chapters.

Introducing DNA into cells

All cells are surrounded by a plasma membrane which is impermeable to large molecules. In the case of plants and fungi there is an additional layer of highly cross-linked and polymerised carbohydrate, the cell wall. DNA is a large molecule, and so the first requirement for gene transfer is a method for introducing the DNA into the cell across this impermeable barrier. If the DNA that is being transferred to the cell is derived from a different organism it is called *foreign* or *heterologous* DNA. The cell that receives the foreign DNA is described as a *transformed cell* and the process is called *transformation*. Unfortunately, this is a rather vague term because it is used to describe any change in the properties of a cell irrespective of the mechanism involved. This means that, for example, when a normal cell becomes a tumour cell, the tumour cell is described as a transformed cell. Clearly not all cells that have been transformed by the introduction of foreign DNA behave like tumour cells. In organisms such as mammals where there could be confusion it is desirable to qualify the term transformation. For example, when the transformation is caused by the introduction of DNA it is often called *DNA-mediated transformation*. Sometimes the word *transfection* is used instead of transformation; this is only strictly correct when the DNA is derived from a virus but this rule is rarely obeyed. In organisms such as yeast there can be no confusion because yeast cells do not become tumour cells and transformation is invariably used.

Some of the general approaches for introducing DNA into cells are outlined in Figure 2.1. The first approach, called *DNA-mediated transfection* or *transformation*, can be used for a wide variety of cells that do not have a cell wall or where the cell wall can be easily removed. The methods were arrived at by trial and error and there is in fact very little real understanding of how they work. It was discovered that cell membranes can be reversibly permeabilised by various chemical treatments, for example, with high molecular weight polymers such as DEA-dextran or PEG (polyethylene glycol) this is often enhanced by divalent cations, particularly calcium ions. When DNA is mixed with these chemicals it is taken up into the cell. Also DNA can be precipitated with calcium phosphate to form visible aggregates and cell membranes treated with these aggregates are permeabilised and take up the DNA. The procedures are often further enhanced by treatments that subject the cells to a reversible osmotic shock; glycerol and dimethylsulphoxide (DMSO) are often used for this. A problem with these DNA-mediated transformation procedures is that many of the cells are damaged and do not recover. Some cells types are so sensitive that they cannot be used. The procedure is also inefficient and many cells that survive the chemical treatment do not contain any transferred DNA. When DNA is successfully taken up by cells it is taken up into the cytoplasm and must be transported to the nucleus before it can be expressed. Much of the DNA is degraded at this stage and therefore even cells that take up DNA may not actually express the new genes. This means that in any experiment only a fraction of the cells that survive actually contain new DNA in the nucleus.

One way of overcoming the problems of DNA-mediated transfection is to inject the DNA directly into the nucleus of the cell, a process called *microinjection*. This requires technical skill and many cells die. However, nearly all those that survive will have new DNA in the nucleus. A third technique is not to purify the cloned DNA but to take the original bacterial clone that harbours the gene, remove the bacterial cell wall with enzymes to release protoplasts and fuse these to the recipient cells using agents such as PEG that enhance membrane fusion. The entire contents of the bacterium are introduced into the cytoplasm and this procedure is known as *protoplast fusion*. It can be more efficient than DNA-mediated transformation but the introduction of extraneous bacterial DNA is often undesirable. The last general technique is to exploit the fact that many microorganisms transfer nucleic acids into eukaryotic cells as a part of their natural life cycle. This applies mostly to viruses but some bacteria transfer parts of their plasmids directly into plant cells. There are many sophistications of these basic procedures and a tremendous amount of research is aimed at improving or devising new techniques to extend the versatility of gene transfer experiments.

Gene transfer vectors

The aim of gene transfer is to analyse or exploit the transferred DNA. Therefore, the DNA is usually introduced into cells in a form that makes

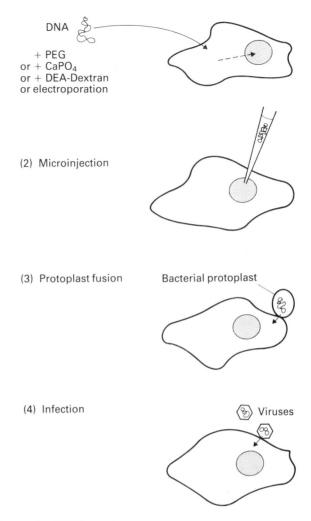

(1) DNA-mediated transformation

DNA

+ PEG
or + CaPO$_4$
or + DEA-Dextran
or electroporation

(2) Microinjection

(3) Protoplast fusion   Bacterial protoplast

(4) Infection   Viruses

Fig. 2.1. Introducing DNA into cells.

these goals easier to achieve. Just as there are specialised DNA molecules, *cloning vectors*, that allow the original cloning of a gene so there are special DNA molecules that facilitate gene transfer experiments. These are *gene transfer vectors* (Figure 2.2).

A common problem with trying to develop gene transfer techniques is knowing when DNA has been successfully transferred into recipient cells. A second problem is distinguishing between cells that contain the new DNA and are therefore interesting to study and those that have survived the transfer procedures but do not contain DNA and therefore should not be studied. In many cases the uninteresting cells are in the majority. To solve this problem the DNA of interest is linked to a gene transfer vector that carries a *marker gene*. When this gene is expressed in the recipient cells it confers readily detectable new properties on the cells. The most useful type

AIMS AND PRINCIPLES OF GENE TRANSFER

A Integrative vector

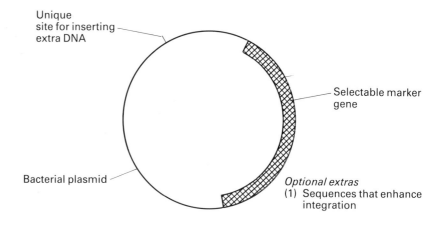

Unique site for inserting extra DNA

Selectable marker gene

Bacterial plasmid

Optional extras
(1) Sequences that enhance integration

B Autonomous vector

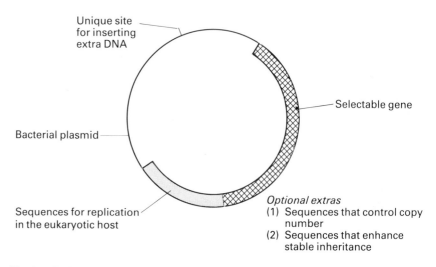

Unique site for inserting extra DNA

Selectable gene

Bacterial plasmid

Sequences for replication in the eukaryotic host

Optional extras
(1) Sequences that control copy number
(2) Sequences that enhance stable inheritance

Fig. 2.2. Basic gene transfer vectors.

of marker gene is one which will allow transformed cells to be selected against a background of untransformed cells, i.e. *a selectable marker gene*. For example, there are cells which depend on special media to provide certain nutrients, because they have a mutation in a key gene. The selectable marker gene could encode the corresponding normal enzyme which will allow growth in the absence of added nutrients. This is a *complementation selectable marker*. All successfully transformed cells will survive when the nutrients are removed whereas untransformed cells will die. Alternatively the selectable marker could carry a drug-resistance gene so that all untransformed cells will die if the drug is added to the culture. This type of marker is a *dominant selectable marker*. The selectable markers both prove that DNA has been transferred and allows the background untransformed cells to be destroyed. The simplest type of gene-transfer vector is therefore, simply a cloning vector such as pBR322 which carries a selectable marker gene.

Once DNA has reached the nucleus it may be expressed and therefore can be detected by phenotype but as the recipient cells divide it will be diluted and become undetectable. For transferred DNA to be inherited by cells as they divide it must either become integrated into the host cell chromosomes or it must be capable of autonomous, i.e. independent replication. The simple vector in Figure 2.2A relies upon integration for stable maintenance and is therefore, called an *integrative vector*. The frequency of integration can be very low and one of the goals of gene transfer research has been to improve this frequency and also to control the process more closely in terms of the site of integration and the number of copies that are integrated. A number of DNA sequences that enhance integration can be included in this vector. These can be derived from viruses or plasmids that normally integrate into genomes or in some cases a chromosomal gene will allow integration by homologous recombination. There are, however, a number of vectors that will replicate autonomously; these are often based on plasmids or viruses that have DNA sequences that act as replication origins that respond to cellular replication proteins and/or virus-or plasmid-encoded proteins. Vectors can be constructed that contain replication signals from several different hosts and these can be used to transfer and maintain genes in all of these hosts. As autonomous vectors can be easily isolated from each host because they are distinct from the chromosome, they can be shuttled between the different host cells very conveniently. They are often called *shuttle vectors*. Autonomous vectors can be present in one copy or multiple copies per cell depending upon the precise replication signals that are used. They are referred to as *single copy* or *multicopy vectors*. In some cases copy number can be varied during the experiment by altering the experimental conditions. In some cases the vectors always replicate in step with cell division; this is a *copy number controlled system*. In other cases there is no control and the vector simply replicates until the cell is filled with plasmid; this is *runaway replication*. These last types of vector are only useful in short-term experiments because the over-replication ultimately kills the cell. A number of DNA sequences can be added to some autonomous vectors that allow them to be stably inherited and to control their copy number.

Expression vectors

There are many situations, particularly when genes are being exploited for biotechnology when the expression of the gene must be modified. The signals that ensure gene expression in prokaryotes and eukaryotes are very different and if a bacterial gene is simply transferred to a mammalian cell it is not expressed except in rare, fortuitous cases. Even the gene expression signals from different eukaryotes may not be interchangeable. Yeast promoters, for example, are not recognised in mammalian cells. Whenever a gene is transferred to a *heterologous cell*, i.e. a cell that is different from the cell in which the DNA originated, and good expression is required, new expression signals must be added which are known to function in the new host cell. It is also very common to wish to transfer and express a cDNA clone and this, of course, will have none of the 5' and 3' flanking control signals that are required for expression. Special vectors that provide the relevant signals have been developed and these are called *expression vectors*. These can be based on integrative or autonomous gene transfer vectors (Figure 2.3). They have a selectable marker and a replication or integration system but in addition there are the sequences derived from genes that are normally expressed in the recipient cell. These are the 5' transcription and translation control signals and the 3' transcription termination signals and in some cases an intron and associated splicing signals. These are often referred to as an *expression cassette*. The new DNA is inserted in the expression vector so that its transcription and translation are controlled by the vector.

There are several different expression configurations, these are outlined in Figure 2.4. The first is a *transcription fusion*. This is where the 5' control signals consist of the entire promoter, the cap site and part or all of the 5' untranslated leader up to but not including the gene's initiating ATG codon. The heterologous coding sequence is then fused in at this point. This fusion gene produces a hybrid mRNA that has the 5' leader sequence from the vector, the coding sequence from the foreign gene and the 3' trailer from the vector. This mRNA will be correctly translated using the initiating ATG from the foreign DNA to produce an authentic foreign protein. It is essential that the vector contains no additional ATG codons in the 5' leader or this would give aberrant translation initiation. The second configuration is a *transcription and translation fusion*. This is where vector 5' control signals consist of the entire promoter, the entire 5' untranslated leader and part of the N-terminal coding region of the gene. This can be as little as the initiating ATG codon up to an extensive stretch of amino acids.

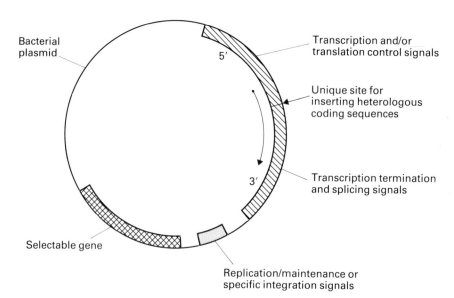

Fig. 2.3. A basic expression vector.

A Transcription fusion

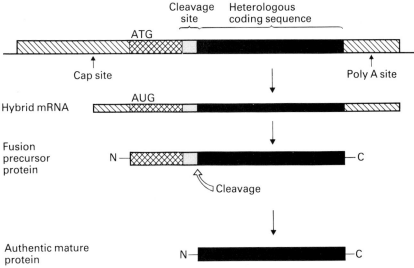

B Transcription and translation fusion

C Cleavage fusions

Fig. 2.4. Expression configurations.

AIMS AND PRINCIPLES OF GENE TRANSFER

This type of vector may be used to express foreign coding sequences which lack their own initiating ATG, e.g. part of a coding region specifying a functional domain or the mature region of a normally secreted protein when the initiating ATG is at the start of the signal sequence which has been removed. It is, of course, essential that the foreign DNA is inserted in such a way as to maintain the correct translational reading frame with respect to the initiating ATG. The fusion gene produces a hybrid mRNA and a hybrid fusion protein that contains additional N-terminal amino acids. This is not an authentic protein. In some cases the fusion confers useful properties on the protein. In other cases, particularly in some applications of biotechnology, a fusion protein is undesirable and even one extra methionine on the N-terminus can perturb the biological properties of the protein. A third configuration is really a special case of a transcription and translation fusion called a *cleavage fusion*. This is where DNA coding for a recognition site for cleavage by a specific endoprotease is inserted at the junction between the vector coding sequences and the foreign coding sequences, resulting in a hybrid mRNA and a hybrid precursor fusion protein which when exposed to the endoprotease is cleaved to release authentic protein. This can occur *in vivo* or *in vitro*. A special case of the cleavage fusion is when the vector coding sequence specifies a signal peptide that directs secretion of the foreign protein with subsequent cleavage removal of the signal by natural processing enzymes.

Virus vectors

Many aspects of the infection and replication strategies of viruses can be harnessed to produce gene-transfer and gene-expression vectors. Viruses can introduce nucleic acids efficiently into cells via specialised surface proteins that interact with cell surface receptors. Many amplify their genomes during replication, so that any foreign genes that they carry will be produced in high copy numbers and therefore expressed at high levels. Some viruses integrate into the genome, allowing the production of stably transformed cells. To convert a viral genome into a gene-transfer/expression vector several criteria must be satisfied. First, as with the development of any vector a suitable cloning site must be present and in many cases this must be engineered artificially. The virus genome must still be capable of replication with the foreign DNA present. If the ability of viruses to introduce nucleic acid into cells is to be exploited then the recombinant virus must still be capable of being assembled into viral particles. This is called *packaging*. Quite often there are constraints on the maximum size of the nucleic acid that can be packaged and so deletions must be made in the virus genome to accommodate the foreign genes. In many cases these deletions disrupt the genes required to make an infectious virus; the virus is therefore termed *defective*. Defective viruses can still be propagated in cells by supplying the missing proteins from a second virus introduced into the cells at the same time. The latter is known as a *helper virus*. It is quite common for the helper virus to have a defect in a different gene, so that each virus is non-infectious alone but they complement each other. An alternative approach is to construct a cell line that expresses the missing protein, which can then be used by the defective virus. These are called *helper cell lines*.

Many viral genomes contain strong gene expression signals to allow the production of the massive amounts of viral proteins required to produce many thousands of viruses from an infected cell. These expression signals can be manipulated to drive the synthesis of foreign proteins.

Distinguishing transferred genes from endogenous genes

In many gene-transfer experiments it is difficult to distinguish between the expression of the transferred gene and that of a similar or identical endogenous gene. If, for example, a human globin gene was transferred into a human cell, the mRNAs and proteins from both genes would be identical in size and would react with the same biochemical probes. To overcome this problem the transferred gene is marked in some way, often by making a hybrid with a second gene known as a *reporter gene* (Figure 2.5). Bacterial genes coding for enzymes such as β-galactosidase or chloramphenicol acetyl transferase (CAT) are particularly useful. These enzymes are easy to assay in cell extracts and so provide a quick biochemical test for gene activity. The reporter gene can be fused to the test gene at any position, depending upon the analysis. If, for example, the experiment only concerns the 5′ control signals then the reporter gene is fused just downstream from the 5′ untranslated leader. If splicing is being analysed the reporter gene is fused into an exon that follows an intron (Figure 2.5B). Another approach is to construct a mini gene, for example by deleting several exons. The resulting transcript will be much smaller and can therefore be distinguished from endogenous transcripts.

Analysing the status and expression of transferred DNA

Once DNA has been successfully introduced into recipient cells the analysis begins. It is usually important to establish the status of the DNA: Is it autonomous or integrated? Has it been rearranged? Is it present in one copy or many copies? The expression of the transferred DNA can then be analysed to answer questions concerning the quantity and quality of the mRNA, and whether the protein is produced, correctly modified and has biological activity. Some of the analyses, particularly the assays for biological activity and the regulation of gene activity will be peculiar to the particular gene but there are several standard approaches which are widely used. These are briefly outlined below.

DNA status

The status of the transferred DNA is analysed by using specific combinations of restriction enzymes followed by the simple technique of *Southern blotting* (Southern, 1975). The general procedure is as follows. Total DNA is isolated from transformed cells, and aliquots are digested with

AIMS AND PRINCIPLES OF GENE TRANSFER

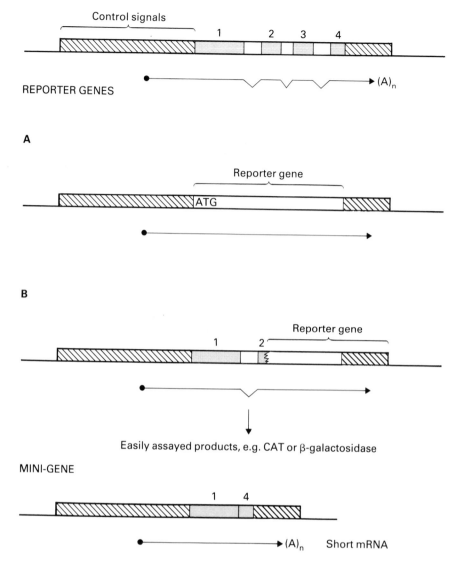

Fig. 2.5. Technique for distinguishing transferred from endogenous genes. The exons are dark numbered boxes.

restriction enzymes. The fragments of digested DNA are then separated by electrophoresis through a gel, usually composed of agarose although polyacrylamide gels can also be used. The distance that DNA migrates through the gel is approximately inversely proportional to the log molecular weight of the DNA. The DNA is then denatured while it is still trapped in the gel by simply immersing the gel in a sodium hydroxide solution. The denatured DNA is transferred from the gel to a sheet of nitrocellulose, exactly preserving the pattern of DNA fragments in the gel. The DNA is irreversibly fixed to the nitrocellulose by baking and can then be analysed.

Analysis is done by hybridising the filter with a radioactively labelled

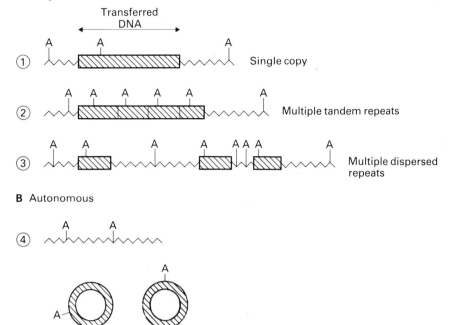

A Integrated

Transferred
DNA

① Single copy

② Multiple tandem repeats

③ Multiple dispersed
repeats

B Autonomous

④

Probe

Fig. 2.6. Some possible DNA configurations in transformed cells. The hatched box represents
the transferred DNA.

single-stranded DNA called a *probe* which is homologous to the DNA of
interest, in this case the foreign vector DNA that has been introduced into
the host cell. The probe will form a stable hybrid with the matching DNA
on the filter. When the hybridisation is finished the filter is washed to
remove any unbound probe and then placed under a piece of X-ray film to
allow the emission from the ^{32}P labelled probe to blacken the film producing a
signal. This process is called *autoradiography*. If the foreign DNA is present
in the DNA from the transformed cell then a signal will be seen. The
pattern and intensity of the signals following different types of restriction
enzyme digestion will reveal the status of the DNA. Some possible arrangements are shown in Figure 2.6 and the pattern of Southern blots
obtained by simple restriction enzyme analysis is shown in Figure 2.7.

   If a single copy is integrated, digesting with an enzyme (A) that cuts
only once in the foreign DNA will produce two fragments that hybridise
with the probe. These contain the foreign DNA and some flanking DNA
from the chromosome at the integration site. If there are multiple tandem
repeats then two flanking fragments are detected but full length foreign
DNA fragments will also be liberated. As these are in multiple copies they
will give an intense signal. If there are multiple copies at different integration
sites then a whole range of flanking fragments will be generated. If the

AIMS AND PRINCIPLES OF GENE TRANSFER

Fig. 2.7. Southern blot profiles of the DNA configurations in Figure 2.6. The panels represent autoradiographs of agarose gels and the bands represent the DNA fragments that would be detected with a radioactively labelled probe that was specific for the transferred DNA.

DNA is autonomous then a band corresponding to linear full-length DNA will be seen but there will be no flanking fragments. Autonomous DNA can also be distinguished from integrated DNA by comparing profiles of undigested DNA. Integrated DNA would be detected as a smudge in high-molecular-weight undigested DNA. Autonomous DNA will migrate as covalently closed circular DNA (this often migrates slightly differently from linear DNA of the same molecular weight). More complex patterns are produced if the DNA has been deleted or rearranged during transfer and multiple digestions with different enzymes must be used to analyse these. It is also possible to use restriction enzymes to prove that autonomous DNA has in fact replicated in the cell rather then being carried over from the original DNA used in the transfection. This uses the fact that DNA replicated in *E. coli* is methylated on adenosine residues whereas DNA that is replicated in eukaryotic cells is methylated on C residues. There are restriction enzymes that distinguish between DNA that is methylated and unmethylated at these residues, e.g. DpnI and MboI recognise the same site, i.e. they are *isoschizomers*, but DpnI only recognises the site when the adenosine is methylated and MboI only recognises the site when it is unmethylated. DNA that has replicated in eukaryotic cells is not methylated at adenosine and is therefore cleaved by MboI and not by DpnI. The reverse is true when the *E. coli* derived, and therefore methylated, DNA is cleaved with the same enzymes.

RNA analysis

There are two main questions that are generally asked about mRNA production in transformed cells. These are how much is produced and whether

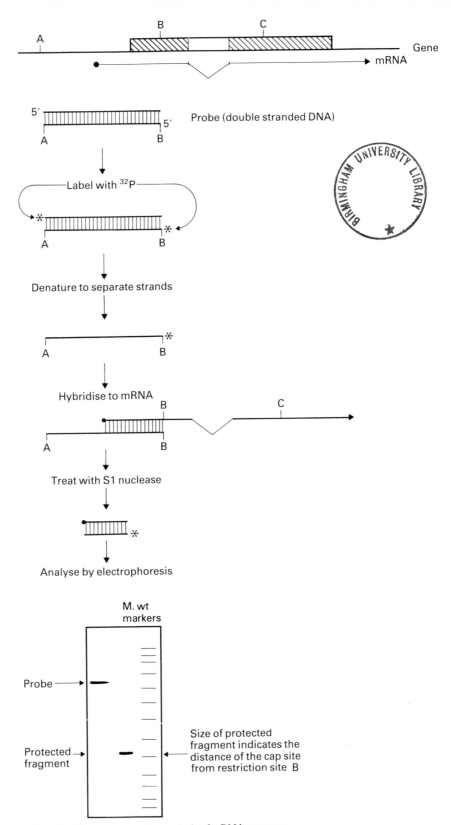

Fig. 2.8. Nuclease protection analysis of mRNA structures.

AIMS AND PRINCIPLES OF GENE TRANSFER

it is authentic, i.e. does it start and stop at the correct place and is it spliced correctly? Levels are usually quantitated by a procedure that is related to Southern blotting and has, therefore, been mischievously called *Northern blotting* (Thomas, 1980). RNA is isolated from transformed cells and subjected to electrophoresis under conditions that prevent secondary structures forming. This separates the RNA according to molecular weight. The RNA is transferred to nitrocellulose and is probed with a DNA probe that will detect the specific mRNA. The intensity of the signal is a reasonable measure of the quantity of RNA and various internal standards can be included to ensure more accurate quantitation.

The major method of characterising the authenticity of RNA is *nuclease protection*. This involves forming a hybrid between the foreign mRNA and a larger complementary nucleic acid probe of precisely known structure. The hybrid is then treated with nucleases that specifically degrade single-stranded nucleic acid but do not affect double-stranded structures. The degree to which the foreign RNA protects the probe indicates the structure of the RNA. The probe can be single-stranded DNA or RNA and a variety of nucleases can be used; the most common are mung bean nuclease or S1 nuclease which digests single-stranded DNA (Green and Roeder, 1980) and RNase A and RNase T1 (Melton *et al.*, 1984) nucleases which digest single-stranded RNA. A simplified procedure for mapping the cap site is shown in Figure 2.8. A probe is prepared using a restriction fragment that spans the presumed cap site. The fragment is labelled *in vitro* with ^{32}P and denatured to separate the strands and these can be independently purified. The single-stranded probe is hybridised with the mRNA and the resulting duplex is then treated with S1 nuclease. This degrades the single-stranded tails and releases a short duplex which is still radioactive because the 5' end of the probe is protected by the main body of the mRNA. This protected fragment can be analysed by electrophoresis and compared with size markers. The size of the protected fragment then shows the distance of the cap site from restriction site B. A similar approach can be used to map the intron−exon junctions using a combination of exonucleases such as exo III and exo VII and S1 nuclease which has exonuclease and endonuclease activities (Berk and Sharp, 1977; Weaver and Weissman, 1979). An alternative method for mapping the 5' ends of RNA is called *primer extension* (Reddy *et al.*, 1979; Figure 2.9). In this procedure a short oligonucleotide is annealed to the mRNA and then extended with reverse transcriptase, which used RNA as a template. The reverse transcript cDNA stops at the 5' end of the mRNA template and the size of the cDNA indicates the distance of the cap site from the primer. The cDNA is labelled either by using a labelled primer or by incorporating labelled nucleotides into the cDNA.

Proteins: levels and biological activity

Occasionally when a total protein extract from a transformed cell is separated by polyacrylamide gel electrophoresis and the proteins are stained with Coomassie blue a new protein encoded by the transferred DNA can be seen directly. More usually a specific probe must be used to identify the protein.

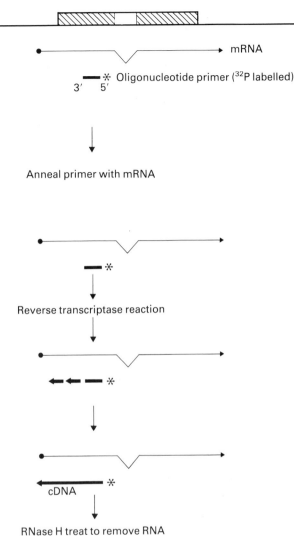

Fig. 2.9. Primer extension mapping of mRNA cap site.

The simplest approach is another blotting technique, this time called *Western blotting* (Towbin *et al.*, 1979). In this case, the proteins from the gel are transferred to nitrocellulose, and this is then probed with an antibody that is specific for the protein. The antibody may be labelled with iodine[125] and a signal is detected by autoradiography, or, more usually, it is detected by a second antibody that is tagged with an enzyme, e.g. horseradish peroxidase which gives a colour reaction when its substrate is added in the presence of suitable reagents (e.g. de Blas and Cherwinski, 1983).

Another immunological technique that is widely used is *immunoprecipitation*. This is where the extract from radioactively labelled cells is simply mixed with specific antibody and the resulting complex is then precipitated, usually by mixing with beads coated with *Staphylococcus aureus* protein A or with formalin-killed *S. aureus* (Kessler, 1975). Protein A binds to the Fc portion of an immunoglobulin which is only exposed when it has reacted with its cognate antigen. The complex is collected by centrifugation and the labelled protein is released and detected by electrophoresis and autoradiography.

If the protein is expected to be displayed on the cell surface then whole cells can be treated with a specific antibody which is labelled with a fluorescent compound such as fluorescin isothiocyanate. The fluorescent cells can then be observed by microscopy and even separated from individual cells which do not express the protein using a fluorescence-activated cell sorter (FACS). This simply detects fluorescent cells passing through the machine and deflects them into a collection chamber.

Summary

This brief account of the basic techniques of gene transfer and gene analysis will allow you to follow the design of the specific experiments that we discuss in later chapters. In Part II we describe in detail the methods for introducing DNA into a whole range of different cells and organisms. In Part III we show you the discoveries that are being made using these techniques. These include new ways to clone genes (Chapter 7); identifying gene expression signals (Chapter 8); isolating and building chromosomes (Chapter 9); discovering how protein structure and function are related (Chapter 10); designing new models and treatments for human diseases (Chapter 11) and finally the exploitation of cloned DNA for biotechnology (Chapter 12). Some useful books that cover background material about basic cloning and recombinant DNA techniques and the principles of gene transfer are listed below.

Background reading

Berger, S.L. and Kimmel, A.R. (1987) A guide to molecular cloning techniques. In *Methods in Enzymology*, Vol. 152. Academic Press, London.

Grossman, L. and Wu, R. (1987) Recombinant DNA Part E. In *Methods in Enzymology*, Vol. 154. Academic Press, London.

Old, R.W. and Primrose, S.B. (1985) *Principles of Gene Manipulation.* Blackwell Scientific Publications, Oxford.

Glover, D.M. (1984) *Gene Cloning: the Mechanics of DNA Manipulation.* Chapman and Hall, London.

Maniatis, T., Fritsch, E.F and Sambrook, J. (1982) *Molecular Cloning: a Laboratory Manual.* Cold Spring Harbor Laboratory, New York.

Wu, R. (1987) Recombinant DNA Part F. In *Methods in Enzymology*, Vol. 155. Academic Press, London.

II   Gene Transfer Systems

3

Gene Transfer into Microbial Eukaryotes

Introduction

Microbial eukaryotes are often called lower eukaryotes but there is little evidence that the strategies that they use for genome replication and expression are any less complex than those of mammalian cells. In fact many of the paradigms of eukaryotic molecular biology are being established in microbial eukaryotes. These organisms are also particularly useful in modern biotechnology because of their simple growth requirements, their ability to grow to very high cell densities and their ability to exploit a wide variety of environments.

Many microbial eukaryotes, such as the yeasts and algae, are unicellular. Some however, such as the slime moulds and filamentous fungi, can form multicellular structures composed of different cell types, i.e. they display some programme of development and differentiation. The simplicity of these systems makes them attractive models for the more complex developmental pathways in mammals.

Gene transfer in *Saccharomyces cerevisiae*

The yeast, *S. cerevisiae* is the most extensively studied microbial eukaryote and it is also amenable to the most sophisticated level of gene manipulation of any of the eukaryotes that are discussed in this book. It is one of the few eukaryotes to possess a natural non-viral plasmid, the 2 μm circle, and this has facilitated the construction of highly tractable gene transfer and gene expression systems. It is also one of the few eukaryotes where genes can be specifically targeted to chromosomal locations and where chromosomal genes can be replaced or mutated by gene transfer techniques. In addition, chromosomal components, i.e. replication origins, centromeres and telomeres, have been isolated from *S. cerevisiae*. This has led to the first *in vitro* reconstruction of a eukaryotic chromosome.

S. cerevisiae is the yeast that is used for making bread, brewing beer and fermenting fruit to produce wine. There is, therefore, an extremely long history of exploitation of this organism by the traditional biotechnology industries. Yeast now has a major role in the new biotechnology industries. In fact one of the first commercial products of recombinant DNA technology applied to eukaryotic cells is a vaccine against hepatitis B virus produced in *S. cerevisiae*.

Growth and culture

S. cerevisiae is unicellular, the cells are roughly elliptical and about 10 μM long. They contain all the organelles and supramolecular structures of a mammalian cell. The yeast cell has a typical outer plasma membrane but it is surrounded by a rigid cell wall that is composed of glucan, phosphomannoproteins and chitin side chains. The interface between the outer surface of the plasma membrane and the cell wall is the *periplasm*. This contains proteins that have been secreted through the plasma membrane.

Most of the fundamental cellular processes in yeast are similar to those in mammalian cells (Hicks and Fox, 1986) but there are some notable differences. For example the secretion and glycosylation pathways are similar except that no complex glycosyl side chains are added in yeast (Kukuruzinska *et al.*, 1987). The cell division cycle is similar to that in higher eukaryotes but the nuclear membrane does not break down during mitosis (Byers, 1981). Several yeast mRNAs are spliced but the 3' splice acceptor sequence is distinct from that in mammalian genes (Langford and Gallwitz, 1983, see Chapter 8).

The *S. cerevisiae* genome is about 13 500 kb, i.e. about four times the size of the *E. coli* genome. It is organised into 17 chromosomes and three small fragments which can be readily separated by pulse field electrophoresis (Schwartz and Cantor, 1984; Carle *et al.*, 1986). Individual yeast chromosomes are small enough to be completely sequenced and this is being done for chromosome III (Palzkill *et al.*, 1986). Less than 10% of the *S. cerevisiae* genome is composed of repetitive DNA. This is much less than most other eukaryotes (Petes, 1980). 50–60% of the genome is transcriptionally active (Hereford and Rosbash, 1977) but recent studies suggest that only 12% of the genome encodes essential genes (Goebi and Petes, 1986).

The life cycle of *S. cerevisiae* is shown in Figure 3.1. Both haploid and diploid cells can be cultured in simple defined media. If a single yeast cell is introduced into 100 ml of medium it will divide to produce 1000 million cells in about 30 hr. The time taken for each cell to divide, i.e. the *doubling time*, is about 1.5 hr in a very rich medium when all complex metabolites are provided. On a less rich medium, growth may be slower because the cells must synthesise complex molecules from precursors. The division of the cells is asymmetric, the new cell or *daughter cell* is formed as a small bud on the surface of the original or *mother cell*. This is why *Saccharomyces* is called a *budding yeast*. Yeast can also be cultured on solid medium in exactly the same way as bacteria to produce colonies from single cells.

Haploid cells of opposite mating type, called *a* or α, can mate to produce diploids. Mating type is controlled by the *MAT* locus on chromosome III. If the *MAT*α allele is expressed in a haploid cell then that cell will be of α mating type and if the *MAT*a allele is expressed the cell will be of *a* mating type. The status of the *MAT* locus controls the expression of α-specific genes in an α haploid and *a* specific genes in an *a* haploid. These genes encode a variety of functions involved in the mating process. Amongst these are the *MF*α and *MF*a genes that encode α-factor and *a*-factor respectively. These are small, secreted, hormone-like peptides that predispose

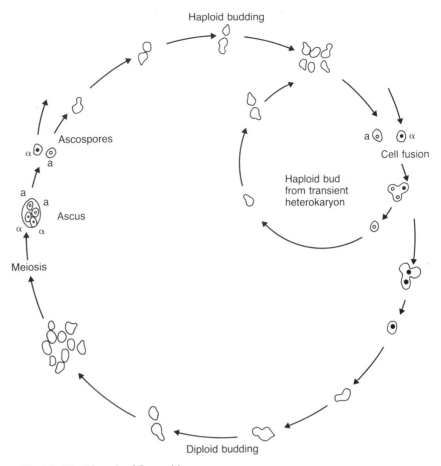

Fig. 3.1. The life cycle of *S. cerevisiae*.

a cells, in the case of α-factor, and α cells, in the case of *a*-factor, to mating. When haploid cells mate *a*/α diploids are created. The presence of both *MATa* and *MAT*α information in the cell switches off the α-specific and *a*-specific genes and leads to the induction of sporulation specific genes. Under appropriate physiological conditions, e.g. nitrogen limitation, these genes encode the functions necessary for meiosis and the production of haploid spores (Herskowitz and Oshema, 1981; Stenberg *et al.*, 1987).

Both the alleles are, in fact, present in a single yeast cell at loci outside of the *MAT* locus. These loci are called *HML* and *HMR* and the *MAT* genes at these loci are silent, i.e. they are not transcribed. One of the alleles is, however, also present at the *MAT* locus where it is expressed. The allele at the *MAT* locus can be altered by gene conversion from either the *HML* or *HMR* locus to change the mating type. In other words the *HML* and *HMR* loci store the information for *a* and α mating types. The alteration of genetic information at the *MAT* locus is called *mating type switching* and it is under complex regulation that involves a gene called *HO* (*homothallism*) (Nasmyth, 1982; Klar, 1987).

Diploid cells can be propagated indefinitely or can be induced to undergo meiosis and sporulation by plating them on nitrogen limiting media.

GENE TRANSFER INTO MICROBIAL EUKARYOTES

Sporulation produces four haploid spores, the direct products of meiosis. These are initially held together in a structure called an ascus and are referred to as a *tetrad*. The individual spores in a tetrad can be teased away from the ascus by microdissection and they will grow into individual isolated colonies. This means that the products of a single meiosis can be analysed directly.

Introducing DNA into yeast

A procedure for transforming yeast was developed independently by Beggs (1978) and Hinnen *et al.* (1978) and is outlined in Figure 3.2. The yeast cells are grown in rich medium to a density of $1 - 2 \times 10^7$ cells ml^{-1} and the cells are harvested by centrifugation. They are then re-suspended in an osmotic stabilising buffer of 1 M sorbitol and treated with cell wall degrading enzymes. These are usually fungal enzymes, e.g. β-glucanase or crude preparation of snail gut called glusulase, is often used. This produces cells without walls, i.e. *spheroplasts*. These are extremely fragile and usually only 10–30% survive the transformation procedure. The spheroplasts are washed in sorbitol to remove the enzyme and then mixed with 1–10 mg DNA in the presence of calcium chloride with a solution of PEG4000 (polyethylene glycol) that in some way enhances DNA uptake. The spheroplasts regenerate their cell walls in special osmotically stabilising regeneration agar and cells that have been successfully transformed produce colonies which can be picked and cultured. The frequency of transformation depends upon the type of vector that is used and ranges from 1–10 per μg of DNA to 10^5 per μg of DNA (see below). The number of DNA molecules taken up per cell is usually between 5 and 30. This means that different plasmids can readily be introduced into the same cell by co-transformation.

An alternative technique for yeast transformation involves treating whole cells with lithium salts followed by PEG4000 treatment (Ito *et al.*, 1983). The advantage of this method is that there is no need to prepare spheroplasts and the transformed colonies grow on the surface of the selection agar rather than embedded in it. Unfortunately, the transformation frequencies tend to be lower with this method and only a single DNA molecule appears to be taken up.

Both of these transformation procedures can be mutagenic, so the preliminary step in any genetic or molecular analysis of a yeast transformant must be to select two independently transformed colonies and confirm that they behave identically. This guards against inadvertently picking a transformant with a mutation in the plasmid that effects the expression of the gene to be analysed. Technical details of yeast transformation can be found in Rothstein (1985).

Selectable markers

S. cerevisiae can be grown as a haploid and therefore it is extremely easy to isolate specific mutant strains, particularly auxotrophs, and then to use the corresponding wild-type gene to complement the defect. Commonly used

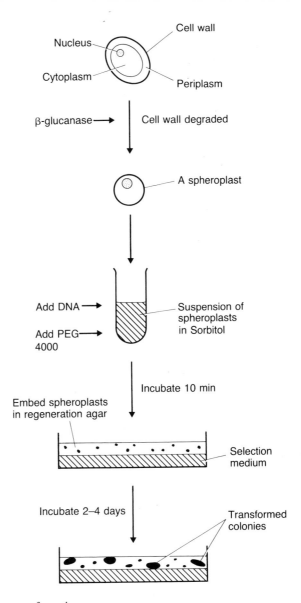

Fig. 3.2. Yeast transformation.

complementing markers are the *LEU2* (Beggs, 1978), *TRP1* (Tschumper and Carbon, 1980), *HIS3* (Struhl *et al.*, 1979) and *URA3* (Struhl *et al.*, 1979) genes. There are also several genes that can be used as dominant selectable markers. The use of dominant selectable markers in eukaryotic cell transformation is discussed in detail in Chapter 4 because they are widely used in mammalian cells where auxotrophic mutations are not as easy to isolate. Those that have been used in yeast are thymidine kinase (McNiel and Friesen, 1981), chloramphenicol acetyl transferase (Jiminez and Davies, 1980), hygromycin B phosphotransferase (Gritz and Davies, 1983) dihydrofolate reductase (Miyajima *et al.*, 1984) and copperthionein (Butt *et al.*, 1984). These confer resistance respectively to folate agonists,

GENE TRANSFER INTO MICROBIAL EUKARYOTES

chloramphenicol, G418, hygromycin B, methotrexate and copper. Dominant selectable markers are particularly useful for transforming industrial strains of *Saccharomyces* because these are often polyploid and therefore it is difficult to isolate auxotrophs.

Gene transfer vectors

A large number of plasmid vectors have been designed to facilitate the genetic manipulation of *S. cerevisiae*. These are reviewed by Parent *et al.* (1985). All of these plasmids are hybrid molecules containing *E. coli* plasmid DNA so that they can be manipulated and prepared in *E. coli*, which is technically easier to handle than *S. cerevisiae*. There are two general types of vector: those that cannot replicate in *S. cerevisiae* and are therefore maintained by integration into the host chromosome, i.e. *integrative vectors* and those that replicate autonomously, i.e. *autonomous vectors*.

SIMPLE INTEGRATIVE VECTORS

A typical integrative vector, called a *yeast integrating plasmid (YIp)* is shown in Figure 3.3. This contains the yeast *HIS3* gene inserted into pBR322 to produce YIpl (Struhl *et al.*, 1979). Following transformation this plasmid will only be maintained when the cell divides if it becomes integrated into the chromosome. This occurs by homologous recombination between the plasmid-borne *HIS3* gene and the chromosomal *HIS3* gene. The chromosomal *HIS3* gene is functionally inactive due to mutation.

The product of the integration event is a chromosome that contains two copies of *HIS3* DNA, one mutant (*his3*) and one wild type (HIS3), with pBR332 DNA integrated in between the repeated *HIS3* sequences. Transformants produced by these types of integrative event are genetically unstable as the integrative plasmid can be lost by recombination across the repeated sequences to excise the transforming DNA. The frequency of transformation with a plasmid such as YIpl is low, about 1−10 transformants per μg of DNA. This low frequency reflects the low efficiency of homologous recombination. The frequency of transformation can however be increased by about 1000-fold by simply cutting the vector within the *S. cerevisiae* DNA with a restriction enzyme (Orr-Weaver *et al.*, 1981). This is because the free ends are recombinogenic.

The frequency of random integration, i.e. non-homologous recombination, is extremely low in *S. cerevisiae* and therefore all cells transformed with an integrative vector are virtually certain to contain only a single copy of the vector inserted by a homologous recombination event.

If a vector is constructed with two separate chromosomal sequences then it can be targeted to one or other of the chromosomal loci simply by making a double-stranded break in one of the sequences on the plasmid. For example, as shown in Figure 3.4, if a vector carrying the *URA3* gene and the *HIS3* gene is cut within the *URA3* sequence then 99.9% of the transformants will have the plasmid integrated into the chromosome at the *URA3* locus. This will occur irrespective of whether the selection is for

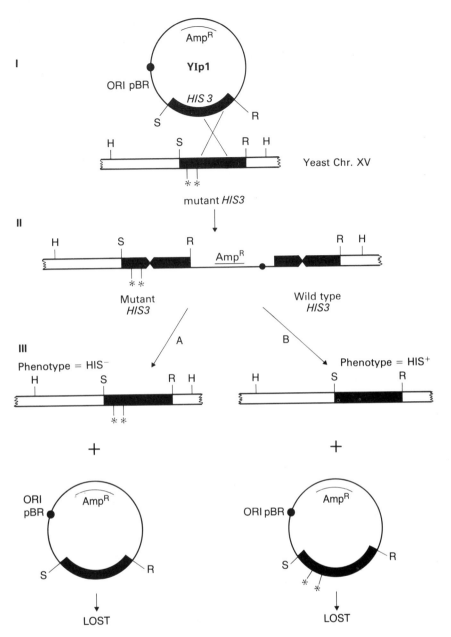

Fig. 3.3. A simple integration vector system. The asterisks indicate mutations. The thin line is pBR322 and the thick lines are yeast DNA. The constriction indicates the crossover point.

histidine or for uracil independence, i.e. selection does not 'drive' the integration. The transformants will still be genetically unstable because of the duplication of the *URA3* DNA. In most cases the gene to be complemented is a double-point mutant to avoid any problems with reversion to wild type that could be confused with transformation to wild type. The point mutations are clustered to avoid separation by recombination which would generate two mutant genes and therefore give no complementation.

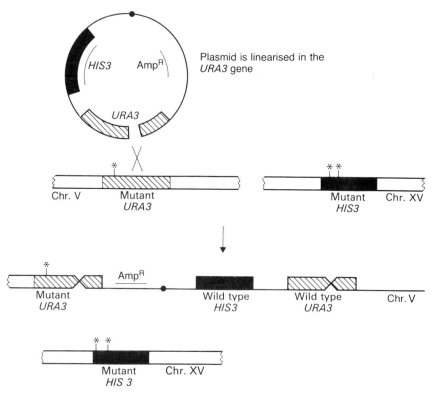

Plasmid is linearised in the *URA3* gene

Both *HIS3* and *URA3* are integrated at the *URA3* locus

Fig. 3.4. Simple targeted integration. See legend to Figure 3.3.

A significant problem with the simple integrating plasmids is their genetic instability. For example HIS$^+$ transformants containing integrated YIpl produce HIS$^-$ cells. After 15 generations of growth in non-selective medium, 1% of the cells are histidine auxotrophs. The genetic instability of the simple integrative vectors reduces their general usefulness, however this instability is an advantage for the technique of *allele exchange* which can be used to swap mutant for wild-type alleles and *vice versa*. When excision of the transforming DNA occurs, the recombination event can occur on either side of the mutations in the chromosomal copy of the gene (Figure 3.3). If, in Figure 3.3 the recombination event occurs on the left of the mutation sites, the original mutant allele of the *HIS3* gene will be left in the chromosome. On the other hand, if the recombination event occurs to the right of the sites of mutation, the wild-type allele of *HIS3* will be left in the chromosome, i.e. the integration and excision events have resulted in an allelic exchange. In both cases the excised plasmid is lost as it is unable to replicate in yeast.

Recently it has been shown that genes can be integrated into the *S. cerevisiae* chromosome at specific targets without producing a duplication. This allows the production of genetically stable transformants. This is discussed later in the section on chromosome engineering.

AUTONOMOUS VECTORS

Vectors that replicate autonomously can transform yeast at a high frequency because they do not require relatively rare recombinations with the chromosome for maintenance in the transformed cells. Two types of autonomous vectors have been developed. One type is based on DNA sequences that have been isolated from the *S. cerevisiae* genome. These function as plasmid replication origins, called *autonomous replicating sequences* (ARS), *centromeres* (CEN) and *telomeres* (TEL). The second type is based on the endogenous yeast plasmid, the 2 μm circle.

(a) Vectors based on chromosomal components
(i) ARS vectors. In the early days of recombinant DNA research a number of yeast genes were cloned by complementation in *E. coli* and these were tested as selectable markers for gene transfer vectors. One fragment containing the *TRP1* gene gave surprising results. Vectors containing this *TRP1* fragment transformed yeast at a very high frequency, about 10^3-10^4 transformants per μg as compared to $1-10$ transformants per μg obtained with other simple integrative vectors carrying the *HIS3* or *LEU2* genes. Southern blot analysis showed that the *TRP1* vector was not integrated into the chromosome but was present as covalently closed circular DNA. This piece of *Saccharomyces* DNA therefore had the ability to replicate autonomously; the vector was called a *yeast replicating plasmid* (YRp) (Struhl *et al.*, 1979). The functional sequence for replication is known as an autonomously replicating sequence, ARS (Kearsey, 1986). The ARS plasmid YRp7 contains a 1.4 kb EcoRI fragment containing the *TRP1* coding sequence and ARS1 (Figure 3.5; see Chapter 9 for more details of ARS structure). Although this plasmid transforms *Saccharomyces* at a high frequency the plasmid is very unstable. In a culture of transformants growing under selective conditions only 20–50%

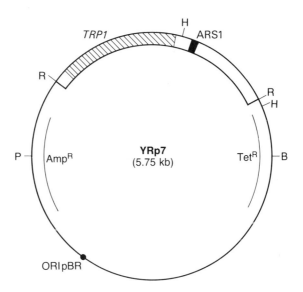

Fig. 3.5. An ARS plasmid: YRp7.

GENE TRANSFER INTO MICROBIAL EUKARYOTES

of the cells are TRP$^+$. If the cells are transferred to non-selective medium then the percentage of TRP$^+$ cells falls to $1-5\%$ within 10 generations. The TRP$^-$ segregants do not contain any plasmid DNA. The high transformation frequency conferred by an ARS allows the direct selection of other ARSs from the *Saccharomyces* genome (Chan and Tye, 1980).

The average copy number of ARS plasmids is low, about $1-5$ per cell, however, only a fraction of the cells contain plasmid and therefore in these cells the copy number must be high, about $20-50$ per cell. The instability of these plasmids despite the very high copy numbers has now been explained by the finding that there is a segregation bias at budding in favour of mother cells (Murray and Szostak, 1983a). In about half the cell divisions both mother and daughter cell receive plasmid but in most of the other divisions only the mother cell received plasmid. The molecular basis for this segregation bias is not yet known but it is proposed that during the cell cycle the circular ARS plasmids become associated with a site or compartment that is destined to segregate predominantly to the mother cell.

The ARS based plasmids such as YRp7 are rarely used as gene transfer vectors because of this instability. Any analysis of gene expression from such plasmids would be complicated by not knowing the precise number of templates per cell.

(ii) ARS/CEN vectors. Clarke and Carbon (1980) isolated an 8 kb fragment from chromosome III of yeast that complemented a centromere linked gene *cdc10*. When this fragment was present on an ARS1 plasmid it imposed stable segregation at mitosis and ordered segregation at meiosis indicating that a functional centromere had been cloned; the sequence was called CEN3 (see Chapter 9 for further details of CEN3). Additional CEN sequences have been isolated from yeast by random cloning and direct selection for sequences that stabilise ARS plasmids (Hsaio and Carbon, 1981). The presence of a CEN sequence on an ARS plasmid not only increases the mitotic stability but it also reduces the average copy number to 1.5 per plasmid bearing cell (Fitzgerald-Hayes *et al.*, 1982).

Gene transfer vectors that have an ARS and a CEN sequence are sometimes referred to as *yeast centromere plasmids* (YCp) (Figure 3.6). They are particularly useful for analysing genes that may be lethal when present in more than one copy per cell. They provide a means of gene analysis at single copy which mimics the situation of a gene in the chromosome but with the advantages of ease of manipulation. The gene can also be readily rescued from the transformant. YCp vectors are not used when high levels of a gene product are required because this can be achieved more readily with the high copy number of 2 μm based vectors (see later).

One problem with using ARS/CEN plasmids as gene transfer vectors is that they are not completely stable. Even under selective conditions the rate of loss is 1% per generation (Fitzgerald-Hayes *et al.*, 1982).

(iii) ARS/CEN/TEL linear vectors. An observation that the telomeric DNA from *Tetrahymena* ribosomal DNA could function in *S. cerevisiae* led to the

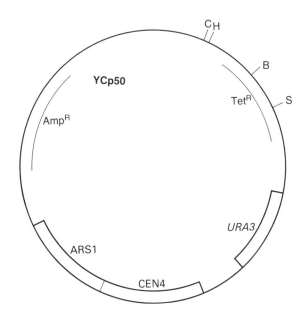

Fig. 3.6. An ARS/CEN plasmid.

cloning of yeast *telomeres* (TEL) (Szostak and Blackburn, 1982). The cloning procedure is discussed in detail in Chapter 9. If telomeres are added to linear DNA containing an ARS it will replicate in yeast as a linear molecule. Such a molecule is sometimes called a *yeast linear plasmid (YLp)* (Figure 3.7). These are still mitotically unstable and are not generally useful as gene transfer vectors. If however, a CEN sequence is added the linear plasmid shows increased mitotic stability at an average copy number of one per cell and shows correct meiotic segregation (Murray and Szostak, 1983b). The combination of ARS, CEN and TEL sequences effectively produces an artificial chromosome. Increasing the size of the artificial chromosome above 50 kb with any extra DNA markedly improves the stability (Hieter *et al.*, 1985b; Murray *et al.*, 1986).

(iv) YAC cloning vectors. Burke *et al.* (1987) have developed a method for using yeast artificial chromosomes (YACs) to clone large DNA fragments of up to 500 kb. This is technologically important because many genes, e.g.

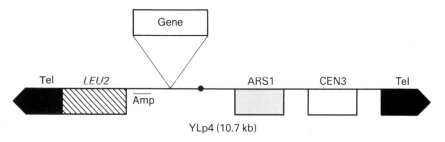

Fig. 3.7. A linear 'minichromosome' vector.

GENE TRANSFER INTO MICROBIAL EUKARYOTES

the human factor VIII gene and *Drosophila* bithorax genes span such distances. These large genes have previously had to be cloned in multiple segments of less than 40 kb which is laborious and some very large loci may never be reconstituted reliably. The new technique is outlined in Figure 3.8. The basic vector YAC2 is a plasmid that is readily propagated in *E. coli*. It can then be cleaved into two fragments that constitute chromosome arms. The left arm has a telomere, a centromere, an ARS and the *TRP1* gene as a selectable marker. The right arm has only a telomere and a different selectable marker, the *URA3* gene. The cleavage site that separates the two arms also divides the yeast *SUP4* gene which encodes an ochre suppressor allele of a tyrosine tRNA. The arms are mixed with exogenous DNA that has been cleaved to produce a population of large, i.e. greater than 100 kb, fragments. The reconstructed chromosomes are introduced into yeast by standard methods. By selecting for *TRP1* and *URA2* only molecules that have both arms are obtained, also the presence of *SUP4* is monitored by using a strain that has an *ade2* ochre mutation, if this is suppressed colonies are white, if not, they are red.

Yeast colonies containing large artificial chromosomes are obtained at about 10^2 colonies per μg and appear to be maintained as stable unrearranged autonomous chromosomes. These YACs can be analysed using the technique of pulsed field gel electrophoresis (PFAGE) which is designed to separate DNA fragments in the 100–1000 kb size range (Schwartz and Cantor, 1984; Carle *et al.*, 1986). This procedure will allow the production of large fragment libraries from complex organisms, e.g. man, that may be representative of the genome. Also yeast is likely to be able to tolerate repetitive DNA and palindromes which are unstable in *E. coli*. Therefore, previously unclonable fragments may be obtained in this yeast system. The large artificial chromosomes can be further manipulated in yeast cells. It may be possible to target genes into their authentic location in a large DNA fragment by relying on the efficiency of homologous recombination in *S. cerevisiae*. The artificial chromosomes can then be re-introduced into mammalian cells. Although they are unlikely to remain autonomous, and will integrate, the targeted gene will be in its authentic context and may be free from position effects that perturb the expression of randomly integrated genes (see Chapters 4, 5 and 11 for discussions of gene targeting and position effects in higher eukaryotic cells). It may also be possible to assemble genes for entire metabolic pathways on YACs to introduce new metabolic versatility into *S. cerevisiae*.

The use of artificial chromosomes to explore fundamental questions about chromosome structure and function is discussed in more detail in Chapter 9.

(b) Vectors based on the 2 μm circle
(i) The 2 μm circle. S. cerevisiae contains a plasmid that has a contour length of 2 μm and as the plasmid confers no detectable phenotype it is simply called the 2 μm circle. It is stably maintained at 50–100 copies per cell and it behaves like a mini circular chromosome. That is, it is found in the nucleus, it has a typical chromatin structure, its replication is generally

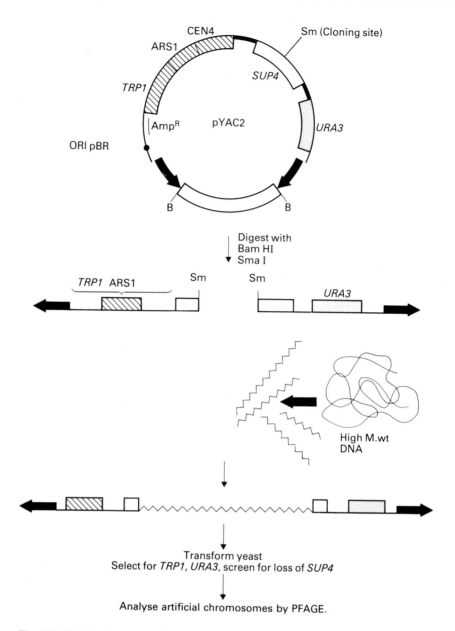

Fig. 3.8. YAC cloning vectors. The telomeres are indicated by arrows.

under the same genetic control as chromosomal DNA replication and it only replicates once per cell cycle in S phase. The plasmid is not essential to *S. cerevisiae* and strains have been isolated which lack 2 μm circle. These are called *cir*° (circle zero) strains as opposed to *cir*⁺ (circle plus) (Dobson *et al.*, 1980). It is 6318 bp in length and has a number of interesting features (Hartley and Donelson, 1980; Broach, 1982), summarised in Figure 3.9.

The most prominent of these features is a precise inverted repeat of a 599 bp sequence. These repeats (IR1 and IR2) split the plasmid into a larger (2.7 kb) and a smaller (2.3 kb) unique sequence. Recombination,

GENE TRANSFER INTO MICROBIAL EUKARYOTES

called 'flipping' between these sequences inverts the unique sequences with respect to each other to produce two molecular forms of the plasmid called A and B; the A form is shown in Figure 3.9 and the B form in Figure 3.10A. Flipping is mediated by a plasmid-encoded recombinase called an FLP protein that recognises a specific target sequence in the repeats (Andrews et al., 1985). The crossover point is always at the same position within the repeats (McCleod et al., 1986).

The major replication origin has been localised to a 75 bp region that spans the junction between the unique sequence and IR1 (Broach et al., 1983a, b; Brewer and Fangman, 1987). The region contains a sequence that is functionally and structurally analogous to a chromosomal ARS and is the site for the assembly of replication complexes in vitro (Jazwinski and Edelman, 1984). A plasmid derivative carrying this region alone behaves like an ARS plasmid, i.e. it has high copy number but is highly unstable. A second cis active region called REP3 or STB is essential to stabilise the replication of the plasmid (Jayaram et al., 1985). This region has an unusual structure. It contains 5.5 copies of a 62 bp repeat; the repeats show 75−90% homology and deletion of one or two of the repeats does not affect the stabilising function. The region containing the origin and REP3 is nucleosome depleted which is typical of origin regions.

Two plasmid-coded proteins, the products of the REP1 and REP2 genes, are required for the stable replication of the 2 μm circle (Jayaram et al., 1983; Kikuchi, 1983). The precise function of these is not known but it is possible that they interact with REP3 to ensure efficient partitioning of the plasmid to both mother and daughter cells at cell division. There is some evidence that the REP1 protein may effect partitioning by interacting with the nuclear matrix (Wu et al., 1986a).

2 μm circle replication is controlled so that each plasmid is only replicated once per cell cycle. If, however, a single copy of a 2 μm plasmid is introduced into a cell it will initially amplify to 50 to 100 copies, i.e. the cell cycle control is bypassed. Recently it has been shown that some of this amplification can be explained by FLP-mediated site-specific recombination of the replicating template (Futcher, 1986; Volkert and Broach, 1986). This serves to re-orientate the replication forks from opposing to concurrent directions. They then chase each other around the plasmid generating large oligomeric structures without re-initiation at the origin. These structures presumably resolve by recombination to produce monomers. REP1 and REP2 are also involved in copy number control although their functions are not fully understood (Sutton and Broach, 1985; Som et al., 1988).

A fourth open reading frame called D is transcribed but as yet no protein has been identified and there is no clear function for this region although insertions within D can lead to reduced plasmid stability (Sutton and Broach, 1985).

Multiple transcripts are produced by the 2 μm circle and the major cap sites and poly A sites have been mapped (Sutton and Broach, 1985). FLP and REP2 are divergently transcribed from a putative divergent promoter (P1) that contains TATA box elements 32 bp and 35 bp upstream of the respective cap sites. D and REP2 are divergently transcribed from a putative

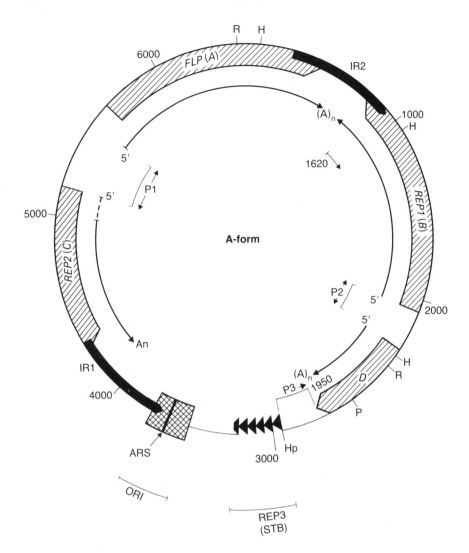

Fig. 3.9. The 2 μm circle.

promoter (P2) that has a TATA box 66 bp upstream from the REP cap site. Two minor transcripts of 1620 and 1950 nucleotides have been detected in the *REP1*/D half of the plasmid and there is a possible third promoter (P3) in the REP3 region.

(ii) Derivatives of the 2 μm circle as vectors. The ability of the 2 μm circle to attain copy numbers of 50−100 per cell and to be stably inherited so that 100% of the cells in a culture contain high copy numbers makes it an excellent basic molecule for the construction of gene transfer vectors.

   The first 2 μm vectors were described by Beggs (1978) and one of these, called pJDB219, is shown in Figure 3.10A. It contains an *E. coli* plasmid,

GENE TRANSFER INTO MICROBIAL EUKARYOTES

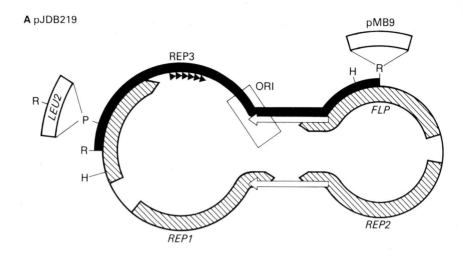

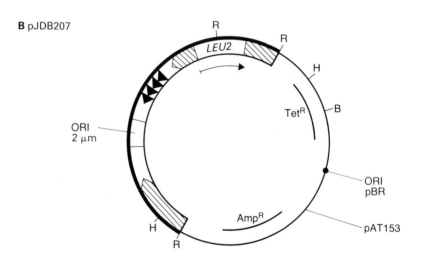

Fig. 3.10. 2 μm based vectors. The thick black line in A is the region that is subcloned to produce pJDB207 in B.

pMB9, inserted at the EcoRI site within the *FLP* gene. This creates a shuttle vector that can replicate in *E. coli* and *S. cerevisiae*. In addition, a fragment of chromosomal DNA carrying the *LEU2* gene is A/T tailed into the PstI site in the D gene to provide a selectable marker. As this gene lacks its own promoter it is assumed to be expressed from the P3 2 μm promoter (Sutton and Broach, 1985). Plasmids such as pJDB219 transform yeast at a high frequency, about 10^4 to 10^5 transformants per μg. The transformants are stable although LEU⁻ cells are occasionally produced. If a *cir°* strain is re-transformed with pJDB219 then the copy number can be as high as 400. This has been explained by a lack of competition with endogenous plasmids for replication factors.

A number of derivatives of the 2 μm circle have been produced to give smaller vectors with fewer restriction sites for ease of manipulation. All of

CHAPTER 3

these derivatives contain the origin and REP3 region but lack one or more of the plasmid-coded proteins. These must be provided in *trans* by the endogenous 2 μm circle and therefore all host strains must be *cir*⁺. A typical derivative, pJDB207, is shown in Figure 3.10B. It contains a double EcoRI fragment from pJDB219 inserted into pAT153. This plasmid is as stably maintained in a *cir*⁺ strain as pJDB219.

Derivatives of the 2 μm circle are usually used when the aim is to produce vectors for the over-expression of proteins. They are also widely used for studying gene expression because the high levels of mRNA achieved by using a high copy number plasmid make the subsequent analyses technically easier. This is particularly true for genes that are expressed poorly or when looking for mutations that reduce the effectiveness of a promoter. When 2 μm based plasmids are, however, used for such analyses the possibility that the high copy number may effectively dilute key regulatory molecules is always acknowledged. Nevertheless, in the majority of studies this has not proved to be a problem, although there are exceptions, e.g. in *ADH2* gene expression (Irani *et al.*, 1987).

Chromosome engineering

The frequency of homologous recombination in *S. cerevisiae* is extremely high whereas non-homologous recombination occurs very rarely. This is the reverse of the situation in many other eukaryotic cells and provides *S. cerevisiae* with a major advantage as a system for analysing eukaryotic cell molecular biology. This is that any gene or functional DNA sequence that has been cloned can be returned to its precise chromosomal location. Specific genes can be mutated *in vitro* and then targeted to the correct chromosomal location and the biological significance of cloned DNA can be verified by specifically disrupting the corresponding chromosomal sequence and observing the resulting phenotype. Yeast chromosomes can therefore be specifically engineered (reviewed by Struhl 1983a; Rothstein, 1985).

Gene targeting

Genes can be targeted to chromosomal localities by simple homologous recombination at DNA breaks within the gene (Figure 3.3). The disadvantage of this procedure is that a duplication is created and these are genetically unstable. In a modification of this procedure (Figure 3.11) a plasmid is digested to specifically excise the yeast chromosomal DNA as a linear fragment. The recombinogenic ends therefore flank the gene as opposed to being within the gene. The linear DNA is used in a standard yeast transformation. Recombination is stimulated by the ends of the fragment of chromosomal DNA resulting in a double crossover to replace the resident gene. In our example, a mutant *HIS3* gene is replaced with the new wild-type *HIS3* gene. No plasmid sequences are introduced and no *S. cerevisiae* sequences are duplicated. This simple procedure is the basis of the more sophisticated techniques of gene disruption and gene replacement.

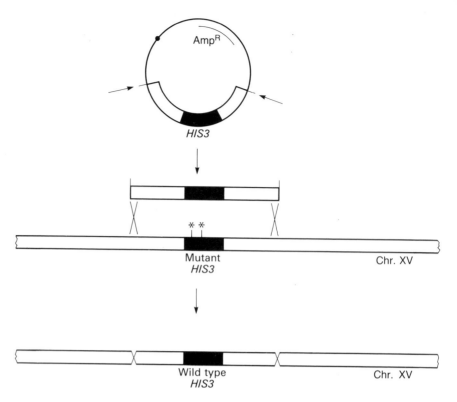

Fig. 3.11. Gene targeting using linear yeast DNA. See legend to Figure 3.3.

Gene disruption

A general strategy for gene disruption is shown in Figure 3.12. The cloned wild-type *HIS3* gene has been manipulated *in vitro* to insert a *URA3* gene into the *HIS3* coding region. This creates a mutant *HIS3* gene. This mutant DNA is then targeted to the chromosomal *HIS3* locus where the double crossover eliminates the wild-type *HIS3* gene and replaces it with the mutated derivative. In other words, the normal *HIS3* locus has been disrupted. This disruption can be monitored by screening transformants for those that have become HIS⁻. A more effective approach is however to directly select for the disrupting DNA. In our example a host strain that is a *URA3* auxotroph is used, transformants are selected on medium lacking uracil and only those that acquire the *URA3*⁺ gene will survive. The only way that the *URA3* gene can be stably inherited is by integration and this will occur at a high frequency at the *HIS3* locus. In this way disruptants are selected in one step. This procedure is widely used to confirm that a specific DNA sequence has been cloned. For example when a mutant centromere fragment was targeted into the yeast genome the resulting transformant lacked centromere functions in one chromosome so confirming that a centromere had, in fact, been cloned (Clarke and Carbon, 1980); see Chapter 9 for more details of this experiment.

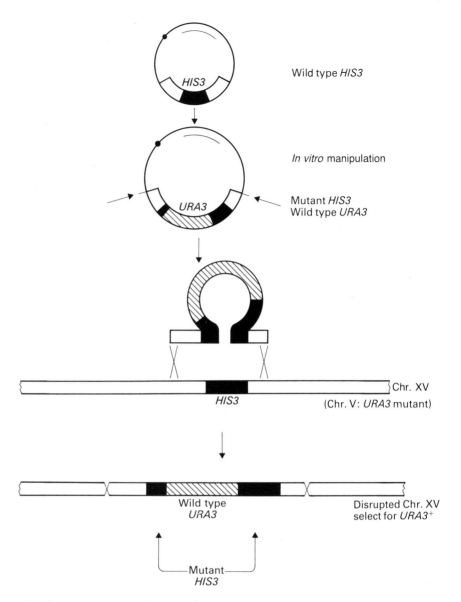

Wild type *HIS3*

In vitro manipulation

Mutant *HIS3*
Wild type *URA3*

Chr. XV
(Chr. V: *URA3* mutant)

Wild type
URA3

Disrupted Chr. XV
select for *URA3*+

Mutant
HIS3

Fig. 3.12. One step gene disruption. See legend to Figure 3.3.

Gene replacement

One of the important uses of gene transfer technology is to study the control of gene expression by making specific mutants *in vitro* and then analysing them back in the living cell. In our discussion of chromosome engineering so far we have considered rather extreme phenotypes, i.e. either replacing a mutant gene with a wild-type gene or the abolition of gene function by gene disruption. In many studies the disrupting mutant gene may have a subtle defect, for example, in a *cis* active regulatory sequence, this could not be used to select or screen for the disruptant if it had little or no effect on the phenotype of the transformant. In order to easily and reliably introduce genes that have no dramatic effect on phenotype into the *S. cerevisiae* a number of simple genetic tricks have been developed.

One approach is outlined in Figure 3.13. It is an extension of the gene disruption technique and is referred to as a *two-step gene replacement strategy*. The aim of the strategy is to replace a wild-type *HIS3* gene with a *HIS3* gene that has been specifically mutated *in vitro*, for example to affect a regulatory sequence involved in general amino acid control. The chromosomal *HIS3* gene is first disrupted as described in Figure 3.12. The disruptant is then used as a recipient in transformation using the mutated *HIS3* gene. This gene will replace the disrupted *HIS3* gene by the standard double crossover recombination and the strain will lose the $URA3^+$ gene. These transformants can be directly selected because strains that are defective in uracil biosynthesis are resistant to the drug 5-fluoroorotic acid (Boeke *et al.*, 1984). The expression of the mutated *HIS3* gene can now be analysed.

Rudolph *et al.* (1985) have exploited the high frequency of co-transformation seen with the yeast spheroplast transformation technique and they have shown that the efficiency of gene replacement is higher if direct selection for the replacing DNA is not imposed immediately. They initially disrupted the *PHO5* gene with *URA3*. They then mixed mutant *PHO5* linear fragments with YEp13, a 2 μm based, *LEU2* selectable plasmid and transformed the disruptant. Transformants were selected for the *LEU2* gene and then replica-plated to screen for loss of the *URA3* gene. About 1−4% of LEU^+ transformants were $URA3^-$. This procedure was also used to target a foreign gene, coding for tissue plasminogen activator into the *PHO5* locus. Many of these procedures are discussed in more detail by Rothstein (1985).

Heterologous gene expression in *S. cerevisiae*

The first study of heterologous gene expression in yeast was by Beggs *et al.* (1980) who inserted a 5.1 kb fragment of rabbit genomic DNA carrying the β-globin gene into the 2 μm vector pJDB219. When the transformants were analysed no β-globin protein was detected, but β-globin RNA was made. The RNA, however, was not authentic: the promoter region of the rabbit β-globin gene was not properly recognised in *S. cerevisiae*, the introns were not processed from the primary transcript and the transcript terminated prematurely in an AT-rich region in the second intron. These results suggested that *S. cerevisiae* would not be a good host system to study the control of expression of non *S. cerevisiae* gene.

More recently it has been shown that some heterologous control signals are recognised in *S. cerevisiae*. For example, transcription initiates correctly in the promoters for maize storage protein genes (Langridge *et al.*, 1984), transcription terminates correctly in the *Drosophila ADE8* gene (Henikoff and Cohen, 1984), the *Drosophila hsp70* gene is induced by heat shock in yeast (DeBanzie *et al.*, 1986) and the translation of soya bean leghemoglobin is regulated by yeast heme interacting with the 5′ untranslated leader (Jensen

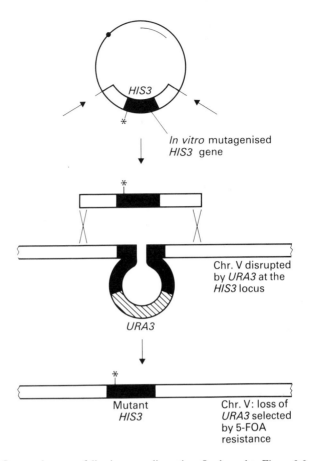

Fig. 3.13. Gene replacement following gene disruption. See legend to Figure 3.3.

et al., 1986). Some regulatory elements may therefore be preserved in all eukaryotic cells. In general, however, in order to ensure efficient expression of heterologous genes in *S. cerevisiae*, it is necessary to use a cDNA and to direct transcription and translation with the appropriate sequences from a yeast gene.

The basic plan of a yeast promoter is shown in Figure 3.14. It has some similarities and some differences from the general plan of a eukaryotic pol II gene outlined in Chapter 1. The mRNA start site is determined in many cases by a TATA box-like element similar to that found in higher eukaryotic promoters. However the distance of the TATA sequence from the mRNA start is less critical than in higher eukaryotes. The DNA sequence around the cap site is also important for specific control and this is called the *initiation* or *I site* (Chen and Struhl, 1985; Hahn *et al.*, 1985; Nagawa and Fink, 1985). The overall region that is involved in determining transcript initiation is conveniently called the *RNA initiation element (RIE)*. In most genes there is a sequence, located usually several hundred nucleotides upstream from the mRNA start, which is essential for any significant transcription to occur. This is called the *upstream activator sequence (UAS)* (Guarente, 1984) and shows many properties of the enhancer sequences of

GENE TRANSFER INTO MICROBIAL EUKARYOTES

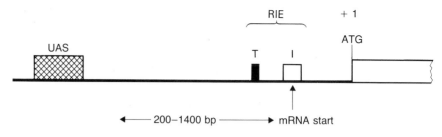

Fig. 3.14. A typical yeast promoter.

higher eukaryotic genes (Serfling *et al.*, 1985). In many cases the UAS is involved in regulating gene expression via the binding of specific regulatory proteins to specific sequences in the UAS (see later).

A number of different promoters have now been manipulated to produce vectors for the expression of foreign genes.

Basic expression vectors

The first demonstration that a foreign protein could be synthesised in *S. cerevisiae* was by Hitzeman *et al.* (1981). They fused the coding sequence for human interferon alpha to the 5′ flanking region of the yeast *ADH1* (alcohol dehydrogenase) gene and inserted the fusion into the ARS plasmid YRp7. They could detect biologically active interferon in extracts from yeast that had been transformed with this expression plasmid. This first expression vector, although important in demonstrating the feasibility of the approach, was not particularly effective, because the *ADH1* expression signals are not very strong and the plasmid was only at an average copy number of one per cell.

The current yeast expression vectors are much more efficient because they are based on 2 μm plasmids and they incorporate a number of improvements. Firstly, they contain very high efficiency promoters. These are usually based on the promoters from genes that code for glycolytic enzymes. The usefulness of glycolytic gene promoters was first demonstrated with the *PGK* gene (Tuite *et al.*, 1982). Examples of other high-efficiency promoters are shown in Table 3.1 (reviewed by Kingsman *et al.*, 1985, 1987). It is also important to provide a transcription terminator close to the end of the foreign coding sequence because long transcripts appear to be unstable (Zaret and Sherman, 1982; Mellor *et al.*, 1983).

A typical high-efficiency expression vector, pMA91 (Mellor *et al.*, 1983), is shown in Figure 3.15. It contains the 5′ flanking region of the *PGK* gene up to nucleotide −2 (Dobson *et al.*, 1982). This is followed by a unique Bgl II expression site and then the 3′ flanking region. Any foreign coding sequence with its own translation initiation and termination sequences can be expressed in this vector. For example, human interferon alpha (Mellor *et al.*, 1985a), calf chymosin (Mellor *et al.*, 1983) and mouse immunoglobulin (Wood *et al.*, 1985). The vector is present at a copy number of 50−100 per cell and generally it is stably maintained in the absence of selection. The levels of heterologous gene expression directed by vectors such as pMA91

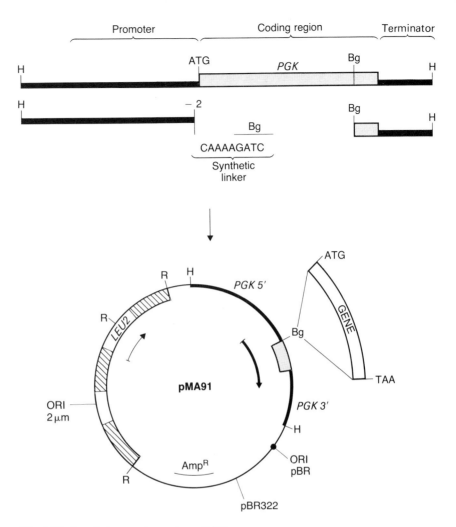

Fig. 3.15. A typical expression vector: pMA91.

usually represent 1–5% of total cell protein and the new protein can easily be seen by polyacrylamide gel electrophoresis (Figure 3.16). The levels of protein are however still lower than predicted from the strength of the promoter and the copy number of the plasmid. The basis for the failure to achieve maximum theoretical efficiency of heterologous gene expression in any yeast vector to date is not known. It may in some cases be due to inefficient transcription (Mellor *et al.*, 1985, 1987).

Regulated expression vectors

Vectors which contain promoters that function continuously are not ideal for every foreign coding sequence because the yeast cells must synthesise foreign proteins throughout the culture period. In some cases this excessive production of foreign proteins increases the doubling time by several hours, which is undesirable in large-scale processes. Also, some proteins are toxic

GENE TRANSFER INTO MICROBIAL EUKARYOTES

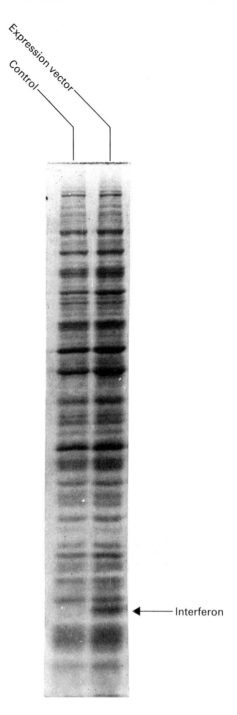

Fig. 3.16. The production of interferon in *S. cerevisiae*. A photograph of yeast proteins separated by polyacrylamide gel electrophoresis and stained with Coomassie blue. Data adapted from Tuite *et al*. (1982).

to yeast such as influenza virus haemagglutinin (Jabbar *et al.*, 1985) and polyomavirus middle T antigen (Belsham *et al.*, 1986). This results in selection for cells that have a reduced plasmid copy number and plasmid rearrangements. These problems can be overcome by using regulated promoters.

The expression of the glycolytic genes is in fact regulated by the carbon source but the induction ratio, i.e. the difference between the lowest and highest levels of expression, is only about 20-fold (Tuite *et al.*, 1982). This means that even under uninduced conditions significant amounts of any toxic protein may be produced. There are several more suitable regulated promoters which have been used to make regulated expression vectors. For example, the transcription of the *PHO5* (acid phosphatase) gene is decreased when the culture medium contains inorganic phosphate and increases when inorganic phosphate is removed. The production of human interferon in a *PHO5* vector was 200-fold higher in low phosphate medium than in high phosphate (Kramer *et al.*, 1984). The most widely used regulated vectors are based on the *GAL* genes (Broach *et al.*, 1983b). The *GAL1* and *GAL10* genes are regulated by galactose with an induction ratio of about 1000 and maximum levels of expression are moderately high with each transcript comprising about 0.25% of total mRNA (St John and Davis, 1981). In the absence of galactose, transcription is barely detectable. The *GAL10* regulated promoter is attractive for industrial process because it is activated by the simple addition of galactose, the *PHO5* promoter is less attractive because phosphate depletion on a large scale is difficult to achieve.

Temperature is a simple and convenient regulator and a general approach is to isolate temperature-sensitive mutations in regulatory genes. For example the activity of the promoter of the *MFα* (mating factor alpha) gene is positively activated albeit indirectly by a group of genes called *SIR* (silent information regulatory) (Nasmyth, 1982). In *SIR* mutants there is no transcription of the *MFα1* gene. If a temperature-sensitive *SIR* mutant (*sir*[ts]) is used then at 37°C when the SIR protein is non-functional the *MFα1* promoter is inactive but at 24°C when the SIR protein regains function, *MFα1* transcription is induced 1000-fold (Brake *et al.*, 1984). This regulation can only be achieved in haploid strains because the *MFα1* gene is always inactive in a diploid (Herskowitz and Oshima, 1981). Temperature-sensitive regulatory mutants have also been used to control *PHO5* transcription (Kramer *et al.*, 1984). Other useful promoters are the *CUP1* (copperthionein) gene which is inducible by 50 times by copper (Karin *et al.*, 1984) and the *HSP70* gene which is induced by heat shock (Finkelstein and Strausberg, 1983).

A list of some of the promoters that have been used in yeast expression vectors and their characteristics are shown in Table 3.1 and reviewed in Kingsman *et al.* (1985).

Translation of heterologous mRNA

There have been suggestions that yields of heterologous proteins in yeast are affected by inefficient translation of heterologous mRNAs. This idea

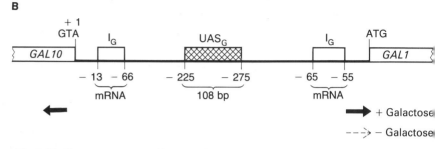

Fig. 3.17. Yeast promoters used in expression vectors. A, *PGK*: high efficiency; poor regulation. B, *GAL10*: moderate efficiency; high induction ratio.

stems from the observation that most efficiently expressed yeast genes show a marked bias in the use of particular codons. For example, the leucine TTG codon is used 91% of the time in the efficiently expressed yeast *PGK* gene and the CTG codon is never used. In a poorly expressed gene such as *TRP1* these two leucine codons are used 35% and 17% of the time. Yeast genes are given a *codon bias index* to reflect the percentage of amino acid residues which show this bias (Bennetzen and Hall, 1982a; Sharp *et al.*, 1986). The codon usage bias of *PGK* and *TRP1* are 0.92 and 0.31, i.e. highly biased and unbiased respectively. Most foreign genes have a low yeast codon usage bias index, for example, the codon usage of human interferon which has an index of 0.32 is compared with the yeast *PGK* gene in Figure 3.18. The abundance of the different tRNA isoacceptor species reflects the codon bias of efficiently expressed yeast genes (Ikemura, 1982). This could mean that translational elongation may be slowed down in genes that have many rare codons because the relevant tRNA will be limiting.

Table 3.1. Some yeast promoters used in expression vectors.

Promoter	Characteristics	Ref.
PGK	Efficient	1
GPD	Efficient	2
ENO	Efficient	3
ADH	Moderately efficient:	4
PHO5	Moderately efficient: regulated by Pi (200)	5
GAL1	Moderately efficient: regulated by galactose (2000)	6
MFα	Moderately efficient: regulated by *SIR* genes (1000)	7
CUP1	Moderately efficient: regulated by copper (50)	8
HSP90	Moderately efficient: regulated by temperature (50)	9

The induction ratios of regulated promoters is indicated in brackets. Key references to the use of these promoters are (1) Tuite *et al.*, 1982; (2) Bitter and Egan, 1984; (3) Innis *et al.*, 1985; (4) Hitzeman *et al.*, 1981; (5) Kramer *et al.*, 1984; (6) Goff *et al.*, 1984; (7) Brake *et al.*, 1984; (8) Etcheverry *et al.*, 1986; (9) Finkelstein and Strausberg, 1983.

CHAPTER 3

There is however no good evidence that this is the case.

Unlike the situation in *E. coli* and mammalian cells (Kozak, 1983a, 1984) there is no stringent sequence requirement for the efficient initiation of translation in yeast (Sherman and Stewart, 1982). The base composition of the initiation region can affect initiation: an A+T rich ATG environment is about three-fold more effective than a G+C rich environment (Kingsman and Kingsman, 1983). Non-specific sequences can also affect elongation, e.g. the translation of a *CYC1* mRNA was reduced to 20% of wild-type levels by a mutation that generated a secondary structure that caused translational pausing (Baim *et al.*, 1985). Sequences in the 3' untranslated region may also influence translation (Zaret and Sherman, 1984).

Post-translational modification and localisation of heterologous proteins

A number of authentic post-translational modifications of foreign proteins have been observed in yeast. For example the human *ras* oncogene protein is fatty acylated (Clark *et al.*, 1985), the human *c-myc* oncogene protein is phosphorylated (Miyamoto *et al.*, 1985) and human *fos* oncogene proteins undergo a number of modifications that alter the electrophoretic mobility (Sambucetti *et al.*, 1986). Proteins which are secreted are usually glycosylated (see later).

LEU: TTA	(5)	1	SER: AGT	(0)	2	ARG: AGA	(89)	5
TTG	(91)	4	AGC	(0)	6	AGG	(0)	4
CTA	(2)	1	TCA	(2)	3	CGA	(0)	0
CTG	(0)	10	TCG	(0)	0	CGG	(0)	0
CTT	(2)	1	TCT	(46)	5	CGT	(10)	0
CTC	(0)	8	TCC	(49)	2	CGC	(0)	0
ALA: GCA	(0)	2	GLY: GGA	(0)	1	VAL: GTA	(0)	0
GCG	(0)	0	GGG	(0)	2	GTG	(0)	7
GCT	(78)	4	GGT	(99)	1	GTT	(50)	1
GCC	(22)	3	GGC	(1)	2	GTC	(50)	2
PRO: CCA	(89)	0	THR: ACA	(3)	3	ILE: ATA	(0)	1
CCG	(0)	0	ACG	(0)	0			
CCT	(11)	3	ACT	(48)	3	ATT	(50)	1
CCC	(0)	2	ACC	(52)	4	ATC	(50)	6
PHE: TTT	(8)	4	CYS: TGT	(100)	3	HIS: CAT	(8)	2
TTC	(92)	6	TGC	(0)	3	CAC	(92)	1
GLN: CAA	(100)	4	ASN: AAT	(3)	2	ASP: GAT	(16)	2
CAG	(0)	8	AAC	(97)	2	GAC	(84)	6
GLU: GAA	(97)	6	TYR: TAT	(2)	1	LYS: AAA	(14)	5
GAG	(3)	8	TAC	(98)	4	AAG	(86)	7
MET: ATG	(100)	5	TRP: TGG	(100)	2	STOP: TAA	0	
						TAG	0	
						TGA	1	

Fig. 3.18. Codon usage bias in *S. cerevisiae*. The figures in parentheses indicate the percentage frequency with which these codons are used in the PGK gene. The numbers after the parentheses show the number of times the codons appear in the human interferon gene.

Proteins normally have a precise localisation within the cell. They are either secreted into the culture medium or periplasm, embedded in membranes as integral membrane proteins, targeted to organelles such as the nucleus or mitochondria or they remain in the cytosol. Localisation appears to be determined by structural features of the proteins (see Chapter 10) and in many cases when foreign proteins are made in yeast these structural features are recognised. Some foreign N-terminal signal sequences will direct these proteins into the secretion pathway (see later). Other examples are correct nuclear localisation of human *c-myc* protein (Miyamoto *et al.*, 1985), the membrane localisation of the Ha-*ras* protein (Clark *et al.*, 1985) and the microsomal localisation of rat cytochrome P450 (Sakaki *et al.*, 1985). A particularly important discovery is that integral membrane proteins such as the influenza virus haemagglutinin and the *Torpedo california* acetyl choline receptor are correctly inserted into the yeast plasma membrane (Jabbar *et al.*, 1985; Fujita *et al.*, 1986). This opens up the exciting possibility of using yeast as a simple system for studying ligand−receptor interactions (Chapter 10).

Manipulating secretion in *S. cerevisiae*

There is tremendous interest in manipulating *S. cerevisiae* to secrete foreign proteins. The reasons for this are (1) to facilitate the production of authentic proteins that lack the N-terminal methionine, (2) to facilitate purification, (3) to prevent the intracellular degradation of unstable proteins, (4) to reduce the intracellular concentration of toxic proteins, (5) to optimise protein folding and (6) to allow post-translational modification.

The pathway of secretion in *S. cerevisiae* has been described by classical genetic analysis of mutants that accumulate secretion intermediates and secretory organelles. The process is essentially the same as in higher eukaryotes (Schekman and Novick, 1982; Schekman, 1985). Different yeast proteins that enter the secretory pathway have a variety of destinations; they may be secreted into the culture medium, e.g. alpha factor, into the periplasm, e.g. invertase and acid phosphatase, into the vaccuole, e.g. carboxypeptidase Y or into the plasma membrane, e.g. permeases. The entry of a protein into the secretion pathway and its ultimate destination is determined by a specific *signal sequence* in the polypeptide. These sequences are usually located at the N-terminus and are usually removed by proteolytic cleavage during localisation (see Chapter 10). If these signal sequences are linked to heterologous coding sequences then the heterologous protein will be directed into the secretory pathway.

Signal sequences

The best-characterised yeast signal sequences are derived from the *MFα1* gene that codes for the mating pheromone, alpha factor and the *SUC2* and *PHO5* genes that code for invertase and acid phosphatase.

The alpha factor signal sequence and processing pathway is shown in Figure 3.19. The alpha factor is synthesised as a 165 amino acid precursor which is processed during secretion to release four copies of a 13 amino acid alpha factor peptide into the culture medium. Secretion and processing is directed by a prepro sequence that consists of a 22 amino acid signal peptide followed by a 63 amino acid pro-segment that ends in the residues Lys, Arg. This is followed by four repeats of the alpha factor peptide, $\alpha F1 - \alpha F4$, which are separated by spacer peptides, S1−S4. These contain two or three dipeptides which are either Glu−Ala or Asp−Ala and these are separated from the alpha factor by Lys−Arg residues. The signal sequence directs the precursor into the endoplasmic reticulum and the pro-segment is glycosylated at three sites during translocation. The precursor is cleaved after the Lys−Arg residues by a cathepsin-B like protease encoded by the *KEX2* gene to liberate the alpha factor with dipeptide extensions at the N-terminus and Lys−Arg at the C-terminus. The extra residues are removed by dipeptidyl amino peptidase encoded by the *STE13* gene and carboxy peptidase respectively (Kurjan and Herskowitz, 1982; Julius *et al.*, 1983, 1984).

Invertase and acid phosphatase are secreted into the periplasm and accumulate in the cell wall with less than 5% being secreted into the culture medium. They have signal sequences of 20 and 17 amino acids respectively and these signals are proteolytically removed by a peptidase during translocation into endoplasmic reticulum (Schauer *et al.*, 1985; Kaiser and Botstein, 1986). The signal sequences show no sequence homology but have three domains that are characteristic of most secreted proteins. These are a charged, i.e. hydrophilic, N-terminus, a central hydrophobic domain and a cleavage site for the signal peptidase adjacent to an alanine residue (Perlman and Halvorsen, 1983).

Glycosylation

Proteins are glycosylated as they are transported through the endoplasmic reticulum. If glycosylation is inhibited, for example in the presence of tunicamycin, then secretion of glycoproteins is reduced but non-glycosylated proteins are still secreted. In addition to enhancing secretion, glycosylation may be necessary for the biological activity of some proteins. The initial stages of glycosylation in yeast are the same as in higher eukaryotes. An oligosaccharide core consisting of N-acetyl glucosamine, glucose and mannose is covalently linked to certain asparagine residues in the nascent polypeptide via an N-glycosidic link (Ballou, 1982; Kukuralinska *et al.*, 1987). Serine and threonine residues can also be glycosylated through an O-glycosidic bond to mannose residues. Although O-glycosylation is rare in *S. cerevisiae* proteins it has been demonstrated on a heterologous protein made in *S. cerevisiae* (Innis *et al.*, 1985; Ernst *et al.*, 1987).

The core oligosaccharide in N-glycoproteins is further modified in the Golgi by the addition of further mannose residues to the core to produce a polymannose backbone of $\alpha 1 - 6$ linked residues up to 150 residues long. The backbone and the core are also modified by the linkage of mannose residues via $\alpha 1 - 3$ linkages to produce complex branching structures. *S. cerevisiae* differs markedly from higher eukaryotes at this stage because only

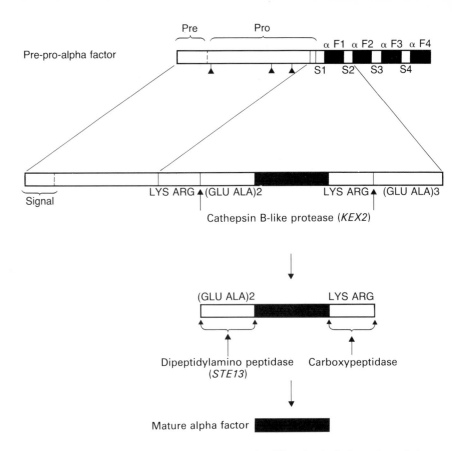

Fig. 3.19. Alpha factor signal sequence and processing. The triangles indicate glycosylation sites.

mannose residues are added as opposed to a whole range of complex sugars that are found for example in mammalian cells (Staneloni and Leloir, 1982).

The secretion of heterologous proteins

A number of foreign proteins have been secreted by *S. cerevisiae* using their own signal sequences with varying degrees of efficiency. An *Aspergillus* glucoamylase is efficiently secreted and the signal sequence is accurately cleaved apparently via the *KEX2* peptidase (Innis *et al.*, 1985). Wheat alpha amylase (Rothstein *et al.*, 1984) and human lysozyme (Jigami *et al.*, 1986) are also secreted and correctly processed to mature proteins. In other cases secretion is very inefficient. For example less than 10% of the pre-interferon produced in yeast is secreted and only 64% of this is correctly processed (Hitzeman *et al.*, 1983b). In some cases the heterologous signal sequence is not recognised at all. For example, human pre-alpha-1-antitrypsin is not secreted (Cabezon *et al.*, 1984). The presence of a heterologous signal sequence can also reduce the overall level of expression of the gene. This may be caused by a block in translation due to inappropriate interaction with the signal recognition protein (Roggenkamp *et al.*, 1985; Kaiser and

Botstein, 1986).

In general the most effective way to secrete foreign proteins is to link the mature coding sequence to a yeast signal sequence. To date the most success has been achieved with alpha factor and invertase. This is because these signal sequences contain all the information that is necessary to direct secretion and the juxtaposition of foreign coding sequences at the cleavage site does not appear to affect translocation and processing. In the case of acid phosphatase, sequences at the N-terminus of the mature protein seem to influence the function of the signal sequence which may, therefore, be less effective in hybrid constructions (Haguenaur-Tsapis and Hinnen, 1984).

The main problem with generating a convenient secretion vector is that firstly the foreign coding sequence must be precisely engineered so that the translational reading frame is maintained with the signal sequence and secondly, the cleavage site must be preserved so that the signal peptide is removed accurately and efficiently. A general approach to producing an alpha factor secretion vector based on a study by Brake et al. (1984) to produce hEGF is outlined in Figure 3.20.

A derivative of the alpha factor gene was produced by in vitro mutagenesis to introduce a unique KpnI site at residue 81 in the pro-sequence. The foreign gene, in this case, human epidermal growth factor (hEGF), is also manipulated in vitro to precisely remove its own signal sequence and replace it with a synthetic oligonucleotide adapter. The adapter begins with a KpnI site and repairs the amino acids of the alpha factor pro-sequence which will be lost after KpnI digestion. When this tailored hEGF gene is inserted into the alpha factor gene an authentic pro-sequence and Lys−Arg cleavage site is recreated. This type of vector allows high levels, about $10 \, \text{mg} \, \text{l}^{-1}$ of authentic mature proteins to be secreted into the culture medium.

Similar approaches are used to link the invertase signal sequence to heterologous coding sequences, for example, to secrete calf prochymosin (Smith et al., 1985) and human interferon (Chang et al., 1986). The invertase secretion pathway appears to become readily saturated as less than 1% of the total protein synthesised is secreted. Smith et al. (1985) mutagenised yeast transformants containing SUC2-prochymosin secretion vectors and identified super-secreting mutants that secreted 80% of the prochymosin that was produced. The analysis of these mutants may lead to the identification of the limiting proteins in the secretion system. These could then be over-expressed to enhance secretion further.

Other useful signal sequences are derived from the killer toxin which is secreted by a similar pathway to the alpha factor (Skipper et al., 1985) and the signals from some fungal enzymes that appear to function well in S. cerevisiae (Innis et al., 1985).

Gene manipulation in other yeasts

Several other yeasts in addition to S. cerevisiae are useful experimental systems, for example Pichia pastoris which is a methylotrophic budding yeast and Schizosaccharomyces pombe which is a fission yeast. Pichia pastoris has the interesting feature that the first enzyme in methanol utilisation,

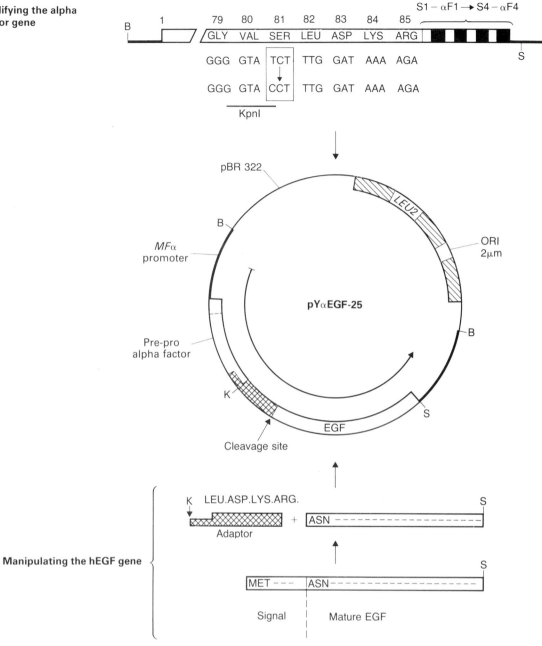

Fig. 3.20. Producing human EGF with an alpha factor secretion vector modifying the alpha factor gene (MFα).

alcohol oxidase, (AO) is expressed as up to 35% of total cell protein and thi protein is compartmentalised into peroxisomes. The expression of AO i tightly regulated, it is switched off during growth on ethanol and glucos and induced by growth on methanol. The ability to grow on a cheap carbo source such as methanol makes *Pichia* attractive for single-cell protei production. In addition the tightly regulated and efficiently expressed AO gene can be manipulated to drive the synthesis of foreign protein.

A transformation system for *Pichia* has been developed along the same lines as for *S. cerevisiae*. The histidinol dehydrogenase gene (*HIS4*) from either *Pichia* or *Saccharomyces* can be used as a selectable marker to complement *HIS4* deficient strains. ARSs can be isolated from *Pichia* (called PARSs). These PARSs allow autonomous replication but when there is other *Pichia* homologous DNA on the plasmid integration often occurs. It appears that there is a higher general recombination frequency in *P. pastoris* as compared to *S. cerevisiae*. Very strong PARSs have been isolated which remain autonomous and strains harbouring these plasmids are extremely vigorous (Cregg *et al.*, 1985). The usefulness of this system for heterologous gene expression has been shown by the high-level production of hepatitis B surface antigens (Cregg *et al.*, 1987).

Schizosaccharomyces pombe is unrelated to *S. cerevisiae*. It divides by producing a septum rather than by budding and displays some characteristics of higher eukaryotic cells that are not seen in the budding yeast. For example, there is a defined G2 phase in the cell cycle and the chromatin condenses which makes *S. pombe* a useful system to study the eukaryotic cell cycle (Nurse, 1985). It has, for example, been possible to use *S. pombe* to identify and isolate mammalian genes which are involved in the cell cycle control (Lee and Nurse, 1987; Draetta *et al.*, 1987). Also some, although by no means all, mammalian introns are processed correctly in *S. pombe* (Kaufer *et al.*, 1985). Despite the unrelatedness of *S. cerevisiae* and *S. pombe* simple *Saccharomyces* plasmids such as pJDB248 will transform *S. pombe* (Beach and Nurse, 1981). ARS sequences have been isolated from *S. pombe* by random cloning and many of these allow the plasmid to replicate as monomers although in general the gene transfer system is unreliable as the plasmids rearrange and recombine (Maundrell *et al.*, 1985). Recently, however, vectors with improved segregation have been produced (Heyer *et al.*, 1986).

Gene transfer in filamentous fungi

Filamentous fungi have been the subject of intensive classical genetic analysis. They have provided basic information about many aspects of eukaryotic cell biology, particularly about chromosome behaviour during mitosis and meiosis, DNA recombination and DNA repair (Fincham *et al.*, 1979). They also have industrial significance, for example in the production of antibiotics, steroids, pigments, enzymes, organic acids, amino and polysaccharides, and in processes such as water purification and lignin degradation. The development of gene transfer techniques in filamentous fungi have lagged behind the yeasts but some general principles and useful systems are now emerging (Bennett and Lazure, 1985; Mishra, 1985). The two most widely studied filamentous fungi are *Neurospora crassa* and *Aspergillus nidulans*. These are both Ascomycetes and therefore related to *S. cerevisiae*. Neither of these filamentous fungi, however, has a well-defined diploid phase that can be propagated in the same way as *S. cerevisiae*. Filamentous fungi grow as multicellular mycelia but they can produce a variety of differentiated cell types that are organised into higher order structures such as the conidiophores which are spore-forming structures. The genomes of these filamentous fungi are about twice the size of the *S. cerevisiae* genome.

Transformation techniques for filamentous fungi are very similar to those developed for *S. cerevisiae* and involve PEG-mediated DNA uptake into spheroplasts. These can be prepared from mycelia (Buxton and Radford, 1984) or conidiophores (e.g. Yelton *et al.*, 1984). As conidiophores are multicellular the transformants are heterokaryons containing transformed and untransformed nuclei which must be separated by crossing with appropriate marker strains. The mycelial spheroplast method appears to be most efficient. Lithium chloride can also be used to allow transformation of intact cells rather than spheroplasts (Dhawale *et al.*, 1984).

Simple *E. coli* plasmids carrying selectable markers such as the *qa-2*$^+$ (dehydroquinase) genes transform *N. crassa* at a frequency of 1000 to 5000 transformants per μg of DNA. The DNA is, however, all integrated but at a number of different sites and in many cases at multiple loci (Case *et al.*, 1979; Grant *et al.*, 1984). There is very efficient non-homologous recombination. The frequency of homologous site integration is usually less than 10%. This is clearly much lower than in *S. cerevisiae* but much higher than in mammalian cells where homologous recombination with transforming DNA is very rare.

There have been several attempts to produce autonomous vectors for *N. crassa*. The most promising seems to be the use of plasmid DNAs that are found in the mitochondria but that are not derived from mitochondrial DNA. A plasmid, pALS1, contains the 4.1 kb mitochondrial plasmid from the Labelle strain of *N. crassa* in the *E. coli* plasmid pBR325 (Stohl and Lambowitz, 1983) although this plasmid replicates in *N. crassa*, the copy number is very low and there is a high frequency of deletion of the Labelle plasmid DNA. Attempts to isolate ARSs by random cloning into a *qa-2*$^+$ plasmid revealed some sequences that enhanced transformation frequency but none of these allowed autonomous replication (Buxton and Radford, 1984). It is possible however, that the *am* gene carries an ARS because some transformants show extreme mitotic instability (Grant *et al.*, 1984).

Similar approaches have been used to transform *A. nidulans* using a variety of complementing genes such as the *N. crassa pyr-4* (orotidine-5′-phosphate decarboxylase) (Ballance *et al.*, 1983) and the homologous *amdS* (acetamidase) gene (Tilburn *et al.*, 1983). The frequency of transformation is very low, usually less than 100 transformants per μg and in all cases the transforming DNA is integrated. As with *N. crassa*, integration occurs at homologous and heterologous sites. There is no efficient autonomous vector for *A. nidulans* although a sequence isolated as an ARS in *S. cerevisiae* enhances transformation frequency (Ballance and Turner, 1985).

Although the frequency of homologous recombination in these filamentous fungi is lower than in *S. cerevisiae* similar techniques for chromosome engineering can be used but they require more extensive screening to identify the correct replacement or disruption (Paietta and Marzluf, 1985; Miller *et al.*, 1985). Transformation systems have also been developed for commercially important filamentous fungi, e.g. *Penecillium chrysogenum* (Cantoval *et al.*, 1987) and *Magnaporthe griseae*, a pathogen of rice (Parsons *et al.*, 1987). Filamentous fungi are now being used as hosts for the expression of foreign genes, e.g. bovine chymosin (Cullen *et al.*, 1987) and interferon-α2 (Gwynne *et al.*, 1987) have been produced in *Aspergillus nidulans*.

Gene transfer in *Dictyostelium discoideum*

The slime mould, *Dictyostelium*, is attractive as a model system for understanding how the sequential expression of genes during development is controlled because it undergoes a simple developmental pathway. *Dictyostelium* amoebae are cultured on agar plates covered with a lawn of killed bacteria (*Klebsiella aerogenes*) as a food source. If they are starved then they enter a developmental programme, the amoebae aggregate to form a cylindrical slug. The cells in this multicellular organism differentiate into two classes, pre-spore and pre-stalk cells, which ultimately develop into the stalk and spores of the fruiting body. These changes in morphology are accompanied by changes in gene expression.

To construct a gene transfer vector for *Dictyostelium* the actin 6 gene was fused to the coding region of the *neo* gene derived from the bacterial transposon Tn5. This confirs resistance to the antibiotic G418 (see Chapter 4). The actin 5' flanking DNA, the 5' untranslated region and the first eight codons of actin 6 were preserved to ensure efficient expression (Nellen *et al.*, 1984). The amoeboid *Dictyostelium* cells are transformed by a modification of the calcium phosphate precipitation technique developed for mammalian cells (see Chapter 4). The actin 6-*neo* vector called pB10 gives 200–2000 G418^R transformants per 10^7 cells and their resistance to G418 was stable for several generations without selection. The expression of the actin 6-*neo* gene was regulated during development in an identical way to the endogenous actin 6 gene. All the vector sequences were present in high-molecular-weight DNA but it was not clear if they were present as tandem integrated repeats or oligomeric structures.

Many developmentally regulated genes have been isolated from *Dictyostelium* including several whose expression is modulated by cyclic AMP. In addition it is easy to isolate mutant organisms that show aberrant development and gene transfer can then be used to isolate regulatory genes by complementation (Kay, 1983).

Gene transfer in *Chlamydomonas reinhardi*

This unicellular green alga has a haploid vegetative phase that undergoes gametogenesis in response to nitrogen deprivation. A single gene with two alleles, mt$^+$ and mt$^-$ controls mating type and the single chloroplast of mt$^+$ cells is always inherited by the haploid products of meiosis. Because of this uniparental inheritance of chloroplasts, *C. reinhardi* is particularly well suited for the genetic analysis of chloroplast genes and for studying organelle gene recombination. There are also many nuclear mutants that define fundamental biological processes such as motility, photosynthesis and circadian rhythms. The organism is also potentially useful as a host system for the efficient and large-scale expression of foreign genes because it uses a free energy source, i.e. sunlight! The alga is readily cultured on agar plates where it produces green colonies.

The first transformation system for *C. reinhardi* used plasmids carrying the *S. cerevisiae ARG4* gene to complement a *C. reinhardi* mutant in *arg7* (argenate succinate lyase, analogous to *ARG4* in *S. cerevisiae*). A low frequency of transformation to phototrophy was obtained. This could not be increased by including autonomous replication sequences derived from *C. reinhardi* (ARCs) DNA in the plasmids. These plasmids remained extra-chromosomal but were not stably maintained (interestingly only one of these ARCs functioned in *S. cerevisiae* (Rochaix *et al.*, 1984). Recently Hasnain *et al.* (1985) have developed a high-efficiency transformation vector. They cloned a 1576 base pair fragment containing the origin of replication from the *S. cerevisiae* 2 μm circle into the mammalian gene transfer vector pSV2-Neo (see Chapter 4). This plasmid, pSV2-Neo-2μ transforms *C. reinhardi* to G418 resistance at a frequency of about 10^2 transformants per 5 μg of DNA. The plasmid was extrachromosomal and stably maintained for at least 230 generations indicating that the 2 μm replication and stabilising sequences appeared to be recognised in this heterologous host. DNA is introduced into the cells by incubating a cell wall deficient (*cwd*) mutant with the plasmid DNA, poly-L-ornithine and $ZnSO_4$.

Summary

Gene manipulation is now possible in a wide range of microbial eukaryotes including yeasts, filamentous fungi, slime moulds and algae. The transfer of genes into the microbial eukaryote *S. cerevisiae* is proving to be central to increasing our understanding of fundamental biological processes, in particular in the areas of chromosome structure and function. The molecular analysis of gene expression may also provide insight into the nature of the eukaryotic transcription initiation complex. *S. cerevisiae* is also a key organism in biotechnology for the large-scale production of valuable proteins. New developments are on the horizon, e.g. transposons may be used to distribute genes throughout chromosomes and YAC vectors will become increasingly sophisticated allowing mammalian cell/yeast shuttle systems to be developed. It is likely that *S. cerevisiae* may soon have to share centre stage with other microbial eukaryotes. *S. pombe*, for example, is being studied increasingly as a model for eukaryotic cell cycle control and filamentous fungi may be developed for the high-level secretion of foreign proteins.

4 Gene Transfer into Animal Cells

Introduction

There are two types of animal cell gene transfer system. These are *transient expression systems* and *stable transformation*. In transient systems the DNA is introduced into cells, gene expression is allowed to occur for 24 to 72 hr and then the cells are lysed and analysed for gene products. In stable transformation, cells that have stably acquired new DNA sequences are selected and propagated to produce a stable, genetically altered, i.e. transformed, cell line. Usually only a fraction of cells that initially take up DNA go on to become stably transformed. There are specialised vectors for the two types of experiment and advantages and limitations of each approach. For an efficient transient system a very high number of cells must take up DNA and the levels of expression must be high to provide sufficient gene product for analysis. For an efficient stable transformation the DNA must be integrated into the chromosome or replicated efficiently so that it is maintained during cell division. There must also be an effective system to select for stably transformed cells.

The growth and properties of animal cells in culture

Establishing a cell culture

Many cells can be maintained in a metabolically active state *in vitro*. A brief account of some of the methods and considerations will be given here but details can be found in Freshney (1983). A tissue is removed from an adult animal or embryo and dispersed into its constituent cells by fine dissection and trypsin treatment. The dispersed cells are put into a culture vessel. This is either a flat sided flask or petri dish made of special plastic or high-quality glass, containing a partially defined nutrient medium supplemented with animal serum. The dispersed cells are then called a *cell culture* or, less correctly, a *tissue culture* (Figure 4.1). A tissue is, of course, composed of many different cell types and this will be reflected initially in the dispersed cell population. The cells settle and adhere to the culture vessel surface to become the *primary cell culture* which closely resembles the cell composition of the original tissue. The adherence of the cells is necessary for their initial

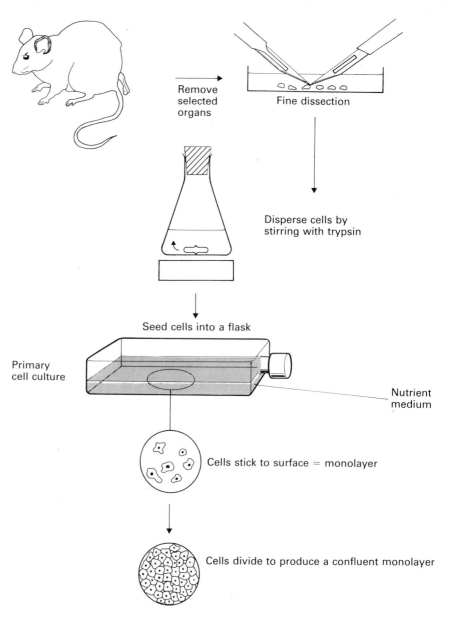

Remove
selected
organs

Fine dissection

Disperse cells by
stirring with trypsin

Seed cells into a flask

Primary
cell culture

Nutrient
medium

Cells stick to surface = monolayer

Cells divide to produce a confluent monolayer

Fig. 4.1. Mammalian cell culture.

growth in culture and there is, at first, a sparse covering of single cells attached to the culture substrate. This is a *monolayer culture*.

If the primary culture is maintained for more than a few hours selection will begin. Some cells simply die, others maintain but do not divide and the remainder are capable of some degree of proliferation. These replication-competent cells will divide until all the surface of the culture vessel has been covered. At this point there is a single layer of cells covering the entire culture vessel surface and the monolayer is said to be *confluent*. Normal cells cease dividing when they have reached a certain cell density and this is usually influenced by the degree of cell-to-cell contact; the cells are said to be *contact inhibited*. The culture now shows its closest functional resemblance to the parent tissue because many aspects of specialised functions are only strongly expressed when there is maximum cell-to-cell contact. At this point the primary culture period is over, there is no more space in the culture vessel and the cells must be sub-cultured. Only the cells which are capable of proliferation can now be propagated and eventually the most vigorous cells will constitute the bulk of the cell population. To sub-culture the cells they are removed from the culture vessel by using trypsin and chelating agents to disrupt cell-to-cell and cell-to-culture surface adhesion and dispersed by gentle pipetting to form a single-cell suspension. An aliquot of these cells is then transferred into a new culture flask. This is called *seeding* and the fraction of the original cells which is used is called the *split ratio*. The process is referred to as *splitting* or *passaging* the culture. The seeding density and split ratio are different for different cell cultures. At the first subculture a *finite cell line* is established. The cells divide and when they approach confluence they are passaged. Usually by the third passage the culture is stable in terms of cell composition and contains hardy, rapidly proliferating cells usually derived from connective tissue fibroblasts or vascular elements. Many cell lines are therefore fibroblastic, but, by modifying culture conditions and cloning sub-populations, the fibroblast dominance can be prevented and specialised cell lines which express specific differentiated functions can be established.

The evolution of a cell culture

Most cell lines have a finite life span; they can only be passaged for 10 to 20 weeks or about 30 to 60 cell divisions. Then they go into senescence and die or give rise to *continuous cell lines* (often referred to simply as a *cell line*) which have altered characteristics. The evolution of a cell culture is depicted in Figure 4.2 and is characteristic and reproducible for individual cell cultures, e.g. normal human fibroblasts invariably go into senescence after 50 generations. Many normal cell cultures never spontaneously form continuous cell lines. Continuous cell lines can be propagated indefinitely, they are usually aneuploid and show some chromosome instability and/or ploidy variations usually between diploid and tetraploid. They display many of the properties of tumour cells, they have reduced nutritional requirements, divide rapidly and grow to high cell densities. Many continuous cell lines will spontaneously yield cells which display the full tumour phenotype and these may eventually dominate the culture, particularly if it is maintained and passaged at a high cell density. These cells lose contact inhibition and so pile up on the culture substrate and they will form tumours when injected into animals. Continuous cell lines are often referred to as *immortalised cells*.

Culturing specialised cells and precursor cells

It is common for *terminally differentiated* or *specialised* cells to be unable to grow in cell culture because loss of proliferative capacity usually accompanies

GENE TRANSFER INTO ANIMAL CELLS

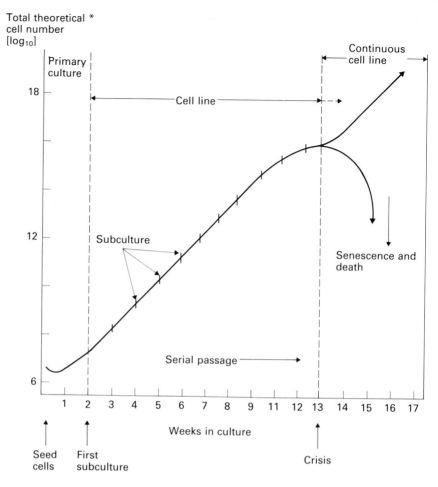

Fig. 4.2. The evolution of a cell culture. (Adapted from figure 2.1 of Freshney, 1983).

terminal differentiation. The only way to study these cells *in vitro* is by using immortalised derivatives. In some cases spontaneous or chemically induced tumours of the specific cell type of interest may have been isolated. It is more effective, however, to immortalise cells directly by infecting them with a tumour virus or transforming them with an oncogene. The tumour virus, SV40, has been used, for example, to generate immortalised cell lines displaying many of the characteristics of differentiated cells such as muscle (Miranda *et al.*, 1983) and macrophages (Nagata *et al.*, 1983).

Specialised cells are derived from precursor cells known as stem cells. Different types of stem cells give rise to different 'families' of specialised cells. Commonly used cell type 'families' are: epithelial cells (e.g. derived from cervix, gastrointestinal tract, kidney tubules, bronchi, trachea), mesenchymal cells (derived from embryonic mesoderm, e.g. connective tissue fibroblasts, adipose cells, cardiac and skeletal muscle, vascular endothelium), neuroectodermal cells (e.g. neurones and glial cells), haemopoietic cells (leucocytes, erythroid cells, macrophages) and gonads.

The pathway of differentiation of stem cell to a specialised cell is usually called the *cell lineage*. Generally cells in one lineage do not transdifferentiate into another lineage. The stem cell has a high proliferative capacity and can develop into any differentiated cell in the particular family. The differentiation pathway proceeds through an intermediate cell stage called the *committed* precursor stem cell which will only differentiate into one specific cell type but which still has proliferative capacity. This process of differentiation is continuously occurring in the haematopoietic cells of an adult animal bone marrow and there is great interest in developing methods for studying these cells at each stage and to control their differentiation *in vitro*. Unfortunately it is difficult to purify these cell populations from animals in sufficient quantities and it is technically demanding to culture bone marrow. It is therefore necessary to immortalise cells at each stage of the lineage. Figure 4.3 shows the lineage of the antibody-producing B lymphocytes in the mouse and the types of continuous cell lines that are available to study the molecular events in differentiation.

As B cells differentiate from committed precursor cells, the antibody genes undergo the rearrangements that were described in Chapter 1. The precursor stem cell does not produce antibodies but proliferates and differentiates into mature B cells that display IgM and IgD on their cell surface. There are three intermediate stages: these are early pre-B cells that do not produce antibody but that have D−J joining in their genomes, pre-B cells that have V−D−J−C joining and produce μ heavy chains and immature B cells that produce membrane bound IgM. When mature B cells are stimulated with antigens or mitogens and specific growth factors called lymphokines they undergo the second phase of differentiation into non-dividing plasma cells that secrete antibodies and memory cells that rapidly differentiate into plasma cells if the animal is treated with a repeat dose of the same antigen. The stem cell can differentiate into any of the blood cell types but once a committed stem cell is formed this differentiates to produce a single family of cells. Immortalised cell lines for each stage of B lymphocyte differentiation are available. One widely used approach is to infect embryo or adult mice with the retrovirus Abelson murine leukaemia virus (A-MLV). Early pre-B cells infected with A-MLV produce a B lymphoma (a tumour of lymphoid tissue) that does not synthesise immunoglobulin whereas infection of adult bone marrow cells produces a B lymphoma that synthesises μ heavy chains. There are also similar spontaneous or chemically induced tumours. These tumour cells can be cultured indefinitely. They retain many of the characteristics of the progenitor cell and some lines will differentiate to the next stage under controlled conditions. Examples of studies using tumour cell lines to analyse different stages in B lymphocyte differentiation can be found in Yancopoulos *et al.* (1984) and Mather *et al.* (1984).

Another particularly useful type of tumour is the teratoma. This is an adult tumour composed of pluripotent embryonic cells that can differentiate into virtually any cell type in culture. These are discussed further in Chapter 5.

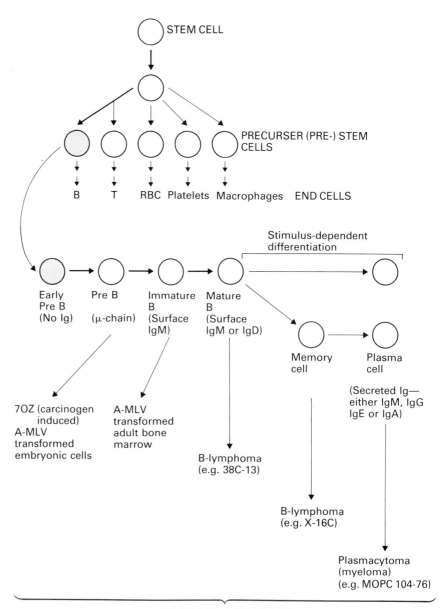

STEM CELL

PRECURSER (PRE-) STEM CELLS

B   T   RBC   Platelets   Macrophages   END CELLS

Stimulus-dependent differentiation

Early Pre B (No Ig)

Pre B (μ-chain)

Immature B (Surface IgM)

Mature B (Surface IgM or IgD)

Memory cell

Plasma cell

(Secreted Ig— either IgM, IgG IgE or IgA)

7OZ (carcinogen induced) A-MLV transformed embryonic cells

A-MLV transformed adult bone marrow

B-lymphoma (e.g. 38C-13)

B-lymphoma (e.g. X-16C)

Plasmacytoma (myeloma) (e.g. MOPC 104-76)

CONTINUOUS CELL LINES

Fig. 4.3. B lymphocyte differentiation.

In all these studies that use tumour cells as models for differentiation there must always be reservations about possible differences from the situation in the animal.

The growth of cells in culture

While many cells require a solid support some cells are non-adhesive, e.g. leukaemia and ascites cells. Others such as lymphoblastoid cells and many

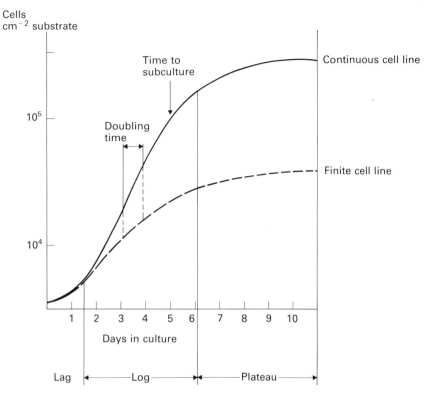

Fig. 4.4. The growth cycle of cells in culture. The doubling times for different cells vary but 24 hours is usual. Notice that the continuous cell line reaches a higher cell density before the plateau than the finite cell line. (Adapted from figure 12.3 of Freshney, 1983).

neoplastic cells do not require a solid substrate for growth. These cells must be, or may be, cultured in suspension in much the same way as microorganisms. Like microorganisms a flask or suspension of cells will go through a growth cycle with a lag, log and plateau phase (Figure 4.4). The properties of the cells will vary throughout this growth cycle, the most reproducible phase being mid to late log.

In cell culture, passage at low density in high serum promotes cells which behave like stem cells whereas passage at high cell density promotes cells which behave more like differentiated cells. In other words the culture conditions can affect the properties of the cell line. Successful experiments with cultured animal cells therefore require great care in maintaining the cells under optimum conditions and require an awareness that a cultured cell is unlikely to have the same properties as the equivalent cell *in vivo*. This is particularly true of continuous cell lines which are hovering on the verge of becoming tumour cells.

Selection systems

The key step in the development of gene transfer systems in mammalian cells was the ability to detect the presence of the transferred DNA in the

GENE TRANSFER INTO ANIMAL CELLS

recipient cells. Once this had been achieved it was possible to refine and develop the transfer techniques. In addition if gene transfer is less than 100% efficient, which is the case for many of the systems that we will describe, it is important to be able to select for stably transformed cells against a background of untransformed cells. In this section we will discuss the biochemical basis of the most commonly used selection systems.

Many of the selection systems exploit aspects of purine and pyrimidine metabolism and so we will first outline these pathways before discussing general methods of selection by complementation or selection for dominant genes.

Purine and pyrimidine biosynthesis

Figure 4.5 gives an outline of nucleotide metabolism and more details can be found in Kornberg (1980). The first key feature is that there are two biosynthetic pathways for purine and pyrimidine biosynthesis that are independent because they involve different enyzmes. These are called the *de novo* and *salvage* pathways. The second key feature is that there are a range of drugs that selectively inhibit one or the other pathway by inactivating specific steps. This means that mutations are relatively easy to obtain because cells that are defective in one pathway can still be cultured provided they can use the alternative pathway. If however, the alternative pathway is blocked by a specific drug the cells will die unless a wild-type gene is introduced to replace the mutant gene. This is the basis of some of the commonly used complementation systems. Alternatively, genes that confer resistance to a drug that inhibits the *de novo* synthesis can be used as dominant selectable markers if salvage precursors are not provided.

Selection by complementation

Selection for thymidine kinase

Most of the initial experiments that established the feasibility of gene transfer into mammalian cells used the thymidine kinase (*tk*) gene selection system.

Cells that lack a functional *tk* gene can be obtained by mutagenesis and selection for resistance to the toxic nucleoside analogue, 5' bromodeoxyuridine (BUdR). This is normally converted to the nucleotide by thymidine kinase and becomes incorporated into DNA where it causes increased sensitivity to UV and fatal misreplication. A derivative of mouse L cells has a mutation in the *tk* gene. These L*tk*$^-$ cells are widely used as they rarely revert to *tk*$^+$ (Kit *et al.*, 1963). Mouse L*tk*$^-$ cells are also very easy to transform and are often used as a recipient even when the selection is not for *tk*. *k*$^-$ derivatives of a number of differentiated cell lines have now been made, e.g. *tk*$^-$ mouse myoblasts (Merrill *et al.*, 1980).

If cells are treated with aminopterin, the *de novo* synthesis of thymidine monophosphate (TMP), guanine monophosphate (GMP) and adenine monophosphate (AMP) is blocked through depletion of the reduced cofactor

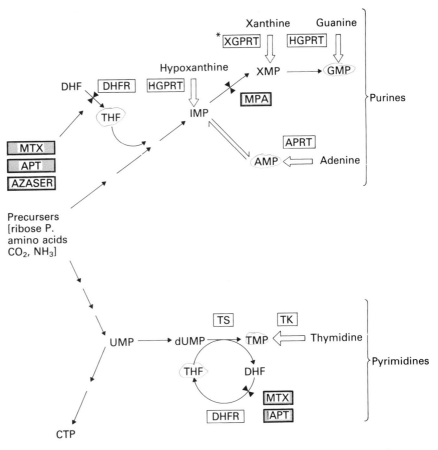

Fig. 4.5. An outline of the *de novo* and salvage pathways of nucleotide biosynthesis. ⋆, XGPRT is not found in mammals. The salvage pathways are indicated by thick open arrows. The important enzymes are boxed, these are HGPRT = hypoxanthine-guanine phosphoribosyl transferase; XGPRT = xanthine-guanine phosphoribosyl transferase; APRT = adenine phosphoribosyl transferase; TK = thymidine kinase; DHFR = dihydrofolate reductase; TS = thymidine synthase. Inhibitors of the pathways are shown in dark boxes; these are: MTX = methotrexate; APT = aminopterin; AZASER = azaserine; MPA = mycophenolic acid. THF = tetrahydrofolate; DHF = dihydrofolate. IMP, AMP, XMP, GMP, UMP are inosine, adenosine, xanthine, guanosine, uridine monophosphates respectively.

tetrahydrofolate (THF). If hypoxanthine is supplied in the medium this bypasses the aminopterin block and purine monophosphates are synthesised. Normally, if thymidine is supplied then thymidylate is produced via the action of thymidine kinase. As this cannot occur in *tk⁻* cells, they will die. If a normal *tk* gene is introduced into these cells they will survive in a medium containing aminopterin, with hypoxanthine and thymidine added as precursors. This is called HAT (hypoxanthine, aminopterine, thymidine) selection and the procedure was first devised by Szybalska and Szybalski (1962). Glycine must also be provided as its synthesis also requires THF. The first source of a *tk* gene was a 3.4 kb Bam HI fragment of herpes simplex virus (HSV) DNA. This was introduced into mouse L*tk⁻* cells and stable *tk⁺* transformants were obtained. Luckily this fragment carries the *tk* gene promoter (McKnight *et al.*, 1984) so that the gene could be expressed (Figure 4.6).

Subsequently *tk* selection was used to detect the transfer of a cellular *tk* gene into *tk*⁻ cells when total cellular DNA was used. The next important observation was that DNA which did not confer a selectable phenotype could be cotransferred at a high frequency with the selectable DNA. For example Huttner *et al.* (1979) introduced a pBR322 recombinant carrying the human β-globin gene with purified HSV *tk* in the ratio 20 000 β-globin sequences to one *tk* sequence. 80% of the stable *tk*⁺ transformants also contained human β-globin sequences. These studies were important because they showed that any DNA could be stably maintained in a recipient cell irrespective of selection.

The frequency of transformation in all these early studies was very low, about 1 in 10^4 to 1 in 10^6 cells that were initially transfected became stable transformants. The powerful HAT selection for transferred *tk*⁺ genes was therefore essential for proving that reliable gene transfer was possible and for permitting the isolation of transformed cells. A detailed review of these early studies can be found in Scangos and Ruddle (1981).

It has now been discovered that the HSV-*tk* promoter is rather weak and requires the addition of an enhancer for maximum activity. As we shall see later, high-level expression of the selectable marker is important for efficient

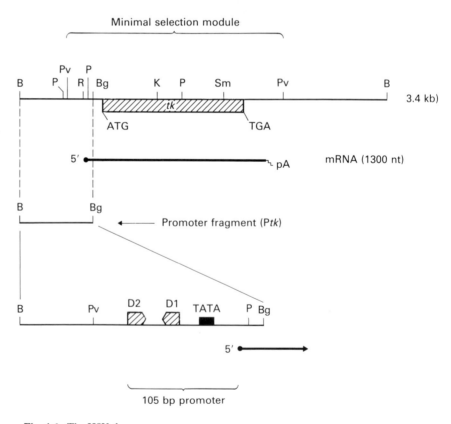

Fig. 4.6. The HSV *tk* gene.

transformation and the weakness of the *tk* promoter may explain some of the low transformation frequencies.

Selection for dihydrofolate reductase

Cells that are defective in the *dhfr* gene can be isolated as resistant to tritiated deoxyuridine which is converted to tritiated thymidine by dihydro-folate reductase (DHFR) and then incorporated into DNA where it causes misreplication. *dhfr* mutants are unable to regenerate THF and will die unless thymidine and purines are supplied. They can be complemented by the wild-type *dhfr* gene. The mouse chromosomal *dhfr* gene is 31 kb long and contains five introns several of which contain repeated sequences that cause instability (Crouse *et al.*, 1982). In order to provide a manageable selectable gene that can be incorporated into gene transfer vectors a *dhfr* cDNA is used. Figure 4.7 shows a typical mouse *dhfr* selection module. The cDNA (Chang *et al.*, 1978) is tailored to remove most of 5' and 3' untranslated regions. Expression is driven by the major late promoter (MLP) of an adenovirus (Ad2) and this also provides a 5' untranslation region and 5' splice donor site. The 3' splice acceptor site is supplied by a fragment from an immunoglobulin gene and the transcription termination and poly-adenylation signals are from the SV40 early transcription and polyadenylation signals are from the SV40 early transcription unit (Kaufman and Sharp, 1982). This rather complex selection module reflects the tendencies to use the most convenient pieces of DNA which are available in a particular laboratory to provide appropriate signals. More recently a more convenient set of selection modules have been produced (see later).

Other complementation systems

The cloned *hgprt* and *aprt* genes have also been used to complement *hgprt*⁻ and *aprt*⁻ cells allowing them to survive in HAT medium (Littlefield, 1963; Willecke *et al.*, 1979; Wigler *et al.*, 1979b) and the human arginosuccinate

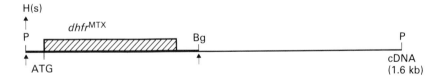

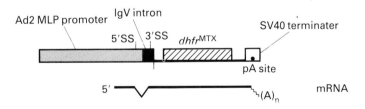

Fig. 4.7. A mouse *dhfr* selection module.

GENE TRANSFER INTO ANIMAL CELLS

synthetase gene has been used to complement a deficient hamster cell line (Hudson *et al.*, 1980). Rodent cells that are deficient in uridine biosynthesis have been complemented with the cloned *cad* (Carbomoylphosphate-aspartate transcarbamylase-dihydroorotase) gene complex from Syrian hamster (de Saint-Vincent *et al.*, 1981) and a galactokinase deficient cell line that is unable to use galactose as a sole carbon source has been complemented with a bacterial galactokinase gene (Schümperli *et al.*, 1982).

Selection for dominant markers

The use of dominant selectable markers overcomes the need to isolate specific mutant cell lines and, therefore, such markers are now used more often than complementing markers. Many of the dominant selectable markers are derived from prokaryotic genes and so they have been manipulated to permit expression in eukaryotic cells.

Selection for G418 resistance

The aminoglycoside antibiotic, G418, is structurally related to the antibacterial antibiotics kanamycin and neomycin. It acts by inhibiting the function of 80S ribosomes and therefore blocks protein synthesis in eukaryotic cells. All aminoglycosides can be inactivated by aminoglycoside 3' phosphotransferases that are encoded by the bacterial transposons, Tn601 (also called Tn903) and Tn5 (Davies and Smith, 1978). These genes, usually referred to as *neo* genes, therefore confer resistance to aminoglycoside antibiotics. Jiminez and Davies (1980) showed that the expression of the Tn601 *neo* gene in a eukaryotic cell, yeast, conferred resistance to G418. Subsequently, the *neo* gene from Tn5 (Beck *et al.*, 1982) (Figure 4.8A) has also been shown to confer resistance in mammalian cells (Colbère-Garapin, 1981; Southern and Berg, 1982).

Mammalian cells treated with G418 stop replicating and eventually die. If the bacterial *neo* gene (Figure 4.8A) is introduced under the control of appropriate expression signals (Figure 4.10) transformed cells become resistant to G418 and can be selected against the background of dying untransformed cells (Southern and Berg, 1982). There is some variation between mammalian cells in sensitivity to G418. Some cells such as HeLa (human) and CV1 (monkey) require treatment with high concentrations, about 400 µg ml^{-1}. It is also important for maximum sensitivity to transform cells at a low cell density as actively growing cells are most sensitive.

The *neo* gene has proved to be one of the most useful selectable markers and has facilitated gene transfer into a wide range of eukaryotic cells.

Selection for mycophenolic acid (MPA) resistance

Bacterial xanthine – guanine phosphoribosyl transferase (XGPRT) can convert xanthine to XMP as well as converting hypoxanthine to IMP and guanine to GMP. Mammalian cells lack this activity and will die if *de novo* guanine synthesis is blocked by mycophenolic acid (MPA) and only xanthine is

A *neo*

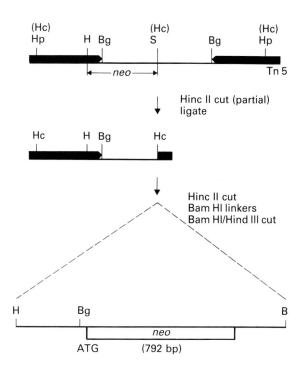

B *Eco-gpt*

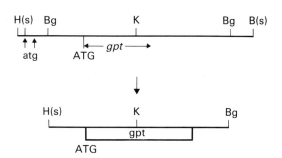

Fig. 4.8. Tailoring bacterial genes for use as dominant selectable markers. In B, the initiating methionine codon for GPT synthesis is shown in upper case and two additional upstream methionine codons are shown in lower case.

supplied as a salvage precursor (Figure 4.5). The *E. coli* XGPRT gene (*Eco-gpt*) has been cloned (Figure 4.8) and when incorporated into a suitable vector it confers resistance to MPA on any mammalian cell (Mulligan and Berg, 1981). A more efficient selection procedure uses aminopterin to block all THF-dependent synthesis and provides xanthine, thymidine, adenine and glycine. A disadvantage of this selection procedure is that the culture medium must be free of guanine and therefore any serum used has to be extensively pre-absorbed to remove guanine.

GENE TRANSFER INTO ANIMAL CELLS

Selection for cadmium resistance

Metallothioneins are small cysteine-rich proteins that chelate metal ions. They are physiologically important in heavy metal detoxification and zinc and copper homeostasis (Hamer, 1986). The over-production of metallothionein confers resistance, in many cells, to heavy metals such as cadmium. The mouse metallothionein gene, Mt-I, can be used, therefore, as a dominant selectable marker to transfer cadmium resistance (Durnham et al., 1980; Pavlakis and Hamer, 1983). It is particularly useful because the promoter functions efficiently in a wide range of cells (Mayo et al., 1982). The cloned gene has been characterised; it is contained in a 2.9 kb Eco RI-Hind III fragment and the coding sequence is split into three exons (Figure 4.9). In addition, the convenient Bgl II site in the leader region can be used for inserting other coding sequences for expression using the Mt gene signals. The Mt-1 promoter has been analysed and is described in more detail in Chapter 8.

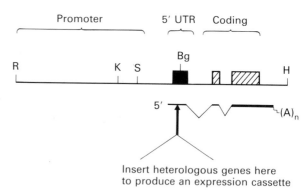

Fig. 4.9. The mouse metallothionein gene, Mt-II. The dark box is the 5′ untranslated region.

Selection for methotrexate resistance

We have mentioned the use of a cloned mouse $dhfr$ gene to complement mutant $dhfr^-$ cells. Certain $dhfr$ genes can, however, be used as dominant selectable markers. These genes encode dihydrofolate reductases which are insensitive to the inhibiting effects of the anti-cancer drug methotrexate (MTX). MTX is an analogue of folate and binds very tightly to normal mammalian dihydrofolate reductase to inhibit its activity. Cells therefore die due to purine and thymidylate starvation (Figure 4.5).

When mouse 3T6 fibroblasts were exposed to MTX some resistant cells emerged and one of these cell lines, 3T6-400 contains a mutant $dhfr$ gene (Haber et al., 1981). The cDNA has been cloned and sequenced and it has a single base change in its coding region as compared to the wild-type gene (Simonsen and Levinson, 1983). The mutant enzyme binds MTX 270 times less efficiently than the normal enzyme and is therefore resistant to concentrations of MTX which kill wild-type cells. Transfer of the mutant $dhfr^{MTX}$ to normal cells therefore confers resistance. A chinese hamster cell line

CHOA29 also contains an MTX-resistant DHFR and total DNA prepared from this cell line can confer resistance on normal cells (Wigler *et al.*, 1980).

Another approach is to use the bacterial *dhfr* gene which is contained on the drug resistance factor R388. This encodes a dihydrofolate reductase which is intrinsically resistant to MTX (Pattishal *et al.*, 1977). Cloned bacterial *dhfr* has been used to transform normal mammalian cells to MTX resistance (O'Hare *et al.*, 1981).

Simple plasmid vectors

In this section we review some of the more widely used vectors that facilitate the detection, selection and expression of heterologous genes.

The pSV plasmids

The pSV plasmids are a family of plasmids based on the DNA tumour virus SV40 (Mulligan and Berg, 1980). This small virus has been studied in great detail and derivatives have been used as cloning vectors, gene transfer vectors and expression vectors. We will be discussing the molecular biology of SV40 later. Here we will just give an overview of the use of the viral gene expression signals to direct the expression of selectable markers. SV40 is a small double-stranded circular DNA virus which contains two divergent transcription units controlled from a single complex promoter/replication region. The transcription units are referred to as early and late to reflect the time of maximal expression during infection. All the viral transcripts contain introns and are polyadenylated.

In the plasmid pSV2, the Pvu II−Hind III fragment, which contains all the promoter signals for early transcription and the mRNA initiation site, but no translation initiation codon, is used to initiate transcription of any coding sequence placed downstream. This is followed by a fragment which contains a 66 bp intron from the transcript for the viral small t protein to provide a splice donor and acceptor site for the heterologous transcript. This is followed by a fragment containing the transcription termination region and polyadenylation site. All these fragments are linked to a pBR322 fragment which contains a prokaryotic replication origin and Amp selectable marker (Figure 4.10). The selectable markers *Eco-gpt*, *neo* (Figure 4.8) and mouse *dhfr* (Figure 4.8) have been inserted as Hind III−Bgl II fragments into pSV2 to produce pSV2-*gpt* (Mulligan and Berg, 1980), pSV2-*neo* (Southern and Berg, 1982) and pSV2-*dhfr* (Subramani *et al.*, 1981) respectively. The SV40 early promoter is used efficiently in a variety of different cells and therefore these plasmids provide a source of dominant selectable markers with a broad host range. More details about the SV40 promoter can be found in the following: McKnight and Tjian, 1986; Maniatis *et al.*, 1987 and Chapter 8).

The pRSV plasmids

The promoter from the retrovirus Rous sarcoma virus (RSV) is more

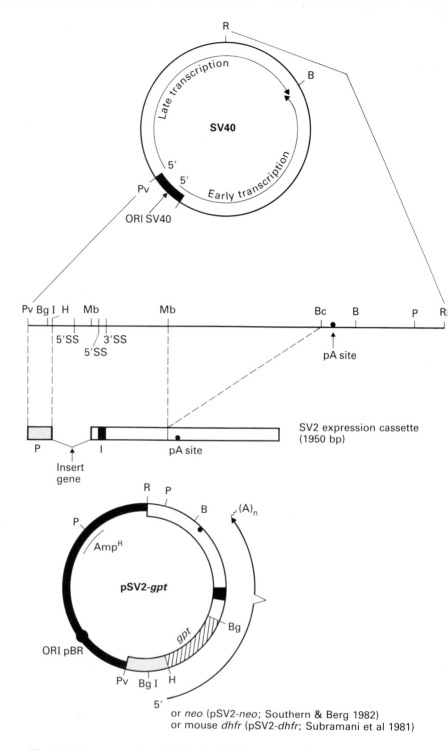

Fig. 4.10. The construction of pSV2 plasmids. The promoter (P), intron (I) and termination region (pA site) from the early region are combined to produce the SV2 expression cassette.

effective than SV40 in some cells (Gorman *et al.*, 1982a). Retroviruses are single-stranded RNA viruses which replicate via a double-stranded DNA intermediate that integrates into the host chromosome where it is called a *provirus* (see later). The provirus has promoter elements located in the long terminal repeats (LTRs) and these have been isolated as a 524 bp fragment and used to replace the SV40 promoter in pSV2 plasmids to produce a pRSV expression module and pRSV-*neo* and pRSV-*gpt* derivatives (Gorman *et al.*, 1983a) (Figure 4.11).

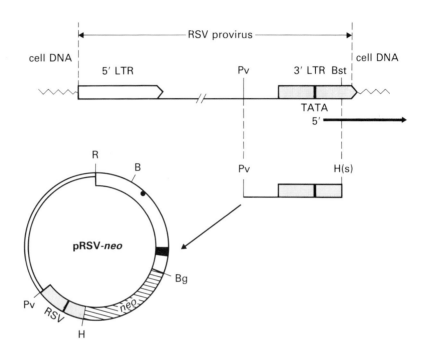

Fig. 4.11. The pRSV plasmids. The 3′ LTR of RSV and the intron and termination region from the SV40 early region (see Figure 4.10) are combined to produce the RSV expression cassette.

Dual selection plasmids

The bacterial genes, *dhfr* and *neo* confer resistance to trimethoprim and kanamycin, respectively, in bacteria and therefore can be used as dual selectable markers in *E. coli*/mammalian cell shuttle vectors. To do this a prokaryotic promoter must be built into the selection module. This can be placed upstream from the eukaryotic promoter because prokaryotic ribosomes will initiate at the correct AUG provided that it is contained within a ribosome binding site, irrespective of whether it is the first AUG in the message. For example, Meneguzzi *et al.* (1984) have constructed a *neo* selection module which contains the original prokaryotic *neo* promoter, followed by the HSV*tk* promoter, the non-coding region and the *tk* gene polyadenylation site. Alternatively there are some promoters such as the avian sarcoma virus LTR which fortuitously function in bacteria and mammalian cells (Mitsialis *et al.*, 1981).

GENE TRANSFER INTO ANIMAL CELLS

CAT plasmids

In the previous sections we have been considering ways of selecting stable transformants. It is also important to have tools to study the transient expression which occurs at 24 to 72 hr after transformation but before selection operates. A series of plasmids which contain the bacterial gene for chloramphenicol acetyl transferase (CAT) have proved useful for monitoring transient gene expression. Eukaryotic cells contain no endogenous CAT activity; therefore the expression of the transferred bacterial *cat* gene can be readily monitored.

The *cat* gene was isolated from the *E. coli* transposon Tn9 and a tailored 773 bp fragment was inserted into pSV2 to produce pSV2-*cat* where the *cat* coding sequence is expressed by the SV40 promoter (Gorman *et al.*, 1982b). The SV40 promoter has also been replaced with the RSV LTR promoter to create pRSV-*cat* and a promoterless derivative pSV0-*cat* can be used with any promoter sequence. A related plasmid, pA10CAT contains the *cat* gene and part of the SV40 promoter region that contains the cap site but which lacks the enhancer (Laimins *et al.*, 1982) (Figure 4.12). There is no CAT expression using this plasmid unless an enhancer is re-introduced. Plasmids such as pA10CAT can therefore be used to assay DNA fragments for enhancer activity. Plasmids of this type have been used to analyse promoter functions and will be discussed further in Chapter 8.

CAT enzyme activity can be assayed by making cell extracts and incubating them with acetyl Co-A and ^{14}C-chloramphenicol. The enzyme acylates the chloramphenicol and this substrate and acylated derivatives are extracted into ethylacetate and separated by thin layer chromatography on silica gel plates. The products are visualised by autoradiography and quantitated by excising the spots from the gel and counting the ^{14}C label. A typical assay is shown in Figure 4.13. The expression of CAT in individual cells in a transformed monolayer can be monitored *in situ* by using immunofluourescent staining using anti-CAT antisera (e.g. Gorman and Howard, 1983).

Amplicons

Wild-type and mutant mammalian *dhfr* genes can undergo amplification up to 1000 copies per cell whether they are present in the chromosome or on a plasmid vector (see Chapter 1 for an introduction to amplification and Chapter 9 for details). When cells are exposed to increasing concentrations of MTX there is a strong selection for cells containing higher numbers of *dhfr* genes. We will be discussing gene amplification in more detail in Chapter 9 but a key point is that the amplicon, i.e. the unit of amplification, is much larger than the selected *dhfr* gene. This means that DNA sequences convalently linked to a *dhfr* gene are co-amplified. This fact has been exploited to increase the copy number of transfected genes. A simple approach has been to insert a foreign gene into a plasmid containing a *dhfr* selection module. The plasmids are then introduced into a *dhfr*⁻ recipient cell and transformants are selected by resistance to MTX. The most widely

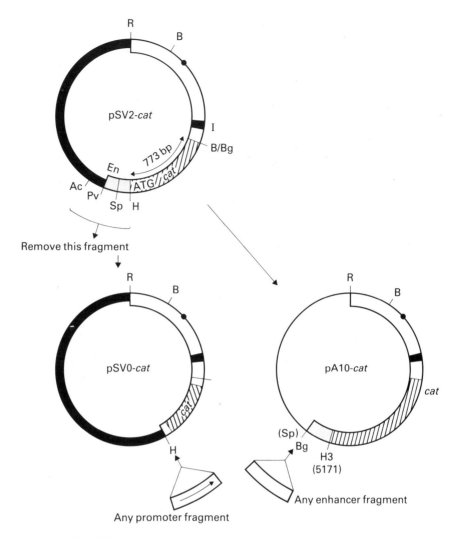

Fig. 4.12. The CAT plasmids.

used host is a *dhfr⁻* chinese hamster ovary (CHO) cell line called CHO-DUK (Urlaub and Chasin, 1980). The initial transformants which contain only one or two copies of the plasmid integrated into the genome are pooled and then cultured in increasing concentration of MTX (0.05 µm to 0.2 mM). Only cells that contain increased copies of the *dhfr* gene will survive because the consequent increased levels of enzyme exceed the concentration of the drug. These cells also express high levels of the unselected genes in the amplicon. The amplicons are usually confined to a single chromosomal site that appears as a homogeneously staining region (Chapter 1), but occasionally there are several sites and karyotypic abnormalities such as tetraploidy and extra minute chromosomes.

An alternative approach is to use co-transformation and to rely upon *in vivo* recombination to convalently link a *dhfr* gene to the unselected gene. For example Christman *et al.* (1982) simply introduced total chromosomal

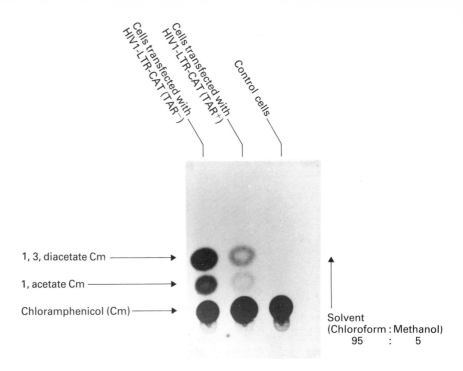

Fig. 4.13. Assaying transient gene expression using CAT enzyme assays. A photograph of an autoradiograph of a thin layer chromatogram. The cell extracts are mixed with ^{14}C chloramphenicol which is acetylated by the CAT enzyme in the extracts to produce the derivatives shown and in some cases 1,3 diacetyl chloramphenicol; these migrate different distances with the solvent. The experiment shows that a fusion between the HIV1 LTR and the CAT gene results in an expression which is significantly increased when the negatively acting TAR site is deleted (see Chapter 11 for more about HIV1). A control extract shows no acetylated chloramphenicol. The data are courtesy of M. Braddock.

DNA from A29 cells that contained $dhfr^{MTX}$ with a plasmid carrying the hepatitis B virus (HBV) genome into mouse 3T3 cells and selected a cell line that secreted HBV particles. Other studies have used co-transformation of pSV2-*dhfr* and pAdD26SVpA to produce cell lines expressing, for example, human α-globin mRNA (Lau *et al.*, 1983), human interferon-gamma (Scahil *et al.*, 1983) and tissue plasminogen activator (Kaufman *et al.*, 1985). Recently Patzer *et al.* (1986) have shown that higher levels of expression are obtained if the *dhfr* coding sequence is not expressed by a strong promoter. This appears to be because poor expression of the selectable marker imposes selection for cells in which integration into a transcriptionally active region of the genome has occurred. A typical amplicon vector is shown in Figure 4.14.

Amplicon vectors are being widely used to generate cell lines that over express interesting and valuable proteins. There are, however, some disadvantages. It requires several months of selection to generate the appropriate cell line and a number of cell lines have to be analysed because there is often about a 50-fold variation in levels of expression of the non-selected gene. This appears to be due to the site of integration (e.g. Meinhoth *et al.*, 1987).

CHAPTER 4

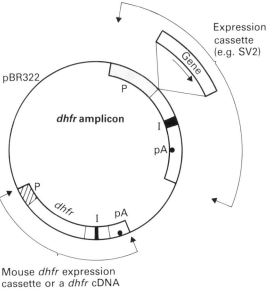

Fig. 4.14. A typical *dhfr* amplicon vector.

The transformed sequences may also rearrange, both during and after the amplification and some clones may lose high-level expression. There is also considerable deletion and rearrangement of the flanking DNA (Federspiel *et al.*, 1984). Ultimately genetically stable lines that continuously express a given protein can be produced. The use of amplicons to increase gene dosage is clearly a valuable approach for producing reasonable quantities of proteins but it is probably not suitable for gene expression analysis. This is because of the time involved in producing cell lines and because any chromosomal location that favours amplification may have special properties. It is also possible that the high copy numbers may titrate transcription factors (see Chapter 8).

It should be noted that in transformations that use mammalian $dhfr^{MTX}$ or *Eco-dhfr* as dominant selectable markers the tranformation frequency is at least 100-fold higher than the background rate of appearance of resistant colonies that result from the amplification of endogenous *dhfr* genes (Simonsen and Levinson, 1983).

Amplicons can also be produced using other selectable markers, e.g. growing cells in increasing concentrations of cadmium selects for amplification of metallothionein genes. Withdrawal of Cd, however, results in slow gene loss and amplified copies are completely lost by 120 days.

Expression vectors

Any plasmid can be turned into an expression vector by adding an expression cassette (see Chapter 2). The major consideration is which promoter to use and there is now a variety to choose from. The SV2 and RSV cassettes are widely used for high-level constitutive expression. Other viral promoters such as the adenovirus major late promoter (MLP) (Kaufman, 1985), the

encephalomyocarditis (EMC) virus late promoter (Pasleau *et al.*, 1985) an
cytomegalovirus (CMV) promoter (Boshart *et al.*, 1985) are also efficien
One of the most useful promoters is that of the mouse metallothionein gene
Mt-I, because it functions in a wide variety of cell types. It can also b
induced with cadmium, although the degree of induction can be low (2–
fold) particularly if high-copy-number vectors are being used.

The over-expression of some proteins can be toxic to cells and in th
case an efficiently regulated promoter must be used to switch off expressio
during cell growth. At present the only efficiently regulated promoter tha
has been well characterised is the mouse mammary tumour virus (MMTV
LTR. This is induced by glucocorticoid hormones, such as dexamethasone
by 50–200 fold (Lee *et al.*, 1981; Klessig *et al.*, 1984).

The efficiency of different promoters may vary between different ce
types (Gorman *et al.*, 1983a) and in some cases a tissue-specific enhance
may be required for maximum expression (e.g. Kriegler and Botchan, 1983
(see Chapter 8).

Introducing DNA into mammalian cells

We have considered how to culture animals cells and how to select fc
transferred DNA or to detect it in a transient expression assay and how t
construct simple plasmid vectors. In this section we will discuss the method
used to introduce the DNA into cells. Some early studies of gene transfe
into mammalian cells used metaphase chromosome preparations as th
source of donor DNA. This procedure, called *chromosome-mediated gen
transfer* (CMGT), is not widely used to analyse gene function and will not b
discussed here; it is reviewed in McBride and Peterson (1980). Introducin
DNA into cells is generally referred to as *DNA-mediated transformation*.

The calcium phosphate precipitation technique

Graham and Van der Eb (1973) devised the calcium phosphate (CaPO$_4$
precipitation technique for introducing DNA into cells. The DNA and .
calcium chloride solution are mixed with a buffered phosphate solution an
a DNA–CaPO$_4$ precipitate is formed. The precipitate is gently pipetted o
to the surface of a monolayer of actively dividing cells which are allowed t
take up the precipitate for several hours, washed and then incubated i
fresh culture medium. For many cells the efficiency of transformation i
increased by giving them a physiological shock with dimethyl sulphoxid
(DMSO) (Stow and Wilkie, 1976) or glycerol (Parker and Stark, 1979).

Under optimum conditions (Gorman *et al.*, 1983a; Chen *et al.*, 1982
nearly all the cells take up the precipitate. However, DNA is only detecte
in the nucleus in a very small proportion, 1–5%, of the cells suggesting tha
one of the factors limiting sucessful transformation is transport to th
nucleus rather than simple entry into the cells (Loyter *et al.*, 1982).

Another factor which affects transformation efficiency is the expressio
of the selectable marker. When the HSV-*tk*, the SV40 early and the RSV 3
LTR promoters were used to drive the expression of selectable markers i

many cell lines, RSV was more efficient than SV40 which was at least 10 times more efficient than HSV-*tk*. (Gorman *et al.*, 1983). In general, the use of promoters that have enhancer sequences or the addition of enhancers to weak promoters, such as HSV-*tk*, increases the stable transformation frequency. Tissue-specific enhancers may be necessary for efficiently transforming specific cell types (e.g. Kriegler and Botchan, 1983 and see Chapters 5 and 8 for more detail). These enhancers may act either in increasing gene expression to allow the cells to survive long enough in selective conditions for stable integration to occur or by enhancing the integration process itself or by increasing expression at the integrated site.

Sodium butyrate treatment of cells at, or shortly after transformation, enhances the transient expression as measured by CAT activity and the stable transformation frequency. Sodium butyrate affects chromatin structure and may stimulate transient expression of newly transformed genes by facilitating their assembly into an active type of chromatin structure (Reeves *et al.*, 1985). Recently, it has also been shown that supercoiled DNA is 100-fold more effective as a transcriptional template than linear DNA (Weintraub *et al.*, 1986).

The DEAE-dextran procedure for transient transformation

Transient expression of genes occurs immediately after DNA uptake and so provides a useful gene analysis system. The success of these short-term experiments to study transcription depends on there being a high enough level of transient expression to generate sufficient gene products for analysis. The gene of interest may not have a particularly powerful promoter so it is important that as many cells as possible express the gene and that the copy number of the gene is as high as possible. Transformation of cells with DNA complexed to the high-molecular-weight polymer diethyl-aminoethyl-dextran (DEAE-dextran, Mr 500 000) is often used to obtain efficient transient expression. If this treatment is coupled with a DMSO shock then up to 80% of transformed cells can express the transferred gene (Lopata *et al.*, 1984). Unfortunately this DEAE-dextran-DNA mediated transient expression does not produce any stable transformants (Sussman and Mulman, 1984). The reasons for this difference between the CaPO4 and DEAE-dextran techniques may relate to the assembly of DNA into chromatin.

The polycation–DMSO technique

The calcium phosphate method is efficient and reproducible but there is a narrow range of optimum conditions. A technically less demanding procedure has now been developed (Kawai and Nishizawa, 1984). This involves using the polycation, polybrene, to increase the adsorption of DNA to the cell surface followed by a brief treatment with 25–30% DMSO to increase membrane permeability and enhance uptake. When transformation efficiency was measured with retroviral DNA using this procedure, focus formation was proportional to DNA input, no carrier DNA was required and stable transformants were produced. The generality of the polybrene–DMSO technique has not been fully tested but it is known to work with chick embryos and mouse fibroblasts.

Microinjection

DNA can be injected directly into the nucleus of cells thereby bypassing the block of transport from the cytoplasm to the nucleus. About 10^{-11} ml of a DNA sample is injected, via glass micropipettes, into cells grown on glass slides. With practice one can inject 500–1000 cells per hour with a successful transfer of material being around 50–100% (Capecchi, 1980). The number of molecules injected can be controlled by making appropriate dilutions of DNA samples and no carrier DNA is added. If the expression of the *tk* gene is increased by adding an enhancer from SV40 or polyomaviruses then one out of five cells becomes stably transformed (Yamaizumi *et al.*, 1983).

Protoplast fusion

Another procedure for stably transforming mammalian cells was devised by Schaffner (1980) and modified by Sandri-Goldin *et al.* (1981). In this technique bacteria which contain the plasmid of interest, amplified to high levels by chloramphenicol treatment, are treated with lysozyme to remove the bacterial cell wall and generate protoplasts. These are fused to mammalian cells with polyethylene glycol and the entire bacterial contents are introduced. Any contaminating whole bacteria are killed by antibiotics in the medium. Frequencies of transformation of mouse Ltk^- cells with HSV-*tk* plasmids of about 5×10^{-2} have been obtained possibly reflecting the efficiency of nuclear localisation due to protection by bacterial components. Although there is a slight advantage in not having to make plasmid preparations this technique is not widely used because of the complications of adding unknown components to the recipient cells and because of the carrier interference mediated by the *E. coli* chromosomal DNA. The carrier effects are discussed in detail in the next section.

Electroporation

High-voltage electric discharges have been shown to induce cells to fuse via their plasma membranes by creating holes or pores in the cell membrane. This is called electroporation (Zimmerman and Vienken, 1983). Electroporation can also be used to promote the uptake of exogenous DNA by cells (Neumann *et al.*, 1982; Potter *et al.*, 1984). A suspension of plasmid DNA and cells is placed in a chamber, which can be made from a plastic cuvette, and subjected to a high voltage, 2.0 to 4.0 kV, shock at 0°C. After 10 min the cells are transferred to fresh medium and grown for two days before imposing selection. The recovery of live cells after this procedure is 60–90%. Treatment of cells with colcemid 18 hr before electroporation enhances transformation efficiencies 3- to 10-fold, possibly because the nuclear membrane in such metaphase-arrested cells is absent or more permeable. Linear DNA is 20-fold more effective than supercoiled DNA, probably due to enhanced integration into the chromosome. 25 to 30 transformants per 10^5 live cells can be obtained using this procedure and it has been used successfully for a number of cell lines including mouse B and T lymphocytes, human lymphoblastoid cells, rat pituitary cells and fibroblasts. It appears to be a reproducible and rapid technique and is successful with a number of cell types that have not been amenable to $CaPO_4$ transformation (Toneguzzo et al., 1986; Chen and Okayama, 1987; Chu et al., 1987).

In addition to routinely used procedures for DNA-mediated transfection that we have described there are some newer procedures that may become widely useful. For example, poly-L-ornithine mediates efficient transformation of some cells (Bond and Wold, 1987), and strontium phosphate has been used to transform primary human cells (Brash et al., 1987). It is also possible to permeabilise cell membranes with lasers (Tao et al., 1987).

The fate of the transferred DNA

The efficiency with which DNA is delivered into the cell depends to some extent upon the technique and the cell type. If DNA is first introduced into the cytoplasm there may be limitations in the transport to the nucleus. Once the DNA enters the nucleus it must be expressed and stably maintained for a transformed cell line to be produced. We will now consider in detail the interactions of the transferred DNA with the host genome. That interaction has important effects on stability and gene expression.

DNA-mediated transformation with carrier DNA

If high-molecular-weight carrier DNA, derived either from the recipient cell or salmon sperm, is mixed with the selectable marker and/or the gene of interest the following temporal sequence of events occurs. During the transformation, the donor DNA becomes fragmented and then randomly religated. This means that plasmids are linearised and completely unrelated sequences from the carrier DNA ligate with each other and with plasmid DNA or any other DNA present. The primary mechanism seems to be end-to-end joining but non-homologous recombination also occurs and the frequency of all these rearrangements is high (Wilson et al., 1982; Miller and Temin, 1983).

The ligation of the donor DNA produces large DNA structures containing several copies of the selected and non-selected DNA embedded in carrier DNA. These structures have been called *transgenomes* or *pekalosomes* and have been estimated to be as large as 2000 kb. A general model for the molecular events which occur during transformation with carrier DNA has been proposed (Perucho et al., 1980; Scangos et al., 1981) and this is summarised in Figure 4.15. The transgenome forms very early after transformation, the plasmid sequences are found associated with high-molecular-weight DNA and the pattern of plasmid DNA interspersed with carrier DNA as shown by restriction enzyme digest is fixed very early. There is probably only a limited time period during which the DNA is in a suitable state for recombination and ligation. A cell may possess several transgenomes

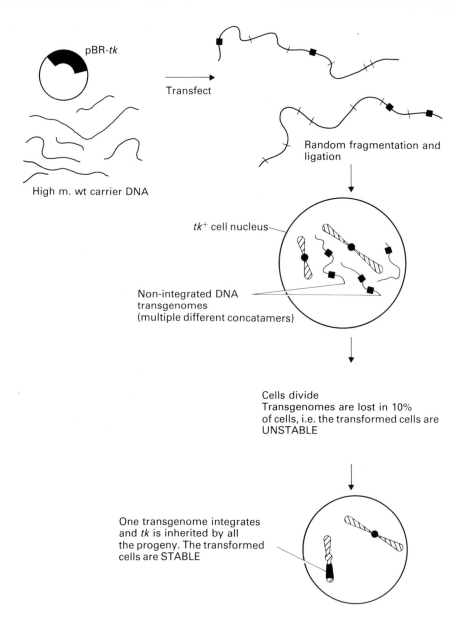

Fig. 4.15. Molecular events in DNA-mediated transfection with carrier DNA.

but normally only one is retained. Cells which retain a transgenome produce, for example, tk^+ colonies. These are, however, unstable transformants. The transgenome initially remains unlinked to any of the recipient chromosomes but it can be maintained during cell division. This maintenance is very inefficient and 10% of the cells always fail to receive a copy of the transgenome and therefore lose the selectable marker. Eventually a transgenome integrates into a chromosome. At this point it is maintained by the normal chromosomal replication and centromere functions and this cell can generate a stably transformed colony. Integration occurs, at a single site, as shown by *in situ* hybridisation to intact cells with a specific labelled probe (e.g. Robbins *et al.*, 1981). In any transformation experiment there will be several stable transformants arising independently and in each of these the transgenome is integrated into a different chromosomal site.

Stable transformants can also lose the selectable marker at a relatively high frequency 10^{-1} to 10^{-4} (Perucho *et al.*, 1980; Graf *et al.*, 1979). This is largely the result of internal deletions in the integrated transgenome which may be mediated by repetitive sequences in the carrier DNA. There is some evidence that the transgenome is not only more active for recombination but also for amplification as transferred *dhfr* genes amplify more readily than the endogenous gene (Wigler *et al.*, 1980).

One advantage of undertaking gene transfer experiments with a mixture of plasmids and high-molecular-weight DNA is that the random linkage of chromosomal DNA to defined sequences for which there are specific molecular probes facilitates the cloning of some genes. This is discussed in detail in Chapter 7. The disadvantages are that there is no way of controlling the genetic environment of the transferred gene and the transgenome appears to be more genetically unstable than the recipient chromosome.

DNA-mediated transformation with pure plasmids

When either circular or linear DNA is microinjected into the nuclei of mouse L*tk*⁻ cells, the DNA is stably integrated into the chromosome and the number of gene copies is roughly proportional to the number of molecules injected. Multiple copies of the donated sequences are found in head-to-tail arrays (concatamers) which are generated by homologous recombination at any site throughout the vector. The concatamers are usually integrated at single chromosomal sites and are stable. When fewer molecules are injected, single plasmids are inserted into the chromosome but there are no preferred sites. Linear plasmids tended to integrate at their ends (e.g. Forger *et al.*, 1982).

In the CaPO₄ transformation technique, using pure plasmids without carrier, plasmids also integrate directly into the chromosome but often several copies are present at different chromosomal locations. For example, pSV2-*neo* generally integrates into mouse L*tk*⁻ cells at 1 to 5 separate sites. Very occasionally multiple copies of head-to-tail concatamers are found at a single location. The *neo* marker integrates into the chromosome very early after transformation and, in contrast to carrier DNA-mediated transformation, the transformants are extremely stable even after prolonged culture in non-selective conditions (Southern and Berg, 1982). In one study, linearisation of the plasmid before transformation increased the frequency with which multiple donor DNA sequences became integrated at a single chromosomal site (Huttner *et al.*, 1981). Using the technique of electroporation only 1–15 copies are integrated.

In some cases the plasmid exists in an extra-chromosomal state. This has been shown for plasmids carrying the human β-globin gene (Huttner *et al.*, 1979; Anderson *et al.*, 1980). It is not known if this is due to autonomous replication or an excision/integration equilibrium.

In both microinjection and CaPO$_4$ transformation it appears that plasmids will undergo extensive recombination and mutation prior to integration (Kucherlapati *et al.*, 1984; Folger *et al.*, 1985a,b). Homologous recombination and gene conversion occur within one hour after the introduction of DNA into the nucleus. Several studies have also shown a high level of mutation in the donor plasmids including deletions, insertions of repetitive chromosomal DNA and point mutations at a level of 1% (Razzaque *et al.*, 1984). These results suggest that any DNA that is stably integrated into a mammalian cell has a high probability of being different from that originally introduced as a result of recombination and mutation. In addition the DNA may be integrated in random permutations so that an entire regulatory or coding region may not be retained. Changes can also occur during the propagation of stable cell lines and this can result in inactivation of the transfected gene. These changes can be deletions, amplification and in many cases methylation (Gebara *et al.*, 1987).

Gene targeting

We have shown in Chapter 3 that it is possible to target genes into precise locations in the yeast chromosome by stimulating recombination with free DNA ends. Although inter-plasmid recombination in mammalian cells is stimulated by double-strand breaks (Kucherlapati *et al.*, 1984) the plasmids integrate into the chromosome randomly. Homologous recombination between plasmid-borne and chromosomal sequences can, however, occur. For example, chromosomally located viral markers can be rescued by infecting with SV40 (Gluzman *et al.*, 1977) but the process is rare. Smithies *et al.* (1985) have shown that a human β-globin gene can be specifically targeted to the normal locus in cultured human cells by using the same approaches as for targeting in *S. cerevisiae*. The frequency of specific targeting was very low, less than one in 1000 transformed cells had the β-globin sequence at the correct site. This can, however, be increased by using microinjection (Thomas *et al.*, 1986). Gene targeting is described in more detail in Chapter 11 as a possible route to gene therapy in man.

Chromatin structure

Replicating and non-replicating plasmids introduced into cells by the CaPO$_4$ or the DEAE-dextran technique are assembled into typical minichromosomes with a 190 bp nucleosome repeat within 24 hr of transformation (Reeves *et al.*, 1985). All the DNA following DEAE-dextran transformation but only 60–70% of the DNA following CaPO$_4$ transformation is assembled into chromatin. This difference may be the basis for the different efficiencies of the two techniques for transient expression and stable transformation although no mechanistic basis has been established. The chromatin structure of the transferred plasmids undoubtedly affects expression and it is likely that many of the plasmids are in an inactive chromatin configuration. This is suggested by the fact that sodium butyrate treatment enhances expression and this is accompanied by increased DNase I sensitivity, decreased levels of acetylated H4 and depletion of H1 on the plasmid minichromosomes. These are all signs of active chromatin (Gorman and Howard, 1983). Once DNA is integrated into the chromosome it is also packaged in chromatin and at some locations may be permanently inactive. Therefore the transferred gene will not be expressed. For example, when retroviruses are integrated at certain sites they are not correctly expressed (e.g. Feinstein *et al.*, 1982). Other properties of the DNA may also be affected. For example the ability of the CAD gene to amplify is dependent upon the chromosomal location (Wahl *et al.*, 1984).

Simple gene transfer systems: a summary

Of all the techniques discussed so far, $CaPO_4$ precipitation is most commonly used to stably transform mammalian cells. Microinjection is more efficient in terms of the number of stable transformants per transformed cell but is not widely used because it requires sophisticated instruments and a skilled operator and it is time consuming. Direct bacterial protoplast fusion is occasionally used. Where possible, the $CaPO_4$-mediated transformation is performed in the absence of high-molecular-weight DNA to avoid trans-genome artefacts. For many cell lines, however, carrier DNA is required to achieve a high transformation efficiency. More recently, it has been shown that DNA can be introduced into some cells, which are refractory to the $CaPO_4$ technique, by electroporation.

Any gene of interest which is not directly selectable can be introduced into mammalian cells by co-transformation with a selectable marker or by insertion into a simple gene transfer vector such as pSV2-*neo*. The techniques that we have described above have some disadvantages. These are: (1) low transformation efficiencies, particularly with finite cell lines. This is due to inefficient transport to the nucleus and unstable maintenance of the DNA; (2) random integration of the transferred gene into the chromosome so that flanking sequences cannot be controlled; (3) low copy number of the intro-duced genes − there are usually only 1 to 10 copies and not all these may be active. Because of these shortcomings more sophisticated vectors have been developed. Most of these vectors use components from eukaryotic viruses and these will now be discussed.

Virus vectors

Eukaryotic viruses which have DNA genomes or a DNA stage in their life cycles have been important for the development of gene transfer vectors for the following reasons:

(1) They contain efficient promoters which are recognised in eukaryotic cells. These can be used to drive the expression of selectable markers as we have already discussed or to express foreign genes.

(2) Many viruses replicate their genomes to a high copy number during an infectious cycle so that any foreign gene inserted into the genome will have a dramatic increase in dosage, and therefore expression, during a transient period of infection.

(3) Some viruses have defined *cis* and *trans* acting elements which control their replication. These can be manipulated to produce replicating plasmids for the long-term maintenance of 'foreign' genes in cells at high copy number.

(4) Some viruses use high-efficiency stable integration into the genome as a normal step in their replication. This can be harnessed to increase the efficiency of stably introducing new genes into chromosomes.

(5) Viruses as infectious agents can introduce genes into host cells very efficiently because of their specialised capsid proteins which recognise cell receptors. *Pseudovirions* can be formed where the virus capsid proteins package a recombinant plasmid to produce a high-efficiency transformation system.

In this section we see how the ability of viruses to infect and replicate in mammalian cells can be exploited to produce gene transfer and gene expression vectors. Potentially any virus can be genetically manipulated provided there is some background information about the infectious cycle and the genome organisation. We will concentrate on the most widely used viral system applications. These are papovaviruses (SV40, polyoma, papillomavirus) and retroviruses. Some viruses that have special applications, such as vaccinia virus, are discussed in Part III.

SV40 and polyomaviruses

These viruses are members of the papovaviruses. Their molecular biology is very similar but they differ in their host range. There are two basic types of vectors that are derived from these viruses. The first type is a *recombinant virus vector*, where foreign DNA is inserted directly into a defective viral genome and this is propagated as a virus in mammalian cells. The second type is a *viral mini-replicon* where only these parts of the viral genome that are required for replication in mammalian cells are used. These vectors are propagated in *E. coli* by including a bacterial plasmid in the hybrid molecule. Both of these types of vector can only be used for transient studies because viral infection or the uncontrolled replication of viral replicons kills the host cell. Before we discuss the development and use of these vectors we will briefly outline the biology of these viruses. We will describe SV40 and comment on any differences shown by polyomavirus.

Life cycle and molecular biology

SV40 infects monkey cells such as CV-1 and AGMK to produce infectious virus and kill the cells. This is a lytic infection and the simian cells are said to be *permissive*. If the virus infects rodent cells, however, no infectious virus is produced, the viral genome integrates into the host chromosome and the cells become transformed, i.e. tumourigenic. The rodent cells are *non-permissive* for replication. Human cells are *semi-permissive*, infectious virus being produced in only 1–2% of human cells that have been exposed to the virus. In some rare cases the virus will integrate into the genome and transform the cells. The polyomaviruses have a different host range, producing infectious virus in mouse cells and transforming rodent, bovine and rabbit cells (Fields, 1981).

The organisation of the SV40 genome and gene products is shown in Figure 4.16. The virus particle is a small icosahedral protein shell comprising three virus proteins (VP), VP1, VP2 and VP3, which encapsidate or package a single copy of the viral genome. The genome is a double-stranded 5243 bp circle that is negatively supercoiled and is organised into chromatin to form a typical minichromosome except that histone H1 is not present in the virus particle. Following infection the genome is transported to the nucleus and transcribed and replicated. As with most viruses there is a strict temporal sequence of gene expression. The genome can be divided into *early* and *late* regions to reflect the time during the infectious cycle when these regions are expressed. The early region codes for two proteins called large T and small t antigens that are produced from two differentially spliced mRNAs (in polyoma there are three differentially spliced early mRNAs encoding three antigens called large, small and middle t). The large T antigen is essential for genome replication. Once a sufficient amount has been produced, DNA

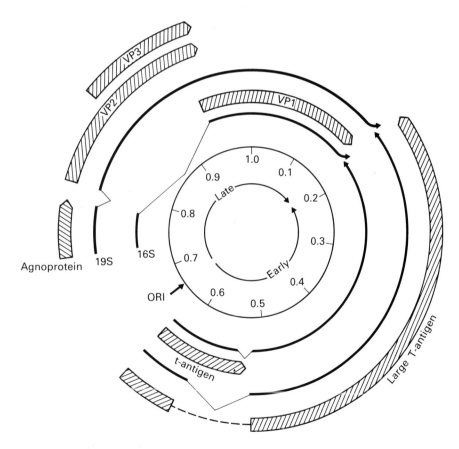

Fig. 4.16. Genetic map of SV40. The map is divided into tenths from the unique EcoRI site as 0.0. The coding regions are hatched and the mRNAs are dark lines with introns indicated as thin V lines. The total genome is 5243 bp.

GENE TRANSFER INTO ANIMAL CELLS

replication begins and early transcription diminishes because T antigen is also autoregulatory. It functions by binding to a specific region in the genome that contains the replication origin and *cis* acting control sequences for gene expression. As DNA replication begins, the late region is expressed to produce differentially spliced mRNAs coding for the capsid proteins. Early and late transcription proceed in opposite directions and the early and late promoters and the replication origin are contained within a 350 bp region of non-coding DNA. This region is described in detail in Chapter 8. The infectious cycle takes about 70 hr, the cells lyse and plaques or infectious centres are produced on a cell monolayer.

In non-permissive cells only the early region is expressed and the viral genome becomes integrated into the host genome. This is a non-specific process. The viral genome is integrated at random sites in the host chromosome, probably by illegitimate recombination, and the crossover can occur anywhere in the viral genome. If the early region is not disrupted then the continued production of T antigens induces the transformed phenotype. In general, only a few copies are integrated and there are often rearrangements and amplifications of both viral and adjacent cellular sequences that continue to occur long after the initial integration (Botchan *et al.*, 1980). Because of the imprecision of integration, low copy numbers and unpredictable alterations in the flanking DNA, these viruses are rarely used as integrative vectors.

Recombinant virus vectors

Recombinant virus vectors have part of the viral genome replaced with the foreign DNA to be studied. The reason for there to be a replacement and not an insertion is that the virus capsid cannot package DNA that is greater than viral genome size, i.e. about 5.3 kb. The final molecule must, however, be larger than 70% of the genome otherwise genomes with duplications are selected because these are packaged more efficiently.

In Figure 4.17 we show a simple experiment to construct an SV40 viral vector for transferring the rat preproinsulin gene into monkey cells (Gruss and Khoury, 1981). Purified SV40 DNA is digested with Hpa I and Bam HI and fragment A, containing an entire early region and functional replication origin, is purified. This is ligated to a Hinc II − Bam HI fragment carrying the complete rat proinsulin gene including the 5′ control regions and an intron. The Bam HI cohesive ends readily join and in this case the blunt ends are left to cyclise *in vivo* (there is no underlying reason for using a linear fragment). The recombinant molecule is introduced into AGMK cells by standard DNA-mediated transformation together with DNA from a helper virus. A widely used helper virus is called tsA$_{58}$. It is a mutant that is unable to make T antigen at 41°C but if T antigen is supplied by the recombinant virus, then tsA$_{58}$ will replicate and produce capsid proteins. After 4−5 days plaques appear in the cell monolayer. These are a mixture of recombinant and helper viruses. There are usually very few plaques because of the inefficiency of the DNA-mediated transformation. The virus is therefore harvested and used to reinfect AGMK cells. This produces

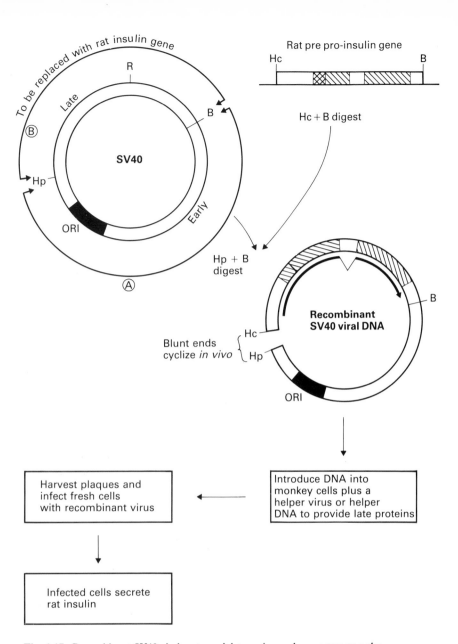

Fig. 4.17. Recombinant SV40 viral vectors. A late region replacement vector: the expression of rat proinsulin from its own signals.

many more plaques and the second lysate forms a high-titre virus stock that is then used for the gene transfer experiment. Fresh AGMK cells are infected and because viral infection is efficient the preproinsulin gene will be introduced into almost all the cells. At about 48−65 hr post infection, when viral genome replication is maximum but before cell lysis, the cells are harvested and extracts prepared for analysis. In this study, rat preproinsulin mRNA was produced using the rat gene's own expression signals and significant amounts of proinsulin polypeptide were produced.

GENE TRANSFER INTO ANIMAL CELLS

If the foreign DNA has no inherent expression signals, e.g. a cDNA, then the SV40 genome must be manipulated more carefully to ensure that viral expression signals are used to express the foreign gene. The first SV40 viral vector, called SVGT5, was in fact designed to express a cDNA. This was a rabbit β-globin cDNA (Mulligan *et al.*, 1979). In this case, part of the late region spanning the VP1 coding region was removed and replaced with the rabbit β-globin cDNA. The cDNA was tailored so that most of the untranslated leader was removed and the initiating ATG was in the same position in the viral genome as the initiating ATG of VP1 would have been. The splice acceptor for the 16S mRNA was retained and so this vector allowed the expression of a rabbit β-globin polypeptide by translation of the substituted 16S RNA (Figure 4.18A). A more sophisticated SV40 viral vector for expressing two separate cDNAs is shown in Figure 4.18B. In this case, cDNAs coding for the α and β polypeptide subunits of human chorionic gonadotrophin (hCG) are inserted into a single SV40 vector so that they replace the VP2 and VP1 coding sequences respectively (Reddy *et al.*, 1985). In both cases the initiating ATG for the hCG polypeptides is in the analogous position on the viral genome to those of VP2 and VP1. Monkey cells infected with this recombinant virus and the tsA_{58} helper produced biologically active dimeric hormone.

The recombinant SV40 viral vectors that we have described so far have parts of the late region removed and so they are called *late region replacement vectors*. Similar approaches can be used to make *early region replacement vectors*. For example Gething and Sambrook (1981) replaced the T antigen coding region with an influenza virus haemagglutinin coding region. The early proteins have to be supplied by a helper virus to produce infectious recombinant viruses. Late region replacement vectors are more widely used because the late promoter is active for a longer period during the infectious cycle and may also be stronger. Because the viral genome is replicated to several hundred thousand copies, and the late promoter is efficient, these vectors can be used to produce significant quantities of protein. For example, 200 µg of influenza virus haemagglutinin from 5×10^6 cells at 62 hr post infection. This transient system is not, however, as suitable for large-scale production as a stable cell line continuously expressing protein.

The advantage of producing an infectious SV40 recombinant virus is that foreign DNA can be introduced into recipient cells with high efficiency, e.g. 100-fold more efficient than DNA-mediated transformation, but, with the increased use of techniques such as electroporation and the efficient DEAE-dextran transient transformation procedure, the use of virus vectors solely to increase transformation efficiency is likely to diminish. The main advantage is therefore in amplifying the template. A disadvantage of this system is that the insert must be a precise size to ensure efficient packaging. Fragments much bigger than 2.5 kb cannot be used.

Viral mini-replicons

The ability of SV40 and polyomaviruses to replicate their genomes very rapidly to very high copy numbers can be exploited by constructing viral mini-replicons.

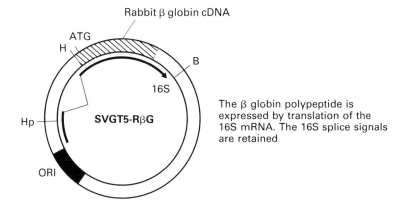

A A VP1 replacement vector

Rabbit β globin cDNA

SVGT5-RβG

The β globin polypeptide is expressed by translation of the 16S mRNA. The 16S splice signals are retained

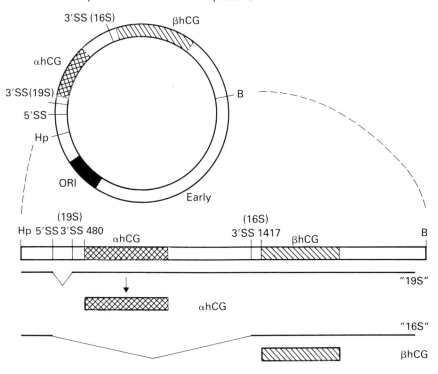

B A VP1 and VP2 double replacement vector for the expression of multisubunit proteins

Fig. 4.18. SV40 late region expression vectors.

A typical polyomavirus-plasmid vector is shown in Figure 4.19. The entire early region and replication origin has been inserted into pBR322 and, in this case, the foreign DNA to be analysed is a mouse immunoglobin heavy-chain gene (Deans *et al.*, 1984). The vector is propagated in *E. coli* and then pure viral-plasmid DNA is introduced into mouse cells by standard DNA-mediated transfection techniques. The vector replicates in mouse cells because the polyoma large T antigen is synthesised using the early promoter and this activates the viral replication origin. During a 4–6 day

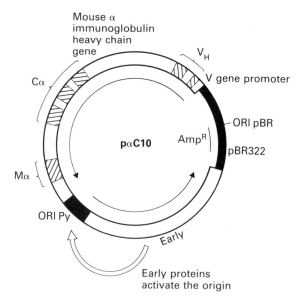

Fig. 4.19. A recombinant polyomavirus plasmid for expression of mouse immunoglobulin genes.

period the vector replicates to 50,000—400,000 copies per cell. This amplification of the template increases the number of transcription units and expression of the foreign gene will give high yields of mRNA and protein for analysis. The system is transient because the massive vector replication eventually kills the cell.

Analogous viral plasmids can be constructed with the SV40 early region for use in simian cells. In these vectors, however, the pBR322 component must be modified to remove a 430 bp sequence called the *poison sequence*. This sequence encompasses a site called *nic* or *bom* that is important in mobilising ColEl type plasmids for transfer between bacteria. For some unknown reason it interferes with replication in simian cells (Lusky and Botchan, 1981).

Instead of constructing vectors that contain the entire early transcription unit it is possible to construct cell lines that permanently express the viral early proteins. The first such lines were obtained by forcing transformation of normally permissive monkey CV-1 cells by using SV40 DNA that was unable to replicate because of a 6 bp deletion around the Bgl I site at the origin of replication (Gluzman, 1981). These lines have SV40 DNA integrated into the genome and they express large T antigen constitutively. They are called COS (CV-1 origin, SV40) cells. Similar cell lines have been produced by transforming human fibroblasts with replication defective SV40 (HFS cells) (Boast *et al.*, 1983). If mini-viral replicons that contain only the viral replication origin are introduced into these cells they replicate to very high copy number using the cellular T antigen. COS cells can also be used to provide the helper functions for early region replacement virus vectors but the level of T antigen is often not sufficient to produce very high titre virus stocks. A typical SV40 viral mini-replicon vector is shown in Figure 4.20.

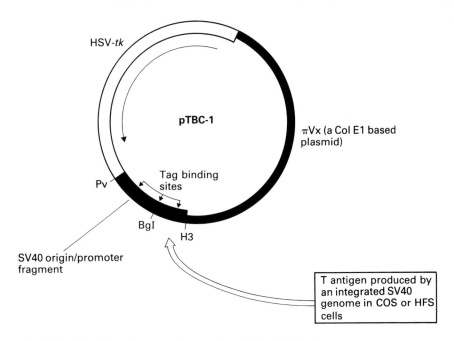

Fig. 4.20. Mini viral replicons used to transform COS and HFS cells.

Retrovirus vectors

There were two main reasons for developing retrovirus vectors, these were:
(1) to increase the frequency of obtaining stable transformants by exploiting
the ability of the virus to integrate into the genome and (2) to exploit the
broad host range of these viruses to introduce DNA into any cell via viral
particles. Retroviral vectors have subsequently been used for a range of
different experiments from cloning genes to acting as markers for stem cell
differentiation. For an extensive review of retroviruses and retrovirus vectors
(see Weiss *et al.*, 1985; Temin, 1986).

An introduction to retrovirology

Retroviruses are RNA viruses that cause a variety of diseases including
tumours. They infect a range of invertebrate and vertebrate hosts, including
man, and they share many properties with some transposable elements
found in *Drosophila* and *S. cerevisiae* (see Chapter 9).

The virus contains two copies of its genomic RNA, a tRNA primer
molecule, reverse transcriptase, RNaseH and integrase. After uptake of a
retrovirus by a host cell the genomic RNA undergoes reverse transcription
to produce a double-stranded DNA form that is integrated into the host
genome to produce a provirus. An abbreviated scheme for replication is
shown in Figure 4.21. A molecule of tRNA hybridises to the negative
strand primer binding site, PBS-ve, and acts as a primer for the reverse
transcriptase to copy the viral RNA template. The enzyme then switches
template to the 3′ end of a viral RNA template, probably the same molecule,

GENE TRANSFER INTO ANIMAL CELLS

1) Viral RNA

2) Strong stop DNA is synthesised

3) RT jumps at r and minus strand DNA is continued

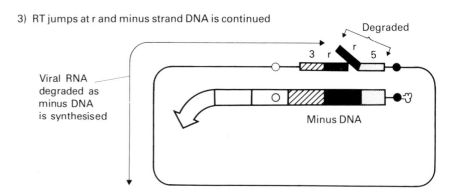

4) Plus strand DNA synthesis begins

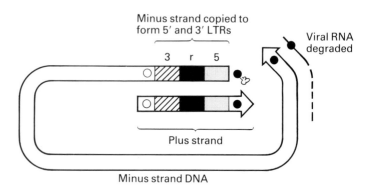

5) RT jumps at PBS −ve, plus and minus strand synthesis
   continues

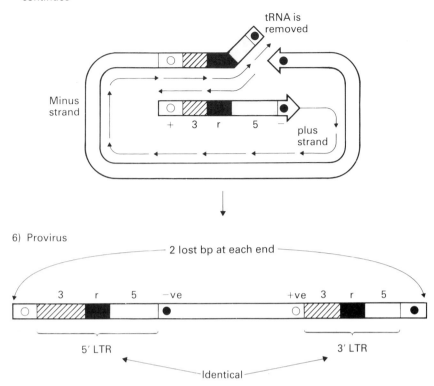

6) Provirus

Fig. 4.21. Retrovirus replication.

and continues to copy the RNA. The RNA template is degraded by RNaseH and the synthesis of the second DNA strand is initiated by priming at the positive strand primer binding site, PBS+ve, site. The first DNA strand is copied to the end and then there is another template jump to continue synthesis. The double-stranded cDNA circularises and integrates into the chromosome to produce the proviral form. The mechanism of integration is unknown but it requires sequences at the ends of the LTRs, and endo-nuclease encoded by the *pol* gene, and it involves a staggered cleavage in the host DNA. The integration is precise with respect to the provirus which is always colinear with the viral RNA. This contrasts with the integration of papovaviruses which occurs anywhere throughout the viral genome. There are no sequence-specific target sites for retroviral integration into the chromosome but integration often occurs in transcriptionally active regions. The provirus is longer than the viral genome because sequences at the 5′ and 3′ end of the viral RNA are duplicated during replication. The integrated provirus directs the synthesis of new viral genomic RNA and proteins to produce new infectious virus.

The organisation of a generalised retroviral provirus is shown in Figure 4.22. It comprises an internal region of a few kb flanked by the long direct terminal repeats (LTRs) of a few hundred bp. The 'full length' RNA is

GENE TRANSFER INTO ANIMAL CELLS

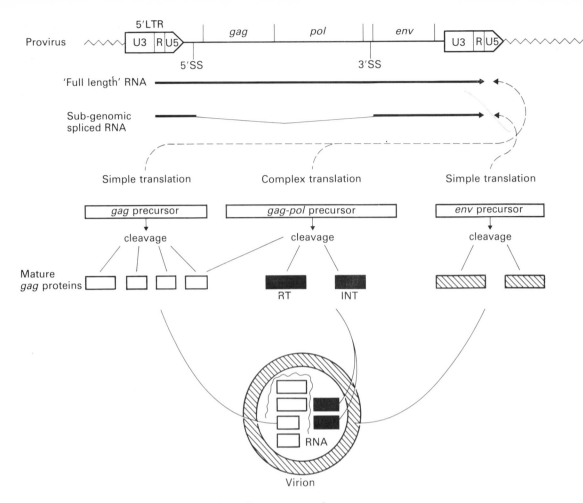

Fig. 4.22. Retrovirus gene expression.

transcribed under the control of promoter elements in the 'left' or 5′ LTR. Transcription initiates at the left end of the R region in the 5′ LTR and terminates at right end of the R region of the 3′ LTR. This creates the terminally redundant genomic RNA (Figure 4.21). This RNA also acts as a message and as a precursor for a sub-genomic spliced message. There are three protein-coding regions in the internal segment of the provirus called *gag* (group specific antigen), *pol* (polymerase) and *env* (envelope). These genes are translated from the two classes of retroviral message. *Gag* and *pol* are translated from the 'full length' RNA. *Gag* translation is simple and produces a precursor protein that is subsequently cleaved to produce the mature viral core proteins. *Pol* translation is more complex and involves translation through *gag*, misreading of the *gag* termination codon by an unknown mechanism and then translation through *pol*. This produces a *gag*–*pol* fusion protein which is also subsequently cleaved to produce more *gag* products but also the protease that mediates the various cleavage events, reverse transcriptase, RNaseH and integrase. *Env* is translated simply from a subgenomic RNA that is produced by a splice that fuses the leader of the RNA to *env*. The *env* protein is cleaved to produce the mature viral surface proteins which determine host range.

A sequence called the *packaging site* (pac or psi, ψ) is located between the 5′ splice donor (5′ SS) site for the *env* message and the initiating ATG for *gag*. It is important for the interaction of the viral RNA with the viral proteins to produce new virus particles. The sub-genomic *env* mRNA lacks this site and is therefore not packaged. The virus particles are assembled in the cytoplasm and released from the cell by budding through the plasma membrane where they acquire a membrane envelope. The viral *env* proteins are embedded in this membrane. Budding does not kill the cells and they can secrete viruses for long periods.

Retroviruses such as Moloney murine leukaemia virus (MoMLV) are referred to as *replication competent* because they contain all the genes necessary to make fully infectious virus particles. Some naturally occurring retroviruses are *replication defective* because they have suffered large deletions of viral DNA and in place of this viral DNA they have acquired cellular DNA sequences. These viruses can be propagated by mixing them with replication competent viruses that act as helper viruses to provide the missing proteins and enzymes. The cellular DNA sequences confer oncogenic (tumourigenic) properties on the viruses that carry them and they are called *oncogenes*. In Figure 4.23 we show a deleted derivative of MoMLV that has acquired the cellular gene *mos*. This virus, Moloney Murine Sarcoma virus (MoMSV), rapidly produces tumours in infected mice. About 30 different cellular genes have now been identified that behave as oncogenes when present in retroviruses (Bishop, 1985a,b). Oncogenes will be discussed in more detail in Chapters 7 and 11 but for the present this information simply tells us that retroviruses can accommodate eukaryotic genomic DNA within

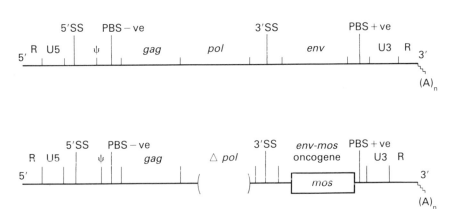

Fig. 4.23. The relationship between replication competent and defective retrovirus genomes.

their genomes and defective viruses that have acquired new sequences can be propagated by complementation with a helper virus.

Infectious recombinant retroviruses

Several groups of researchers have developed gene transfer systems based on retroviruses (Shimotohno and Temin, 1981; Tabin *et al.*, 1982; Miller *et al.*, 1983; Joyner and Bernstein, 1983; Cepko *et al.*, 1984). In Figure 4.24 we show the simplest type of retrovirus gene transfer system. A retroviral provirus is manipulated so that all or most of the *gag*, *pol* and *env* genes are removed but all the *cis* active regions that are required for replication, integration and RNA synthesis are retained. These are the 5' and 3' LTRs, PBS-ve, PBS+ve and psi. A selectable gene is inserted into this vector such that the initiating ATG is in the same place as the ATG for *gag*. This means that transcription will be directed by the 5' LTR and translation will initiate at the correct ATG for expressing the selectable marker. These manipulated proviruses are propagated as plasmids in *E. coli* and then the DNA is used in a standard DNA-mediated transfection of suitable recipient cells and stable transformants are selected. These transformants will be expressing the selectable marker and producing RNA with the structure shown in Figure 4.24.

The next step is to package this RNA into a retrovirus particle to produce an *infectious recombinant retrovirus*. Clearly the retroviral vector is completely defective and will produce no enzymes or proteins for infectious particle production. These must therefore be supplied. The simplest way of achieving this is to super-infect the transformed cells with a helper virus. The *gag* and *env* proteins produced by the helper virus will recognise the psi packaging site on the recombinant transcript and a virus particle will be formed. The particle will also contain reverse transcriptase produced by the helper virus and a tRNA primer hybridised to the PBS-ve site. The particle will, therefore, be able to infect a second host cell, replicate its RNA using the particle associated reverse transcriptase and primer and integrate into the host chromosome as a provirus. The recombinant provirus will not, however, be able to produce further infectious viruses unless helper functions are present. It is often referred to as a pseudovirus, i.e. it has all the right proteins for initial infection but a defective genome. The process of packaging one viral genome with proteins from a second virus is called *pseudotyping*. This is illustrated in Figure 4.25.

Once the recombinant RNA is packaged into virus particles it is a much more efficient gene transfer system than the original plasmid that has all the limitations of DNA-mediated transformation. The virus particles can infect any cell that has suitable receptors and efficiencies of successful gene transfer can approach 100%. The second major point is that following viral infection any recombinant DNA is sure to integrate as a provirus. Although the site of integration into the chromosome is still largely random the structure of the provirus and therefore the DNA immediately adjacent to the introduced gene is known with precision. The copy number of proviruses is usually only one per genome and so there are no complications of multiple insertions.

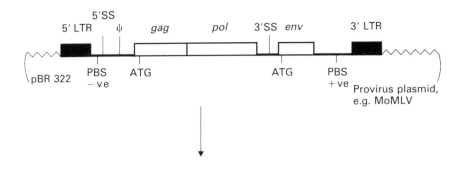

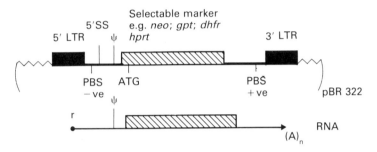

Fig. 4.24. Construction of a simple retrovirus vector, e.g. pMLV-TK (Tabin *et al.*, 1982)
(MLV + HSV *tk*); pSNV-TK (Shimotohno and Temin, 1981) (SNV + HSV *tk*);
N2 (Keller *et al.*, 1985) (MLV + *neo*); pMSV*gpt* (Mann *et al.*, 1983) (MSV +
Ecogpt); pLPL (Miller *et al.*, 1983) (MSV + human *hprt*); pLDL (Miller *et al.*,
1984b) (MLV + *dhfr*MTX).

The infected transformed cell will continue to produce recombinant
retrovirus and it is often referred to as a *producer cell*. There are other ways
of making a recombinant retrovirus producer cell. For example the cell can
be infected with wild-type virus before transformation or the retroviral
vector plasmid and a wild-type proviral plasmid can be cotransformed (e.g.
Tabin *et al.*, 1982).

It is very important to produce a large number of recombinant viruses,
i.e. a high-titre virus stock, to ensure that the maximum number of cells can
be infected at the next stage.

Helper-virus-free retrovirus vectors

The systems that we have described so far have the disadvantage that
infectious helper virus will also be produced. This means that the culture
supernatants from producer cell lines will always contain a mixture of
recombinant and wild-type viruses. This helper virus may compete with the
recombinant virus for cell receptors and therefore reduce the number of
cells that are infected by the recombinant. It also limits the general usefulness
of the system for generating recombinant retroviruses for use in whole
animals (Chapter 5) and man (Chapter 11). This is because infectious virus
could be pathogenic.

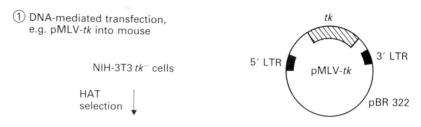

① DNA-mediated transfection,
   e.g. pMLV-*tk* into mouse

NIH-3T3 *tk⁻* cells

HAT
selection

Rare *tk⁺* cells have pMLV-*tk* in the genome

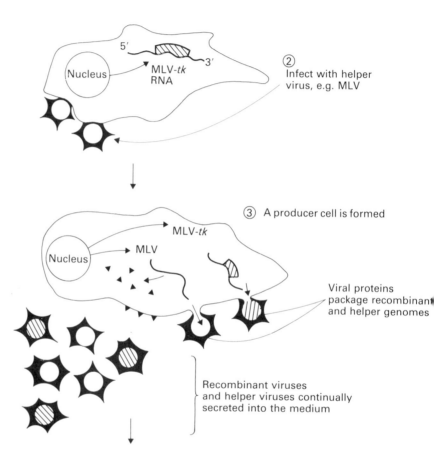

② Infect with helper
   virus, e.g. MLV

③ A producer cell is formed

Viral proteins
package recombinant
and helper genomes

Recombinant viruses
and helper viruses continually
secreted into the medium

Infect fresh cells with virus particles
to give high frequency transformation to *tk⁺*

Fig. 4.25. Production of infectious recombinant retroviruses: the use of a helper virus.

The next stage in the development of retrovirus gene transfer system
was to produce recombinant virus stocks that were completely free from
helper virus. To do this cell lines were constructed that contained retroviral
genomes that produced all the proteins and enzymes but could not form
infectious particles themselves. This was done by deleting the packaging
site from a helper virus genome and using this viral DNA to transform a

cell line. For example Mann *et al.* (1983) deleted 350 nucleotides from an infectious proviral DNA clone of MoMLV to remove the packaging site. This plasmid was called pMOV and cell lines that had pMOV stably integrated into the genome were constructed. One of these lines is called ψ2 (see Figure 4.26A). When defective recombinant retrovirus vectors such as pMLV-*gpt* are introduced into ψ2 cells, fully infectious recombinant virus is produced. This virus transforms cells to *gpt*$^+$ but does not produce reverse transcriptase or infectious virus particles. Therefore it is not contaminated with helper virus. In these ψ2 cells the helper virus produces all the necessary proteins but its own RNA will not be packaged into infectious particles.

Broad host range retrovirus vectors

In the ψ2 type of system the pseudoviruses formed will have the same host range as MoMLV and unfortunately this is rather limited. MoMLV will only infect mouse and closely related rodent cells. This type of host range is called *ecotropic*. Other retroviruses can infect cells from the species of origin and unrelated species. These are referred to as *amphotropic*. These cell tropisms are largely determined by the viral envelope proteins, particularly gp70, and reflect the ability of the virus to interact with specific cell surface receptors. The amphotropic murine retroviruses are not very well studied at the molecular level and are thus not suitable as vectors. The approach, therefore, has been to replace the *env* gene of an ecotropic virus such as MLV with the *env* gene of an amphotropic virus such as 4070A (Figure 4.26B). A number of cell lines containing broad host range, but packaging deficient, helper viruses have been constructed such as ψ-AM (Cone and Mulligan, 1984) and PA12 (Miller *et al.*, 1985a). The recombinant retroviruses produced from these cells have been shown to infect a wide range of rodent, avian and human cells including primary cell cultures and embryos (Chapter 5). The recombinant viruses are produced in high yields with lysates containing 10^5 to 10^7 recombinant virus ml^{-1}.

Although the cell lines are designed to be free of helper virus, fully infectious virus can occasionally be generated by recombination between the vector DNA and the helper-virus DNA or even with endogenous retrovirus-like sequences which are present in most eukaryotic genomes. The lysates produced in these systems are therefore tested for the presence of infectious virus.

Recently improved helper cell lines have been constructed to prevent the production of infectious virus (e.g. Miller and Buttermore, 1986). These cell lines, such as PA317, contain a helper virus genome that is multiply defective. For example, the helper virus in plasmid pPAM3 (Figure 4.26C) has the packaging site deleted and the 3′ LTR is replaced with the SV40 termination region. It would require two recombination events to generate an infectious genome from the helper and the vector. So far, no infectious virus has been detected when propagating retroviruses on PAM317 cells. The general scheme for using packaging cell lines is shown in Figure 4.27.

GENE TRANSFER INTO ANIMAL CELLS

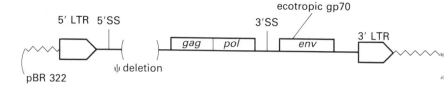

A pMOV ψ⁻ (Helper cell line: ψ2)

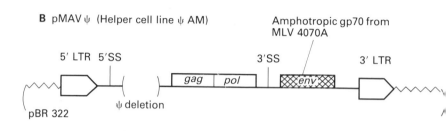

B pMAV ψ (Helper cell line ψ AM)

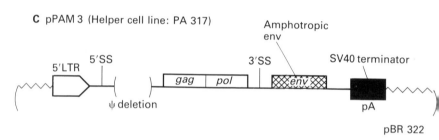

C pPAM 3 (Helper cell line: PA 317)

Fig. 4.26. Packaging-defective MLV helper plasmids. The helper cell line derived from these plasmids is indicated in brackets.

Retrovirus expression vectors

The retrovirus vectors that we have described so far only contain a selectable marker. The aim, however, is to use them to introduce and express a number of genes in a range of eukaryotic cells, i.e. to produce retroviral expression vectors. It is possible to put more than one gene into a retrovirus vector. The maximum size of the insert that can be propagated is about 6 kb but large inserts of up to 20 kb can be included if these are spliced down to a smaller size before packaging (see Chapter 7). The vector pLPL contains the human *hprt* gene as a selectable marker and has a unique Hpa I site (Miller *et al.*, 1983). A rat growth hormone mini-gene was constructed by fusing the normal growth hormone promoter region on to a growth hormone cDNA and this was inserted into the Hpa I site of pLPL to produce pLPGHL (Figure 4.28). High-titre infectious recombinant retroviruses were produced and used to infect rat *hprt⁻* cells and these were then shown to be producing growth hormone. The expression was inducible by dexamethasone and the growth hormone was secreted. In this study the non-selected gene was expressed from its own promoter and responded to

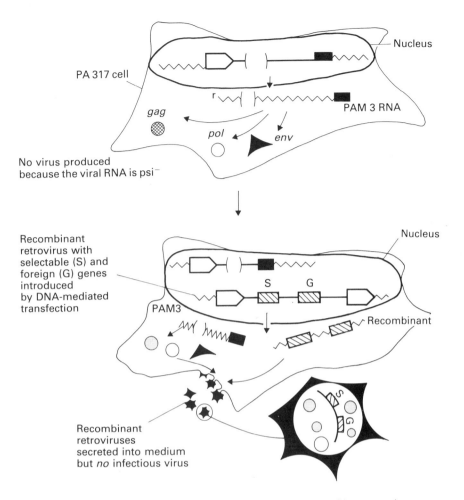

Fig. 4.27. The use of packaging cell lines to generate infectious recombinant retroviruses.

the normal controls (Miller *et al.*, 1984b). The correct expression and developmental regulation of human globin genes inserted into retrovirus vectors has also been shown (Cone *et al.*, 1987; Karlsson *et al.*, 1987). In another study the retroviral LTR was used to drive the expression of both the selectable marker and the second gene by exploiting the splicing potential of the retrovirus genome (Hellerman *et al.*, 1984). A derivative of the vector pMSV-*gpt* was constructed to contain a cDNA for human parathyroid hormone (PTH) as well as the Eco-*gpt* gene. The plasmid was rearranged *in vitro* so that the retrovirus 3' splice acceptor site preceded the Eco-*gpt* gene. Two transcripts are produced: a full-length transcript that is translated to produce PTH and a spliced transcript that is translated to produce GPT. Another configuration is to incorporate a second promoter within the vector to drive the expression of a foreign cDNA. The *tk* gene promoter is particularly useful as it functions in a wide range of cells (Magli *et al.*, 1987).

GENE TRANSFER INTO ANIMAL CELLS

A pLPGHL:rGH gene expression from its own promoter

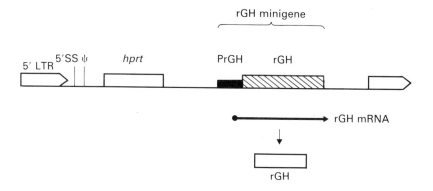

B pMSV-hPTH-*gpt*:hPTH expression from an LTR promoter

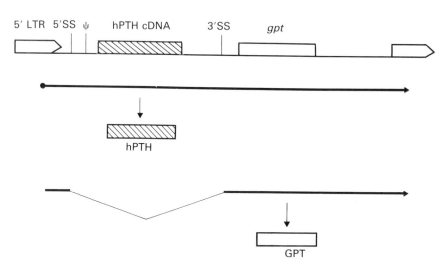

Fig. 4.28. Retroviral expression vectors. The genes expressed are rat growth hormone (rGH) and human parathyroid hormone (hPTH).

There is increasing evidence that the precise structure of the vector can affect expression in some cell types (reviewed by Temin, 1986). In particular the 5′ LTR is susceptible to suppression, probably by cellular factors, in some undifferentiated cells, e.g. EC cells (Stewart *et al.*, 1985) and MEL cells (Cone *et al.*, 1987). There is also some evidence for competition between multiple promoters. For example, Emerman and Temin (1984a) inserted the HSV-*tk* gene and the mouse α-globin genes with their own promoters into an avian spleen necrosis virus vector. If the *tk* expression was selected, deletions were found in the α-globin DNA, largely in the promoter region, in all of the *tk*$^+$ clones. It is suggested that the α-globin promoter sequences suppress the *tk* gene promoter activity and therefore there is selective pressure for mutants that have lost this promoter. Promoters placed within the retrovirus also vary in their efficiency, e.g. the *tk* promoter is very active whereas the SV40 promoter is also suppressed in cells that suppress the 5' LTR (Stewart *et al.*, 1987).

There are a number of other factors that must be considered when using retroviral vectors. The inserted DNA should not contain transcription termination signals as this would disturb the production of full-length viral transcripts that are required to generate the provirus (Shimotohno and Temin, 1981). In addition, if vectors with the 5' splice site are used it is important that any inserted sequences do not contain or create an inappropriate 3' splice acceptor site that will interfere with the expression of the inserted DNA. Also if the aim of an experiment is to study the function of introns these vectors cannot be used because intron-containing genes are processed (Shimotohno and Temin, 1982) (Chapter 7).

Although retrovirus vectors are valuable in creating a known and constant DNA sequence environment flanking any transferred gene, this environment may not be ideal for studying gene expression. This is because retroviral LTRs are highly transcriptionally active. They are strong promoters, driving transcripts to 6–10% of total cell RNA and they contain enhancers in the LTR. This activity may interfere with gene expression or it may allow expression to occur by creating a transcriptionally active chromatin domain. For example, the rat growth hormone (rGH) mini-gene is not expressed when introduced into cells by DNA-mediated transformation but is expressed in a retrovirus vector (Miller *et al.*, 1984b). The randomness of retroviral integration can also result in variable expression possibly because of influences from the surrounding chromatin structure (Feinstein *et al.*, 1982). Proviral deletions and rearrangements have been observed and loss of expression can occur with a frequency of $10^{-4} - 10^{-5}$ (Jolley *et al.*, 1986). In addition, replication via an RNA intermediate is error-prone and mutations have been detected in genes that have been transferred in retroviral vectors (Temin, 1986).

Despite some of these limitations retrovirus vectors have a wide variety of uses and we will be discussing the use of specific retrovirus vectors for gene isolation, for producing cDNA mini-genes, as insertional mutagens for isolating and studying developmentally important genes (Chapter 7), as vectors for gene replacement therapy in humans (Chapter 11), and as markers for following hematopoiesis (Chapter 5). Retroviruses are not widely used in biotechnological applications for producing high levels of protein because of their low copy number.

Viruses as autonomous stable vectors

In many studies it is important to produce a stable cell line that is expressing a gene of interest. All the stable vectors that we have discussed so far integrate randomly into the host genome and suffer from the general disadvantage that the neighbouring genomic environment may affect expression. In addition, the integrative vectors, with the exception of the amplicons, are only present in one or a few copies per cell. The alternative is to maintain a gene extrachromosomally. Unfortunately, the viral replicons and recombinant viruses, discussed above, that allow this also kill the host cell and are therefore only useful for studying transient expression of foreign genes. In this section we discuss vectors that have been derived from DNA viruses that normally maintain their genomes as episomes at a controlled copy number. These are papillomaviruses, Epstein–Barr virus (EBV), BK viruses and vectors produced by regulation of SV40 copy number. These viruses promise to be the basis of autonomous, multicopy stable vectors for mammalian cells.

Papillomaviruses

Papillomaviruses induce benign epithelial tumours (warts and papillomas) in a number of vertebrates including man. There is also evidence that they are associated with some human cervical cancers. Infectious virus is only produced in fully differentiated squamous epithelial cells and to date these viruses have not been propagated in cell culture. Several papillomaviruses will however transform cells in culture and are oncogenic in hamsters. The key feature of viral transformation in terms of vector production is that the genome is maintained episomally at 50–200 copies per cell in dividing cells.

The papillomaviruses have a small, about 7.9 kb, double-stranded circular genome and they appear to be related to viruses like SV40 and polyoma. Bovine papillomavirus (BPV1) has been most extensively studied. It has been completely sequenced and the genome has been divided into early and late regions (Figure 4.29) (Chen et al., 1982). The early region, contained on a 5.5 kb Bam HI to Hind III fragment comprising 69% of the genome is sufficient to cause transformation. The fragment is called BPV_{69T}. There is a region that is probably an origin of replication (Waldeck et al., 1984) and also contains transcriptional control signals. In addition there are two cis acting sequences, called PMS1 and PMS2, which are required for stable plasmid maintenance (Lusky and Botchan, 1984). There are eight open reading frames and the transcription of the early region is complex. There are at least three separate promoters (Heilman et al., 1982; Sarver et al., 1984; Stenlund et al., 1985), at least two enhancers (Lusky et al., 1983; Spalholz et al., 1985) and there are a large number of differentially spliced mRNAs (Stenlund et al., 1985; Yang et al., 1985). The full repertoire of viral proteins and their functions has yet to be established (reviewed by Giri and Danos, 1986). To date it is thought that E6 (Androphy et al., 1985) and E5 (Yang et al., 1985) are transforming proteins, E2 is a transcriptional trans activator (Spalholz et al., 1985) and E1, E6 and E7 are involved in replication and copy number control (Berg et al., 1986a, b; Lusky and Botchan, 1986; Roberts and Weintraub, 1986).

Although the understanding of the molecular biology of papillomaviruses is incomplete they have been manipulated to produce vectors. In initial studies, the BPV_{69T} fragment was inserted into a bacterial plasmid containing a foreign gene with its own promoter and the vector was selected by its ability to morphologically transform mouse cells. For example, a BPV vector expressing rat insulin was described by Sarver et al. (1982; Figure

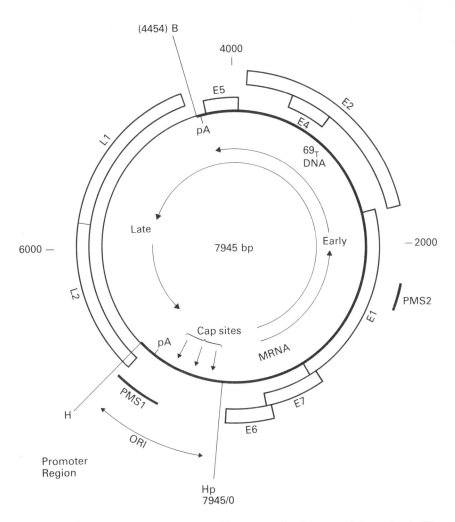

(4454) B

4000

E5

E2

E4

pA

L1

69 DNA

Late

6000 —

7945 bp

Early

— 2000

L2

PMS2

Cap sites

E1

pA

MRNA

PMS1

E7

H

E6

ORI

Promoter
Region

Hp
7945/0

Fig. 4.29. Bovine papilloma virus genome. The open reading frames are designated early (E)
1 to 8 and late (L) 1 and 2. PMS = plasmid maintenance sequences.

4.30). In most cases it was essential to excise the bacterial plasmid DNA
before transfection because of an inhibitory effect of the bacterial sequences.
The resulting vectors cyclise in mouse cells and are maintained as multicopy
episomes. In many cases, however, they rearrange or acquire new DNA
sequences from the chromosome or the carrier DNA and the levels of
expression are highly variable. These vectors also cannot be rescued from
the mouse cells.

The next development was to produce shuttle vectors. The first of these
was derived from a fortuitous observation by Di Maio *et al.* (1982) that when
a 2.7 kb fragment of a human β-globin gene is included in the vector, the
E. coli plasmid DNA can be retained. In this case the vector pBPV-BVl was
maintained at 10–30 copies per cell, it was readily rescued by transforming

GENE TRANSFER INTO ANIMAL CELLS

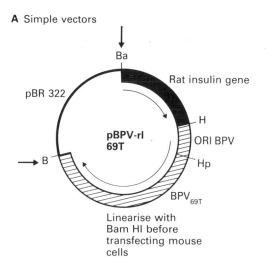

A Simple vectors

Ba

Rat insulin gene

pBR 322

pBPV-rl
69T

H

ORI BPV

Hp

B

BPV₆₉ₜ

Linearise with
Bam HI before
transfecting mouse
cells

B Shuttle vectors

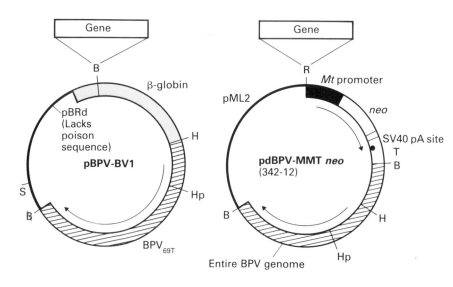

Gene

Gene

B

R

Mt promoter

β-globin

pML2

neo

pBRd
(Lacks
poison
sequence)

H

SV40 pA site

pBPV-BV1

pdBPV-MMT neo
(342-12)

T
B

S

B

Hp

B

H

BPV₆₉ₜ

Entire BPV genome

Hp

No need to linearise before transfection

Fig. 4.30. BPV vectors.

E. coli, it was identical to the input plasmid and using this vector human β-
interferon was produced at similar levels in a number of cell lines and could
be induced 400-fold by poly (I)−poly (C), (Zinn *et al.*, 1983; Figure 4.30).

When foreign genes have been inserted into BPV shuttle vectors they
can have variable effects on episomal maintenance. For example, when an
8.5 kb fragment of human DNA encoding the heavy chain of an HLA
human histocompatibility antigen was inserted into a vector similar to
pBPV-BVl it integrated into the host chromosome apparently as tandem

head-to-tail oligomers (Di Maio *et al.*, 1984). An interferon gene inserted into the same vector had been maintained episomally (Zinn *et al.*, 1983). When BPV vectors integrate they are often still present at a high copy number so that yields of foreign protein can be high but the advantages of a shuttle vector are lost. The promoter that is used in expression cassettes can also influence vector status. When a SV40 72 bp enhancer element was introduced into a mouse mammary tumour virus (MMTV) LTR-BPV vector, rearrangements and integration occurred (Ostrowski *et al.*, 1983). The host cell can also determine episomal maintenance or integration. Sambrook *et al.* (1985) used pBPV-BV1 to express influenza virus haemagglutinin on the surface of mouse cells. In MME cells the vector integrated at 1−5 copies per cell, in NIH3T3 cells it integrated as multicopy oligomers whereas in C127 cells the vector was maintained as a monomeric multicopy episome. Mouse C127 cells are the most widely used host cell for ensuring autonomous maintenance but integration can occur even in these cells. There is also variation in the length of time that a cell line can be passaged and still express a foreign gene in a BPV vector. Fukunaga *et al.* (1984) have passaged mouse C127 cells for four months and they continue to secrete human interferon-gamma from an episomal BPV vector. Cells containing the gene for hepatitis B surface antigen on a full-length BPV vector, on the other hand, stopped producing antigen after six passages (Denniston *et al.*, 1984).

In order to study the expression of foreign genes from their own signals, transcription should oppose BPV transcription to avoid aberrant read-through (Di Maio *et al.*, 1984). Faithful initiation and correct regulation of expression have now been demonstrated for a number of genes in BPV vectors, e.g. metallothionein (Pavlakis and Hamer, 1983); MMTV-LTR (Ostrowski *et al.*, 1983); human β-interferon (e.g. Zinn *et al.*, 1983). The idiosyncracies of BPV vectors and their use are discussed in detail by Campo (1985).

Epstein−Barr virus (EBV)

EBV is a human herpes virus that causes infectious mononucleosis (glandular fever) and is associated with two human cancers, Burkitt's lymphoma and nasopharyngeal carcinoma. The genome is double-stranded DNA of about 172 kb that has multiple direct repeats of about 0.5 kb at each end (Baer *et al.*, 1984). The genome is linear in the virus particle but it exists as a circular episome in about 1−100 copies in the nucleus of infected cells.

A plasmid vector has been constructed consisting of a selectable marker (either G418 or hygromycin B resistance), the EBV origin of replication (ORI P) and a 2.9 kb fragment of EBV genome that codes for a replication protein, EBNA-1. This plasmid replicates autonomously in a wide range of human cells from different lineages, i.e. fibroblasts, epithelial cells and lymphoid cells and also in dog but not rodent cells. The copy numbers are low, about 2−4 copies per cell, and the plasmid is unstable in the absence of selection, the loss rate being about 5% per cell generation (Yates *et al.*, 1985). EBV vectors will be useful for analysing aspects of gene expression such as chromatin structure or interactions with regulatory proteins that

GENE TRANSFER INTO ANIMAL CELLS

may be affected by chromosomal integration or high copy numbers. It has also been shown that a shuttle vector based on EBV can carry as much as 35 kb of DNA which is expressed and can be readily rescued (Kioussis *et al.*, 1987). It is possible that EBV shuttle vectors may be useful for the direct cloning of mammalian genes (see Chapter 7).

BK virus

BK virus is 80% homologous to SV40 but it has a different host range. It efficiently infects human cells and transforms hamster, rat, mouse, rabbit and monkey cells. In many cells the BKV genome exists as a full-length episome although some rearrangements and deletions can occur. A derivative of BKV which lacks part of the late region and incorporates a *tk* selectable marker and the *E. coli* vector pML2 persists episomally in three different human cell lines (Milanese *et al.*, 1984). The copy number varies from 20 to several hundred. T antigen is required and may limit replication by auto-regulation. This autonomous viral replicon may be particularly useful for manipulating human cells.

SV40 replicons in 'regulated' COS cells

COS cells transformed with SV40 replicons usually die because of massive plasmid replication. It is possible to produce stable lines by regulating the synthesis of T antigen which controls replication. One approach has been to establish COS-type cells with a temperature-sensitive T antigen controlled by the RSV promoter. The cells are transformed with the SV40 replicon which is allowed to replicate for a limited time before the temperature is raised to inactivate T. The copy number of the transfected plasmids can be regulated by temperature modulation of T antigen (Rio *et al.*, 1985). An alternative approach is to place T antigen expression under the control of an inducible promoter such as that of human *Mt*-IIA (Gerard and Gluzman, 1985). The cell lines (BMT and CMT) express low basal levels of T antigen and are efficiently transformed to G418[R] with SV40 replicons which are maintained at low copy number episomally without selection. After heavy metal induction the plasmid copy number rises by 10- to 15-fold, allowing increased expression of any foreign gene in the vector.

Other mammalian virus vectors

As more is learned about the molecular biology of different viruses they become amenable to genetic manipulation. We cannot describe all the possible virus vectors that are being used but a few others with special uses will be discussed in later chapters. In particular, in Chapter 12, the use of vaccinia virus vectors for the production of new live vaccines is discussed. Adenoviruses which are medium-sized DNA viruses are also used for stable and transient expression (Van Dorven *et al.*, 1984; Davis *et al.*, 1985; Mansour *et al.*, 1985).

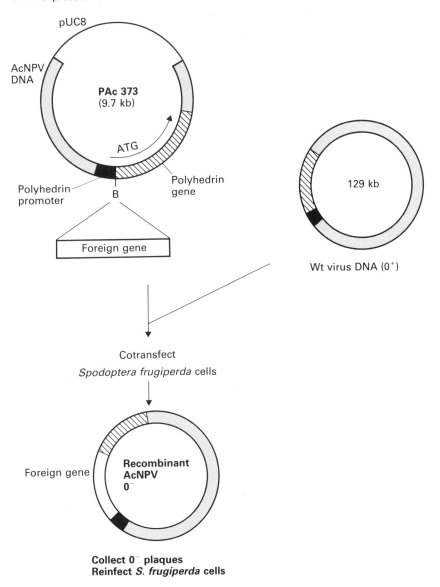

Fig. 4.31. Molecular engineering of AcNPV.

Gene transfer into insect cells

The insect Baculovirus, *Autographa californica* nuclear polyhedrosis virus (AcNPV), has been engineered to produce a vector for insect cells (Smith *et al.*, 1983a; Miller *et al.*, 1987). During a normal infection the virus produces inclusion bodies in the nucleus. These consist of virus particles embedded in a virus-coded protein called *polyhedrin*. The polyhedrin gene has an extremely efficient promoter but is itself not essential for viral

GENE TRANSFER INTO ANIMAL CELLS

Table 4.1. General features of some mammalian host/vector systems.
(A) Stable systems

Vector	Transfer method	Transfer efficiency	DNA status	Comments
Simple plasmids	DMT-CaPO$_4$ DMT-microinjection DMT-electroporation DMT-polybrene-DMSO DMT-protoplast fusion	Low Mod. High High High	Integrated randomly, often rearranged, ligated into large transgenome if carrier DNA is used, usually 1–5 copies	Useful for a wide range of cell lines. CaPO$_4$ may not be effective for primary cell lines
dhfr amplicons	DMT-(as above)	(As above)	Integrated concatemers or extra minute chromosomes. High copy numbers, up to 1000/cell. Some rearrangements	Takes time to amplify by MTX selection. Best in DHFR⁻ cells and if *dhfr* expression is weak. May be genetically unstable. Expression of co-amplified genes can vary widely
Retroviruses	DMT-(as above) Infection	(As above) High	Usually single copy	Broad host range including primary cells. Expression of transferred genes can be affected by the retrovirus and flanking DNA. Max. insert about 6 kb. Cannot be used to study splicing
BPV1	DMT-(as above)	(As above)	Usually stable episome but depends on particular construction and cell line. Copy number 50–200. Integration and rearrangements can occur	Restricted host range — rodent and bovine. Copy number controlled
EBV mini-replicon	DMT-(as above)	(As above)	Stable episome. Copy number 1–5	Broad host range
BK mini-replicon	DMT-(as above)	(As above)	Stable episome. Variable copy number	Human cells only
SV40 mini-replicon	DMT-(as above)	(As above)	Stable episome	Only in cell lines such as BMT that regulate T-antigen production

(B) Transient systems

Vector	Transfer method	Transfer efficiency	Comments
Simple plasmids	DMT-DEAE dextran	Mod.	Expression of transferred DNA is influenced by topology and chromatin structure. Sodium butyrate enhances expression
	DMT-polybrene-DMSO	Mod.	
SV40 and polyoma	Infection	High	Used in monkey and rodent cells respectively. Kills cells in 3–10 days. Very high copy number
SV40 Mini-replicon	DMT-(as above)	(As above)	Only in COS or HFS cells. Kills cells in 3–10 days. Very high copy number

replication. It is, therefore, an ideal source of expression signals for the construction of an expression vector.

AcNPV has a complex double-stranded DNA circular genome of about 130 kbp and therefore it cannot be used as a direct cloning vector. The approaches used to manipulate the virus therefore involve two stages. The first is the construction of an expression and recombination plasmid. The second is the *in vivo* homologous recombination between the wild-type viral genome and the plasmid, to produce a recombinant virus.

A typical AcNPV expression vector, pAc373, is shown in Figure 4.31. An EcoRI fragment containing the polyhedrin gene and flanking DNA was sub-cloned into pUC8 and manipulated to place a unique Bam HI site between the polyhedrin cap site and the ATG. Any foreign gene inserted at this site will be expressed from the polyhedrin promoter. The insertion can be readily detected because no polyhedrin will be produced by the disrupted gene. Recombinant pAc373 derivatives are cotransfected into *Spodoptera frugiperda* cells with wild-type AcNPV DNA. Homologous recombination replaces the wild-type polyhedrin region in the virus with the recombinant DNA. The resulting recombinant viruses will replicate but will not form occlusion bodies. Occlusion minus (o^-) plaques are readily detected and the virus is purified and cultured to produce a recombinant virus stock. When this is used to reinfect *S. frugiperda* cells the foreign gene is expressed transiently until the cells lyse about 72 hr post infection. During this time very high yields of proteins can be obtained. Several proteins have now been produced in this system, e.g. interferon β (Smith *et al.*, 1983b), influenza virus haemag-glutinin (Kuroda *et al.*, 1986), interleukin 2 (Smith *et al.*, 1985) and an oncogene, *c-myc*, protein (Miyamoto *et al.*, 1985).

The biotechnological applications of insect virus vectors are discussed more fully in Chapter 12.

Summary

A wide variety of procedures and vectors have been developed for manipulating cultured animal cells. There is no universal gene transfer and gene expression vector. The choice of vector depends to a large extent upon the goals of the particular experiment. If the aim is to produce massive quantities of a protein, then a high-efficiency promoter is used to drive expression and this should be combined with a high copy number of the gene. The promoter must however be matched with the host cell. For example, a tissue-specific enhancer may be required. In addition if the product is toxic, a regulated promoter will be used. The choice of host cell will depend upon whether cell-specific modifications to the protein, such as proteolytic processing or glycosylation, are necessary. Transient systems are suitable for laboratory-scale production but large-scale commercial production of proteins is often best suited by stable continuously expressing cell lines.

If the aim of the experiment is to study gene expression then it is most appropriate to use host cells which normally express the gene of interest. However the decision must be made whether to use a transient system or a stable system. Transient systems are simple, rapid and reproducible but may not be physiologically relevant. For example, the topology and chromatin structure of the template may be atypical and the copy number of the template will be abnormally high. This may titrate key regulatory factors. Stable systems take longer to generate and there is often a high degree of variability in copy number and expression levels between cell lines. Once a suitable line has been made, however, expression can be studied for long periods.

One of the main factors that has limited the use of gene transfer technology in studying fundamental aspects of animal molecular biology is that it is extremely difficult to introduce DNA into highly specialised cells. Retrovirus vectors and electroporation may, however, overcome this. Many of the advantages and disadvantages of the different gene transfer systems that we have introduced will be re-examined when we discuss the applications in more detail in later chapters. A brief summary of the key features of some of the systems is given in Table 4.1.

5 Gene Transfer into Whole Animals

Introduction

Cells in culture are unlikely to mimic all aspects of behaviour of the same cells in the animal. In addition, it is likely that the factors which determine the activity or inactivity of genes to achieve correct development and differentiation may only be expressed in embryos and stem cells. The definitive way of analysing gene expression therefore, is to transfer the genes back into an intact animal and to follow expression during development and differentiation. This is now possible in a number of different organisms and there are three general approaches.

The first is to introduce DNA directly into a fertilised egg. If the egg has not yet divided then there is a good chance that any introduced DNA will be inherited by all the cells of the embryo. This includes both the somatic and germ line cells. The embryo develops into a viable adult and the expression of the transferred genes can be analysed in all the tissues. Furthermore, because the introduced genes are incorporated into the germ line they will be transmitted to any progeny. Organisms that have been genetically manipulated in this way are referred to as *transgenic* animals. The second approach is to create *chimaeric organisms* by introducing genetically distinct cells into an embryo. The tissues of the resulting adult are derived from both the original embryo and the introduced cells and the organism is therefore a chimaera. An extension of this technique is to introduce DNA into cells before they, in turn, are introduced into the embryo. The resulting organism is therefore a *transgenic chimaera* and may produce fully transgenic progeny if the introduced cells contribute to the formation of the germ line. The third approach is to manipulate the somatic cells of an adult organism. This largely involves manipulating bone marrow cells because they can be easily removed from the animal and then returned and they are a self-renewing population of cells. Manipulating adult cells allows questions to be asked about the status of gene expression in the various differentiated cells and provides a method of introducing DNA into adult animals. This technique is the basis for gene therapy in man and is discussed further in Chapter 11.

In this chapter we concentrate on the major experimental organisms. These are the mouse and the fruit fly, *Drosophila melanogaster*. One of the primary goals of current molecular biology, and in fact a major reason for creating transgenic organisms, is to understand the process of development. This is to discover what happens during progression from the single celled zygote to the complex multicellular adult organism. Experimental embryologists have described the stages in embryogenesis and have shown that in many cases cells become limited to a specific developmental programme very early in embryogenesis. There is then a progressive *restriction* in developmental potential until finally a cell is *committed* to differentiate into one specialised cell type. Some of the factors that influence how a cell will differentiate appear to be established in the egg itself; other factors come into play during embryogenesis. Many aspects of development are a consequence of differential gene expression with specific regulatory steps occurring at any of the levels outlined in Chapter 1. A detailed consideration of embryogenesis and development is far beyond the scope of this book but we have given a brief outline of the salient features of each of the organisms discussed in this chapter. This should allow you to appreciate the relative merits of the different organisms for studying these fundamental biological questions. Books by Slack (1983) and Graham and Wareing (1984) provide more detailed accounts. In addition comments by Stent (1985) and Lawrence (1985) show how molecular biology and experimental embryology may or may not provide a definitive explanation of development.

Gene manipulation in the mouse

A guide to mouse embryology

In this section we give a very brief account of the development of the mouse embryo for further details see (Slack, 1983; Hogan *et al.*, 1986 and Figure 5.1).

Ovulation occurs in response to mating, the fertilised egg is first located in the oviduct and the early stages of development occur here; this is the *pre-implantation stage*. In the fertilised egg or zygote, there are the nuclei of the oocyte and sperm, called the female and male *pronuclei* respectively. The first *cleavage*, i.e. *cell division* occurs after 24 hr and then the second and third cleavage occur at 12 hr intervals to produce an eight cell *pre-embryo*. The pre-embryo becomes more compact and division continues to the 16−32 cell stage during which time the pre-embryo is called a *morula*. At this stage at about 3 days after fertilisation the 'pre-embryo' begins to, '*take shape*'. A fluid filled cavity called the *blastocoel* is formed and the pre-embryo is now called a *blastocyst*. It is composed of an outer cell layer called the *trophectoderm* and a clump of cells called the *inner cell mass (ICM)* attached to one point inside. At this stage the pre-embryo travels to the uterus and around 4 to 5 days after fertilisation it becomes implanted in the uterus. The pre-embryo is now referred to as *post-implantation*. During this stage, further sub regions of determined cells are formed. Some of these cells produce the baby mouse, i.e. these cells form the *embryo* proper, and some produce the tissues that maintain the embryo, i.e. the placenta and yolk sac; these are called *extra-embryonic* tissue. The trophoectoderm develops

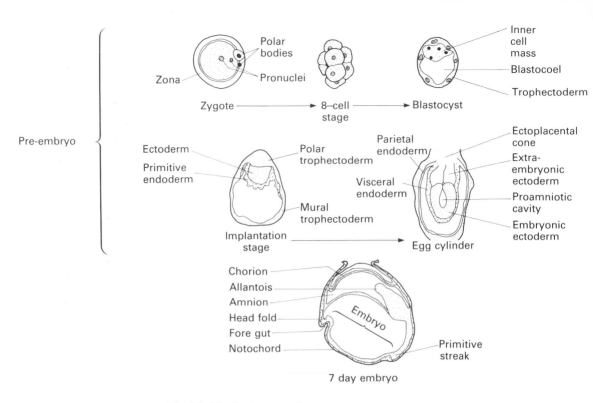

Fig. 5.1. The development of the mouse embryo. (Adapted from figure 6.1 of Slack, 1983.)

into entirely extra-embryonic tissue whereas the inner cell mass develops into the *primitive endoderm* and the *primitive ectoderm*. The primitive endoderm produces extraembryonic tissues (the paretal and visceral endoderm) and the primitive ectoderm produces some extra-embryonic tissue but also comprises the cells that will form the embryo. At about 5 days post-fertilisation, the embryo cells proliferate to form a structure called the *egg cylinder* containing 500 to 600 cells. By 7 days after fertilisation the embryo is 'polarised' with the *primitive streak* running between the posterior and anterior poles and the *endoderm*, *ectoderm* and *mesoderm* of the embryo have been formed.

In Figure 5.2 you can see how the basic body plan of a vertebrate is built up from different regions of the developing embryo. Most cell types in the adult can be traced back along a defined lineage that originates in one of the three regions of the early embryo, i.e. the ectoderm, the mesoderm and the endoderm. The basic body plan is not, however, fully defined in the early embryo. There is a hierarchy of irreversible decisions after each of which the cells are said to be in a different *determined state*. At each decision they become less versatile in the properties that they can display. Experimental manipulation of embryos suggests that there are two types of process which may account for the spatial distribution of cell types in the embryo. The first is *cytoplasmic localisation* which supposes that cytoplasmic components are asymmetrically distributed in a cell such that daughter cells receive different components and hence develop differently. The second is

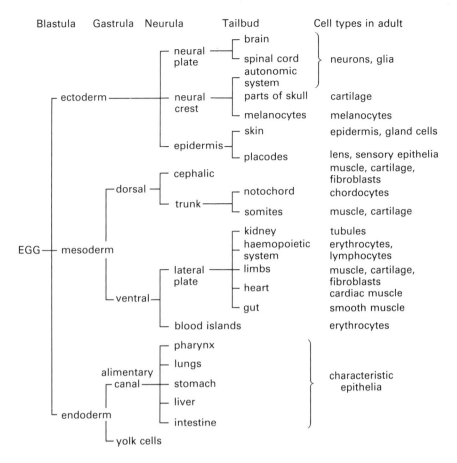

Fig. 5.2. The body plan of a vertebrate. (Adapted from figure 1.1 of Slack, 1983.)

induction which supposes that 'chemical signals' from one region of the embryo induce other cells to become differentially determined. The biochemical basis of determination is completely unknown but the response to 'developmental stimuli' must presumably be mediated by differential gene expression at some level.

Transgenic mice

The ability to produce transgenic mice has opened up a whole new approach to studying fundamental problems of development and disease in mammals. The expression of transferred genes can be analysed in the correct tissue and at the correct developmental stage (Chapter 8) or transferred DNA can be used as an insertional mutagen to identify developmentally important genes and to aid in their subsequent cloning (Chapter 7). Introducing various oncogenes into embryos produces adult animals with tissue specific tumours providing the most tractable models for human cancers that have ever been available (Chapter 11). In addition, as oncogenes can immortalise specialised cells these can now be cultured for the production of specialised proteins with pharmaceutical applications (Chapter 12). Finally a range of physiological interactions can be analysed and models for a number of diseases, e.g. growth defects, can be established (Chapter 11).

In this section we will describe the techniques used to produce transgenic mice and also some of the features of the expression and inheritence of the transferred genes. Further details will be found in Part III. The subject is reviewed by Palmiter and Brinster (1985, 1986), Gordon and Ruddle (1985) and Hogan *et al.* (1986).

Microinjecting DNA into fertilised eggs

The first successful transfer of DNA by microinjection into a fertilised egg was reported by Gordon *et al.* (1980) and the procedure they developed is outlined in Figure 5.3.

To produce the fertilised eggs female mice are hormonally induced to superovulate and then mated. They are killed 22 hr later and their oviducts are removed into a simple buffered salt solution. The oviducts are dissected to release the fertilised eggs which are washed and stored at 37°C in culture medium. The fertilised eggs can be identified by observing the two pronuclei under a dissecting microscope.

Microinjection needles are prepared by 'drawing out' fine bore capillary tubing. One needle is used to hold the fertilised egg in place and a second finer needle is used to introduce the DNA directly into the male pronucleus which is larger than the female pronucleus and therefore easier to manipulate. DNA is introduced until the pronucleus just doubles in size and the volume transferred is about $1-2$ pl. About $40-60$ eggs can be injected in $1-2$ hr and they are stored in culture medium until all the manipulations are complete. Many eggs lyse as a result of this procedure.

In order for the fertilised eggs to develop into baby mice they must be reintroduced into the female reproductive tract. A second group of female mice are made pseudo-pregnant by mating with vasectomised males. This means they will be in the correct hormonal state to allow the introduced embryo to implant but none of their own eggs will form embryos. About $10-20$ genetically manipulated embryos are introduced into the oviduct of a pseudo-pregnant host by a minor surgical procedure. Live offspring are delivered $18-21$ days later although only about 50% of the transferred eggs produce live births even when control, i.e. uninjected eggs are transferred.

The presence of foreign DNA in the offspring can be determined by a '*tail blot*'. A small amount of blood is collected by snipping the end of the tail, DNA is prepared and analysed by restriction digestion and Southern blotting using an appropriate probe. Transgenic mice have been produced at a range of efficiencies from $3-50$% of the mice which are born. If linear DNA free from prokaryotic sequences is used, then 25% of the surviving injected embryos are transgenic. These initial transgenic offspring are referred to as the *founder animals* or in genetic terms, the G0.

THE STATUS AND INHERITANCE OF MICROINJECTED DNA

When either plasmids or linear ·DNA fragments are microinjected into

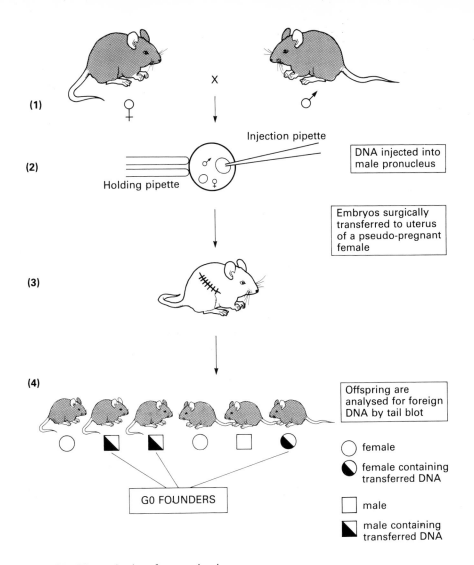

Fig. 5.3. The production of transgenic mice.

mouse embryos the status of the transferred DNA is somewhat unpredictable just as we have discussed for gene transfer into cultured cells. In the majority of cases the DNA is integrated randomly at a single chromosomal site. There are often multiple copies arranged as head to tail concatemers comprising 2−200 tandem repeats. Single copies are also found but often these have rearranged. Very rarely, the DNA appears to remain extra-chromosomal (e.g. Rassoulzadegan *et al.*, 1986) but the basis for this is unknown. Integration occurs at the one cell stage so that all the cells in the transgenic animal contain foreign DNA integrated into the same site in one of their chromosomes. Frequently however integration is delayed until a later multicellular stage and this produces animals whose tissues are not genetically identical, these are called *mosaics*. Mosaic animals are only readily detected when they are bred. For example, if the progeny have a larger number of copies of the foreign DNA than detected in the blood sample taken from the parent this suggests that some of the cells in the parent lacked foreign DNA. Mosaic animals may also fail to transmit the foreign DNA because it has not integrated in the germ line cells (reviewed by Gordon and Ruddle, 1985).

In the majority of cases the foreign DNA is inherited as a simple Mendelian trait and stable lines of transgenic mice can be established. G0 animals are mated with normal animals, usually one of the original parents, i.e. a back cross, and 50% of the progency will be *hemizygous*, that is heterozygous for the foreign DNA. These animals are referred to as G1 and the transgenic line can be maintained perpetually by mating hemizygous offspring with normal animals to produce G2, G3, etc. mice. The inheritance of the introduced gene is confirmed by tail blots at each generation. When transgenic males are mated with transgenic females (e.g. mating between siblings, which is often called *inter se* mating) then 25% of the progeny should be homozygous for the chromosome carrying the foreign DNA. A typical analysis of transgenic mice is provided by a study by Wagner *et al.* (1983) and is illustrated in Figures 5.4 and 5.5. A human growth hormone (hGH) was injected into 229 embryos and six transgenic G0 mice were produced from 20 live births. The Southern blot analysis showed that the pattern of integration was highly variable and in mouse hGH-3 and hGH-6 it was too complex for analysis. Mouse hGH-5 had less than one copy of hGH per cell indicating that it was probably a mosaic and this was confirmed by back-crossing the G0 with one of the parents. Mouse hGH-5 only transmitted hGH to 10% of the progeny; therefore only 10% of its germ-line cells must have contained hGH DNA. The other five mice were fully transgenic and transmitted hGH to about 50% of the progeny. The progeny (G1) were hemizygous for hGH.

Figure 5.5 shows a pedigree analysis of two of the transgenic lines following the inheritence of hGH in back-crosses and *inter se* matings. Mouse hGH-2 showed simple Mendelian inheritance of hGH. It produced 40% hemizygotes in a back-cross and 30% hGH homozygotes in *inter se* crosses. Mouse hGH-3 however fails to produce any homozyotes in a number of *inter se* matings showing that the insertion in this strain produced a recessive lethal mutation. About 20% of all transgenic mice have such mutations and they are often alterations in genes that are important in development which result in the production of deformed embryos. In other cases when homozygotes are produced they have some phenotypic abnormality. This phenomenon of gene mutation by transferred DNA is called *insertional mutagenesis* and it may be an important tool for the identification and subsequent cloning of mouse genes that are important in development (see Chapter 7). It has, however, been shown that insertion of foreign genes is often associated with extensive rearrangements and deletions of flanking DNA and identification of the gene that produces the mutation may therefore be complicated (Covarrubias *et al.*, 1987). There is also some suggestion that integration may not be random and there may be 'hot spots' for integration into chromosomal regions that are naturally recombinogenic, e.g. fragile sites and immunoglobulin genes (Covarrubias *et al.*, 1986).

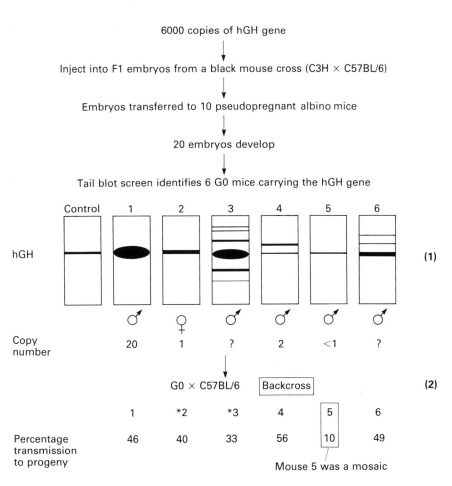

Fig. 5.4. Analysing the status and inheritance of foreign DNA in transgenic mouse lines. (Based on data from Wagner *et al.*, 1983.) Asterisks, see Figure 5.5.

FOREIGN GENE EXPRESSION IN TRANSGENIC MICE

In many cases the transferred DNA is expressed in the tissues of transgenic mice. The quality and quantity of expression is however dependent upon the particular gene and possibly on the precise structure of the injected DNA (reviewed by Palmiter and Brinster, 1986). We will discuss several studies to illustrate the design and interpretation of transgenic mouse experiments. Several of the plasmids used in these studies are shown in Figure 5.6.

(a) Mouse metallothionein promoter — hybrid genes

A number of key studies have been undertaken with fusion genes comprising the promoter region from the mouse metallothionein gene ($Mt-1$, see Figure 4.8) linked to a number of different coding regions.

The mouse $Mt-1$ promoter was chosen because $Mt-1$ is a *house keeping gene* that is, it is expressed in nearly all tissues. It is also inducible by heavy

GENE TRANSFER INTO WHOLE ANIMALS

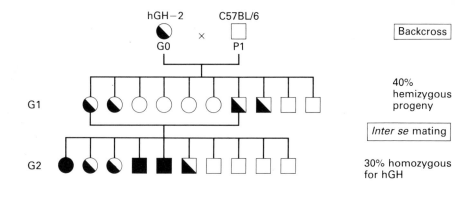

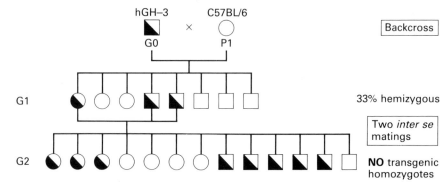

Fig. 5.5. A pedigree analysis of the inheritance of the hGH gene in two transgenic lines of mice. (Based on data from Wagner *et al.*, 1983.)

metals in most of these tissues although the extent of induction varies. The highest levels are achieved in the liver (Durnam and Palmiter, 1981). The earliest analyses used a plasmid pMK that has the $Mt-1$ promoter fused to the HSV$-tk$ coding region. Of ten transgenic mice seven expressed high levels of thymidine kinase in the liver. Furthermore the levels of expression were increased by feeding cadmium to the animals. This was one of the earliest convincing demonstrations that a foreign gene could be expressed in transgenic mice.

Subsequent studies by this research group provided the first dramatic illustration of the power of the transgenic mouse as an experimental system (Palmiter *et al.*, 1982a, 1983). They fused the $Mt-1$ promoter to either the rat growth hormone (rGH) or the human growth hormone (hGH) genes and injected the plasmids into mouse eggs. The hGH plasmid (pMT$-$hGH) is shown in Figure 5.6. The progeny mice were reared on water supplemented with 5000 ppm of $ZnSO_4$. Over two thirds of the transgenic animals grew more than 18% larger than their litter mates and several animals were twice as big. This increase in size was a direct consequence of elevated expression of growth hormone. The $Mt-$hGH genes were induced up to 170 fold with zinc treatment and circulating growth hormone levels were up to 800 times higher than normal levels.

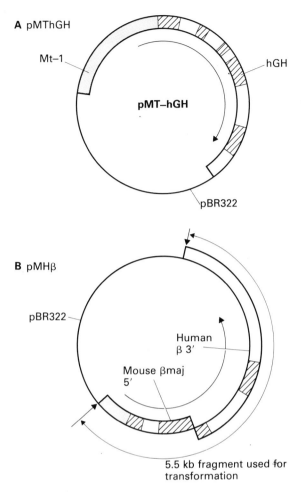

A pMThGH

Mt−1

hGH

pMT−hGH

pBR322

B pMHβ

pBR322

Human
β 3′

Mouse βmaj
5′

5.5 kb fragment used for
transformation

Fig. 5.6. Some plasmids used to study gene expression in transgenic mice. A, From Palmiter
et al. (1983). B, From Chada *et al.* (1985).

These studies showed that a foreign gene could produce a significant
change in the phenotype of a transgenic animal. Furthermore the $Mt-1$
promoter was responding to heavy metal induction. There are, however,
some peculiar features of expression directed by the $Mt-1$ promoter in the
transferred fusion genes compared to the expression of the endogenous
metallothionein gene (Palmiter *et al.*, 1982b). The expression of the transferred gene is not induced by glucocorticoid treatment under conditions
when the endogenous gene is optimally induced. Also, the induction by
heavy metals is not completely normal because the ratio of $Mt-$hGH
mRNA to endogenous $Mt-1$ mRNA varies between tissues, implying that
the exogenous and endogenous genes are responding differentially to the
same stimuli (Table 5.1). The absolute levels of expression of the transferred
gene are variable producing circulating growth hormone at levels from 10 to
64 000 ng^{-1} ml and the level of expression is unrelated to copy number,
which ranged from 0.9 to 455 copies per cell. These data are interpreted as
showing that there is a hierarchy of $Mt-$hGH mRNA production in different

GENE TRANSFER INTO WHOLE ANIMALS

Table 5.1. Variable induction of *Mt*-hGH gene expression in transgenic mice (adapted from Palmiter *et al.*, 1983).

Animal	Ratio *Mt*-hGH mRNA to *Mt*-1 mRNA			
	Liver	Intestine	Brain	Heart
167−2	0.56	0.07	0.02	0.16
186−1	0.26	0.01	0.29	0.58
186−5	0.43	0.06	0.01	0.83

tissues which is due to differential transcription and differential mRNA stability. Superimposed on this hierarchy there are strong position effects that show up as extraordinarily high or low levels. These position effects might be due to integration near a gene that is expressed in a comparable tissue specific manner.

The nature of the coding region used to tag the promoter also influences expression. For example in *Mt*−*tk* transgenic mice the levels of *tk* never exceeded 0.01 to 0.1% of the endogenous *Mt*−1 mRNA. This may reflect the instability of the *tk* mRNA. In several *Mt*−rGH and *Mt*−hGH transgenic mice a completely novel tissue specificity of expression was detected. High levels of fusion mRNA were detected in the large neurones in the brain which are not normally sites of either excessive growth hormone or metallothionein expression (Swanson *et al.*, 1985). The levels and regulation of expression of the *Mt*−GH genes in transgenic mice were heritable traits. However, *Mt*−*tk* offspring showed marked variations in expression both between siblings and when compared to the parent. The *Mt*−*tk* DNA was extensively methylated but this did not entirely correlate with expression and the reasons for this variable inheritance of expression have not yet been resolved (Palmiter *et al.*, 1982b).

(b) The expression of intact genes

(i) Globin genes. In one of the first transgenic mouse experiments Costantini and Lacy (1982) introduced 100 to 1000 copies of a lambda Charon 4 clone carrying a 19 kb region of the rabbit β−globin locus into fertilised mouse eggs. No expression of the rabbit β−globin genes could be detected in the erythroid tissues of the seven transgenic animals that were studied. In two mouse lines rabbit β−globin transcripts were detected in specific but inappropriate tissues, skeletal muscle in one and testis in the other. The pattern of expression was inherited in these lines and it was suggested that the chromosomal site of integration was influencing expression (Lacy *et al.*, 1983).

These early results were disappointing but it now seems likely that the inefficient expression was due to the presence of the bacteriophage DNA or due to attempting to express a rabbit gene in the mouse. Chada *et al.* (1985) have now produced transgenic mice with the plasmid pMHB (Figure 5.6) that has the 5′ portion of the mouse β−major globin gene including 1.2 kb of 5′ flanking DNA fused to the 3′ portion of the human β−globin gene including 3.8 kb of 3′ flanking DNA. The eggs were injected either with supercoiled plasmid or with a purified linear fragment containing only the β−globin DNA. The hybrid gene is used to distinguish the transferred gene from the endogenous mouse β^{maj}−globin gene. After treating transgenic mice with phenylhydrazine to enrich for erythropoietic cells in the spleen, various tissues were analysed by quantitive Sl nuclease protection. In six mice, including all three that had been produced with supercoiled plasmid, there was little or no expression of β−globin. In four mice β−globin was expressed and the major sites of expression were the blood, bone marrow and spleen i.e. expression was tissue specific (Table 5.2A). The transcripts were correctly initiated and spliced but no globin protein was detected. The maximum level of expression was also only 2−4% of the normal endogenous levels. Further studies have now shown that transgenic lines that display tissue specific expression also show the correct pattern of regulation of the hybrid β−globin gene during development (Magram et al., 1985). Homozygous transgenic males from line 46 and 49 (Table 5.2A) were mated with normal (C57Bl/6) females and the resulting hemizygous transgenic animals were analysed at various stages of gestation. Both the endogenous and foreign globin genes were silent in early embryos and were first expressed in foetal liver erythroblasts and expression continued in adult bone marrow. The control of expression of the hybrid gene was not entirely normal

Table 5.2. The expression of a mouse−human β-globin hybrid gene in transgenic mice. (A) Tissue specific expression (Chada et al., 1985).

Mouse line	DNA	Copy no.	% total β-globin RNA contributed by transferred gene			
			Blood	Bone marrow	Spleen	Others
46	5 kb−Linear	1	1.60	0.60	0.60	0
49	5 kb−Linear	1	2.00	1.20	0.02	Te <0.001
75A	5 kb−Linear	6	0.02	0.02	0.015	H,K <0.001
77	5 kb−Linear	3	0.08	0.08	0.10	0
43	5 kb−Linear	1				
47	5 kb−Linear	1				
53	5 kb−Linear	1	No expression detected.			
11	5 kb−Plasmid	9				
16	5 kb−Plasmid	180				
15	5 kb−Plasmid	3	Very low expression in all tissues.			

⋆ Other tissues tested were testes (Te), heart (H), kidney (K), lung, brain, liver, skeletal muscle.

(B) Development stage-specific expression (Magram et al., 1985).

Mouse line	Fraction of total RNA represented by β-globin RNA			Foetal: adult ratio
	Foetal blood	Foetal liver	Adult bone marrow	
46	0	3×10^{-6}	1.5×10^{-5}	1:5−20
49	0	5×10^{-7}	10^{-5}	1:5−20
77	0	2×10^{-5}	2×10^{-6}	10:1
endogenous	0	10^{-3}	10^{-3}	1:1

because the ratios of hybrid β-globin levels in foetal and adult tissues differed both from the normal ratios and between transgenic animals. This study is summarised in Table 5.2B.

Different globin genes have now been introduced into mice and some interesting features of developmental switching are being described. For example, simian primates (including human) express a distinct foetal type of globin chain in foetal erythroid cells encoded by the γ^A and γ^G genes. In mice there are genes related to γ-globin genes but they are expressed only in embryos. When human γ-globin genes are introduced into mice they revert to an embryonic pattern of expression (Chada et al., 1985; Kollias et al., 1986). These data show that some aspects of tissue specific and stage specific gene expression can be studied in transgenic mice. Recently sequences have been isolated from the far upstream regions of the β-globin locus that prevent position effects and allow quantitative studies (Grosveld et al., 1987).

(ii) Immunoglobulin genes. Brinster et al. (1983) have established transgenic lines by injecting a functionally rearranged mouse immunoglobulin κ gene. The gene was expressed in very high yields, about 3000 molecules per cell, and equivalent levels were found in all transgenic mice and their offspring. The gene was only expressed in the lymphoid cells and not in testis, liver, kidney, heart, muscle, brain or thyroid gland (Storb et al., 1984). Similar results were obtained with a functionally rearranged μ heavy chain gene (Grosschedl et al., 1984). These data show that rearranged immunoglobulin genes contain all the information necessary for tissue specific expression, irrespective of chromosomal integration site, and to allow stable and normal levels of expression throughout successive generations. It is proposed that these genes contain tissue specific enhancers which ensure efficient expression in appropriate cells (see Chapter 8 for further discussion of enhancers).

(iii) Pancreatic elastase: tissue specific expression. Although tissue specific expression at normal levels was demonstrated with immunoglobulin genes, these are somewhat unusual in being derived from spatially distinct regions of the chromosome by rearrangement. The first 'normal' chromosomal gene to show normal levels of expression in the correct tissue in a transgenic mouse was rat pancreatic elastase (Swift et al., 1984). Different transgenic mice containing a 23 kb segment of rat DNA carrying the elastase gene all express elastase at above 10 000 molecules per cell. The levels also correlate with the copy number of the transferred gene. Some other tissues contained elastase transcripts but always at less than 30 molecules per cell (Table 5.3A). In an extension of these studies Ornitz et al. (1985) have fused the elastase gene promoter from -205 to $+8$ to an hGH gene and analysed 18 transgenic animals. In 12 animals there was tissue specific expression whereas in six animals there was litle or no expression in any tissue (Table 5.3B). These results suggest that a very small region of the promoter can confer tissue specific expression but that the small promoter region can be influenced by chromosomal location. Recently, a tissue specific enhancer has been identified in the elastase I gene by analysing the expression of elastase growth hormone fusion genes in transgenic mice (Hammer et al., 1987).

Table 5.3. Tissue-specific expression of rat elastase in transgenic mice (adapted from Swift *et al.*, 1984; Ornitz *et al.*, 1985).

(A)   A 23 kb rat elastase genomic clone is transferred.

Genes per cell	Molecules of elastase mRNA per cell				
	Pancreas	Intestine	Kidney	Liver	Spleen
2	10 000	2.8	4.0	10	0.1
100	120 000	2	<0.2	3	30
9	18 000	<0.2	<0.1	<0.6	5
7	45 000	<2	<1	300	30
2	10 000	75	<0.3	<1	1.6
★120	55	10	<0.5	1	14
0	11	<0.5	<0.3	0.1	<0.02

★ This animal was mosiac.

(B)   A rat elastase promoter–hGH hybrid gene is transferred.

Genes per cell	Molecules of elastase-hGH RNA per cell				
	Pancreas	Intestine	Kidney	Liver	Spleen
1	0				
6	0				
1.2	1 460	No significant RNA levels detected in any other tissue.			
199	9 760				
136	11 900				
2.3	15 400				
4.2	39 400				

(iv) Other genes. The number of studies of transgenic mice will increase exponentially in the near future but at the moment a limited number of genes have been tried (reviewed by Palmiter and Brinster, 1986). Many genes show normal tissue specific and stage specific expression but anomalies have been observed. Often the levels of expression are low and in many cases some transgenic offspring show no expression at all. For example, neither the rat nor human growth hormone genes with their own 5' and 3' flanking region are expressed in transgenic mice whereas both are expressed as *Mt*−1 fusion genes (Hammer *et al.*, 1984). Although the intact chicken transferrin gene is expressed predominantly in the liver of transgenic mice, levels are very variable (McKnight *et al.*, 1983). Similarly a 4.7 kb genome fragment carrying the rat myosin gene was expressed largely in skeletal muscle in two transgenic mice analysed, but the levels showed a ten fold variation (Shani, 1985).

Viral infection

Cells of the pre-implantation mouse embryo can be infected with a retrovirus such as Moloney murine leukaemia virus (MoMuLV) either by co-cultivating 4–8 cell pre-embryos with virus producer cells or by injecting virus into the blastocyst *in vitro*. The infected pre-embryos are returned to pseudo-pregnant recipients and allowed to develop. Pre-embryos can also be infected *in utero* at the 8 day stage. The retrovirus integrates as a single copy and if germ line cells are infected then the integrated provirus is transmitted as a Mendelian trait (Jaenisch *et al.*, 1981). The probability of germ line infection is highest when very early pre-embryos are used. This approach can be used to transfer any foreign gene into embryos using infectious recombinant retroviral vectors. There are, however, disadvantages. Firstly, the offspring are usually *genetic mosaics*. This is because more than one cell in the embryo may be infected and the virus will be integrated at a different position or infection will occur at later embryonic stages so that some cell lineages may not be transformed. Outbreeding of the founder animal is necessary to establish pure lines that have only a single copy of the retrovirus in one of the chromosome homologues (i.e. hemizygous). An alternative approach is to inject the retroviral provirus DNA into the cytoplasm of a single cell egg. The resulting virus integrates into the genome at a single chromosomal site (Harbers *et al.*, 1981).

There have been problems in obtaining expression of foreign genes in retroviral vectors that are inserted into early embryonic cells. Jahner *et al.* (1982) have shown that MoMLV becomes methylated *de novo* after infection of pre-implantation stages and this blocks subsequent expression. Flanking host sequences are also methylated and inactivated (Jahner and Jaenisch, 1985). In addition, later in development the integrated genome is expressed but in a tissue specific manner that is different for different integrants. The chromosomal position of integration therefore has a marked effect on gene expression (Harbers *et al.*, 1981; Jaenisch *et al.*, 1981).

Several studies have now shown that the LTR is inactivated after embryo infection (Stewart *et al.*, 1987). The SV40 promoter was also inactive after embryo infection. It has been proposed that there might be *trans*-acting negative regulatory factors in undifferentiated cells that specifically inhibit transcription from LTR and SV40 promoters (Gorman *et al.*, 1985). These appear to interact with the enhancer region of the LTR (Cone *et al.*, 1987). If, however, the *tk* promoter is used to express genes within retrovirus vectors then, for example, *tk-neo* is expressed in all tissues. Tissue specific and developmentally regulated expression of transferred β-globin genes inserted into retroviral vectors and expressed with their own promoters has also been shown (Soriuno *et al.*, 1986; Karlsson *et al.*, 1987; Cone *et al.*, 1987). The precise design of the retroviral vector, therefore, influences subsequent expression (reviewed by Temin, 1986).

Retrovirus vectors are also a useful experimental tool for creating mutations and then isolating the mutated gene. This application is discussed in Chapter 7. In addition, an analysis of the activation of retroviruses themselves at different developmental stages may provide insight into the control of gene activity.

Cell transfer into embryos

If cells from the morula stage from two different strains of mice are mixed and then reintroduced into a foster mother, then both cell types contribute to the tissues in the developing embryo. This gives rise to an adult whose somatic tissues and germ cells are comprised of two genetically distinct cell types. This is a *chimaera*. It is also possible to produce chimaeras by injecting certain cultured cells into the early pre-embryo at the blastocyst stage. In order to contribute to all tissues in the adult the cells must be *pluripotent* and there are two types of cultured cells that can be used for this type of experiment; these are *embryonal carcinoma (EC)* cells and *embryonal stem (ES)* cells. It is possible to produce *transgenic chimaeras* by introducing genes into the cells *in vitro* and then using these cells to colonise host (reviewed by Robertson, 1986).

Embryonal carcinoma (EC) cells

Embryonal carcinoma (EC) cells are undifferentiated cells that are found in tumours called *teratocarcinomas* that are derived from primordial germinal cells. These tumours are distinct from other tumours because they contain a variety of cell types such as muscle, hair, nerve and pigment cells as well as the undifferentiated progenitor cell, that is, the EC cell. A single EC cell can form all the cell types found in the teratocarcinoma. There are also specific lines of mice that have a high frequency of spontaneous teratocarcinomas. Teratocarcinomas are produced experimentally by grafting early mouse embryos to sites, such as the kidney capsules, that are outside the uterus of a syngeneic (i.e. immunologically identical) host. The embryo becomes disorganised and a teratocarcinoma develops. The teratocarcinoma is successfully transplanted to select for the undifferentiated EC cells and permanent cell lines can be produced. Undifferentiated EC cells can be maintained as monolayers in tissue culture flasks, although in some cases a feeder layer of inactivated cells, e.g. by mitomycin treatment, has to be used to allow the cells to be cultured. Removal of the feeder cells or addition of retinoic acid induces the cells to differentiate into *embryoid bodies* that resemble early stage pre embryos. Different EC cell lines do not display uniform properties and many are karyotypically abnormal, however, euploid lines can be established, e.g. METT 1. When METT 1 cells are injected into the blastocyst stage of a normal mouse embryo they contribute to all the tissues of the adult including the germ line. This means that the progeny of these chimaeric mice can inherit the genotype of the teratocarcinoma cell rather than the host embryo (e.g. see Stewart and Mintz, 1981).

There are some problems with producing chimaeric mice with EC cells. Firstly, the efficiency with which they colonise the host blastocyst is very low; only 11–12% of animals resulting from EC embryo manipulations are chimaeric. Secondly, if the EC cells are aneuploid they will not form germ cells. The ability of EC cells to reach the germ cell pool is also very low and the METT 1 line is the only line maintained *in vitro* that has produced females with EC derived oocytes.

Embryonal stem (ES) cells

Normal pre-implantation pre-embryos can be grown *in vitro*, but once they reach the stage where they would normally implant in the uterus then the blastocyst must attach to a substrate for growth to continue. Attachment to a substrate *in vitro* disrupts the organisation of the embryo and the inner cell mas (ICM) becomes exposed. This can be disaggregated and cell colonies that exhibit a stable stem cell morphology can be identified and established as permanent cell cultures of pluripotent stem cell (e.g. Evans and Kaufman, 1981). These cells have to be carefully cultured in highly supplemented media and on a layer of X-irradiated inactivated feeder fibroblasts to minimise the selection of abnormal cells. ES cells are euploid and colonize embryos more efficiently than EC cells. They also contribute to the germ line.

Transgenic chimaeras

MANIPULATING THE PRE-IMPLANTATION EMBRYO

There is great interest in EC/ and ES/ embryo chimaeras because there are distinct experimental advantages in first transforming cells *in vitro* and then using them to produce transgenic embryos. Clonal cell population can be established *in vitro* in which the copy number of the transferred DNA is known and cells can be selected for the expression of the required phenotype. Gene expression signals can be analysed initially in EC/ES cells by standard *in vitro* mutagenesis to identify key regions. These key regions can then be analysed further in transgenic mice. It is clearly faster to analyse genes in cultured cells than by making transgenic mouse lines for every mutant gene. It is anticipated that EC/ES cells will be a reliable model for gene expression in early embryo development and there are already some promising studies. For example, when EC cell line F9 is induced to differentiate by retinoic acid treatment it forms embryoid bodies where cells resemble the visceral endoderm and alpha foetoprotein (AFP) is synthesised. Scott *et al.* (1984) co-transformed undifferentiated F9 cells with a plasmid carrying an α-foetoprotein mini-gene and an HSV-*tk* plasmid and selected for HAT[R] colonies. The HAT[R] cells were induced with retinoic acid and differentiation into endoderm was accompanied by expression of the AFP mini-gene in a manner quantitatively identical to that of the endogenous α-foetoprotein gene.

One problem with the EC cell system is that chimaeras will be produced and therefore not all the cells in the animal will contain the transferred gene and as a result screening for gene expression cannot be comprehensive. The chimaera must be bred to produce non-chimaeric progeny derived entirely from the EC cell genotype. This relies on the EC cells having contributed to the germ line in the chimaera and as this occurs in less than 20% of the animals, experiments have to be somewhat large to guarantee success. EC cells are not particularly easy to transform but recently retrovirus vectors have been used with reasonable efficiency to introduce DNA into ES cells and then these cells have contributed to germ cell and somatic cell lineages in transgenic mice (Robertson *et al.*, 1986).

A typical procedure for producing transgenic chimaeric mice, based on Stewart *et al.* (1985), is shown in Figure 5.7.

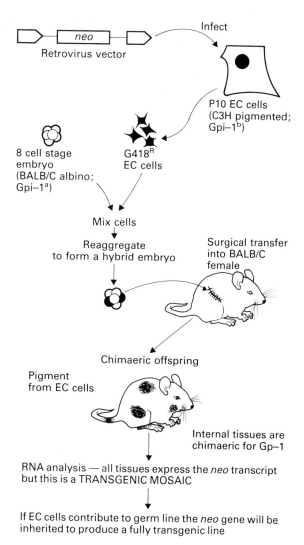

Fig. 5.7. The production of a transgenic chimaera. Genetically distinct cells from an embryo and an EC cell culture are mixed. In this case they carry different alleles of the gene for the enzyme glucose phosphate isomerase (Gpi) and different pigment markers. The hybrid embryo develops in the surrogate mother to produce chimaeric progeny whose tissues are derived from a mixture of the two cell types. If the EC cells have contributed to the germ line then the *neo* marker will be transmitted to all the progeny. Of course, any other gene that had been linked to the *neo* gene would also be transmitted.

MANIPULATING THE POST IMPLANTATION EMBRYO

The manipulation of post implantation embryos is also important for addressing the fundamental questions about the patterns of cell migration in the early stages of morphogenesis. For example, Jaenisch (1985) has shown that cultured neural crest cells can be microinjected into 9 day mouse embryos and can migrate over considerable distance and participate in normal development. Neural crest cells (NC) are the progenitors of pigment cells. When NC cells from pigmented mice are introduced into albino mice the chimaeric animals show very precise patterns of symmetrical pigmentation which can be interpreted as the result of migration along the normal pathways of neural crest cell migration. This type of approach has tremendous implications for studying cell to cell interactions. Cellular genes coding for cell surface antigens that have been implicated in cell adhesion and morphogenesis could be introduced into inappropriate cells by standard gene transfer techniques and the effects on migration and homing in the embryo can be evaluated. Alternatively, plasmids producing specific anti-sense RNAs could be transferred to block translation of the corresponding antigen RNA to determine if this perturbs normal morphogenesis (Weintraub, 1985; see Chapter 8 for a discussion of anti-sense RNA).

An appraisal of the transgenic mouse as an experimental system

The creation and analysis of transgenic mice requires considerable resources and therefore it must be clear that they provide a more valid experimental system than the more manageable cultured cells. For some gene systems this has already proved to be true, genes such as the immunoglobulin, elastase and α-foetoprotein genes appear to display their full expression potential in transgenic animals. Full expression potential means that the genes are expressed (1) predominantly in the correct tissues, (2) at the correct developmental or differentiated stage, (3) at the correct levels and (4) the expression characteristics are fully heritable traits. The majority of genes do not fulfil all these criteria in transgenic animals. In some cases there is a consistent property which probably points to a relevant biological characteristic of the gene or promoter being analysed. For example, Mt-1 promoter fusion with the human hprt gene is consistently expressed at high levels in the central nervous system which is atypical for metallothionein expression (Stout et al., 1985). It has recently been discovered that many genes including hprt contain additional expression signals within the transcribed region (see Chapter 8). This result, obtained in transgenic mice, is therefore reflecting a characteristic of the hprt gene. For other genes, however, the expression potential is variable and unpredictable. The favoured explanation is that there is a position effect, i.e. neighbouring genes or the general chromatin environment are influencing expression. It appears that there are dominance hierarchies between control signals such that certain signals can augment or suppress adjacent signals. For example, Townes et al. (1985) constructed plasmids that either contain the Mt-hGH gene adjacent to the entire human β-globin gene or on a separate plasmid and used these to produce transgenic mice. The human β-globin gene alone was expressed predominantly in erythroid tissues, the Mt-hGH gene was expressed predominantly in the liver, but when the two genes were adjacent the human β-globin gene was expressed with the same tissue specificity as the Mt-hGH gene. Clearly, if 'weak' promoters integrate near 'dominant' promoters they may be subject to different controls. One would propose that genes such as elastase and immunoglobulin contain dominant control signals that override any position effects.

It has been suggested that the failure to transmit expression characteristics also points to a fundamental characteristic of genes, namely that they contain *organiser sequences* that establish the heritability of expression (Palmiter *et al.*, 1984). As yet, there is no evidence for this concept. An interesting recent discovery is that in some cases when transferred DNA is inherited from the founder animal to subsequent generations the methylation status of the transferred gene in each generation depends upon whether it was inherited from a paternal chromosome or a maternal chromosome (Reck *et al.*, 1987; Sapienza *et al.*, 1987). This is called *genomic imprinting*. It is not shown by all transferred DNA, i.e. it may depend upon the site of insertion. This differential methylation could affect the expression of the transferred DNA. Clearly, studying genes that lack the full expression potential in transgenic mice may identify some fundamental features of the control of gene expression. Experiments aimed at elucidating the mechanism of differential gene activity in the normal animal by studying transgenic animals where expression is not quite right must, however, be interpreted with some caution. It is likely that many preliminary observations and analyses will be made in cultured cells and then hypotheses will be tested in transgenic mice. Improvements in EC cell manipulation will be beneficial in this respect. The ability to address questions about mammalian gene expression in a whole mammalian system is extremely exciting. As more data accumulates about the validity of the systems we suspect that the onus will be on other 'model systems' to justify their value.

Manipulating somatic cells

It is possible to remove bone marrow cells from an animal, genetically manipulate them *in vitro* and then return them. There are two reasons for undertaking such experiments. The first is to study haematopoiesis, i.e. the production of the lymphoid (immune) and myeloid (blood) cells in the adult animal. The second is to explore the possibility of gene therapy in man via reconstituting the bone marrow with cells that produce essential gene products.

The general hierarchy of haematopoiesis as established by histological techniques is outlined in Figure 5.8. It is proposed that there is a single pluripotent stem cell (Sp) that generates a progenitor stem cell that is committed to develop into either the lymphoid lineage (SL) or the myeloid lineage (Sm). These progenitor cells ultimately give rise to all the differentiated cell types. The differentiated cells have a finite life span and must be perpetually renewed from primitive stem cells. The process occurs in the bone marrow which therefore contains all the cell types from pluripotent stem cells to fully differentiated cells. Pluripotent stem cells are only a minor component, less than 0.01%, of the bone marrow population. They can be assayed, however, because when bone marrow cells are injected into X-irradiated mice the Sp cells establish colonies in the spleen. The cells are therefore called CFU-S (colony forming units − spleen). Other cells give rise to colonies *in vitro* in semi-solid medium such as methyl cellulose. The colonies are distinguishable and contain either mixed cells of only one lineage or a single differentiated cell type depending upon the stage of commitment (see legend to Figure 5.8).

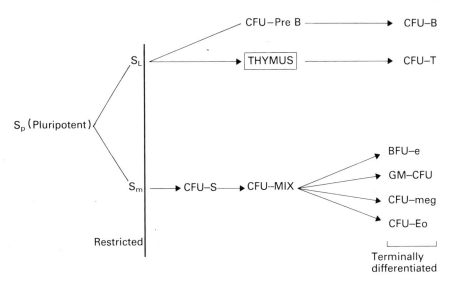

Fig. 5.8. S_p = pluripotent stem cell. S_L = committed stem cell for the lymphoid lineage. S_m = committed stem cell for the myeloid lineage. CFU = colony forming unit. BFU = burst forming unit. S = spleen. MIX = progenitor cell of the myeloid lineage capable of differentiating into any myeloid cell. B = B cells (antibody producers). T = T cells (cellular immunity). e = erythroid cells. GM = granulocytes and monocytes. meg = megakaryocytes. Eo = eosinophils.

The ability to transfer genes into Sp cells allows an analysis of their expression in cells at all stages of commitment in haematopoiesis and should provide insight into the general aspects of gene activity in mammalian development. The major technical problem is how to ensure gene transfer to cells that are so poorly represented in a mixed cell population. As there are only about five CFU-S in 10^5 cells, the transformation frequency must be high enough to ensure a reasonable probability that one of these CFU-S cells will be transformed. This has been achieved by exploiting the retrovirus vectors that we described in Chapter 4.

The general approaches for manipulating haematopoietic cells with retroviruses were established by Joyner *et al.* (1983); Miller *et al.* (1984b) and Williams *et al.* (1984) and more recently the techniques have been improved (Dick *et al.*, 1985; Eglitis *et al.*, 1985; Keller *et al.*, 1985). A typical procedure is shown in Figure 5.9. First a retrovirus producer cell line is created. For example, ψ2 cells are transformed with a retroviral vector (e.g. Figure 4.27) and G418^R colonies are isolated. These are screened for virus production, because there is always some variation, and a cell line that consistently produces a high titre of recombinant retrovirus is chosen. It is important to use a virus stock that has more than 10^4 virus ml^{-1} to ensure efficient infection of the bone marrow cells. The producer cells are seeded into a flask and allowed to reach sub-confluence. The medium is then changed and 24 hour later about 5 x 10^6 bone marrow cells are introduced into the flask. The bone marrow cells are obtained from the femurs of mice that have been treated for 2−5 days previously with 5−fluorouracil (5−FU) to enrich for primitive precursor cells (the 5-FU is toxic for proliferating

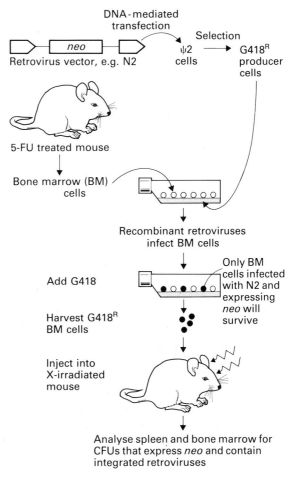

DNA-mediated
transfection

Selection

neo

Retrovirus vector, e.g. N2   ψ2   →   G418^R
cells   producer
cells

5-FU treated mouse

Bone marrow (BM)
cells

Recombinant retroviruses
infect BM cells

Add G418

Only BM
cells infected
with N2 and
expressing
neo will
survive

Harvest G418^R
BM cells

Inject into
X-irradiated
mouse

Analyse spleen and bone marrow for
CFUs that express *neo* and contain
integrated retroviruses

Fig. 5.9. Manipulating pluripotent haematopoietic stem cells.

cells but spares non-cycling stem cells). After co-cultivating the bone marrow
cells with the retrovirus producer cells for 24 hr to allow infection, G418 is
added to the medium to pre-select for bone marrow cells that contain the
neo gene. After 48 hr the non adherent bone marrow cells are easily collected
and they are injected into mice that have been X-irradiated to destroy all
endogenous bone marrow cells. The spleens of the reconstituted mice are
examined for CFU-S any time from 12 days after innoculation and for
G418^R colony forming cells *in vitro*.

These studies have shown that up to 85% of the primitive stem cells can
be transformed with retrovirus vectors. These produce colonies which contain
G418^R cells in the spleen. The CFU-S that are present at 6–17 weeks after
innoculation have been introduced into secondary recipients and can reconstitute the entire haematopoietic system. This is demonstrated by the fact
that haematopoietic cells of all lineages isolated from these secondary recipients are resistant to G418, showing that cells with the capacity for perpetual
renewal must have been transformed.

Because retroviruses integrate at only one site and their integrated structure is precise and stably maintained they act as a clonal marker. All cells that have been derived from the same transformed precursor will contain a unique restriction fragment containing the retrovirus genome. Appropriate Southern blot analysis of the different colonies will reveal the relationship between the lineages. Studies have now confirmed that cells with distinct lymphoid and myeloid phenotypes arise from a common precursor (Lemischka et al., 1986). In addition, because the retroviruses can also integrate into cells at later stages of differentiation, unique restriction fragments can also be followed to identify the potential for trans-differentiation at different stages of commitment. The life span and capacity for self renewal of each of the cell types can also be studied. It is also possible to study cell lineages during development (Soriano and Jaenisch, 1986; Palmiter et al., 1987).

These studies also show that gene therapy via transforming bone marrow cells should be possible because the transformed cells are capable of self renewal. Retroviral vectors can be used to introduce non-selected genes into pluripotent stem cells from human bone marrow (This is discussed in more detail in Chapter 11).

Gene manipulation in *Drosophila*

The fruit fly *Drosophila melanogaster* is an extremely important organism for studying eukaryotic gene expression. It has a short life cycle of only two weeks from fertilised egg to fertile adult, making it suitable for extensive genetic analysis. In addition the precise banding patterns in the polytene chromosomes of the salivary gland have allowed a detailed cytogenetic description of the genome. The genetic map of *Drosophila* is the most comprehensive of any eukaryote, so that genes involved in most of the fundamental biological processes have been identified. *Drosophila* is particularly important for studying early embryonic development. This is because this occurs entirely outside the animal and takes only 22 hr. A large number of eggs can be screened for early embryonic lethal mutations and a number of key genes involved in development have been identified. Such mutations in mammalian embryos would be aborted before the investigator could identify them.

Because of the key discoveries about embryogenesis that are being made with *Drosophila*, we will devote some space to a simplified discussion of *Drosophila* development. We will then discuss the methods of gene transfer into *Drosophila* embryos. This was made possible by two developments. First, the cloning of *Drosophila* genes that could act as selectable markers and second, the manipulation of a *Drosophila* transposon, the P-element, to increase the efficiency of stable integration of transferred DNA.

A guide to *Drosophila* development

The body plan of an insect is made up of a set of tandemly arranged segments each of which develops into the specialised structures of the adult. The number of segments and their specialised potential are established very early in embryogenesis; this is the *pre-pattern* for development. Although there is no apparent correlation between the segmental pattern of the insect body plan and the mammalian body plan which has repeating structure called somites (Hogan, 1985), recent results suggest that the same fundamental controls of differential gene expression may be operating in all eukaryotes (Ruddle *et al.*, 1985).

An outline of the development of *Drosophila* is shown in Figure 5.10. Eggs are laid shortly after fertilisation, the fertilised egg is asymmetric in the distribution of cytoplasm and this is controlled by maternal genes active in the ovaries. In particular a region at the posterior end called the pole plasm produces the germ line precursor or pole cells. The male and female pronuclei fuse to form the zygotic nucleus which undergoes a series of divisions but no cell membranes are formed. The early embryo is therefore a syncitium of about 6000 nuclei. At about 1.5 hr after egg laying (AEL) after the ninth nuclear division, pole cells are formed and the nuclei migrate to the periphery where cell membranes develop to form the cellular blastoderm stage. At this point the pattern of subsequent development has been established. Cells in a specific location within the embryo will develop into specific lineages that form specialised structures in the adult. At 2.5 hr AEL, this pre-pattern begins to be 'interpreted' via a series of cellular movements beginning with gastrulation and by 8–10 hr the segments are visible. The precise number of segments is not known because some segments in the head and terminal abdomens are difficult to distinguish. There are at least three head segments, three thoracic segments and 8–10 abdominal segments. By about 12 hr the head and terminal abdominal segments have undergone some re-organisation to produce specialised structures such as spiracles and sense organs. The larva hatches at 22 hr.

The larvae undergo three moults as they increase in size; the stages are called first, second and third *larval instars*. The third instar pupariates and undergoes metamorphosis to produce the adult fly. The surface structures of the adult, that is, the cuticle and the appendages (antennae, legs, wings, etc.) are all of epidermal origin and they are derived from structures called *imaginal discs*. These are clusters of about 40 cells laid down in early embryogenesis which expand to about 40 000 cells during the larval phase. Pairs of imaginal discs develop into a specific specialised adult body segment and appendage. Each segment in the larva becomes the corresponding segment in the adult fly. The segments are further divided into compartments, the most significant of these is the anterior–posterior compartments. The imaginal discs are also compartmentalised and the commitment of a cell lineage to a specific compartment occurs very early in embryonic development. Within a compartment however the cell lineages are not restricted, that is to say the same cuticular structures such as bristles can be formed from different cell lineages.

Many genes have now been identified which when mutated cause disruptions in the number of segments, or the identity of segments, or the compartmentalisation of segments. These are called *homeotic genes*. Some of the mutants produce bizarre adult flies. For example, defects in the gene *antennapoeadia* cause the antennae to be replaced by a pair of middle legs.

GENE TRANSFER INTO WHOLE ANIMALS

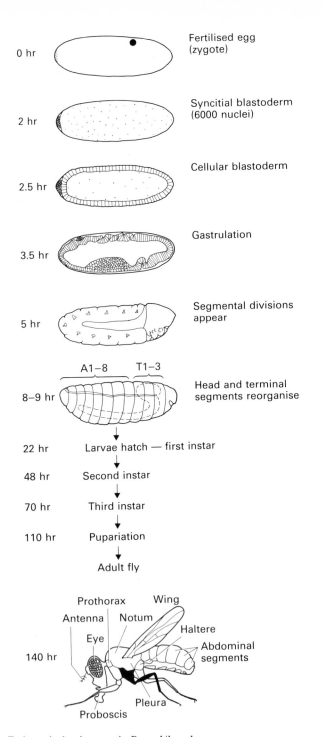

0 hr — Fertilised egg (zygote)

2 hr — Syncitial blastoderm (6000 nuclei)

2.5 hr — Cellular blastoderm

3.5 hr — Gastrulation

5 hr — Segmental divisions appear

8–9 hr — Head and terminal segments reorganise

A1–8   T1–3

22 hr — Larvae hatch — first instar

48 hr — Second instar

70 hr — Third instar

110 hr — Pupariation

Adult fly

140 hr

Prothorax Wing Antenna Notum Eye Haltere Abdominal segments Pleura Proboscis

Fig. 5.10. Embryonic development in *Drosophila melanogaster*.

The working idea is that the expression of a particular homeotic gene in a cell will determine how it responds to positional signals. There is a hierarchy of homeotic genes that probably control a number of other genes to establish

176   CHAPTER 5

both the segment identity and the ability of the imaginal disc to respond correctly to the segment identity.

Three developmentally important loci have been identified each containing several genes: these are the *bithorax* complex, BX-C, the *antennapaedia* complex, ANT-C, and the *engrailed* complex, EN-C. The genetic organisation of these complexes and the way some of the genes influence segmentation are shown in Figure 5.11. EN-C determines the anterior/posterior decision, *engrailed* mutants lack posterior compartments. Many of the genes of the BX-C and ANT-C complexes are expressed in different segments. They function to suppress the structures characteristic of more anterior segments and to promote the development of appropriate posterior structures. Thus mutations resulting in loss of function transform one or more segments to resemble a more anterior segment. For example, BX-C determines the identity of all segments posterior to T2, the ground state of development of these segments is T2 and the expression of individual BX-C genes 'advances' the state to T3, Al, etc.; mutants 'revert' to the previous state of development. Similarly, ANT-C genes control segment identity in the head and anterior thorax. Some genes in the complexes only affect one or a few segments, others such as *Ftz* affect all segments, mutants in *Ftz* lose the segment boundaries to create a larva with fushi tarazu ('not enough segments' in Japanese). Several of the genes in the complex contain a 180 bp sequence in their coding regions that is highly conserved (Scott and Weiner, 1984). This is called the *homeo box*. It has the characteristic helix-turn-helix structure and high content of basic amino acids typical of DNA binding proteins. It is proposed that the genes containing homeo boxes are the controlling genes that regulate other genes in the complex. It may be significant that the genes that affect specific segments lie in the same order on the physical map as the segments in the fly.

Recently there has been the exciting discovery that some genes in other organisms, man, mouse, *Xenopus laevis* and *S. cerevisiae* (Shepherd *et al.*, 1984) contain homeo box regions. The mouse homeo box genes are summarised in the Table 5.4. The important points are that the genes are developmentally regulated, expression being maximal between 10–17 days of gestation (Colberg-Poley *et al.*, 1985; Hart *et al.*, 1985; Hauser *et al.*, 1985); the homeo boxes are clustered suggesting that they may be complex loci; in some cases they map in regions that have been implicated by classical genetic analysis as important in development and in the case of Mo-en-1 there is homology outside of the homeo box with *Drosophila engrailed*. These data suggest that understanding the functions of genes with homeo boxes will give insight into fundamental aspects of development (reviewed by Gehring, 1985; Manley and Levine, 1985; Ruddle, 1985; Gehring, 1987; Martin, 1987).

DNA transfer into *Drosophila* embryos

The key developments that led to efficient gene transfer into *Drosophila* embryos were (1) the isolation and characterisation of transposons called P

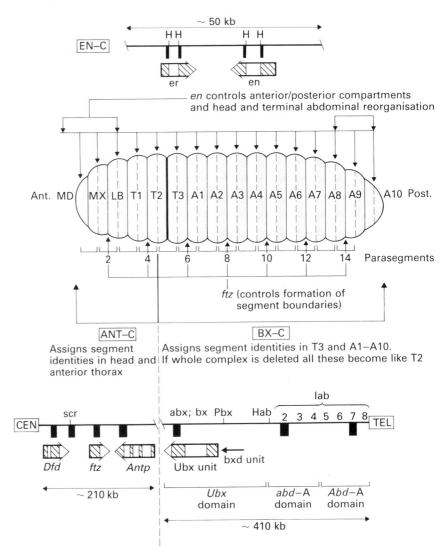

Fig. 5.11. The molecular genetics of development in *Drosophila*. The figure shows a representation of the larva with segment boundaries shown as solid vertical lines and parasegment boundaries shown as dotted lines. The segments are MD = mandibular, MX = metathorax, LB = labial, T1 – T3 = thoracic and A1 – A3 = abdominal. The major homeotic gene complexes are shown, these are engrailed (EN-C) antennapedia (ANT-C) and bithorax (BX-C). An outline transcript map of the complexes is also shown with some of the key genes indicated. The homeo boxes are shaded dark and the coding regions are hatched.

elements and 2 the cloning of *Drosophila* genes such as *ry* (xanthine dehydrogenase, XDH) that could be used as transformation markers.

P-elements and gene transfer vectors

P elements are a class of transposons found only in *Drosophila*. They are present in 30−50 copies per genome and both full length and partial elements are found. Only certain strains, called P strains, carry these

Table 5.4. Homeo box genes in mice.

Mouse		
Hox-1.1		
Hox-1.2		
Hox-1.3	Chromosome 6	
Hox-1.4		
Hox-1.5		
Hox-1.6		
Hox-1.1		
Hox-2.2		
Hox-2.3	Chromosome 11	
Hox-2.4		
Hox-2.5		
Hox-2.6		
Hox-3	Chromosome 15	
Hox-4	Chromosome 12	
En-1	Chromosome 1	Homology with *Drosophila*
En-2	Chromosome 5	*engrailed* gene

elements. Other strains that lack the elements are called M strains. When a P male is mated with an M female the P elements are activated to transpose. This occurs only in the germ-line cells of the resulting embryo and has a number of genetic consequences. Insertions of P elements cause mutations, and facilitates recombination and rearrangements. In addition, any defective P elements in the M strain are activated to excise and this may also be mutagenic. The embryos develop into F1 flies that contain many mutations and they are often sterile, a phenomenon called P–M hybrid dysgenesis (reviewed by Engels, 1983). The simplest explanation is that P elements normally repress their own transposition but that dilution of the repressor in the P/M zygote allows the element encoded transposase to function; the M strain lacks any transposase or repressor because it contains no P elements or only defective elements.

A number of P-elements have been cloned and sequenced (O'Hare and Rubin, 1983; Figure 5.12A). The full length element is 2.9 kb and it contains four open reading frames (ORFS) in different phases on the same strand. The element is expressed in germ-line cells via a single transcription unit that has three splice points. The translation product is a fusion of all four ORFs to produce an 87 kD protein that is the putative transposase. The third splice point is only processed in germ cells and not in somatic cells which accounts for the tissue specificity of transposition (Karess and Rubin, 1984; Laski *et al.*, 1986; Rio *et al.*, 1986). At the ends of the element there are inverted repeats of 31 bp, the shorter elements all retain these inverted repeats but lack parts of the coding region.

Spradling and Rubin (1983a) were able to produce hybrid dysgenesis by simply injecting an intact P-element cloned into pBR322 into an M/M embryo. They used a genetic trick to facilitate their assay for successful P-element transfer. The M strain carried two defective P elements in the *singed* locus producing the signed weak (Sn^w) phenotype. In a P/M mating this locus is highly mutable generating $Sn+$ or a more extreme singed

GENE TRANSFER INTO WHOLE ANIMALS

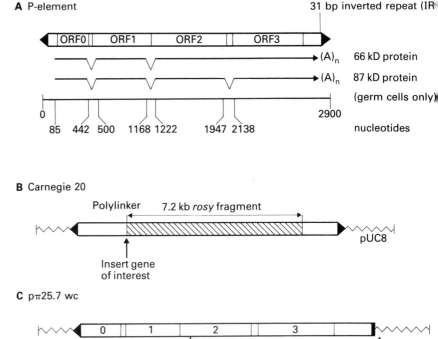

A P-element

31 bp inverted repeat (IR)

| ORF0 | ORF1 | ORF2 | ORF3 |

→ (A)$_n$ 66 kD protein

→ (A)$_n$ 87 kD protein

(germ cells only)

0   2900

85   442   500   1168 1222   1947 2138   nucleotides

B Carnegie 20

Polylinker   7.2 kb *rosy* fragment

pUC8

Insert gene
of interest

C pπ25.7 wc

| 0 | 1 | 2 | 3 |

Produces 87 kD
transposase

Last 23 bp of IR is lost

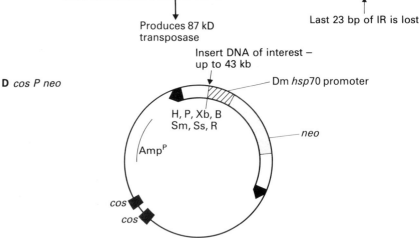

Insert DNA of interest –
up to 43 kb

D *cos P neo*

Dm *hsp*70 promoter

H, P, Xb, B
Sm, Ss, R

neo

Amp^P

cos
cos

Fig. 5.12. P-elements and their derivatives as vectors. A, P-element (Laski *et al.*, 1986). B, A
defective P-element carrying the *rosy* selectable master (Rubin and Sprading,
1983b). C, A transposition ('wings clipped') helper P-element (Karess and Rubin,
1984). D, P-element cosmid vector (Steller and Pirrotta, 1985).

phenotype Sn^e due to P-element excision. The progeny of the microinjected
Sn^w/Sn^w embryos were examined and the presence of $Sn+$ or Sn^e was
evidence that the injected P-element was functional. New P-elements were
found in the chromosomes, whereas no pBR322 sequences were detected.
This showed that the P-element had integrated by transposition rather than
by recombination. This study confirmed that P-elements could be stably
integrated into the chromosomes following microinjection and opened the

way for the development of P-element vectors (Spradling and Rubin, 1983b; Rubin and Spradling 1983a,b).

A simple vector, Carnegie 20, is shown in Figure 5.12B. Most of the P-element coding region has been replaced with a 7.2 kb fragment containing the gene for XDH (rosy, ry). In addition, there is a poly-linker for introducing additional DNA and the vector is genetically designated p [ry], the square brackets denoting the genetic locus that is being transferred. This molecule is clearly defective and will only transpose if a helper element is co-injected. The most useful type of helper is one that will provide the transposase but will not itself transpose. This prevents extreme hybrid dysgenesis from occurring in the resulting transgenic *Drosophila*. Such helpers are referred to as *wings clipped* because they lack the terminus of one of the inverted repeats so that they are not a substrate for the transposase but they have the full coding capacity. For example pπ25.7wc (Karess and Rubin, 1984; Figure 5.12C) lacks the last 23 bp of the inverted repeat.

The *rosy* gene is particularly useful as a scorable marker for successful transformation because it can be detected in flies that are homozygous for a rosy mutation (ry$^-$). This is because XDH is involved in producing eye pigment and ry$^-$/ry$^-$ flies have brown eyes whereas wild-type eyes are brick red. Very low levels (1% of wild-type) of XDH produced anywhere in the fly are sufficient to produce wild-type eye colour. The progeny of transformed embryos can therefore be picked out from untransformed embryos even if the expression of the transformed DNA is very poor. In the early studies with this system transformant offspring were produced in less than 0.5% of the total progeny of an injected fly. If all the progeny had to be screened by Southern hybridisation, success may have been missed, and so the ability to see the change in eye colour was very important.

There appears to be no limit on the size of DNA that can be inserted in a P-element vector. Recently Haenlin *et al.* (1985) have introduced a 43 kb region of the genome containing the *fs* (1) k10 morphogenetic locus into embryos. The efficiency of transformation was however reduced. This group have also constructed a combination cosmid−P−element vector (Steller and Pirotta, 1985) to facilitate both cloning large DNA fragments and then screening them for biological activity by transferring them into *Drosophila* (Figure 5.12D). The use of P elements to facilitate gene isolation and identification is discussed in Chapter 7.

The *rosy* marker allows transformed flies to be detected but not selected, i.e., large numbers of flies may have to be screened to find the transformants. There are now two selectable markers that allow only transformed flies to survive. One of these is the *Drosophila Adh* (alcohol dehydrogenase) gene (Goldberg *et al.*, 1983). Flies which are *Adh$^-$* are killed by 6% ethanol and a single copy of the wild-type *Adh* gene is sufficient to confer resistance. Flies transformed with P-elements carrying the *Adh* gene can be selected by growing them in medium containing ethanol. The bacterial neomycin resistance gene (*neo*) can also be used, the expression is driven with a *Drosophila* promoter such as the heat shock promoter from *hsp70*. Transformants are selected by feeding larvae with food containing 1 mg ml^{-1} of G418 which is normally toxic (Steller and Pirotta, 1985).

The advantage of using these selectable markers is that flies do not have to be inspected individually to detect transformants. The disadvantage is that some transformation events may be missed, if the levels of expression are too low to confer resistance. This could bias the analysis particularly in questions about position effects on expression. The other disadvantage is that in any experiment to construct a transgenic strain the site of integration must be analysed and suitable transgenic stocks established. These genetic analyses are easier with a visible marker to score.

THE PRODUCTION OF TRANSGENIC *DROSOPHILA*

A typical protocol is shown in Figure 5.13. The P-element vector (e.g. p[ry]) and helper element are injected into an appropriate genetically marked embryo, in this case an ry^- embryo. The embryo must be of the M type to allow transposition of the injected P-element. The embryos are left to hatch and develop into adult flies. In these G0 flies, the P-element vector will have integrated into the chromosomes only in the germ line cells. These cells will be the only ones expressing the transferred gene and therefore the G0 flies may not display the transformed phenotype. Sufficient XDH may, however, be produced to affect eye colour detectably when the *ry* marker is used. If the *Adh* or *neo* markers are used selection is not imposed on the G0 flies as insufficient gene product is being made. The germ line of the G0 flies may also be mosaic. There are 30–50 pole cells that are the germ line precursors and not all these may contain a P-element. The next step is, therefore, to mate these G0 flies to produce Gl offspring. Only those Gl progeny that are derived from P element transformed germ cells will be transgenic. Usually only 10% of the G0 offspring are transgenic and it can be as low as 1%. Selection is imposed at the Gl stage if the appropriate P element has been used. A stock of transformed flies is obtained by appropriate genetic crosses with parent strains or siblings and the site of insertion is mapped genetically by *in situ* hybridisation to polytene chromosomes. The production of transgenic *Drosophila* can be rather inefficient as there are losses at each stage, the viability of the embryos increasing with the skill of the experimenter, but there are still problems of infertility in the G0 flies (Karess, 1985).

THE STATUS AND EXPRESSION OF GENES TRANSFERRED BY P-ELEMENTS

Following microinjection, the P-element vector transposes from the plasmid using the transposase provided in *trans* by the helper element. Usually only a single element becomes integrated but where multiple insertions occur these can be separated by standard genetic crosses to establish transgenic lines with single insertions. Spradling and Rubin (1983b) have analysed the integration, inheritence and expression of p[ry⁻]. They established 36 transformed lines. The sites of integration were very diverse but only one insertion was found in heterochromatin. About 30% of the insertions were associated with recessive lethality indicating that the P-element vectors act

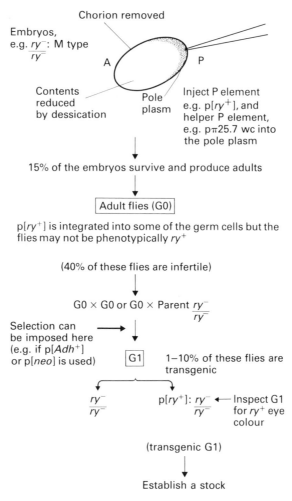

Fig. 5.13. The production of transgenic *Drosophila*.

as insertional mutagens. There was no evidence for any rearrangements in the transferred P-element vector. In the absence of any complete P-elements in the genome, the transferred defective P element is genetically stable throughout successive generations. The transferred XDH gene was expressed in all the 36 transformants. Furthermore the enzyme was expressed in the same tissues as wild-type XDH, that is, in the adult Malpighian tubules and larval fat body cells. The position of the insertion had some effect as the specific activity of XDH ranged from 30 to 130% of wild-type levels. In this study 20/36 showed no influence of position, six had elevated activity and ten had reduced activity.

One case of a biologically relevant position effect was the fact that *rosy* genes inserted into the X chromosome showed dosage compensation in males. The expression of the gene was approximately doubled. Occasional instances of extreme position effects on expression have been detected. For example 10% of flies transformed with the *white* locus did not display wild-type eye colour. Several of these insertions were, however, close to either the telomeres or centromeres and so expression may have been reduced [by] the proximity to heterochromatin (Hazelrigg *et al.*, 1984). Nevertheles[s] several studies have shown that most genes are expressed correctly wh[en] introduced into *Drosophila* by P-element mediated transformation. For e[x]ample, the transformed dopa decarboxylase (*Ddc*) is under the same tempo[ral] control of expression as the wild-type gene, i.e. it is expressed only duri[ng] times of cuticle synthesis and predominantly in the hypodermal tiss[ue] (Scholnick *et al.*, 1983). An 11.8 kb chromosomal fragment containing t[he] *Adh* gene is expressed only in the correct tissues, i.e. larval and adult [fat] body and mid-gut. The levels of ADH are also correctly modulated betwe[en] tissues with high specific activities in the larval fat body and lower activiti[es] in the larval mid-gut. Finally, the *Adh* gene has two promoters; one is use[d] primarily in larval tissues and the other in adult tissues. The correct dev[el]opmental switch in promoter use also occurs with the transferred ge[ne] (Goldberg *et al.*, 1983). Tissue specific regulation of yolk protein genes h[as] also been demonstrated and the control signals are being identified using t[he] P element system (e.g. Garabedian *et al.*, 1986). P elements are also bei[ng] used to study specific gene amplification (see Chapter 9).

An appraisal of the P-element system

The P-elements have a number of advantages as gene transfer vectors: th[ey] ensure efficient integration and they do not rearrange so the transferr[ed] gene is always in a defined environment. Fly strains with single site insertio[ns] are readily produced so there are no gene dosage problems and large DN[A] fragments up to 50 kb can be transferred. The transferred genes also sh[ow] accurate tissue and developmental stage specific expression and only min[or] variations in the absolute levels of expression. A further advantage is th[at] the transferred P-elements can be readily mobilised simply by introducing [a] transposase producing P-element into a transgenic embryo. This allo[ws] many different chromosomal positions to be tested if necessary.

One interesting idea is to adapt the P-element system to function [in] mammalian cells (suggested by Rio *et al.*, 1986). This is possible becau[se] the transposase can be expressed under the control of a mammalian promo[ter] and the 31 bp inverted repeats may be all that is required for efficie[nt] integration of any DNA that they flank. This system would allow very lar[ge] fragments of chromosomes to be integrated which may reduce the positi[on] effects that are observed in mammalian cells.

Gene manipulation in *Xenopus laevis*

The African clawed toad (a frog), *Xenopus laevis* has been a favouri[te] organism for developmental biologists for many years. This is because t[he] fertilised egg is large, 1–2 mm and embryogenesis occurs entirely outsi[de] the body of the female and takes only 48 hr from egg laying to larv[al] hatching. The egg is therefore amenable to microdissection and interferen[ce] to follow the fates of cells during embryogenesis. Some key experimen[ts] that demonstrated genetic pluripotency of the somatic nucleus involv[ed] transplanting the nuclei from intestinal cells from tadpoles into enucleated embryos and showing that adult frogs were produced (Gurdon, 1974). In addition *Xenopus* oocytes and unfertilised eggs were the first eukaryotic cells to be manipulated by introducing foreign DNA (e.g. Mertz and Gurdon, 1977) and the analysis of *Xenopus* 5S RNA expression in *Xenopus* oocytes produced the first definition of a eukaryotic promoter (Brown, 1982).

An outline of *Xenopus* development is shown in Figure 5.14. Normal development is shown on the right and four stages that are amenable to manipulation are shown on the left (reviewed by Gurdon and Melton, 1981). Oocytes can be dissected from the ovaries of adult females in large numbers and the mature oocytes are recognised by their size and asymmetric pigmentation. DNA can be injected into the germinal vesicle (nucleus) which although not readily visible is large enough (400 μm) to make this a fairly reliable procedure (Figure 5.14A). The oocytes can be maintained *in vitro* for several days and they continue to transcribe the injected DNA. RNA can also be introduced into the cytoplasm and post-transcriptional processing and translation can be studied. The oocytes can be treated with progesterone *in vitro* and they will develop into eggs where the nuclear membrane ruptures and meiosis proceeds to the second metaphase. Alternatively eggs can be collected after laying naturally or following induction of ovulation by injecting females with chorionic gonadotrophin. Following removal of the protective jelly, eggs can be injected (Figure 5.14B). The procedure is easy because the nuclear contents are now dispersed throughout the cell. Transcription and translation can be studied at this stage.

Transcription is usually more reproducible in eggs than in oocytes (Mertz and Gurdon, 1977). Unfertilised eggs are also used to analyse DNA replication whereas DNA injected into the germinal vesicle of oocytes does not replicate. If the eggs are pricked with a fine needle to mimic fertilisation (referred to as activation) replication increases. The eggs are often UV-irradiated to reduce any endogenous DNA synthesis directed by the female chromosomes (e.g. Harland and Laskey, 1980). The disadvantages of using eggs to study transcription and translation is that in the absence of fertilisation they rapidly deteriorate, whereas oocytes can be maintained for longer periods *in vitro*. Eggs can be fertilised *in vitro* and then injected with DNA to attempt to produce transgenic frogs (Figure 5.14C). For example, Bendig and Williams (1983) have injected cloned *Xenopus* adult α1 and β1-globin genes into fertilised eggs. The DNA was rapidly integrated and it replicated and persisted through to the tadpole stage. The genes were not, however, correctly expressed. Adult globin was produced at low levels throughout embryogenesis whereas it is normally not expressed until metamorphosis.

It is also possible to produce clones of transgenic frogs very rapidly in this system as shown by Etkin and Roberts (1983). Foreign DNA is injected into fertilised eggs and then gastrula cells which are pluripotent are dissected and the nuclei are transplanted into enucleated eggs which then develop into identical cloned frogs (Figure 5.14D). It is not yet clear whether the expression of transferred genes in transgenic frogs will mimic any aspects of normal development.

GENE TRANSFER INTO WHOLE ANIMALS

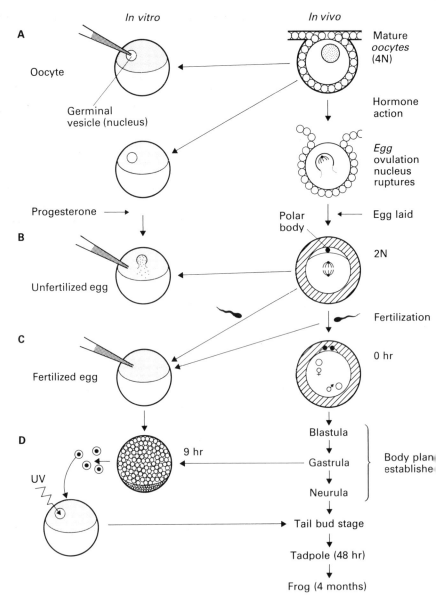

Fig. 5.14. Gene transfer in *Xenopus laevis*.

One problem with *Xenopus* as a model organism is that there is no good genetic analysis. This is largely because the life cycle is long and it takes at least four months to produce an adult frog. This may restrict the usefulness of this system for studying gene expression during all stages of development. It is, however, possible to study early development, e.g. Krieg and Melton (1987) have studied the expression of genes expressed in early *Xenopus* embryos by injecting them into fertilised eggs and analysing transcription at the early gastrula stage.

Caenorhabditis elegans: a model for studying lineage restrictions

The nematode *C. elegans* provides a contrast to mouse and *Drosophila* development because it develops from a small number of cells with a rigid cell lineage that is established very early in development. The lineage of all the adult organs has been discribed precisely and there is no variation between animals (Sulston *et al.*, 1983). Specific genes that specify cell fates, e.g. *lin*−12, are being identified (Greenwald *et al.*, 1983) and recently a gene transfer system has been established which will increase the power of genetic analysis of cell fate determination (Stinchcomb *et al.*, 1985a). DNA is introduced directly into the gonads of the nematodes which are then allowed to lay eggs. The progeny are propagated for two generations before analysis. About 10% of the injected worms give rise to progeny. Injected supercoiled molecules form high molecular weight tandem head to tail arrays and linear molecules form head to head and head to tail arrays. These high-molecular weight arrays remain extra-chromosomal and are heritable. About 50% of the transformed animals carry DNA in all their cells and the remainder are mosaics. The transferred DNA is therefore reasonably stable and preliminary evidence indicates that foreign genes are expressed. Stability might be improved by incorporating segregator sequences that were isolated from *C. elegans* DNA by assaying for CEN activities in *S. cerevisiae* (Stinchcomb *et al.*, 1985b; Fire, 1986). Analysis of development in trans-genic nematodes may provide definitive evidence for the importance of differential gene activity in development. Furthermore, it may identify the mechanisms that lead to the restriction in developmental potential that results in specific cell lineages.

Summary

DNA can be introduced directly into fertilised eggs of mice by microinjection or into embryos by retroviral infection. The DNA is usually inherited by any progeny and in some cases is expressed in a tissue-specific and developmentally stage specific manner. Retroviruses are less reliable for allowing expression of transferred genes than simple plasmid injection. The factors that affect expression of transferred DNA are not fully characterised but the site of chromosomal integration has a marked effect. This is less apparent with genes such as immunoglobulin genes that carry strong tissue specific enhancer elements. Transgenic mouse lines are very important tools for studying the control of gene expression but they are still technically and logistically difficult to generate for all but the best equipped laboratories. A second approach is to transfer DNA into pluripotent teratocarcinoma cells in culture and then mix these with normal embryo cells to produce transgenic chimaeras. DNA can also be transferred to mouse haematopoeitic stem cells. These are then used to reconstitute the stem cell population in adult mice.

Gene manipulation in *Drosophila* is now highly sophisticated using vectors based on the P-element transposon and the transferred genes rarely display position effects.

Organisms such as *X. laevis* and *C. elegans*, which are key model systems for developmental processes, can also be genetically manipulated.

6 Gene Transfer into Plants

Introduction

The potential applications of gene manipulation in plants are enormous, ranging from managing cereal crops to ensure adequate world food supplies to producing new horticultural varieties for pleasure. Plant breeding techniques involving sexual and vegetative propagation are well established, the aim being to introduce genetic diversity into plant populations, to select 'superior' plants carrying genes for desired traits and to maintain the range of plant varieties. The application of these classical techniques has produced significant achievements such as yield improvements in the major crops, wheat and maize. However, this takes a long time. It usually takes, for example, 4–8 years to produce a new variety of wheat by sexual propagation (reviewed by Borlaug, 1983). The addition of molecular biology and gene transfer techniques to the traditional techniques will increase their versatility and accelerate the production of new plant varieties (Cocking and Davey, 1987). Also, the gene pool available for traditional plant breeding is in fact diminishing because fewer plant species and varieties are being cultivated. Only eight species of cereals supply about 50% of the total world food calories. Gene manipulation technology will allow genes to cross between species and will therefore contribute to increasing genetic diversity in the world plant resource (Barton and Brill, 1983).

Gene manipulation in plants has lagged behind the eukaryotic systems that we have discussed previously. It is only recently that significant numbers of plant genes have been analysed at the molecular level and therefore transformation markers and expression cassettes have not been readily available. In addition, techniques for the genetic manipulation of plant cells and tissues *in vitro* were still only being developed in the late 1970s. The initial development of gene transfer techniques in plants relied upon the understanding and exploitation of a natural plant-pathogenic bacterium, *Agrobacterium tumefaciens*, which introduces plasmid DNA into the plant genome. Agrobacteria infect dicotyledonous plants (dicots) and for this reason, together with the fact that there are many 'dicots' that are easy to culture *in vitro* from single cells, most molecular genetic research has focused on dicots. The major food crops are however all monocotyledons (monocots), which are not readily infected by *Agrobacterium* and which on the whole cannot be cultured *in vitro* from single cells. Some of the more agriculturally important dicots such as the legumes are also difficult to culture *in vitro* from single cells. Much research is aimed at manipulating these agriculturally important plants and novel gene transfer methods and vectors are being developed. In this chapter we give an outline of the goals of gene manipulation in plants and then give a brief review of *in vitro* techniques for culturing plant tissues and cells. We then discuss gene transfer systems for plant cells and organelles. Additional details about traditional and molecular genetic analysis of plants can be found in Mantell *et al.* (1985) and Barton and Brill (1983). Plant genome structure and function is reviewed by Sorensen (1984) and Heidecker and Messing (1986) and plant gene expression by Kuhlemeier *et al.* (1987).

Some targets of gene manipulation in plants

A number of clear goals of gene manipulation in plants have been defined but the time-scale for achieving some of these may be tens of years. This is largely because so little is known about the fundamental biochemistry and genetic basis of many of the characteristics to be modified.

Plants are a major component of the human diet but the nutritional qualities of plants are sub-optimal. Also there are many areas of the world where environmental conditions are unfavourable for cultivation and the crop yields represent less than 5% conversion of solar energy into plant biomass. Much research effort is therefore aimed at improving plant productivity by manipulating photosynthesis, nitrogen fixation, seed storage proteins and herbicide and disease resistance. Plants are also a source of a large number of chemicals and biochemicals. In this section we will outline some of the targets of plant gene manipulation and comment on the potential problems in achieving these aims. In Chapter 12 we will discuss some of the successes.

Photosynthesis

The reactions of photosynthesis occur in the chloroplast which contain a number of pigment molecules which can readily absorb photons. The major pigments are the chlorophylls, a and b, and there are accessory pigments called *carotenoids*. The pigment molecules are found in complexes, with various binding proteins, in specialised structures called *thylakoid membranes*. The pigments form clusters called *photosystems* within which the majority of the chlorophyll is complexed with proteins called chlorophyll a/b (Cab) proteins. All the pigments absorb photons in a process known as *light harvesting* and each photosystem contains a pigment/protein complex known as the *photochemical reaction centre or photoreceptor*. The best characterised photoreceptor is *phytochrome* which absorbs red light. There are two different photosystems. Photosystem I has a high chlorophyll a: chlorophyll b ratio and is maximally excited at relatively long wavelengths (700 nm, red light). Photosystem II has more chlorophyll b and is activated at wavelengths below 680 nm. The two photosystems form the *light harvesting complex* and cooperate in the generation of reducing power from light. When plants are illuminated the chlorophylls become excited and this 'excited state' is transmitted between pigment molecules until the photochemical reaction centre is activated and releases an electron. The 'energy-rich' electron is passed down a chain of electron carriers to generate NADPH and ATP. The electron donated by the photochemical reaction centre of photosystem I is replaced by an electron derived from the light activation of photosystem II, which in turn is replenished by an electron derived from the light mediated cleavage of water. The initial phase of photosynthesis, therefore, generates oxygen in addition to NADPH and ATP.

The next stage is *carbon dioxide fixation* which is the synthesis of glucose from atmospheric carbon dioxide and ribulose 1−5 diphosphate via the *Calvin cycle* of reactions using the photochemically generated energy (ATP) and reducing power (NADPH). The first enzyme in carbon dioxide fixation is *ribulose bisphosphate carboxylase (rubisco)* and this is extremely inefficient and rate limiting. This is because it has a low affinity for CO_2 and because it can also catalyse an entirely separate reaction. This is the interaction of oxygen with ribulose 1−5 diphosphate to produce glyoxalate in a process

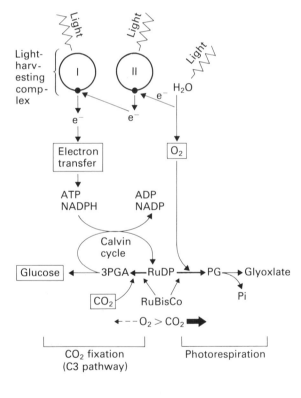

RuBisCo = Ribulose 1−5 bisphosphate carboxylase
RuDP   = Ribulose 1−5 diphosphate
3PGA   = 3−phosphoglycerate
PG   = Phosphoglycolate
I, II   = Photosystems I and II
●   = Photochemical reaction centre

Fig. 6.1. An outline of photosynthesis and photorespiration.

GENE TRANSFER INTO PLANTS

known as *photorespiration*. This appears to be a wasteful process. It consumes NADPH, reduces the amount of precursors for the Calvin cycle and releases low-energy inorganic phosphate rather than ATP. The oxygen produced in light harvesting therefore competes with carbon dioxide for Rubisco and in low carbon dioxide the major pathway is photorespiration. The reactions of photosynthesis and photorespiration are illustrated superficially in Figure 6.1. Many of the genes for the photosynthetic apparatus have been cloned and the regulatory response of gene expression to light is being analysed (Tobin and Silverthorne, 1985; Kuhlemeier *et al.*, 1987). The biophysical properties of the key enzyme, rubisco, are also becoming increasingly well characterised.

One of the targets of gene manipulation is the photosynthetic enzyme rubisco. The aims are (1) to increase the affinity for CO_2 and (2) to abolish or reduce the competing reactions of photorespiration. Rubisco is a 16 subunit enzyme: there are eight large (55 kD) polypeptides (rbc-L) and eight small (16 kD) polypeptides (rbc-S). The rbc-L proteins are encoded by the chloroplast genome and the rbc-S proteins are encoded by a nuclear gene family. Rubisco from different plant sources varies in kinetic constants and therefore one strategy would be to introduce genes from different organisms to produce heterosubunit enzymes. Another approach is to modify the enzyme activity by site directed mutagenesis to increase the CO_2 affinity and replace the normal gene with a more effective mutant gene. Other more long term aims would be to exchange photosystem components to optimise electron flow rate through photosystems I and II. Any manipulation of photosynthesis requires the ability to engineer both nuclear and chloroplast genes and to express these genes specifically in leaves and in response to light.

Nitrogen fixation and assimilation

Most crop plants need large amounts of soluble nitrogen for optimal growth and this is generally supplied by chemical fertilisers. However, leguminous plants such as peas, beans or clover can convert atmospheric nitrogen into ammonia, obviating the need for expensive fertilisers. This process of *nitrogen fixation* requires a symbiotic interaction with bacteria of the *Rhizobium* spp. The bacteria infect the plant roots and stimulate the production of nodules where nitrogen fixation occurs. The nodule is a specialised plant organ that develops from the root hairs in response to a highly specific cell to cell interaction between Rhizobia and root cortex cells. The nodule creates the anaerobic environment that is required for nitrogen fixation. Nodulation is controlled by a few bacterial genes, the *nod* genes and requires at least thirty plant proteins called 'nodulins'. Inside the nodule the Rhizobia develop into pleomorphic 'bacteroids' that fix nitrogen. The process of nitrogen fixation requires both bacterial and plant proteins. The bacterial proteins are encoded by the *nif* genes. There are seventeen *nif* genes contained in eight operons that encode the oxygen-sensitive nitrogenase complex and components of the electron transfer chain. The functions of many of the plant proteins are still unclear but one of the major proteins in the nodule is plant leghaemoglobin which binds oxygen to protect the nitrogenase complex from inactivation. Plant uricase, xanthine dehydrogenase and glutamine synthetase, which are involved in ammonium metabolism, are also present (Figure 6.2; Kondorosi and Kondorosi, 1986; Downie and Johnston, 1986).

Genetic conversion of a plant cell for nitrogen fixation would require: (1) the appropriate expression of all the bacterial *nif* genes in a plant cell; (2) the appropriate processing and assembly of the nitrogenase complex; (3) an anaerobic environment; (4) a supply of ATP; (5) a supply of NADPH. This seems a daunting task. One approach might be to transform chloroplasts with the *nif* plasmid so that the normal prokaryotic signals will be used rather than place all 17 genes under plant nuclear promoter control. Oxygen inactivation of the nitrogenase would however still be a problem. An alternative and perhaps more likely approach is to engineer the nodulation genes from a leguminous plant into the cells of other plants so that they can

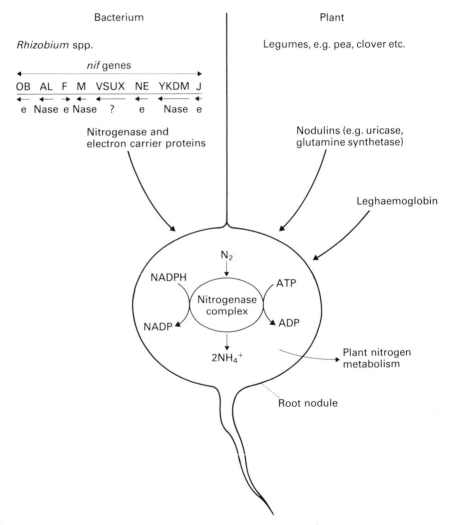

Fig. 6.2. Nitrogen fixation. Nase = nitrogenase enzymes and regulatory genes; e = electron carrier protein genes.

GENE TRANSFER INTO PLANTS

respond appropriately to invasion by nitrogen fixing bacteria. This would require, however, a deeper understanding of the factors that influence the association between nitrogen fixing bacteria and plant roots. A number of plant nodulin genes have been cloned and a model system for studying nodulation has been established with the legume *Lotus corniculatus* (birds foot trefoil) (Jensen *et al.*, 1986).

Nitrogen fixation is restricted to the legumes. Other higher plants rely upon assimilating nitrogen from nitrates in the soil. This is achieved by the reduction of nitrate to ammonia in a two-step process. First there is reduction to nitrite by nitrate reductase and then reduction of nitrite to ammonia by nitrite reductase. The key enzyme is *nitrate reductase*. This is highly regulated, the levels are increased by light, cytokinin and nitrate but it is still rate limiting. There are many situations, for example, during seed production, where the demand for nitrogen exceeds the supply from nitrate reduction. It might be possible to increase the efficiency of a plant's use of nitrates by manipulating the catalytic activity of nitrate reductase or by increasing the levels by modifying gene expression signals. The cDNA for nitrate reductase from the squash *Cucurbita maxima* has now been cloned which opens the way to cloning the gene and manipulating both expression and enzyme activity (Crawford *et al.*, 1986).

Increasing the nutritional value of seeds

Plant seeds are rich in protein and can provide a major component of the protein requirements in the animal diet. Humans cannot synthesis ten amino acids and it is these essential amino acids that must be provided. Many seeds are deficient in some of these. For example corn is low in lysine and tryptophan and legumes are deficient in methionine and cysteine. It may be useful, therefore, to manipulate the genes for seed storage proteins to increase the content of specific amino acids (Messing, 1983; Jaynes *et al.*, 1986). A number of the relevant genes have been cloned and characterised, e.g. phaseolin (French bean, Sengupta-Gopalan *et al.*, 1985) legumin (pea, Lycett *et al.*, 1984) and zein (maize, Messing, 1983) and these can be transferred and expressed in plant cells using the techniques discussed later in the chapter. There are, however, a number of fundamental biological problems to be overcome. These can be illustrated by considering zein, the storage protein in maize (Figure 6.3). The protein has a highly ordered structure, and different zein genes have four highly conserved domains suggesting that there has been rigorous selection for the organisation of the protein. The structure may be necessary for the stability and packaging of the protein in the seed. Increasing the lysine content may therefore interfere with protein architecture and packaging. The zein mRNA also has a highly ordered secondary structure which is thought to contribute to mRNA stability (Heidecher and Messing, 1987) and this may be affected by the codon substitution required to increase lysine content. The most significant problem is, however, that zein proteins are encoded by five related multigene families possibly comprising 100 genes which map to at least three different chromosomal locations. The transfer of a single modified zein gene will not therefore

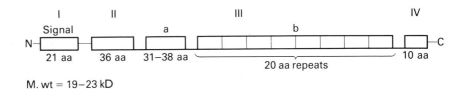

Fig. 6.3. Conserved domains in zein: the storage protein of corn.

significantly affect the nutritional quality of the corn. This means that methods must be developed to inactivate endogenous genes. One approach that is being investigated is to block their activities by introducing a gene that produces large amounts of RNA from the opposite DNA strand. This 'antisense' RNA may block endogenous expression by forming a hybrid with the authentic mRNA (Ecker and Davis, 1986; see Chapters 8 and 12 for further applications of antisense RNA technology).

The expression of any transferred genes destined to improve seed quality would have to be under precise developmental control as the zein proteins are produced 18 to 32 days after pollination in response to complex hormonal signals. It is also essential that the transferred gene is expressed in the part of the plant that is usually eaten, e.g. expression in leaves and not seeds would be useless. Another potential problem relates to the fact that some storage proteins have a dual function, for example, patatin the storage protein in the potato, is an esterase in addition to being a storage protein. Other seed proteins are proteinase inhibitors or have antifungal or antibacterial activity (Rosahl et al., 1987). Clearly it might be undesirable to lose these traits in the process of improving another. An alternative approach is to introduce a totally synthetic gene that has been designed to contain high numbers of codons for essential amino acids. Such a high essential amino acid encoding (HEAAE) gene has been introduced into potato tubers and the nutritional quality of the tubers is being characterised (Jaynes et al., 1986). It will, however, be important to ensure that the amino acid pools are sufficient to produce large quantities of new proteins with a different amino acid content without imbalancing metabolism which could have a deleterious effect on seed yield.

Resistance to pests, pathogens and herbicides

A significant proportion of potential crop yield is lost due to disease or by being consumed by pests. There is therefore economic advantage in producing disease and pest-resistant plants. Many plants have effective broad spectrum resistance mechanisms but these have to be induced by prior infection or damage. One aim is to understand the induction processes so that resistance can be expressed 'constitutively' (Sequira, 1983; Dean and Kuć, 1985). Another aim is to introduce the genes for bacterial toxins into plants. For example Bacillus thuringiensis produces a toxic protein that effectively kills the larvae and caterpillar stages of various pests as they feed on leaves covered in the toxin. This toxin has been crystallised and can be sprayed on

GENE TRANSFER INTO PLANTS

to plants but this is expensive and therefore genetically modified toxin-producing plants are an advantage.

Herbicides are used to kill weeds but they must either be selective in action or sprayed very locally to avoid damaging crop plants. Weed control would be more effective if crops were resistant to herbicides. Natural resistance occurs in some weeds and identifying the basis for this allows the introduction of resistance genes into crop plants. In some cases the resistance mechanism is already known. For example, atrazine resistance is conferred by the presence of a 32 kD photosystem II protein that has a single amino acid change that prevents the herbicide binding to the protein and so photosynthesis is unaffected (Hirschberg and McIntosh, 1983). In another case, resistance to glyphosphate is conferred by over-production or alteration of the specificity of the enzyme 3-phosphoshikimate 1-carboxyvinyl transferase. The transfer of genes that confer resistance to pests and to some herbicides has been achieved for a number of plants and is discussed in Chapter 12.

Production of secondary metabolites

Plants provide 25% of the world's pharmaceuticals and they also produce fine chemicals, biochemicals, cosmetics and food additives (Table 6.1). At present these are mostly purified from entire plants but progress is being made in defining culture conditions for optimising production *in vitro* (see Mantell *et al.*, 1985). It may be possible to genetically manipulate these pathways to increase the efficiency of synthesis of known compounds. A more exciting prospect is manipulation to produce entirely novel chemicals with beneficial properties. This is an extremely long-term goal and will require extensive research into basic plant biochemistry before it can be tackled.

Others

Plants are always subject to environmental stress, e.g. drought and temperature stress, and sub-optimal land is cultivated out of necessity in many areas of the world. Crops that are tolerant to drought, high salt, mineral deficiency, etc. would be extremely valuable to Third World countries. When more is known about the genetic control of response to environmental stress it may be possible to transfer the relevant genes between species. A start has already been made in understanding salt tolerance which is correlated with the induction of new mRNAs (Ramagopal, 1987). Cloning and characterisation of these genes is underway.

Another key area is that of plant fertility. It may be useful to restrict the spread of plants that have been modified by recombinant DNA techniques and this can be done by ensuring male sterility. Male sterility is also useful to prevent cross-pollination in classical breeding programmes. Recently genes that control sexual compatibility in plants have been characterised (Anderson *et al.*, 1986) and it may therefore be possible to manipulate the fertility of transgenic plants.

Table 6.1 Some important product from plants

Product	Use
Enzymes	
e.g. papain	Brewing (chill proofing);
bromelain proteases	meat tenderising; leather
ficin	tanning; flour improvement
Pharmaceuticals	
Codeine	Analgesis
Digoxin	Cardiotonic
Diosgenin	Anti-fertility
Quinine	Anti-malaria
Scopalamine	Anti-hypertensive
Vincristine	Anti-leukaemia
Agriculture and food	
Pyrethrin	Insecticide
Quinine	Bitter flavour
Thaumatin	Sweet flavour
Cosmetic	
Jasmine	Fragrance
Jojoba	Waxes

Culturing plants and plant cells

Plants can be propagated sexually, via self-fertilisation or cross-fertilisation and the production of seeds, or asexually, from cuttings or explants. The ability to produce complete fertile plants from cuttings from any part of an adult plant, e.g. shoot, stem or leaves. It indicates that some non-dividing specialised cells of the adult plant are totipotent. That is, they have not been restricted to one pattern of gene expression but can 'trans-differentiate' into other specialised structures (Wareing, 1984; Walbot, 1985). This is a major difference in development between plants and animals.

In many cases a single cell can be induced to differentiate into a whole plant. This is of fundamental interest to the researcher interested in how gene expression is controlled during differentiation but it also has important practical consequences. It means that a single plant can be the progenitor of many thousands of genetically identical plants, i.e. *clones*. For example the commercial production of oil palms now involves over 12 000 plants generated from just 30 progenitor plants. The genetic manipulation of a single plant cell would therefore be sufficient to rapidly produce a new plant variety. It also means that foreign DNA does not have to be introduced into the germ line in order to obtain a fully developed fertile plant with the new DNA in all its cells. The production of transgenic plants may, therefore, be simpler than the production of transgenic animals. The disadvantage of this toti-potency of plant cells, however, is that isolated cells *in vitro* often rapidly lose the differentiated state of the tissue of origin. This means that rapid transient assay systems for transferred gene expression are unlikely to provide meaningful data on specific regulation of gene expression, unlike the situation with animal cells.

Asexual micropropagation techniques involve excising small pieces of plant tissue called *explants* and allowing them to increase in size on nutrient medium that maintains the potential for differentiation. Fragments of these explants can be serially sub-cultured and by changing the composition of the medium (see below) they can be induced to develop into whole plants. Diploid fertile plants can be produced from most parts of a plant. In some plants, pollen can be cultured to produce callus which differentiates into haploid plants.

There are two routes to culturing undifferentiated plant tissue *in vitro*. These are: (1) *callus* culture and (2) *protoplast* culture. Callus is an undifferentiated mass of cells that grows out from a site of wounding and protoplasts are single cells that lack their cell wall. The state of differentiation can be maintained or altered by varying the ratios of *phytohormones* (plant hormones) in the culture medium. The major phytohormones are the *auxins* (e.g. indole-3-acetic acid (IAA), the *cytokinins* (e.g., kinetin) and *gibberelins*. Callus derived from plants or protoplasts can be induced to differentiate into fertile adult plants. The main plant species used for experimental gene manipulation are *Nicotiana tabacum* (tobacco) and *Petunia hybrida*. A general scheme showing the experimental manipulation of a plant such as *Nicotiana* is shown in Figure 6.4. Sexual and asexual micropropagation techniques are outside the scope of this book and details can be found in Mantell *et al.* (1985). Some of the details of *in vitro* techniques relevant to gene manipulation are discussed below. Practical details can be found in Reinert and Yeoman (1982).

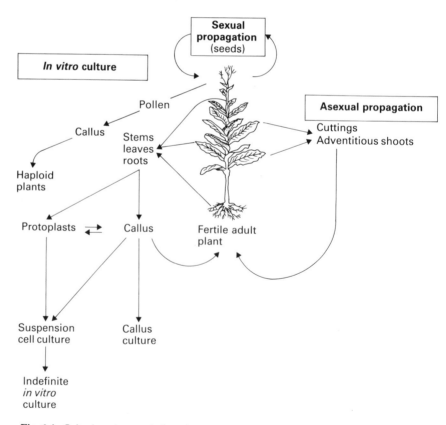

Fig. 6.4. Culturing plants and plant tissue.

Callus culture

When excised tissues from many plants are placed on medium containing low concentrations of phytohormones, a mass of disorganised undifferentiated cells grows at the cut surfaces. This is callus tissue and it can be propagated indefinitely by simply excising small pieces and placing them on fresh medium where they increase in size. If the concentrations of phytohormones are altered then the callus can be induced to differentiate. A very high auxin-to-cytokinin ratio induces root formation and a high cytokinin-to-auxin ratio induces shoots and in a more restricted number of species, flowering parts. This is referred to as *organogenesis*. Alternatively, a structure called an embryoid may develop, a process called *embryogenesis*. If the hormonal conditions are correctly balanced then an entire plant can be produced; this is referred to as *regeneration*. In many cases the adult plant is fertile and produces viable seeds. The concentration and ratios of hormones are determined empirically for different plants and different explanted tissues which will vary in their endogenous hormone levels. Not all plants will form callus cultures and not all calli will regenerate. An outline of callus culture is given in Figure 6.5.

Protoplast culture

The cell walls can be removed form plant cells mechanically by treating with a mixture of cellulase and pectinase enzymes. Provided that the medium is osmotically balanced this releases *protoplasts* which have all the properties of an intact cell except that the plasma membrane is now accessible to chemical and physical agents (Cocking, 1960; Lichtenstein and Draper, 1985). Plant protoplasts can be treated in much the same way as animal cells. Somatic hybrids can be produced by fusing protoplasts from different species and DNA can be readily introduced by calcium phosphate or PEG-mediated transfection, by infection or by electroporation.

Although protoplasts can be produced from a range of plant tissues, the mesophyll cells of leaves are most commonly used. This is because it is easy to disrupt leaves into single cells and because leaf cells have a high regeneration potential allowing mature plants to be derived from single protoplasts. The mesophyll cells are also homogeneous in their physiological properties. Protoplasts can be maintained in osmotically balanced liquid medium or in soft agar in large-scale or small-scale cultures such as microdrops. The cell wall is resynthesised after about 2–10 days and cell division occurs. The cells can be maintained for long periods as a cell culture by repeated sub-culturing about every 2 weeks. Plant cells tend to remain associated after cell division and the cultures usually contain aggregates rather than single cells. Protoplasts can also be produced by enzyme treatment of callus or plant cell cultures.

Individual cells or cell aggregates derived from protoplasts can be induced to form callus cultures by growing in the appropriate hormone supplemented media. Single cells are often plated on to a filter paper disc that is placed over a layer of plant cells to encourage growth. This is called *feeder layer technique*. It is necessary because single plant cells require a minimum cell

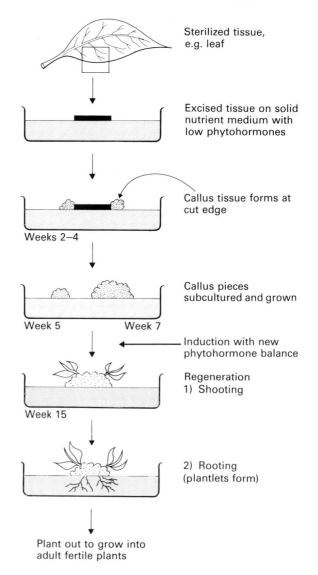

Sterilized tissue, e.g. leaf

Excised tissue on solid nutrient medium with low phytohormones

Callus tissue forms at cut edge

Weeks 2–4

Callus pieces subcultured and grown

Week 5   Week 7

Induction with new phytohormone balance

Regeneration
1) Shooting

Week 15

2) Rooting (plantlets form)

Plant out to grow into adult fertile plants

Fig. 6.5. Callus culture.

density to allow growth and cell division. Aggregates or small calli induced in liquid culture are plated in soft agar. The resulting calli can be subcultured or induced to regenerate to produce mature plants.

The ability of protoplasts to regenerate fertile adult plants is highly variable. When plant tissues are maintained for extended periods *in vitro* either as callus or suspension culture they loose the potential for regeneration. This is because genetic changes occur *in vitro*. There are often abnormal mitoses resulting in polyploidy and aneuploidy and various other karyotypic abnormalities are observed (Chaleff, 1983). For this reason genetic manipulation procedures are usually designed to minimise the *in vitro* culture period. Furthermore, in many plants, abnormal mitoses also occur in some tissues *in vivo* so they are naturally polyploid. This results in a heterogenous cell culture containing genetically different cells which can complicate any analyses and reduce regeneration potential. It is generally difficult to regenerate monocot plants from protoplasts (Ozias-Akins and Lörz, 1984). There has been success with some wild, i.e. non-cultivated, diploid strains of corn (Sondahl et al., 1984) and regeneration of rice has been achieved (Uchimaya et al., 1986). These studies may mark the beginning of improved in vitro manipulation of crop plants.

One important feature of plant cell culture is that by culturing pollen it is possible to produce haploid protoplasts and then, via callus culture, to regenerate haploid plants. This allows chromosomal mutations to be isolated and may allow the interaction between host cell genome and any transferred genes to be analysed more easily. This may make the identification of regulatory genes in plants an easier task than identifying similar genes in cultured animal cells which are diploid. An outline of some of the techniques of protoplast culture is given in Figure 6.6.

Gene transfer

A breakthrough in gene manipulation in plants came by characterising and exploiting plasmids carried by the bacterial plant pathogens, *Agrobacterium tumefaciens* and *Agrobacterium rhizogenes*. These provide natural gene transfer, gene expression and selection systems. More recently electroporation has been used to introduce DNA directly into plant protoplasts and this may replace the *Agrobacterium* systems for some applications. There has also been much effort to develop viral vectors to increase the efficiency of gene transfer. Several plants viruses have provided powerful promoters for producing plant gene expression vectors (see articles in Weissbach and Weissbach, 1986).

Exploiting *Agrobacteria*

The *Agrobacteria* are gram-negative soil bacteria that can infect over 330 genera of dicotyledonous plants after they have been wounded. This infection produces aberrant tissue growths. Most research has been carried out with *A. tumefaciens* which produces *crown gall disease* but recently *A. rhizogenes* which produces *hairy root disease* has also been exploited. The pathogenic effects are mediated by bacterial plasmids and these have been manipulated to produce vectors (Lichtenstein and Draper, 1985; Klee et al., 1987).

The interaction of A. tumefaciens with plants

A. tumefaciens adsorbs to the surfaces of cells exposed at wound sites, usually at the root crowns, and causes the wound tissue to proliferate to produce a neo-plastic growth called a crown gall tumour. The bacterium is referred to as oncogenic or virulent and this property is conferred by a plasmid called the *Ti (Tumour inducing)* plasmid. Bacteria which do not have the plasmid do not produce tumours but the plasmid is readily transferred to such non-virulent strains by conjugation. During the course of the infection,

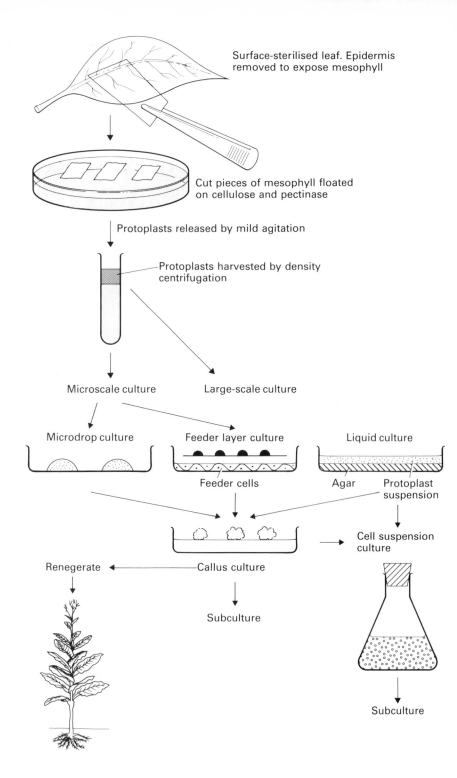

Surface-sterilised leaf. Epidermis removed to expose mesophyll

Cut pieces of mesophyll floated on cellulose and pectinase

Protoplasts released by mild agitation

Protoplasts harvested by density centrifugation

Microscale culture   Large-scale culture

Microdrop culture   Feeder layer culture   Liquid culture

Feeder cells   Agar   Protoplast suspension

Cell suspension culture

Renegerate ◄——— Callus culture

Subculture

Subculture

Fig. 6.6. Protoplast culture.

part of the Ti plasmid is introduced into the plant cell and becomes integrated into the plant cell genome. This is the *T-DNA (transferred DNA)*. Expression of the genes contained in the T-DNA confers new properties on the plant.

T-DNA codes for enzymes that synthesise auxin and cythokinin which disturb the normal balance of plant growth and result in the proliferation of undifferentiated wound tissue that forms the crown gall. The plant cells also produce unusual nitrogenous compounds called *opines* via opine synthase genes which are present on the T-DNA. There are at least six different classes of opines and some tumours may synthesise more than one. This reflects differences in the opine synthase genes carried by the T-DNA of different Ti plasmids. Some of the opines and relevant Ti plasmids are summarised in Table 6.2. The most widely studied Ti plasmids are the nopaline and octopine type plasmids. Ti plasmid-containing bacteria can metabolise opines as a sole source of carbon and nitrogen because the relevant catabolic enzymes are also encoded by the plasmid. In addition opines stimulate the conjugal transfer of Ti plasmids ensuring that all bacteria in a population have this metabolic potential. Once the tumour has been initiated the bacteria can be removed and the crown gall will continue to grow. The gall can be excised and cultured *in vitro* and unlike all other plant tissues it grows in the total absence of exogenous phytohormones. The key features of a crown gall tumour are then, (1) hormone-independent growth *in vitro*; (2) the production of opines; and (3) the presence of T-DNA in the plant cell genome. The gall continues to grow as an undifferentiated callus and unlike other plant tissues it cannot be induced to regenerate into adult plants by phytohormone treatment. There are, however, two classes of defective Ti plasmids that show partial regeneration: *Shooty mutants*, (*tms⁻*), which regenerate shoots and can be propagated by grafting on to adult plants, and *rooty mutants*, (*tmr⁻*), which regenerate roots. The interactions between *Agrobacterium* and plants are summarised in Figure 6.7 and are reviewed in Caplan *et al.* (1983), Nester *et al.* (1984) and Hookyas and Schilperoot (1984).

Ti plasmids and T-DNA

The Ti plasmids are large plasmids of about 150 to 200 kb. In addition to the T-DNA, they contain genes that are essential for the establishment of a

Table 6.2. Opines, opine synthases and Ti plasmids.

Opine	Opine synthase gene	Ti plasmid
Nopaline	*nos*	pTiT37; pTiC58
Octapine	*ocs*	pTiA6NC; pTiAch5; pTiB653; pTiAg162 (limited host range)
Agrocinopine	*acs*	
Mannopine	*mas*	

GENE TRANSFER INTO PLANTS

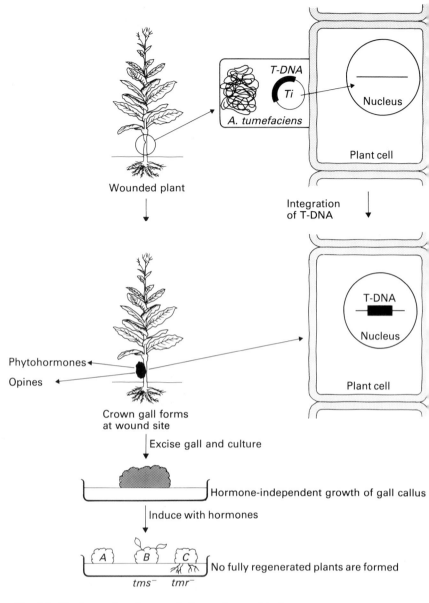

Fig. 6.7. The interaction of oncogenic *A. tumefaciens* with plants.

tumour, the *vir* (virulence) genes. They also encode functions for replication (*ori*, *rep*) and conjugal transfer (*tra* genes) of the plasmids (Figure 6.8) (Joos *et al.*, 1983; Hoekema *et al.*, 1983).

A more detailed map of the T-DNAs of an octopine- and nopaline-type plasmid are shown in Figure 6.9 (Willmitzer *et al.*, 1983; Winter *et al.*, 1984). The nopaline-type T-DNA is 23 kb in length and is flanked by 2 bp, almost perfect, direct repeats called the left and right border (*LB* and *RB*) sequences. Three transcriptional units called 1, 2 and 4 are the *onc* (tumour) genes. Deletion of 1 and 2 produces the shooty tumours and these

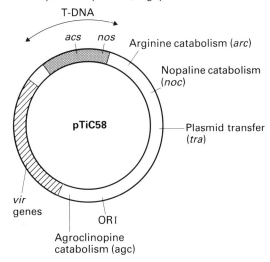

A Nopaline Ti plasmid, e.g. pTiC58

T-DNA

acs *nos*

Arginine catabolism (*arc*)

Nopaline catabolism (*noc*)

pTiC58

Plasmid transfer (*tra*)

vir genes

ORI

Agroclinopine catabolism (agc)

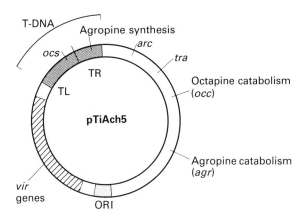

B Octapine Ti plasmid, e.g. pTiAch5

T-DNA

Agropine synthesis

ocs

arc

tra

TR

TL

Octapine catabolism (*occ*)

pTiAch5

Agropine catabolism (*agr*)

vir genes

ORI

Fig. 6.8. Ti plasmids. See Figure 6.9 for gene designation.

genes, called *tms1* or *IaaM* and *tms2* or *IaaH*, respectively code for the enzymes tryptophan monooxygenase and indole acetamide hydrolase. These are involved in the synthesis of auxin which would normally inhibit shoot production. Deletion of region 4 produces rooty tumours and this gene, called *tmr* or *ipt*, encodes isopentenyl transferase that is involved in the synthesis of cytokinin which normally inhibits root formation. If both regions are deleted the plasmid becomes non-transforming, or Onc⁻. The octapine T-DNA has a slightly different organisation. It comprises two adjacent independently acting segments: the left T-DNA (TL-DNA) of 13.6 kb and the right T-DNA (TR-DNA) of approximately 7 kb. The TL-DNA shares a 9 kb region of homology with the right half of the nopaline T-DNA that encompasses the tumour genes. Octapine TL-DNA is also flanked by 25 bp direct repeats that have significant homology with the

GENE TRANSFER INTO PLANTS

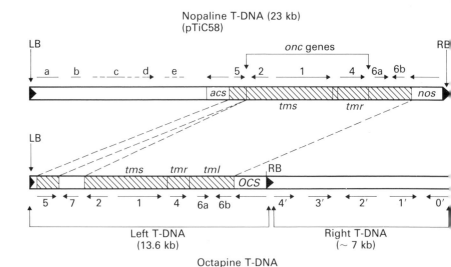

Fig. 6.9. T-DNA. The shared genes between the octapine and nopaline T-DNAs are hatched. *acs* = agrocinopine synthase; *nos* = nopaline synthase; *ocs* = octapine synthase. Tumour morphology loci: *tms* = shooty tumour, *tmr* = rooty tumour, *tml* = large tumour. The approximate size and direction of the transcripts is indicated by the arrows. LB and RB are left and right borders respectively.

nopaline T-DNA border sequences (Figure 6.10; Holsters *et al.*, 1983). The octapine TL-DNA is necessary for tumour formation, whereas the TR DNA has no tumour inducing functions but codes for enzymes involved in agropine biosynthesis (Saloman *et al.*, 1984). A number of different opine synthase genes are found in T-DNA and the best characterised are *nos* (De Greve *et al.*, 1982a) and *ocs* (Depicker *et al.*, 1982). Although the T-DNA is found on a bacterial plasmid the promoters for the T-DNA transcripts are essentially eukaryotic so that the genes can be expressed in the plant cells. It is possible that the T-DNA is derived from a plant genome, although the intimate relationship between bacteria and plant may have resulted in selection for plant-compatible promoters. The T-DNA genes provide a good source of expression signals for constructing plant vectors and the synthesis of opines can be used as a transformation marker.

The mechanism of T-DNA transfer is not fully understood but it appears to involve a process that is very similar to conjugation between bacteria. Usually the only part of the Ti plasmid that is transferred is the T DNA itself. The transfer process is activated by plant phenols, e.g. aceto syringone (AS) that are released as a result of wounding and transfer is controlled by bacterial genes located on the chromosome (*chv* genes) and on the Ti plasmid (*vir* genes) (Stachel and Zambryski, 1986; Stachel and Nestor, 1986). The T-DNA is nicked by a endonuclease encoded by the *vir* D gene on the same strand at both borders (Yanofsky *et al.*, 1986). It is transferred as a single strand, called the *T-strand*, in a single direction with the RB sequence leading (Stachel *et al.*, 1986). The polarity of transfer explains earlier observations that the RB must be present for efficient

opaline	L	T	G	G	C	A	G	G	A	T	A	T	A	T	T	G	T	G	G	T	G	T	A	A	A	C
	R	T	G	A	C	A	G	G	A	T	A	T	A	T			G	G	C	G	G	G	T	A	A	A	C
ctapine	L	C	G	G	C	A	G	G	A	T	A	T	A	T			C	A	A	T	T	G	T	A	A	A	T
-TDNA)	R	T	G	G	C	A	G	G	A	T	A	T	A	A			A	C	C	G	T	T	G	T	A	A	T	T

Fig. 6.10. T-DNA border sequences. The nucleotide differences are boxed.

transfer to occur (e.g. Wang *et al.*, 1984). The RB of the nopalene plasmid pTiT37 can, in fact, transfer at least 20 kb of adjacent DNA in the absence of an LB sequence. The LB sequence usually serves to delimit the transferred region by virtue of the nick. In the absence of LB, this end must be dictated by random nicking. The LB sequence has little or no transfer activity alone (Jen and Chilton, 1986). The integration of T-DNA is only approximate but the sites of integration are always within or very close to the border sequences and are more uniform at the RB sequence. It is not clear whether this pattern of integration reflects the structure of the T-DNA or whether the border sequences direct integration in some way.

Transformation markers

Plant cells only synthesise opines when they are transformed with T-DNA and therefore opine synthesis can be used as a marker for transformation (e.g. Zambryski *et al.*, 1983). Octapine and nopaline synthase is readily assayed by the conversion of arginine to opines which are detected by paper electrophoresis (Otten and Schilperoort, 1978). The opine synthase genes were extremely important as transformation markers in early gene transfer studies but they have largely been replaced by dominant selectable markers (Fraley *et al.*, 1983; Herrera-Estrella *et al.*, 1983a, b; Rogers *et al.*, 1987). For example (Figure 6.11), the *nos* gene promoter and terminator have been used to express the neomycin phosphotransferase gene, *neo*, and a prokaryotic dihydrofolate reductase gene, *dhfr*, so that transformed plant cells can be selected by growth in the presence of kanamycin or methotrexate respectively. Resistance to hygromycin B has also been used for plants such as *Arabidopsis* which are not very sensitive to killing by neomycin (van den Elzen *et al.*, 1985). The T-DNA promoters are useful because they function in a wide range of plants and they are efficient and constitutive in most plant tissues.

Velten *et al.* (1984) have produced a dual plant promoter fragment derived from the transcript $1'-2'$ region of the right T-DNA of the octopine plasmid pTiAch5. These genes encode enzymes in agropine biosynthesis. A 479 bp bidirectional promoter fragment can drive both a selectable marker such as *neo* and a non-selected foreign gene and it is more efficient than the *nos* promoter (Figure 6.12). The ability of Onc⁻ Ti-induced callus to proliferate in the absence of phytohormones can also be used as a selectable marker. This is, however, limited to experiments where full regeneration into plants is not required.

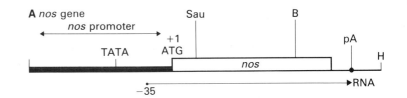

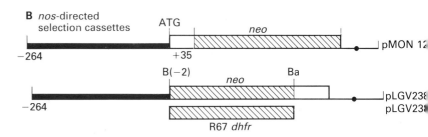

Fig. 6.11. Transformation markers.

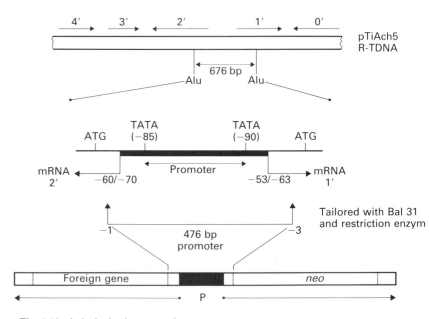

Fig. 6.12. A dual selection expression cassette.

Disarmed (non-oncogenic) Ti vectors

The Ti plasmids are not generally suitable as gene transfer vectors becau
the resulting tumours cannot be regenerated into fertile plants because
the hormone imbalance in the tissue. Analyses of Ti plasmids has show
however, that while the border regions of the T-DNA are important f
transfer of the T-DNA into the plant cell genome, other regions can
deleted. For example the shooty and rooty mutants lack one or other of tl
transforming genes and still produce tumours but with altered regeneratic

potential. Zambryski *et al.* (1983) systematically modified a nopaline Ti plasmid, pTiC58, to produce a Ti plasmid that could still transfer T-DNA into plant cell genomes but no longer caused tumours. This involved deleting all of the T-DNA except for the left and right border sequences and the *nos* gene and replacing this with a region of pBR322 (Figure 6.13). The resulting plasmid, pGV3850, is Onc⁻, i.e. non-oncogenic or *disarmed*. *Agrobacterium*-carrying pGV3850 were used to infect discs of carrot or potato or wounded *Petunia* and *Nicotiana* plants. No crown gall-like tumour was produced and callus tissue only grew when explants of wound tissue or discs were cultured in the presence of phytohormones. This indicated that the internal deletion had removed the genes required for hormone independent growth. The resulting callus tissue however, was producing nopaline showing that the T-DNA had still been transferred because the *nos* gene was being expressed. Furthermore, nopaline-positive tobacco stem callus could now be regenerated into fertile adult plants by hormone induction. This study opened the way for the production of transgenic plants where foreign genes could be transmitted through seeds.

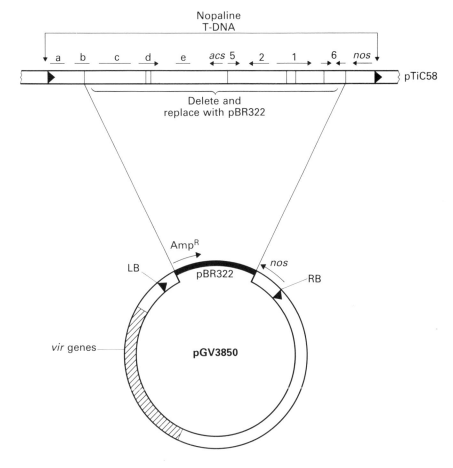

Fig. 6.13. A disarmed Ti vector.

GENE TRANSFER INTO PLANTS

Disarmed Ti plasmids cannot be used as primary cloning vectors becaus[e] like their parental Ti plasmids, they are so large that there are no uniq[ue] restriction enzyme sites. Foreign genes are therefore inserted into the[se] vectors by *in vivo* recombination between plasmids to produce *cointegrate*[s]. The formation of a cointegrate vector is illustrated in Fig. 6.14 using t[he] disarmed vector pGV3850. This vector is particularly useful because t[he] pBR322 DNA in the T-DNA provides a region of homology with mo[st] other cloning vectors. In this example the foreign gene is inserted into t[he] vector pLGVneo1103 (De Block *et al.*, 1984). This carries a plant-selectab[le] marker, the *nos-neo* hybrid gene form pLGV23-*neo* (Figure 6.11), and [a] bacterial selectable marker, the *kan* gene from Tn903. We will refer to th[is] type of plasmid as the *donor vector*. Recombination between the homologo[us] pBR322 DNA in the two plasmids produces the cointegrate which no[w] carries a plant selectable marker and the foreign gene, flanked by T-DN[A] border sequences. When *Agrobacterium* carrying this cointegrate infec[ts] plant cells the recombinant T-DNA will become integrated into the genom[e] and transformed cells are selected by resistance, conferred by the *neo* gen[e] to kanamycin.

Cointegrates have to be produced by transferring the donor vect[or] which is in *E. coli* to *Agrobacterium* by conjugation. The ability to transf[er] plasmids between different species of Gram-negative bacteria depends upo[n] several properties of the plasmids. Both *cis* acting sites on the plasmid an[d] transacting plasmid-encoded proteins are required for conjugal transfer. I[n] addition many plasmids are only able to replicate in a narrow host range [of] bacteria and/or they are *incompatible* with another plasmid in the same cel[l].

The easiest way to achieve the end result of an *Agrobacterium* carrying [a] disarmed vector and a donor vector that can recombine is to set up *triparental mating* (Van Haute *et al.*, 1983). A typical procedure to produc[e] the cointegrate described in Figure 6.14 is outlined in Figure 6.15. The thre[e] parental bacterial strains are *E. coli* transformed with the donor vector, *E. coli* containing the plasmids pGJ28 and R64drd11 and *A. tumefaciens* co[n]taining the disarmed vector pGV3850. Vectors based on pBR322 are ColE[1] derivatives and they are, therefore, not capable of transferring to anoth[er] bacterium, i.e. they are not *self-transmissible*, but they can be mobilised b[y] helper plasmids (Van Haute *et al.*, 1983). Transfer functions encoded by *tr[a]* genes can be provided by a self-transmissible plasmid such as R64drd11[.] Additional proteins called *mob* proteins are also required for transfer. Thes[e] are ColE1 encoded but the genes have been deleted from pBR322. The *mo[b]* proteins can be provided by a ColE1 helper plasmid such as pGJ28. The[y] interact with a specific site called *bom* which is retained in pBR322. Whe[n] the three bacterial strains are mixed R64drd11 mobilises itself and pGJ2[8] into the *E. coli* strain containing pLGVneo1103 (Figure 6.15, step (a)[)]. Conjugal transfer between *E. coli* strains is very efficient and 50% of th[e] recipients contain all three plasmids. Recombination between these plasmid[s] is prevented by a mutation in the recipient's *recA* gene. All three plasmid[s] are then transferred to the *Agrobacterium* recipient, pLGVneo1103 is mobilise[d] by the helper *tra* genes of R64drd11 and the *mob* functions of pGJ28 (Figur[e] 6.12, step (b). None of the *E. coli* plasmids can replicate in *Agrobacteriu[m]*

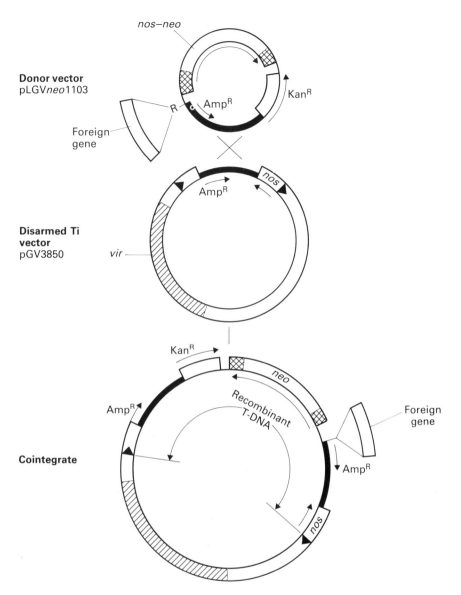

Fig. 6.14. Producing a recombinant T-DNA by cointegrate formation. The donor vector recombines with the disarmed vector by homologous recombination across the pBR322 sequences (dark line) to produce the cointegrate.

because they are not broad host range and therefore the only way pLGVneo1103 can be maintained is by recombination with pGV3850 to produce a cointegrate. Strains carrying cointegrates are selected by growth on minimal medium containing kanamycin as they are prototrophic and the *kan* gene is provided by the donor vector. The bacterial donor parents are selected against (*contraselected*) as they are auxotrophic and/or sensitive to kanamycin (Figure 6.15, step (c)). The structure of the cointegrate is usually checked by an appropriate restriction digest and Southern blot to detect specific diagnostic fragments. The cointegrate containing *Agrobacterium* strain

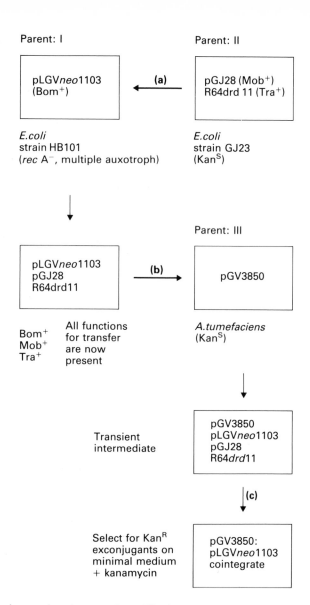

Fig. 6.15. A triparented mating to produce a Ti cointegrate.

is always grown in kanamycin to select against deletions of pLGVneo1103 across the duplicated pBR322 sequences. Deletions do not occur once inside plant cells. This may reflect the fact that plant genomes contain many tandem repeats and may, therefore, tolerate repeats these in plasmids. In some systems, cointegrate formation is mediated by including a piece of T-DNA in the donor vector to recombine with T-DNA in the disarmed vector e.g. pMON145 (Morelli *et al.*, 1985).

Binary Ti vector systems

The production of cointegrates is a simple procedure but because *in vivo* recombination is required the precise structure of the recombinant T-DNA is not known with the certainty of vectors constructed *in vitro*. In addition 'odd ball' events can occur *in vivo* and the gross structure of the cointegrates must always be checked. These disadvantages can be overcome by using binary vector systems. In order to transfer DNA into plant cells, both the T-DNA and the virulence genes are required. The *vir* functions can however be provided in *trans* by a helper Ti plasmid. Two plasmids are therefore introduced into *Agrobacterium*. One is a *disarmed helper Ti plasmid* carrying the *vir* genes and the other is a *mini T-DNA* donor vector carrying foreign genes. In this system both plasmids must be capable of replication in *Agrobacterium* to avoid the need for cointegrate formation. The T-DNA is therefore cloned into a broad host range plasmid such as RK2. Such plasmids are large and they must be manipulated to produce cloning vectors. This involves making a deletion that usually removes the *tra* genes and disrupts the *mob* functions and therefore they, in turn, require a helper plasmid to mobilise them from *E. coli* into *A. tumefaciens*.

A typical binary vector system is shown in Figure 6.16 (Bevan, 1984; Hoekema *et al.*, 1983). The helper Ti plasmid, pAL4404, is a derivative of the octopine plasmid pTiAch5 that has all the T-DNA region deleted. The mini T-DNA donor vector, Bin19, contains a truncated T-DNA from the nopaline plasmid pTiT37 comprising the LB and RB sequences and a *nos-neo* hybrid gene as a selectable marker. A *lac* polylinker is inserted in the mini T-DNA to provide unique cloning sites and a colour test for insertion. The linker carries the alpha-complimentary region of β-galactosidase which will produce blue colonies on X-GAL indicator plates and white colonies if it is disrupted by an insertion. The T-DNA is inserted into a derivative of the broad host range plasmid RK2 which has been modified by the addition of the kanamycin resistance gene from *Streptococcus* and a 4 kb deletion. The plasmid can be mobilised to *Agrobacterium* with a helper plasmid pRK2013 by a mixed plate mating between *E. coli*: pRK2013, *E. coli*: Bin19 and *A. tumefaciens*: pAL4404. Alternatively the mini T-DNA can be transformed into *A. tumefaciens* directly.

The design of mini-T-DNA donor vectors is being modified as more information is obtained about the mechanism of T-DNA transfer. It is now clear that only the RB sequence needs to be present and a rational design is to construct a donor vector that has RB, then the unselected foreign gene immediately to the left of RB to ensure transfer and then the selectable marker. Any plant that expresses the selectable marker must have received the non-selected DNA (Jen and Chilton, 1986).

A binary system without T-DNA!

The discovery that the process of T-DNA transfer shows similarities with bacterial conjugation suggested functional analogies between mobility (*mob*) genes and *virD* and between the origin of conjugational DNA transfer (*oriT*) and the T-DNA borders (Lichtenstein, 1987). This similarity has now been confirmed by the demonstration of the direct transfer of a non-Ti plasmid from *A. tumefaciens* to plant cells (Buchanan-Wollaston *et al.*, 1987). The

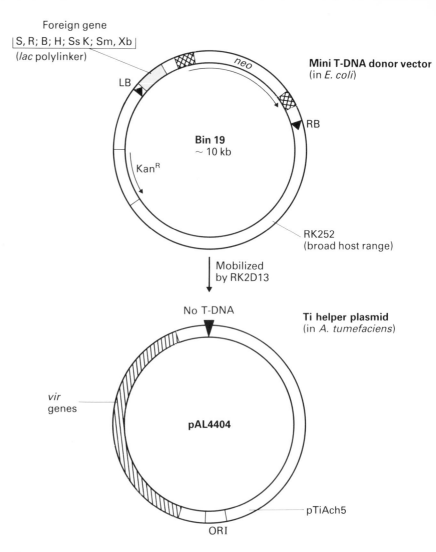

Foreign gene
| S, R; B; H; Ss K; Sm, Xb |
(*lac* polylinker)

LB

neo

Mini T-DNA donor vector
(in *E. coli*)

Bin 19
~ 10 kb

Kan^R

RB

RK252
(broad host range)

Mobilized
by RK2D13

No T-DNA

Ti helper plasmid
(in *A. tumefaciens*)

vir
genes

pAL4404

pTiAch5

ORI

Fig. 6.16. A binary Ti vector system. Bin 19 (Bevan, 1984). pAL 4404 (Hoekema *et al.*, 1983).

plasmid pJP181 (Fig. 6.1) contains a plant selectable marker (*nos–neo*), an *E. coli* replication origin and selectable marker (*Cm^R*) derived from pACYC184 and a region of the wide host range plasmid, RSF1010, that includes *oriT* and *mob*. This plasmid, pJP181, was mobilised into the helper strain of *A. tumefaciens* that carries pAL4404 (Fig. 6.16) in a triparental mating using RK2013 as a helper to provide *tra* gene functions. When the recipient *A. tumefaciens* was used to infect *Nicotiana* leaf discs the *nos-neo* marker was transferred. The plasmid was integrated into the plant genome and was inherited as a single dominant selectable marker. This study shows that bacterial *mob* and *oriT* functions can substitute for T-DNA to effect DNA transfer. *A. tumefaciens* functions are, however, essential to provide the specific cell-to-cell contact needed to allow transfer. This opens up the

CHAPTER 6

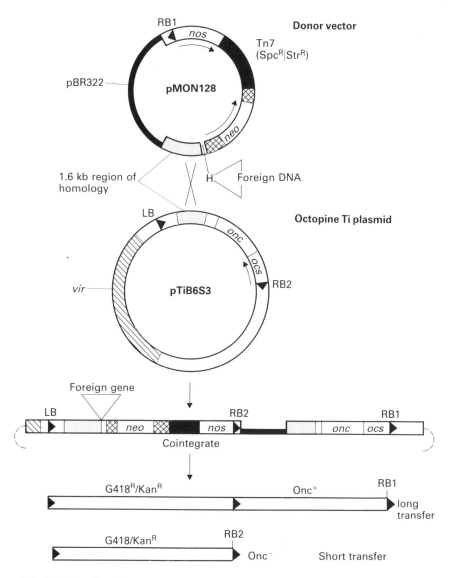

Fig. 6.17. A split end vector (SEV) system. The T-DNA can be transferred using either RB1 or RB2 to produce a long or short T-DNA respectively.

possibility of constructing vectors from a number of plasmids from Gram-negative bacteria.

Split end vectors

The deletion of the tumour-inducing genes in disarmed and binary Ti vector systems has one disadvantage and that is that the transformed tissues grow much more slowly than Onc$^+$ tumour tissue. This means there is a significant delay in monitoring the potential success of the transfer and expression of foreign genes. To circumvent this problem the *split end* vector (*SEV*) system was developed (Horsch *et al.*, 1984, 1985). This system relies on the formation of cointegrates that can transfer *onc*⁺ or *onc*⁻ recombinant T-DNA (Fig. 6.18). A typical donor vector is pMON128 which carries a *nos*−*neo* selectable marker, a single RB sequence and bacterial selectable markers. The recipient is a fully oncogenic Ti plasmid, pTiB6S3 (virtually identical to pTiAch5). A cointegrate is produced that contains two RB sequences and both of these are potentially functional in the plant cell. If RB1 happens to be used then a 'long' T-DNA is transferred and this will produce transformed cells that are Onc⁺ and proliferate rapidly in the absence of phytohormones. This rapidly provides callus tissue that can be analysed for the presence and expression of foreign genes and will indicate if the experiment is worth pursuing. If RB2 happens to be used then only a 'short' T-DNA will be transferred. Transformants can be selected by resistance and can be distinguished from Onc⁺ transformants by slow growth. Fertile adult plants can be generated from these calli. In any one plant cell either RB1 or RB2 is used and the experimenter selects Onc⁺ and Onc⁻ calli for further analysis.

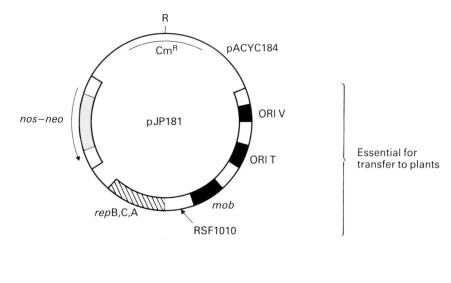

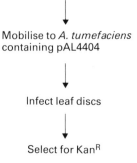

Fig. 6.18. Transfer of a non-Ti plasmid into plants.

A. rhizogenes and Ri plasmid vectors

When *A. rhizogenes* infects plants the production of adventitious roots rather than undifferentiated tumours is induced at the site of infection. This is mediated by *Ri (root inducing)* plasmids which are closely related to Ti plasmids (Huffman *et al.*, 1984). Vectors based on the Ri plasmids have been produced and these have a major advantage because fully *onc*[+] vectors can be used. This is because the roots induced by Ri plasmids in many plants retain the full potential for phytohormone-induced differentiation into fertile adult plants (Tepfer, 1984). A cointegrate vector system involving an Ri plasmid, pRiA4, is shown in Figure 6.19. The plasmid pRiA4 is about 250 kb in length and is as yet poorly characterised. Regions containing T-DNA which have distinct left and right components and the virulence genes have however been located. The vector pCGN529 contains a fragment of TL DNA to allow recombination with the TL DNA of pRiA4 to produce a cointegrate by standard procedures. It also contains a selectable marker, in this case, the *neo* gene driven by the promoter from the mannopine synthase gene (*mas-neo*) and an expression cassette. The cointegrate pRiA4: pCGN529 induces root formation in transformed tissues. These can be excised and cultured in the presence of kanamycin to select for cells carrying

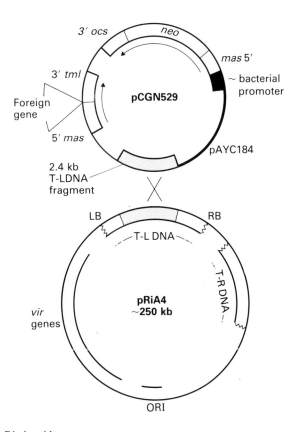

Fig. 6.19. An Ri plasmid vector system.

GENE TRANSFER INTO PLANTS

recombinant T-DNA and simultaneously regenerated by using the appropriate phytohormone balance in the medium. Ri vectors therefore allow single-step selection and regeneration in some species (Comai *et al.*, 1985). Ri vectors are particularly useful for studying nodulation (e.g. Jensen *et al.*, 1986) and manipulating root cultures for secondary metabolite production (Flores *et al.*, 1987).

Experimental procedures for producing transgenic plants

When an *Agrobacterium* strain carrying the desired Ti or Ri cointegrate or binary vectors has been constructed there are a number of approaches to producing transgenic plants. One consideration is whether an extended period of callus or protoplast culture is desirable because karyotypic changes may occur. A second consideration is that the time from the initial transformation to the production of fertile plants should be as short as possible. There are three general approaches: (1) infection of wounded plants or explants: (2) co-cultivation with protoplasts; (3) leaf disc transformation. In all these procedures, steps are taken to minimise contamination by bacteria, e.g. sterilised seeds may be grown in sterilised soil. After T-DNA transfer the inducing *Agrobacteria* are killed with antibiotics such as carbenicillin. The procedures are illustrated in Figures 6.20, 6.21 and 6.22 and much technical information can be found in Lichtenstein and Draper (1985) and Rogers *et al.* (1987).

INFECTION OF WOUNDED PLANTS

Seedlings are decapitated and the freshly cut surface is innoculated with an overnight culture of *Agrobacterium*. An Onc$^+$ Ti vector will produce a $3-10$ mm diameter tumour at the cut surface in $3-6$ weeks. This is excised and grown as a callus culture in the absence of phytohormones. After a further three weeks there is sufficient material to analyse gene transfer but the callus cannot be regenerated. If a SEV system is used the tumour will be a mosaic of short and long T-DNA transformed cells. In this case the callus can be dispersed and cultured under selection for kanamycin resistance, for example, and then induced with phytohormones when a proportion of the calli will regenerate. If an Onc$^-$ vector is used the stem tip is excised after $2-3$ weeks and placed on callus-inducing medium. After $2-4$ weeks the resulting callus is dispersed and microcalli are plated into selective callus medium or non-selective medium to screen for opine synthesis. The transformed calli are picked off, grown and then transferred to an appropriate hormone-balanced medium to induce regeneration. This procedure is laborious and it takes $12-16$ weeks to produce plantlets.

CO-CULTIVATION WITH PROTOPLASTS

Agrobacterium will infect protoplasts as they are reforming the cell wall and undergoing cell division because this mimics the wound response. The protoplasts are incubated for $24-40$ hr in a suspension of *Agrobacteria* at

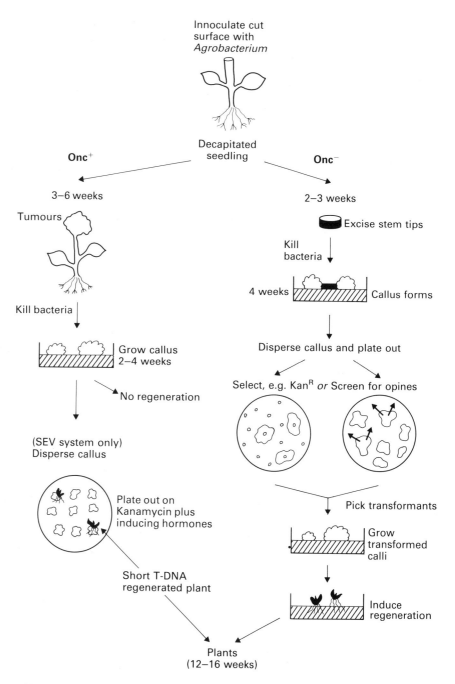

Fig. 6.20. Gene transfer by *Agrobacterium* infection of wounded plants.

about 100 bacteria per protoplast. The protoplasts are plated out either immediately or after forming microcalli and then selection is imposed. The transformed calli can then be regenerated. The advantage of this procedure is that plants are produced from single transformed cells so providing a genetically uniform collection of transgenic plants. The disadvantages are

GENE TRANSFER INTO PLANTS

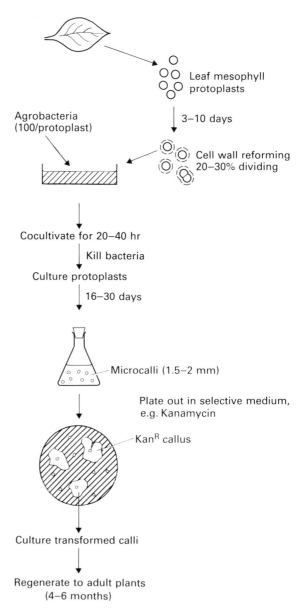

Fig. 6.21. Gene transfer by cocultivating *Agrobacterium* with protoplasts.

that only very robust protoplasts will survive co-cultivation and therefore the procedure is only suitable for a limited number of species. It can take up to six months to produce a plant by this procedure.

LEAF DISC TRANSFORMATION

Discs are punched from a young leaf and co-cultivated with *Agrobacteria* with gentle shaking to ensure infection at the cut surfaces (Horsch *et al.*, 1985). The discs are transferred to filter paper on feeder plates for two days

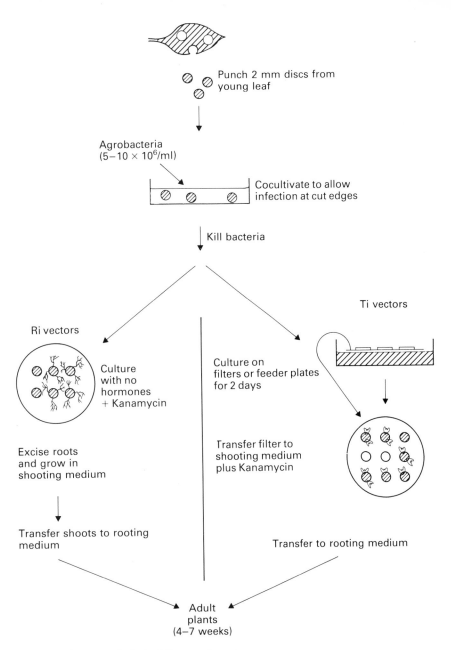

Fig. 6.22. Gene transfer by leaf disc transformation.

and then the paper is transferred to 'shooting medium' containing kanamycin. This imposes one-step selection and regeneration. The shoots are transferred to rooting medium and then planted out in soil. Plantlets can be produced in 4–7 weeks using this procedure. Leaves are particularly useful because they are genetically uniform and they have a high regeneration capacity. The technique has been used for a number of species including *Petunia*, *Nicotiana* and tomato. Improved transformation frequencies can be obtained

GENE TRANSFER INTO PLANTS

by mixing tobacco cells with the discs. This probably works by increasing the concentration of wound factors that induce the *vir* genes.

A similar technique is used with Ri plasmid vectors. In this case adventitious roots are induced at the cut surface. These are excised and put on selective shooting medium. The shoots are then rooted and plantlets are produced in about four weeks. The leaf disc transformation does not involve any undifferentiated tissue stage, i.e. callus or protoplast, and therefore the success of regeneration of genetically normal plants is high. It is possible to use discs of any explanted tissue and usually the best regenerating tissue is chosen for each species, e.g. for tomatoes, cotyledons are used (Klee *et al.*, 1987). The disc method is technically simple and produces transgenic plants very rapidly and is therefore the favoured procedure at present.

Agrobacterium infection of monocots

Agrobacterium normally only infects dicotyledonous plants. However, Hernalsteens *et al.* (1984) have been able to induce tumours in a monocot, *Asparagus officinalis*. The tumours could be propagated *in vitro* without hormones and nopaline and agrocinopine were produced. Hooykaas-van Slogteren *et al.* (1984) infected *Chlorophytum* and *Narcissus* and although tumours were not produced, tissue at the wound site synthesised opines and swellings were produced. These studies showed that *Agrobacterium* can initiate an infection in monocots but this does not happen readily and may be difficult to detect if tumours are not produced. Schafer *et al.* (1987) reasoned that monocots may lack some of the specific wound response factors that are required to initiate *Agrobacterium* infection via induction of the *vir* genes. They, therefore, pre-treated *A. tumefaciens* with wound exudate from tubers of the potato, a normally susceptible dicot. The Agrobacteria were then mixed with bulbil discs from the yam (*Dioscorea bulbifera*) which is an agriculturally important monocot. The infection resulted in swelling of the discs and typical crown gall-type tumours were produced after ten weeks. Untreated Agrobacteria had no effect and wounded yam tissue failed to induce Agrobacteria. T-DNA was detected integrated into the plant genome. The use of pre-induced Agrobacteria may, therefore, extend the host range of this gene transfer system.

Plant viruses as vectors

Vectors based on viruses are desirable because of the high efficiency of gene transfer that can be obtained by infection and because of the amplification of transferred genes that occurs via viral genome replication. Also, many viral infections are systemic so that genes can be introduced into all cells in the plant. In addition there are some advantages of developing a system that does not involve integration into the plant genome which could disrupt essential genes. At present there is no really useful vector but three viruses are being studied: these are cauliflower mosiac virus, brome mosaic virus and gemini viruses.

Cauliflower mosaic virus

Cauliflower mosiac virus (CMV) was initially considered as a potential vector because it has a small (8 kb) double-stranded DNA genome that might facilitate manipulation *in vitro*. The organisation of the genome (Figure 6.23) is, however, extremely compact. There are only two non-coding regions, IR1 and IR2 (intergenic regions) and most insertions into the genome result in loss of infectivity. In addition, the viral capsid is small (50 nm in diameter) and any significant increase in genome size above 300 bp disrupts packaging (Gronenborn *et al.*, 1981). The only non essential genes are genes VII and II which are very small (Figure 6.22) and so there is not much scope for deletion and a selectable marker and non-selected gene cannot be included in the same genome. The replication of the virus is also complex as it involves a reverse transcriptase step (Pfeiffer and Hohn, 1983). This increases the potential for recombination via strand switching and therefore it is not feasible to create a defective-virus/helper-virus system. This is because the infectious virus is readily generated by recombination and this will kill the plant. The involvement of an RNA stage in the life cycle also poses doubts about the stability of expression vectors. This is because enzymes that copy RNA tend to be error prone because they lack

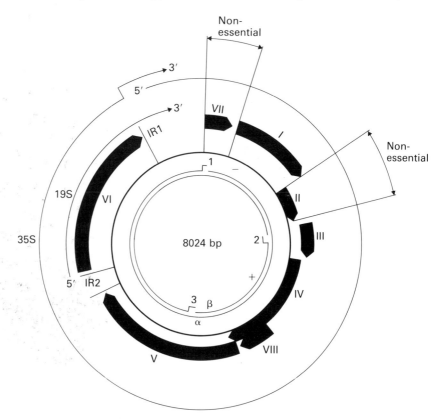

Fig. 6.23. A functional map of the cauliflower mosaic virus genome. The coding regions are shown as dark boxes. The different reading frames are indicated by the position of the boxes. The thin lines in the centre are the DNA and the lines around the outside are the 35S and 19S transcripts.

GENE TRANSFER INTO PLANTS

the proof reading activities of DNA polymerases. In one study, however, CaMV carrying the *dhfr* gene replacing gene II coding sequence was transferred to turnip plants and was still expressed after four weeks in the absence of selection (Brisson *et al.*, 1984). This indicates that medium term expression studies are possible with RNA/DNA viruses.

The major interest in CaMV has switched to exploiting the two viral promoters. During infection, virus particles accumulate in the cytoplasm embedded in a protein matrix to form an inclusion body. The matrix is composed of polypeptide VI and the gene VI promoter is very efficient. The IR2 region has, therefore, been manipulated to drive the expression of other genes. For example, in the vectors pABD1 and pKR612B1 the bacterial *neo* gene is under IR2 control (Figure 6.24); Paszkowski *et al.*, 1984; Balazs *et al.*, 1985). The 35S RNA promoter has also been manipulated to drive the *neo* gene in the vector CaMV-*neo* (Fromm *et al.*, 1986). Selection/expression cassettes based on CaMV expression signals are being used to construct simple plasmid vectors for plant protoplast transformation (see later).

Brome mosiac virus vectors

Brome mosaic virus is a small multicomponent RNA virus that infects a number of monocotyledonous plants including cereals (Ahlquist *et al.*, 1987). There are three separate viral genomic RNAs which are also messenger RNAs and which are packaged into separate virions. During infection, a sub-genomic RNA, RNA4, is derived from RNA3. This sub-genomic RNA encodes the gene for the virion coat protein which is produced in milligram quantities in infected cells. It is simple to clone cDNA corresponding to each of the viral RNAs, and when these are used as templates in an *in vitro* transcription reaction the resulting mixture of RNAs can be injected into whole barley plants and isolated barley protoplasts to produce a typical BMV infection (Ahlquist *et al.*, 1984; French *et al.*, 1986). Replication and expression of the genome only requires RNA1 and RNA2. French *et al.* (1986) have now constructed a vector based on BMV for the expression of foreign genes in monocots (Figure 6.25). A derivative of RNA3 cDNA was

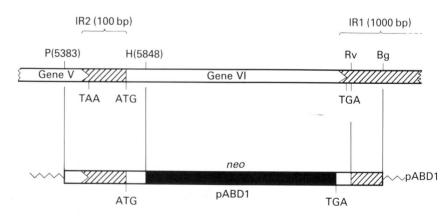

Fig. 6.24. An expression vector based on CaMV.

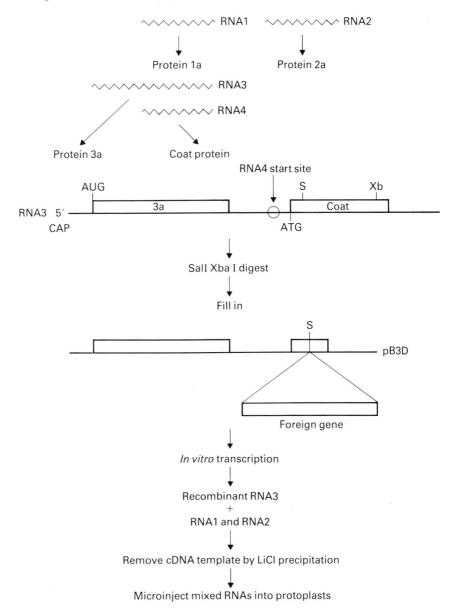

BMV genome

RNA1   RNA2

Protein 1a   Protein 2a

RNA3

RNA4

Protein 3a   Coat protein

RNA4 start site

AUG   S   Xb

RNA3 5' ——————[3a   ]———○——[Coat   ]———————

CAP   ATG

Sal I Xba I digest

Fill in

S

————[   ]————[ ]———— pB3D

Foreign gene

In vitro transcription

Recombinant RNA3
+
RNA1 and RNA2

Remove cDNA template by LiCl precipitation

Microinject mixed RNAs into protoplasts

Fig. 6.25. A Brome mosaic virus vector.

constructed with a 0.5 kb deletion in the coat protein gene to produce the vector pBD3. Any coding sequence can be inserted at the unique Sal I site in the deleted coat gene and should be expressed using the coat gene AUG provided that the correct reading frame is maintained. The system has been tested by inserting the bacterial CAT gene, and high levels of CAT activity were detected in protoplasts 20 hr after innoculation. These levels were about 20 fold higher than those reported with promoters from the nopaline synthase and Rubisco genes suggesting that BMV vectors may be useful

GENE TRANSFER INTO PLANTS

when high yields of heterologous proteins must be synthesised. The system is currently being tested for long-term stability of expression to show whether mutations occur via RNA replication.

Agrobacterium-mediated virus infection

It has now been shown that plant virus genomes that are cloned into Ti plasmids can be introduced into plants from *Agrobacterium* and they will then initiate a typical infection. This *Agrobacterium*-mediated virus infection has been called *Agroinfection*. It has been used to provide a sensitive assay for T-DNA transfer into monocots. Successful transfer of a single viral genome can be detected as a result of the massive amplification during viral replication (Grimsley *et al.*, 1987). A vector was constructed that contained the cloned double-stranded copy of the maize streak virus (MSV) genome. This is a gemini virus that has a single-stranded DNA genome that replicates in the plant nucleus via DNA intermediates that are autonomous. The naked viral DNA and cloned DNA are not infectious and are transmitted in nature by insects. A head-to-tail dimer of the MSV genome was inserted into a mini-T-DNA vector between border sequences. The plasmid was mobilised into *A. tumefaciens* in a triparental mating and the host Agrobacteria were innoculated into leaf wounds in maize seedlings. The maize plants developed all the symptoms of MSV infection. This study showed that Agrobacteria can transfer DNA to the *Graminae* in addition to the monocots we have already discussed. The study also shows that gemini viruses might provide the basis for useful vectors. The genomes can be manipulated via cDNA − there is no packaging constraint and helper systems for replication without disease could be developed. Successful agroinfection of another gemini virus, tomato golden mosaic virus (TGMV) to *Petunia* has also been demonstrated (Rogers *et al.*, 1986).

Direct DNA transfer into plant cells

DNA can be introduced directly into plant protoplasts by techniques that are similar to those used for animal and yeast cells. The advantage of this procedure is that it can be used on any plants from which protoplasts can be obtained and it is the most widely used procedure for introducing DNA into monocots. Paszkowski *et al.* (1984) introduced pABDI (Figure 6.22) into *Nicotiana tabacum* protoplasts by treating them with PEG 6000. The protoplasts were cultured by embedding them in agar beads to facilitate changing the selective medium and after six weeks of culture in kanamycin-containing medium, resistant calli had developed which could be sub-cultured by standard procedures. The transformation frequency was extremely low, 10^{-5} to 10^{-6}.

Similar results were obtained when Ti plasmid DNA was introduced into tobacco leaf protoplasts by either PEG6000 treatment or by liposome-mediated transformation (Kreus *et al.*, 1982; Deshayes *et al.*, 1985). The transformation frequencies in these studies were much lower than by infecting protoplasts, which gives 1−3% transformants, or by wound infection, which

CHAPTER 6

produces 20–50% transformed calli. More recently the technique of electroporation has been used and recent modifications have increased the frequency of successful transformation to 2% (Fromm et al., 1985, 1986; Shillito et al., 1985). In this procedure plant protoplasts are electrically permeabilised in the presence of vector DNA by discharging about 350 V through a protoplast suspension. The protoplasts are cultured for two weeks and then selection is imposed. After a further two weeks microcalli can be transferred to solid selective medium and between five and six weeks after electroporation there is sufficient material for analysis (Figure 6.26). When pCaMV-neo was introduced into maize protoplasts, 161 kanamycin-resistant protoplasts were obtained from 2×10^6 initial protoplasts. As only 1% of the culture survived electroporation, this is a transformation frequency of about 1%. The transformed calli contained 1–5 copies of CaMV-neo integrated per haploid genome and all expressed neomycin phosphototransferase (Fromm et al., 1986). Direct DNA-mediated transfer by electroporation is particularly useful for transient expression studies to rapidly assess whether a particular plasmid construction will be expressed in particular plant cells. Gene expression can be monitored 48 hr after electroporation (Fromm et al., 1985). The technique is only limited by the quality of the protoplasts and for many Graminae this is still a problem (Ozias-Akins and Lörz, 1984).

It is possible that the inclusion of sequences that increase the efficiency of DNA integration into the genome may further improve these direct transfer techniques. One possibility is to manipulate plant transposons in much the same way as described for P-element vectors in Drosophila. Two potential transposon vectors are the Ac transposons of maize and the mutator element Mu-1 (reviewed by Freeling, 1984). At present, however, there is not enough known about the genetic organisation of these elements to produce vectors.

Although it is now possible to introduce DNA into monocot protoplasts, it is still not possible to regenerate entire plants from most commercially important monocots, the exception being rice (Uchimaya et al., 1986). Several new approaches may be relevant here. Ohta (1986) has shown that maize can be transformed by mixing DNA with pollen grains which are then used to fertilise plants. The DNA was found in the embryo and endosperm and appeared to be transmitted to the next generation. It is also possible to inject DNA directly into the floral side shoots (tillers) of rye plants to produce pollen that contains foreign DNA. This appears to be taken up by pre-meiotic cells that give rise to the pollen which then cross-fertilises to produce transgenic seedlings (De La Pena et al., 1987). Cocking (1985) has been able to produce protoplasts in situ by enzymatically treating the root hairs from a number of species including Zea mays (maize) and Triticum aestivum (wheat). This opens the possibility of transforming with agents that can spread systemically and deliver foreign genes to all parts of the plant. It might also be possible to regenerate transformed root tips which have not de-differentiated by in vitro culture and Rathore and Goldsworthy (1985) have been able to induce significant shooting of wheat roots by electrical treatment. Another approach has been to shoot genes into epidermal cells using 4 μm tungsten particles coated with DNA. This has

GENE TRANSFER INTO PLANTS

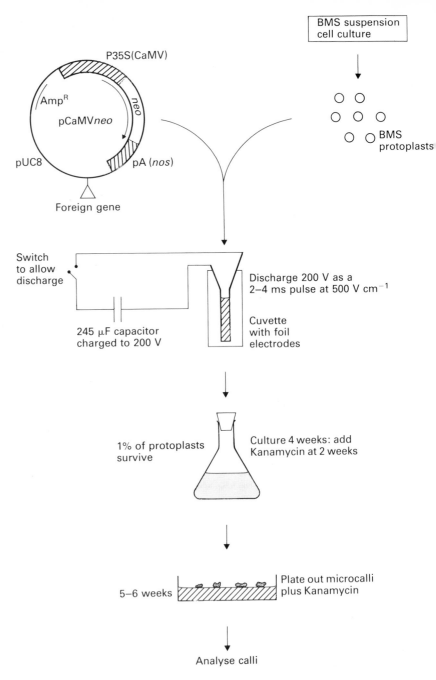

Fig. 6.26. Introducing DNA into maize protoplasts by electroporation. BMS = black Mexican sweetcorn.

only been attempted with onion cells which are easily obtained and large but the frequencies of 23–44% were the highest obtained for any monocot (Klein *et al.*, 1987). It seems likely that the routine production of transgenic monocots may not be far off.

Status and expression of transferred genes

Several studies have analysed the inheritence and expression of recombinant T-DNA. For example *Nicotiana* plants regenerated from pGV3850:*neo* transformed calli were completely homogeneous for T-DNA incorporation (De Block *et al.*, 1984). 3500 calli derived from leaf protoplasts from the transgenic plant, shoot, stem and root explants were all kanamycin-resistant. The plants were self-fertilised and produced a normal seed yield with 100% germination. Seeds obtained from a self-cross showed a 3:1 ratio or kanamycin resistance to sensitivity showing that the original plant was hemizygous and *kan* was segregating as a simple dominant Mendelian trait. This simple pattern of inheritance indicates a single site of insertion and this has been confirmed by Southern blotting in a number of different studies. The copy number is usually low, between $1-5$ copies per cell. Similar results have been obtained in *Petunia* (Horsch *et al.*, 1984) and by the leaf disc co-cultivation technique (Horsch *et al.*, 1985). When DNA is transferred directly into protoplasts using non Ti-vectors similar patterns of inheritance are found. There is usually only a single random site of insertion and $3-5$ copies of the gene. The DNA is transmitted through meiosis as a simple Mendelian trait. Occasionally, however, more complex patterns of integration are seen with this technique indicating multiple insertion sites (e.g. Paszkowski *et al.*, 1984).

The expression of a number of genes in transgenic plants has been analysed and both tissue specific and developmental-stage-specific expression have been obtained. Transgenic plants have, for example, been produced using a number of genes involved in photosynthesis. In some cases the entire gene has been transferred and in others the 5′ flanking DNA has been fused to a marker gene such as *neo* or *cat*. For example, the transgenic plants have been produced with *cab* genes from pea (Simpson *et al.*, 1985), wheat (Lamppa *et al.*, 1985) and *Petunia* (Jones *et al.*, 1985) and pea *rbc-s* (Nagy *et al.*, 1985), reviewed by Kuhlemeier *et al.* (1986b). Two vectors for the expression of photosynthesis genes are shown in Figure 6.27. Using these vectors, genes were expressed predominantly in the leaves with a small amount in the stem and no expression in roots. In addition the leaf-specific expression was inducible by light by about 50-fold. These studies showed that faithful tissue-specific gene expression occurs in transgenic plants. Some quantitative anomalies, however, have been described. For example, when the pea *rbc-s* gene was expressed in transgenic *Nicotiana*, the levels of expression relative to the *nos*−*neo* transcript on the same vector (Figure 6.25A) varied by 25-fold between different plants and the maximum level of expression was only $0.2-10\%$ of that in peas. Similarly, when the *Petunia cab* gene promoter was fused to *ocs* (Figure 6.25B) there was more than 200-fold variation in levels of expression of *cab* and there was independent variation in nopaline synthase levels directed by the co-transferred *nos* gene (Jones *et al.*, 1985). These data are probably due to position effects that interfere with regulated promoters, such as the *cab* gene promoter, more severely than with constitutive Ti DNA promoters. Unlike the situation in animal cells, chromosomal position in plants only appears to affect the

A   The expression of pea rbc-s

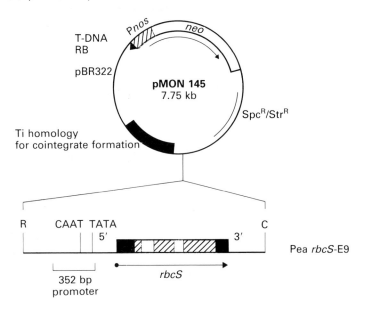

B   Cointegrate T-DNA for transferring petunia *Cabb/ocs* fusions

Fig. 6.27. Heterologous gene expression. P*nos* = the nopaline synthase gene promoter; P*nos*/Kan[R] is the promoter driving the Kan[R] gene.

quantity and not the control of transcription. It is also becoming clear that not all genes from monocots are expressed in dicots, for example, the wheat *cab* gene is expressed in a tissue-specific fashion in *Nicotiana* (Lamppa *et al.*, 1985) but the wheat *rbc-s* gene is not expressed in any tissue (Keith and Chua, 1986).

Correct developmental regulation of gene expression has been demonstrated by transferring the gene for β-phaseolin (the major storage protein of the French bean) into tobacco stem explants. The seeds of the regenerated plants contained 1000-fold more β-phaseolin than in other tissues and the genes were switched on only in embryonic tissue during seed development. This differed from the endogeneous tobacco storage proteins which are expressed in embryonic and endoderm tissue, but was identical to the expression in French beans. The phaseolin was glycosylated and processed and packaged into protein bodies as 3% of total seed protein in a similar fashion to the natural situation (Sengupta-Gopalan *et al.*, 1985).

Manipulating organelles

Choloroplasts

Many modifications of plant characteristics require altering the properties of chloroplasts. Resistance to the herbicide atrazine is conferred by a single amino acid change in a photosystem II chloroplast protein and improving photosynthetic efficiency will require changes in the chloroplast genotypes.

The gene transfer systems that we have discussed so far have all resulted in the nuclear transcription of genes. The chloroplasts are however essentially prokaryotic organelles. They each contain 30−200 copies of small circular genome of about 100 kb which codes for a number of chloroplast proteins, ribosomal RNAs and tRNAs and aminoacylating enzymes. There are about 100 chloroplast directed polypeptides but nuclear genes also contribute proteins, e.g. the small subunits of rubisco which is the major protein in chloroplasts. The nuclear encoded proteins are transported into the chloroplasts via specific signals in the polypeptide. The signals for gene expression in chloroplasts are prokaryotic. Transcription is directed by promoter regions that contain a 'Pribnow box' and −35 sequence and the transcripts are not polyadenylated. The ribosomes are 70S and the chloroplast tRNAs are distinct from the cytoplasmic tRNAs. In addition, gene expression in chloroplasts is sensitive to antibiotics such chloramphenicol which do not affect 'eukaryotic' protein synthesis.

In order to obtain gene expression in chloroplasts it is therefore necessary to use prokaryotic signals. De Block *et al.* (1985) have recent recently shown that the Ti vectors can be used to introduce genes into chloroplasts. They fused the *nos* promoter to the bacterial gene for chloramphenicol acetyl transferase, *cat*, and produced a cointegrate with pGV3850. This was transferred to *Nicotiana* protoplasts by co-cultivation and calli resistant to chloramphenicol were obtained. The *nos* promoter is recognised in both plant and bacterial cells and the *cat* gene was engineered to retain the Shine and Dalgarno sequence to allow translation in a prokaryotic system. The vector DNA was detected in purified chloroplasts, but not all the chloroplasts were transformed. Repeated sub-culturing of the callus without selection resulted in loss of chloramphenicol resistance. Resistance was transmitted maternally but could not be transmitted via the pollen. This is the expected pattern of inheritance of an organelle gene.

Another approach is to manipulate the nuclear genes to produce proteins that can then be transported into the chloroplast. The small subunit of rubisco is synthesised as a precursor protein with a 57 amino acid N-terminal *transit peptide* which is removed by cleavage as the protein is transported into the chloroplast (see Chapter 10 for more details). Van den

Broek *et al.* (1985) have fused the coding region for this peptide to the *n* coding region and introduced the chimaeric gene into tobacco using a ' vector. The neomycin phosphotransferase is transported into the chloroplas the transit peptide is accurately removed and neomycin resistant plants a produced.

Mitochondria

Plant mitochondria are more complex than those in animal cells. Th genomes can be circular or linear and can be as large as 2500 kb. addition, the mitochondria of many plant species contain small autonomo plasmids. Gene expression in plant mitochondria is poorly characterised b both prokaryotic and eukaryotic features have been identified. There ha been suggestions that it might be possible to manipulate the mitochondri plasmids to allow gene transfer into isolated mitochondria. These cou then be introduced into plant protoplasts by PEG-mediated uptake and t modified protoplasts could be regenerated into plants (Sondahl *et al.*, 1984

Boutry *et al.* (1987) have succeeded in targeting a recombinant gene the mitochondria of transgenic *Nicotiana plumbaginifolia*. They fused t pre-sequence from the gene encoding the β subunit of mitochondrial AT synthase, *atp2*−1 to the bacterial *cat* gene. Expression was driven by t CaMV 35S promoter and the hybrid gene was introduced into tobacco ce using an *Agrobacterium* binary vector. CAT activity was obtained predom nantly in the mitochondria and none was detected in chloroplasts. Th study shows that protein can be specifically directed to mitochondria ar opens the way for both studying mitochondria specific traits such as ma sterility and for manipulating these traits.

Summary

The ability to introduce new genes directly into plants so that they a inherited as simple Mendelian traits is revolutionising plant breeding. Th is largely because entirely new characteristics can be introduced whic would not be possible by conventional breeding, for example, the introductic of genes from viruses, bacteria and other plant species. Gene transfer is als accelerating the pace of discovery about fundamental processes such nitrogen fixation and photosynthesis.

The original *Agrobacterium* vectors have been refined and these ne vectors and new techniques such as electroporation are increasing the ran of plant species that can be manipulated. This, coupled with improvemen in *in vitro* culture of plant tissues and calli and plant regeneration ha resulted in the successful manipulation of several agriculturally importa legumes and cereals.

CHAPTER 6

III The Use and Exploitation of Gene Transfer Technology

7 Gene Isolation and Identification

Introduction

The most common method of cloning genes is by screening a genomic or cDNA library in *E. coli* (Old and Primrose, 1985). This can, however, be a time consuming process. Cloning in *E. coli* also relies on having a specific test for the gene. This may be a nucleic acid probe or an assay for gene function. There are however, many situations where there is insufficient information about the gene product to develop a reliable screening technique in *E. coli*. This lack of information could be as fundamental as not even knowing if there is a gene involved in a particular process! For example the malignant phenotype of a tumour cell could be initiated via complex genetic and environmental interactions. How would you clone the 'tumour genes' in *E. coli*? Similarly the regulation of gene expression in eukaryotes presumably involves regulatory genes whose products may be rare. How may these regulatory genes be cloned? Many eukaryotic genes are organised as multigene families, each member gene encoding a 'functionally distinct but related protein. How do you make the correct gene–protein assignments? In this chapter we use some selected examples to show how the ability to transfer DNA into eukaryotic cells can be used to facilitate the isolation of eukaryotic genes.

DNA tagging

DNA tagging is a general procedure that can be applied to cloning potentially any gene from any organism. It simply relies upon the fact that the gene to be cloned is physically linked to a piece of DNA, the tag, that can be readily identified. In some cases, there is a natural tag. This is where DNA from different species can be recognised by the presence of species-specific, repetitive DNA. In other cases the tag is introduced; for example, a retrovirus will insert randomly into genomes and so any flanking chromosomal DNA will be tagged with a retrovirus genome. Transposons and simple plasmids are also used as tags. In some cases the tag itself causes a mutation which can indicate that it has disrupted an interesting gene − this is *insertional mutagenesis*. The gene is therefore identified and tagged at the same time. A useful procedure is to use a tag that can function as a plasmid in *E. coli*. This means that the tagged DNA is very easy to clone by simply transforming

E. coli — this is called *rescue*. Some of these techniques are now described in more detail.

Cloning the chicken thymidine kinase gene by plasmid rescue

The chicken *tk* (thymidine kinase) gene was the first eukaryotic gene to be isolated by using gene transfer techniques (Perucho *et al.*, 1980). Chicken chromosomal DNA was first digested with a range of different enzymes and the resulting DNA was used to transform mouse L*tk*⁻ cells to HAT resistance. This preliminary study showed that Hind III- and Eco RI-digested DNA was capable of transferring the *tk*⁺ gene to L*tk*⁻ cells and therefore these enzymes were used for the subsequent cloning. Hind III digested chicken DNA was ligated to an equal amount of Hind III digested pBR322 and the mix was used to transform L*tk*⁻ cells. Twelve L*tk*⁺ clones were obtained. Southern blot analysis showed that eight clones contained 2−20 copies of pBR322 DNA. The vector DNA was in vast excess of the *tk* gene and therefore many of these copies of pBR322 would have represented random multiple insertions into the genome and would not necessarily be linked to the *tk* gene. A second round of transformation and selection was therefore undertaken. This time, total DNA from a *tk*⁺ transformant was digested with EcoRi before transforming mouse L*tk*⁻ cells. Several *tk*⁺ clones from this second round now contained only a single copy of pBR322 DNA and the co-transformation of *tk*⁺ and vector DNA indicated that these sequences must be closely linked. DNA from one of these secondary transformants was digested with EcoRI or BamHI to release linear DNA fragments which were then ligated to produce circular molecules which were used to transform *E. coli*. The majority of the chromosomal DNA clearly cannot transform *E. coli*, but chromosomal DNA that was linked to pBR322 could transform *E. coli* to ampicillin resistance using the replication and selection sequences on the vector. Two amp^R colonies were isolated and one of these carried plasmids that contained a 9.2 kb EcoRI fragment of chicken DNA. This plasmid transformed mouse L*tk*⁻ to *tk*⁺ at a high frequency and the L*tk*⁺ transformants contained a new enzyme activity that had the isoelectric point of chicken thymidine kinase and not mouse thymidine kinase. The chicken *tk* gene had therefore been cloned (Figure 7.1).

This study was successful for several reasons: (1) the *tk* gene was small and so a number of different restriction enzymes could be used to manipulate the DNA without destroying the gene, (2) the recipient cells have a very low reversion frequency and so the rare *tk*⁺ transformants were readily detected, and (3) a very high efficiency *E. coli* transformation rescue procedure was used which gave more than 10^7 transformants per µg of pure plasmid. This ensured that the single copy pBR322 that was present at a concentration equivalent to about 10^{-6} µg of pure plasmid could be rescued. A variation of this procedure is to introduce a cosmid vector that allows the rescue of very large segments of chromosomal DNA (e.g. Brand *et al.*, 1985).

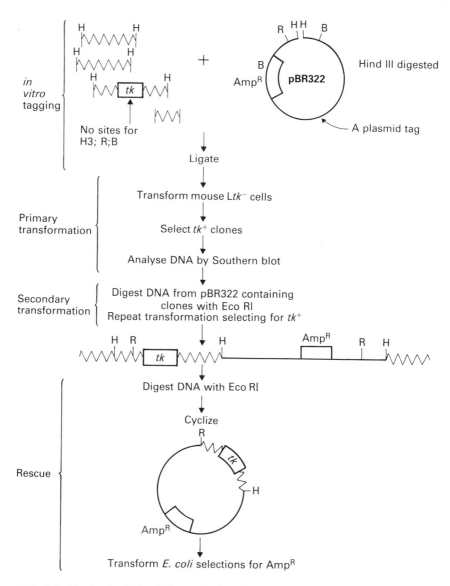

Fig. 7.1. Cloning the chicken TK gene by plasmid mouse.

Isolating oncogenes by suppressor rescue

Total DNA isolated from a variety of animal and human tumours will induce neo-plastic transformation after DNA mediated transformation of non tumour cells such as mouse NIH3T3 cells (Shih *et al.*, 1981; Perucho *et al.*, 1980). This finding indicated that the tumour phenotype was likely to be determined by one or a few genes and opened the way to cloning these genes. This was first achieved by Goldfarb *et al.* (1982), using a gene rescue technique. Total DNA from a human bladder carcinoma cell line (T24) was digested with Hind III, which did not destroy the transforming activity, and then ligated with equimolar amounts of a 1.1 kb DNA fragment which

GENE ISOLATION AND IDENTIFICATION

carried the *E. coli supF* gene, which encodes a tRNA amber suppressor. The DNA was used to transform NIH3T3 cells selecting for morphologically transformed foci. DNA from these primary transformants was then used to re-transform NIH3T3 cells in a second round of selection for transformed foci. These foci were pooled and used to prepare large DNA fragments that were cloned into a bacteriophage lambda vector carrying an amber mutation, *Sam*7, in the lysis gene, S. This phage will normally only grow on a *supF E. coli* host, i.e. one which will suppress the lysis gene mutation. If, however, a fragment carrying the *supF* gene is inserted into the phage it will now produce plaques on a non-suppressor, sup^0, host.

This system provides, therefore, a simple direct means of selecting for DNA fragments containing the *SupF* gene. In this particular case, it permitted the isolation of fragments enriched for the presence of an oncogene. The detailed procedure is shown in Figure 7.2. DNA from the second round NIH3T3 transformants was partially digested with Sau3a to produce fragments of 7−18 kb suitable for inserting into the *Sam*7 lambda vector to produce a phage library. After *in vitro* packaging the phage S were plated on to an *E. coli Sup^0* or *SupF* host strains. Any plaques which form on the sup^0 host must contain the *supF* gene in the recombinant phage. DNA from four such *supF* phages was capable of morphologically transforming NIH3T3 cells at a high frequency showing that the 'tumour-inducing' oncogene had also been cloned.

DNA sequences homologous to this transforming DNA from the T24 bladder carcinoma were also detected, by Southern blotting, in a variety of non-transformed cell lines. Subsequent studies have shown that the T24 transforming DNA carries an oncogene called *Ha-ras* which had previously been identified in the retrovirus, Harvey murine sarcoma virus. The *Ha-ras* oncogene in the T24 cells (*cHa-ras*) and in the retrovirus (*Ha-ras*) are mutant alleles of a normal cellular gene which does not have transforming activity. A number of different oncogenes have now been cloned using gene transfer and rescue procedures although the *ras* oncogenes are most readily identified by this technique. Oncogenes are discussed in more detail in Chapter 11 and reviewed by Varmus (1984).

Cloning DNA repair genes using a species specific tag

The procedures we have discussed so far have involved *in vitro* ligation of the tag DNA and the chromosomal DNA. It is possible however to use a naturally occurring DNA sequence to follow the fate of the transferred DNA and to clone a gene. For example, the human Alu repeat DNA is randomly dispersed throughout the genome such that on average there is one sequence every every 5 to 10 kb (Chapter 1). Any human gene of interest is likely therefore to be linked to an Alu I repeat. A human DNA excision repair gene was cloned by exploiting the Alu I repeats (Rubin *et al.*, 1985). A chinese hamster ovary cell line (CHO-UV20) that was sensitive to ultraviolet (UV) and a variety of DNA-damaging chemicals such as mitomcyin C (MM-C) was used as the recipient cell. The aim was to complement the deficiency with human DNA and to select transformed

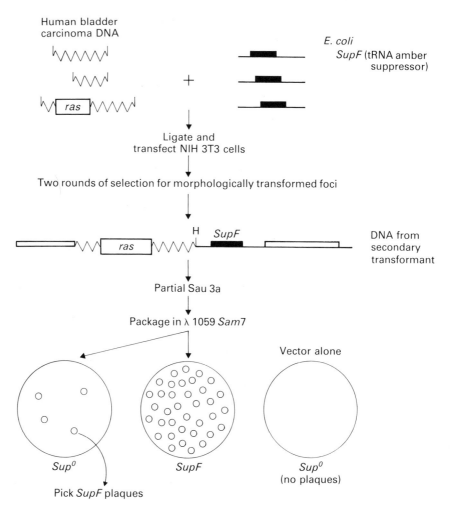

Fig. 7.2. Cloning the *ras* oncogene by suppressor rescue.

cells that were resistant to MM-C. High molecular-weight DNA from HeLa cells was transferred to CHO-UV20 cells in the presence of pSV2-*gpt* (Figure 4.5) and initial transformants were selected on the basis of resistance to MM-C and DNA from these primary transformants was used in a second round of transformation. A genomic library was prepared in λ gtWES.AB with DNA from a secondary transformant. The library was screened using a standard plaque hybridisation with nick-translated total human DNA as a source of Alu I repeats. Two recombinant phages were identified that contained DNA fragments able to transform CHO-UV20 cells to MM-C resistance at high frequency and which presumably contained a human DNA repair gene. An independent approach using cosmid rescue resulted in recovery of the same gene, which has been called human excision repair gene ERCC-1, and this has recently been shown to have significant homology with a yeast excision repair gene, RAD10 (van Duin *et al.*, 1986).

GENE ISOLATION AND IDENTIFICATION

Gene cloning by insertional mutagenesis

This *in vivo* procedure involves creating a mutant phenotype by inserting a DNA tag into or near the gene. Theoretically any gene can be cloned by this approach: it only requires that the mutation can be scored and that insertion of the DNA tag can occur anywhere in the genome. It is not even necessary to know much detail about the genes to be cloned. For example one of the major goals of the approach is to clone genes that are required for normal development, a very general phenotype. In this section we describe the use of retroviruses, plasmids and transposons as insertional mutagens for gene cloning.

Cloning a mouse collagen gene following retroviral mutagenesis

Retroviruses and retroviral vectors become integrated into the germ line of mice following infection of early embryos (Chapter 5) (Jaenisch *et al.*, 1981; van der Putten *et al.*, 1985). This allows the production of transgenic mice carrying a provirus integrated at a single chromosomal location and in some cases the integration may disrupt a gene. For example Jaenisch *et al.* (1983) produced thirteen mouse substrains transmitting MLV at a single locus, designated *Mov*. When the heterozygous parent strains were bred homozygous progeny were obtained in the correct Mendelian frequency in 12 strains, but in one strain, Mov13, no viable homozygotes were obtained. This indicated that MLV had inserted into an essential gene. The mutated gene was isolated by making a genomic library from Mov13 DNA and probing with an MLV-specific probe (Harbers *et al.*, 1984). Molecular analysis showed that the virus was integrated into the first intron of the $\alpha 1(I)$ collagen gene and inhibits transcription (Hartung *et al.*, 1986). Clearly in this case early embryonic death was a consequence of the phenotypic effect of loss of a gene for a major structural component rather than a developmental abnormality affecting the differentiation and spatial organisation of cells (Lohler *et al.*, 1984).

Some naturally occurring mutations such as the dilute (d) coat colour mutation are due to the loss of an ecotropic retroviral provirus. Furthermore, the natural germ-line acquisition of ecotropic murine leukaemia viruses is high. Following crosses between particular strains of mice up to 75% of the progeny may contain 1−20 new insertions (Jenkins and Copeland, 1985). These studies indicate that the frequency of retroviral integration either by gene transfer or natural infection is high enough to generate many mutations that can be cloned by probing for the particular retroviral tag.

Retroviral shuttle vectors for isolating proviruses and linked genomic DNA

In Chapter 4 we discussed the general strategies of designing retrovirus vectors for introducing selectable and non selectable genes into mammalian cells. Cepko *et al.* (1984) have described a more sophisticated type of vector that allows the rapid isolation of chromosomal sequences flanking the site of proviral insertion. The vector pZIP-NeoSV(X)1 (SVX) is derived from

MoMLV and contains the *neo* gene from Tn5 for selecting G418 resistance in mammalian cells and kanamycin resistance in *E. coli*. The *neo* gene is expressed in mammalian cells via a sub-genomic spliced mRNA using the retroviral splicing signals and in *E. coli* via the prokaryotic promoter (PPR) that is present on the fragment. In addition the vector contains the SV40 replication origin (SVORI) and the pBR322 origin (Figure 7.3). Infectious SVX pseudovirions can be produced by using appropriate helper viruses or helper cell lines and used to transform a range of cell types. Transformants are selected for G418 resistance and in each independent G418[R] colony the provirus will be integrated at different locations. The integrated provirus can then be rescued from these transformants using a number of procedures that rely on the functions of the SV40 and pBR322 replication origins. One of the most efficient procedures is to fuse the SVX transformed cells with COS cells. This is done by exposing the mixed cultures of COS cells and SVX transformants to PEG 1000. The SV40 large T antigen that is produced in COS cells 'activates' the SV40 origin in the SVX provirus and causes 'onion skin' replication of the proviral DNA and flanking DNA in the chromosome (Figure 7.4). This amplification process promotes recombination and proviral sequences are readily excised as circles by intrastrand recombination. In the majority of cases excision is by homologous recombination across the LTRs (long terminal repeats) to generate an SVX plasmid with a single LTR. In 10−15% of the cases illegitimate recombination occurs in the flanking chromosomal DNA. This generates larger plasmids that have a piece of chromosomal DNA inserted between two LTRs. These SVX plasmids can be rescued by using plasmid enriched DNA preparations to transform *E. coli* selecting for resistance to kanamycin. The plasmids will replicate in *E. coli* using the pBR322 origin. The plasmids carrying chromosomal inserts can be enriched further by size selecting them by agarose gel electrophoresis before transformation. Once the sequences flanking SVX have been isolated by these procedures the undisrupted gene can be cloned

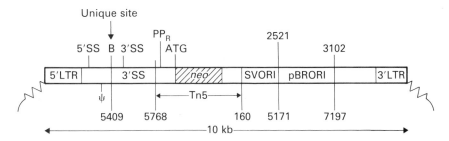

Fig. 7.3. The retroviral shuttle vector pZIP-Neo SV(X)1 (SVX). The vector has two replication origins derived from SV40 (SVORI) and pBR322 (PBRORI), the *neo* cassette with the prokaryotic promoter (PP$_R$) and retroviral promoter, splicing and packaging signals. The numbers below the line are the nucleotide coordinates of the MoMLV and SV40 genomes and those above the line are for pBR322. See Genetic Maps 1984 for coordinates in SV40 MoMLV and Sutcliffe (1978) for pBR322. 1, The 3′ SS fragment is a BglII-XbaI fragment from MuMLV containing the 3′ accepter site. 2, The Bam HI site is a conversion of the PstI site at 566 in MoMLV. 3, The initiating ATG for gag is normally at 621. 4, 1.4 kb *neo* fragment is the HindIII-SalI fragment from Tn 5 (see Figure 4.8A).

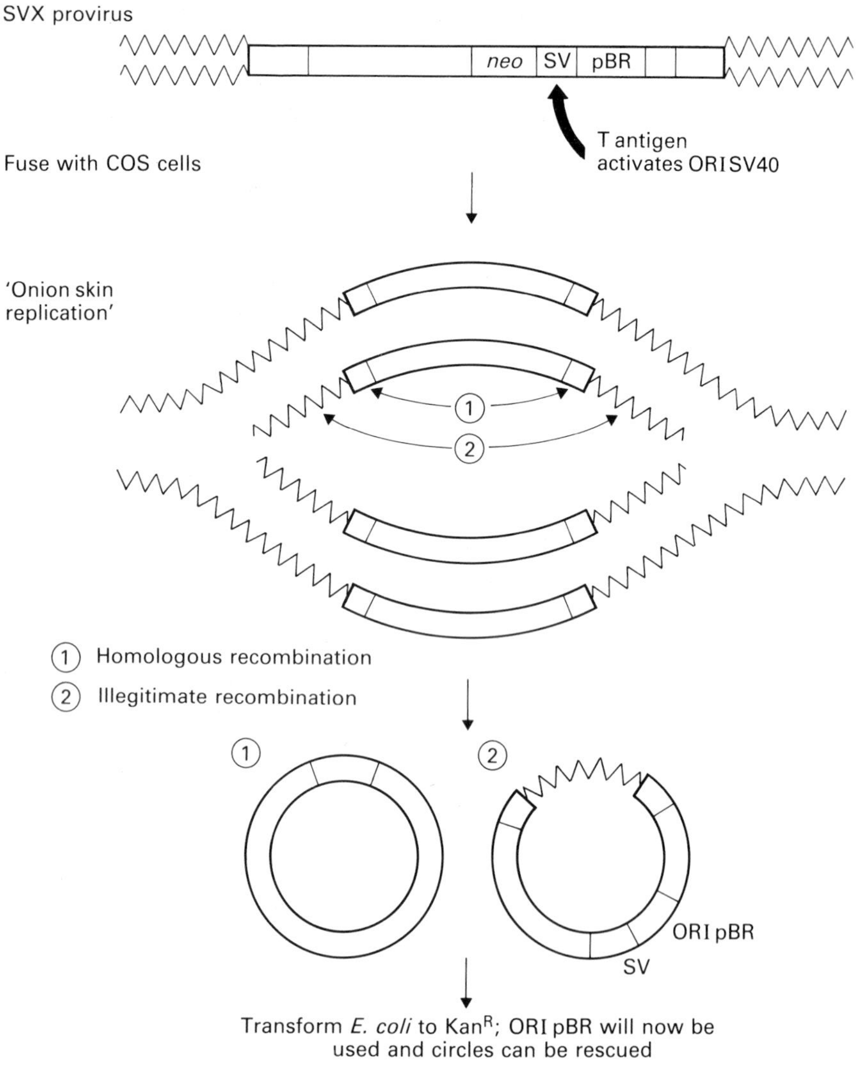

SVX provirus

Fuse with COS cells

T antigen
activates ORI SV40

'Onion skin
replication'

(1) Homologous recombination

(2) Illegitimate recombination

ORI pBR

SV

Transform *E. coli* to Kan^R; ORI pBR will now be
used and circles can be rescued

Fig. 7.4. Rescue of genomic DNA in SVX.

by using the SVX flanking sequences as a probe against an appropriate genomic bank.

This system for retroviral vector rescue can be useful in other circumstances. Firstly the system can be used to easily check the integrity of the retroviral vector. This may be important if it is being used as the basis of a gene transfer system. The retroviral vector and any inserted DNA can be rapidly recovered from any transformant and analysed. This will show if any changes have occurred in the vector that might have affected expression and led to false interpretations in gene analysis studies. Secondly, the chromosomal location of genes transferred by this vector can be analysed to indicate whether it may have any effect on expression of transferred genes. For example the chromatin structure and methylation can be examined.

Finally, the system can be used to study the biology of the retroviruses themselves. For example retrovirus replication should be error prone because the reverse transcriptase has a high rate of misincorporation *in vitro* and an ability to switch templates, suggesting the potential for duplication and deletions. The frequency of these during *in vivo* infection can be analysed by rescuing and sequencing proviruses.

In a variation of this rescue technique SVX pseudovirions can be used to infect COS cells. In this case, the proviruses that are generated as integration intermediates are amplified. This allows the rescue of plasmids that have two LTRs linked head to head, providing material to analyse integration.

Cloning genes involved in mammalian limb morphogenesis

As we have seen in Chapter 5 about 20% of transgenic mice, produced by injecting DNA into early embryos, have recessive mutations as a result of the insertion of foreign DNA (see Figure 5.5). One of these insertion mutations has been shown to affect male fertility (Palmiter *et al.*, 1984) and another affects the pattern of limb formation (Woychik *et al.*, 1984). This second study represents a key approach to cloning a developmentally significant mammalian gene (Figure 7.5). The mutation resulted from the insertion of a mouse mammary tumour virus (MMTV) LTR-mouse *c-myc* fusion gene in pBR322, which was being used as part of a separate study of oncogenes. Homozygous progeny of this transgenic strain had severe deformities of the fore and hind limbs involving fusions of bones and missing digits. The mutation was inherited as a typical autosomal recessive trait and co-segregated with the *c-myc* sequences.

The mutant allele was cloned by preparing a cosmid library from total liver DNA of the mutant strain and selecting for the Amp^R gene of the pBR322 vector which had been used to create the transgenic line. The wild type gene was then obtained by using one of the recombinant cosmids to screen a normal mouse library. The wild type and mutant alleles differed by a 1 kb deletion in the mutant DNA. The region was mapped to band C1 on chromosome 2 by hybridisation to a panel of mouse/hamster somatic cell hybrids. Several mutations affecting the development of fore and hind limbs had previously been identified by classical genetic analysis and these map to the same region of chromosome 2. The mutation in the transgenic mouse line failed to complement the classical limb deficiency (ld) mutations when heterozygous strains were crossed suggesting that the mutations were allelic. A different limb bud mutation (lst) that increases the number of bones has also been mapped to this region of chromosome 2. Insertional mutagens have therefore provided the first cloned DNA from what may be an important cluster of genes involved in specifying mammalian limb morphogenesis.

Transposon tagging

Many spontaneous mutations in organisms such as *S. cerevisiae*, *Drosophila*, *C. elegans* and maize arise from the insertion of a transposon. If the transposon has been cloned then the mutant gene is easily isolated by

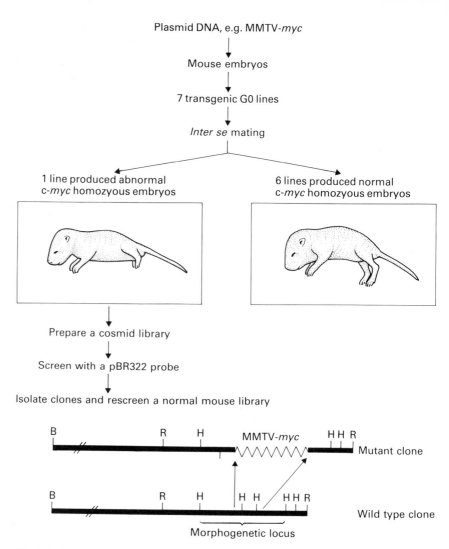

Fig. 7.5. Cloning a mammalian morphogenetic locus. The abnormal embryos on the left have changes in the skeletal structure producing shortened limbs. The MMTV-*myc* DNA has inserted into the limb morphogenetic locus, disrupting it and producing a deletion.

probing a genomic library with the transposon. This *transposon tagging* has been used to clone, for example, the white eye colour locus in *Drosophila* by probing with the *copia* element (Bingham *et al.*, 1981), the *Drosophila* DNA polymerase gene (RpII) by probing for a P-element (Searles *et al.*, 1982) and the homeotic gene lin-12 of *C. elegans* by probing for Tcl (Greenwald, 1985). The natural frequency of transposition of most transposons is rather low, the exception being P-element transposition in dysgenic hybrids (see Chapter 5). This means that natural transposon tagging is not a generally useful method for gene cloning. To improve the technique transposons are being manipulated in *in vitro* to increase their activity and then reintroduced into cells.

The *Drosophila* P-element transposase can be expressed as a cDNA to circumvent the requirement for germ-line specific splicing (Chapter 5). This allows the element to transpose in any cell type. It also opens up the possibility of using P-elements in mammalian cells. The aim is to drive transposase expression with a mammalian cell promoter and introduce this construction together with a defective P-element vector into mammalian cells or embryos. This should give a high frequency of insertional mutagenesis. At present it is not clear if this will be an advantage over retroviral mutagenesis which also occurs at high frequency. Retroviruses also have the advantage that there is usually only a single insertion site. If there are multiple insertions then genetic linkage between the mutation and a single insertion site must be established by the appropriate genetic crosses. This is trivial in an organism such as *Drosophila* with a two week breeding cycle but difficult in the mouse where each cycle takes several months. It is also likely that the limiting factor in cloning mouse genes is not creating the insertional mutations but analysing them (see Jackson, 1986). The transposition frequency of the yeast Ty element can be increasd by exchanging the Ty promoter with the inducible *GAL1-10* promoter (Boeke *et al.*, 1985) but to date this has not been used as an aid to gene cloning.

Gene isolation using transformation bioassays

There are many examples where a region of DNA has been cloned using criteria independent of gene function, e.g. by *chromosome walking* or *chromosome dissection*, and proof that a particular gene has been cloned is required. One approach is to look for biological activity following gene transfer.

Cloning a biorhythm gene from *Drosophila*

A number of *Drosophila* mutations have been identified that affect circadian rhythms and the periodicity in the male courtship song. These map to a single chromosomal locus designated *per*. This region of the chromosome has been cloned by two procedures illustrated in Figure 7.6. The first approach used a *chromosome walk*, jumping form the cloned *Notch* gene across a deletion that brought Notch just distal to *per*. This resulted in the isolation of a 90 kb contiguous stretch of DNA within which the *per* locus could be cytogenetically localised to a 25 kb interval (Bargiello *et al.*, 1984). The second approach was to microdissect polytene chromosomes to produce a *sub-genomic bank* of the 3B1.2 band which contained the *per* locus (Pirrotta *et al.*, 1983). The *per* gene(s) could not be further identified because nothing was known about the gene product. To overcome this, a behavioural assay for *per*[+] DNA was developed to define the functional regions of the locus. Fragments from the locus were sub-cloned into an *Adh*[+] P-element vector (Figure 5.12), pPA-1 and introduced into *per*[0] embryos by microinjection in the presence of the helper P-element pπ25.1. G1 transgenic flies were selected on the basis of resistance to 6% ethanol. The resulting transformed flies were analysed for their circadian and courtship song rhythms. Some of the sub-segments (shaded in Figure 7.6)

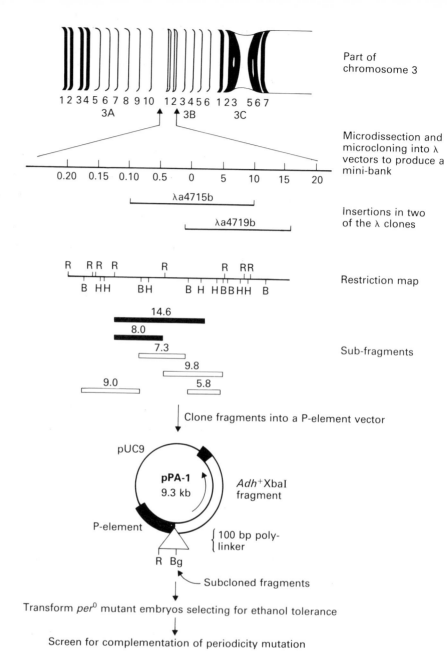

Fig. 7.6. Cloning *Drosophila* biorhythm genes. Part of chromosome III with the characteristic banding pattern is shown at the top. This was dissected and a partial restriction map derived from overlapping cosmid clones is shown. The clones that carried the *per* locus are shaded.

restored rhythmicity in *per*[0] flies although none of the transformants was completely normal. Preliminary data implicated a 0.9 kb transcript in the biological activity of the *per* locus (Zehring *et al.*, 1984; Citri *et al.*, 1987). This codes for a polypeptide that has an unusual repeating amino acid structure. The number of repeats is directly responsible for the timing and

CHAPTER 7

duration of the rhythms (Baylies *et al.*, 1987). A fascinating discovery arising from this is that the levels of this transcript vary throughout the day: it is more abundant at 2 p.m. than 2 a.m. Planning when to make an RNA preparation is therefore critical in this type of system! This study was the first identification, by bioassay, of a metazoan gene about which nothing was known of the gene product.

Matching gene product to gene in a multigene family

Transplantation antigens that mediate graft rejection, and differentiation antigens in the mouse are encoded by the major histocompatibility complex (MHC), H-2 on chromosome 17. This comprises a multigene family whose members encode cell-surface glycoproteins that are involved in the recognition and immune response to foreign antigens. The complex can be divided into three types of genes on the basis of structural and functional similarities. These are called classes I, II and III (see Chapter 10 for more detail about the properties of class I and class II antigens). Class I cDNA probes hybridise to 30—40 different genomic clones that span 837 kb of DNA which is organised into 13 different gene clusters (Hood *et al.*, 1982). The class I genes are closely related. Serological analyses have identified four groups of transplantation antigens, K, D, L and R and a set of hematopoietic (Qa) and T cell (Tla) differentiation antigens (Flavell *et al.*, 1986). Different inbred mouse strains have serologically distinct arrays of class I antigens referred to as a haplotype. Gene transfer has been used as a rapid method of identifying the gene which encodes each serologically identified class of surface proteins. For example class I clones from a BALB/C mouse (H-2^d haplotype) can be transferred to mouse L cells (H-2^k haplotype) where they are expressed and displayed as new antigens on the L cell surface. The different antigens are readily distinguished with monoclonal antibodies. A DNA clone, called 27.5, was co-transferred to mouse Ltk^- cells with an HSV-tk plasmid and HAT^R cells were selected. These were screened for the expression of class I antigens by radioimmunoassay using a panel of monoclonal antibodies. The cells were positive for the L^d antigen, therefore providing an unequivocal gene assignment (Goodenow *et al.*, 1982, see Fig. 7.7).

Cloning the human transferrin receptor gene by immuno-selection

If the gene for a cell surface antigen is to be cloned and specific antibody to that antigen is available it is possible to select transformants that express the antigen using a fluorescence activated cell sorter (FACS). This approach has been used, for example, to clone the human transferrin receptor gene (Kühn *et al.*, 1984) (Fig. 7.8). Mouse Ltk^- cells were co-transformed with total human DNA and plasmid pTK DNA and then HAT^R transformants were selected. (The source of the human DNA was an acute lymphoblastic T cell leukaemia cell line, MOLT-4). Pools of 10 000 HAT^R colonies were stained with antitransferrin receptor monoclonal mouse antibody and fluorescein conjugated goat anti-mouse immunoglobulin. The stained cells were

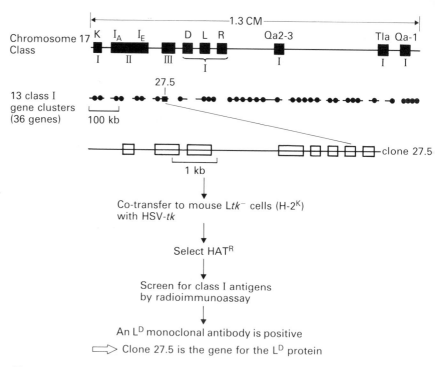

Fig. 7.7. Identifying a mouse MHC class I gene by gene transfer. The MHC region covering 1.3 cM (centi-Morgans) is shown. The expanded class I region, DLR and a cosmid clone (27.5) derived from this cluster is shown.

then sorted by FACS. In the first round there was no obvious peak of positively stained cells but cells with above background fluorescence were cultured and re-analysed. In the third round of FACS a clearly positive cell population was identified. DNA from the positive cell line was probed with nick-translated human DNA to detect highly species specific repetitive DNA and was found to contain 3000 kb of human DNA. Six secondary transformant cell lines were derived from the primary transformant DNA. This represented an overall frequency of one transferrin receptor positive transformant per 12 000 HAT^R cells. The secondary transformants contained 30–90 kb of human DNA. A phage library was constructed from secondary transformant DNA and screened with nick-translated human DNA. Subsequent analyses indicated that the transferrin receptor gene spans 31 kb and produces a 4.9 kb mRNA. The receptor synthesised by the L cell transformants had all the biochemical characterstics of the human transferrin receptor.

This study demonstrates that genes can be cloned directly in eukaryotic cells provided that there is a powerful enough selection. Several points emerge: (1) an efficient transformation system is required as the frequency of gene isolation was only 1 per 12 000 clones, (2) large fragments may have to be transferred to ensure expression, (3) the transferred gene must be expressed reasonably efficiently to ensure detection. Similar approaches have been to clone T cell differentiation antigens (Kavathas and Herzenberg, 1983; Kühn *et al.*, 1983).

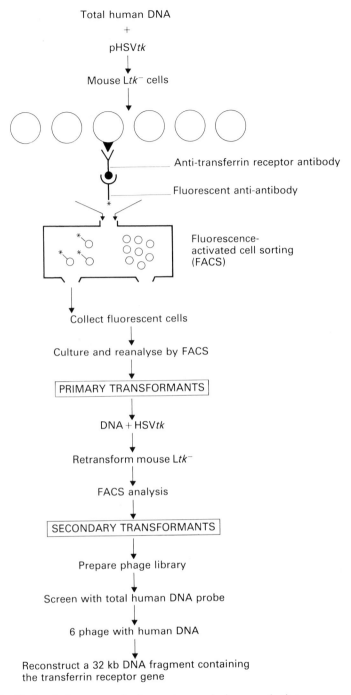

Fig. 7.8. Cloning the human transferrin receptor gene by immunoselection.

Gene transfer and cDNA subtraction

Gene transfer can be used as a method of enriching for a specific mRNA to facilitate cDNA cloning. The general approach is to prepare ^{32}P labelled

GENE ISOLATION AND IDENTIFICATION

cDNA from cells transformed with the desired gene and hybridise this to an excess of poly A RNA from untransformed cells. Single stranded cDNA is then isolated by HAP chromatography. This removes, or subtracts, all the common sequences in the two cells. The isolated cDNA will be enriched for the transferred gene which has no homologue in the untransformed cell poly-A RNA. It can then be used as a probe against a standard cDNA library. This approach has been used to clone the gene for the human T cell antigens T8 (Leu−2) and T4 (Kavathas et al., 1984; Littman et al., 1985).

Vectors for direct cloning in eukaryotic cells

We have seen how genes that are expressed in transformants can be readily cloned provided that there is a selection for the specific phenotype, e.g. immunoselection or tumour formation. These procedures depend upon efficient expression of the transferred gene and this may not occur if all the expression signals are not recognised in the recipient cell or if the gene is disrupted during transfer. In addition, all the procedures that we have described so far have been multistep processes. There has been a considerable amount of effort directed at producing vectors that can be used to select genes directly in mammalian cells. The essential features would be (1) guaranteed expression of the cloned genes, (2) high-frequency transformation and (3) easy re-isolation of the required recombinant vector. Such a system has been developed by Okayama and Berg (1983). This system is composed of two plasmids. One of these, pcDV1, is used to provide an SV40 poly-adenylation sequence and oligo (dT) tail that is used as a primer for cDNA synthesis using mammalian cell mRNA as a template (Figure 7.9A, B). The cDNA is then tailed with oligo (dC). The second plasmid, pL1, provides a linker fragment comprising the SV40 early promoter, the SV40 origin of replication, the 16S splice junction and a PstI expression site. The vector is tailed with oligo (dG) at this PstI site and a linker fragment is generated by Hind III cleavage (Figure 7.9C). The linker fragment and the vector primed cDNA are ligated together via a Hind III site and the oligo (dC):oligo(dG) tails. The mRNA is removed by digestion with RNaseH and the second DNA strand is synthesised by DNA polymerase I followed by ligation to produce the final vector pcD-X (Figure 7.9D), where X is the cDNA of any mRNA in the original template population. This vector has the SV40 promoter and splice junctions and poly-A site flanking the cDNA and the vector is designed so that the orientation of the cDNA is correct for expression from the SV40 promoter. In addition the cDNA cloning method ensures high-efficiency cloning of full-length cDNAs (Okayama and Berg, 1982).

To test this vector, total mRNA from a human fibroblast cell line that had been transformed with HGPRT was used to make a cDNA library in pcD1. The library contained HGPRT cDNA clones at a frequency of 2 x 10^{-5} as shown by standard hybridisation screening of the library in E. coli. When, however, HGPRT negative cells were transformed with this library no positive clones were isolated (Jolly et al., 1983).

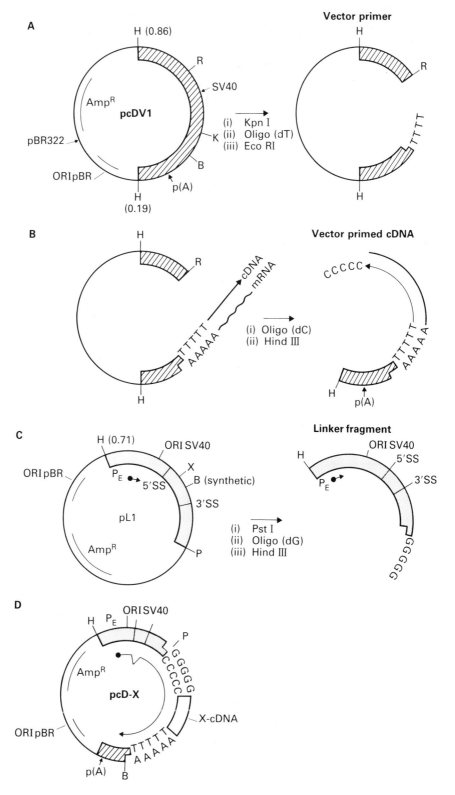

Fig. 7.9 Okayama and Berg procedure for direct cDNA cloning in mammalian cells. The hatching and dark shading allow the different plasmids to be identified and the regions that are being manipulated to be followed.

The failure to clone genes by direct transformation into mammalian cells probably reflects the low concentration of the desired full-length cDNA. At a gene frequency of 10^{-5} to 10^{-6} the mammalian cell transformation system must produce at least 10^{-5} to 10^{-6} transformants to be sure of detecting the gene. The use of a viral vector or electroporation may therefore improve the efficiency of direct cDNA cloning (Chapter 4). Recently an Okayama and Berg human cDNA library has been used to transform cell cycle mutants of *S. pombe* and a *cdc2* homologue was isolated (Lee and Nurse, 1987).

Using retrovirus vectors to generate 'instant' cDNA

When DNA fragments carrying introns are inserted into retroviruses and propagated as viruses a precise deletion of the intervening sequence occurs (Shimotohno and Temin, 1982). This is probably due to splicing of the RNA just before packaging. The SVX shuttle vector (Figure 7.3) can be used to exploit this property of retroviruses to rescue 'instant' cDNA copies from genomic DNA containing intervening sequences. This could be particularly useful for viral transcription units which may display complex splicing patterns, e.g. SV40, adenovirus and BPV-1. These are difficult to infer from sequence analysis and difficult to analyse if some of the spliced transcripts are rare, making conventional cDNA cloning difficult. To test the SVX system Cepko *et al.* (1984) inserted the adenovirus 5 E1A promoter and coding region but no enhancer or poly-A site into the unique BamH1 site of SVX (Fig. 7.10). This region directs the synthesis of three distinct E1 mRNAs, 9S, 12S and 13S, in adenovirus-infected HeLa cells from a single transcript that is processed using three different 5' splice donor sites. The recombinant plasmid was transfected into psi2 cells and after 18 hr a transient harvest was made and the supernatent was used to infect mouse NIH3T3 cells and G418^R colonies were selected. The E1A-retrovirus genome was rescued from different transformants by COS cell fusion. Surprisingly only the 13S E1A mRNA copy was found. However when unspliced E1A SVX virus was passaged through HeLa cells, virus containing precise DNA copies of all three E1A mRNAs were obtained. The failure to detect the 12S and 9S by passage in ψ2 cells was not a cell type difference because adenovirus 5 infects N1H3T3 cells (the precursor to ψ2) and produces all three mRNAs.

Recently this system has been used to construct an instant cDNA of the mammalian oncogene, *int*-1 (Rijsenvijk *et al.*, 1986). It seems likely that many products of splicing can be detected using this system although some cDNAs, such as the E1A cDNAS may not always be produced.

Combining classical and molecular genetics

Identifying regulatory genes

The classical approach to analysing the genetic basis of cellular functions is to mutagenise the cells, select mutant cells that no longer perform the

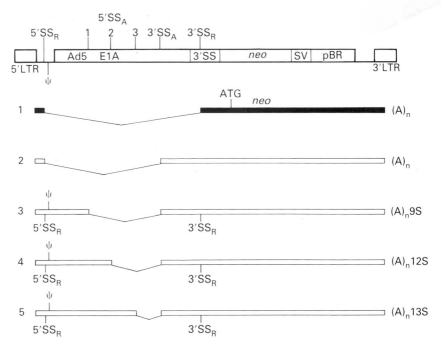

Fig. 7.10. 'Instant' cDNA cloning in SVX. The adenovirus E1A gene has three 5' splice sites (5', SS_A 1−3) and a single 3' splice site (3', SS_A). The retrovirus vector has a single 5 splice site (5', SS_R) and a 3' splice site (3', SS_R). A more detailed map of SVX is shown in Figure 7.3. Potential transcripts are shown. Transcripts 1 and 2 will not be packaged. Transcripts 3−5 will be packaged and reverse transcription will form proviruses with the same structure. Proviruses produced from transcriptase 3−5 will still express *neo* from a subgenomic transcript produced using the retroviral splice junctions. These can be rescued by fusion with COS cells to clone the individual cDNAs.

function correctly and then to map the position of the mutation by appropriate genetic crosses. The molecular genetic approach is to use the biochemical knowledge about the cellular function to clone the gene. This approach is largely limited to genes that encode proteins whose biochemical function is readily assayed or whose amino acid sequence is at least partially available.

One of the problems facing the eukaryotic molecular biologist is understanding how gene expression is controlled. There is a lot of information on *cis*-acting sequences but very little on the regulatory genes and gene products. In microbial systems, regulatory genes have been discovered by screening mutants that are altered in the expression of the gene of interest. In mammalian systems it is difficult to isolate such mutants (1) because the organism is diploid and (2) there is no powerful selection. One new approach described by Sitoyama *et al.* (1985) is to combine the classical and molecular genetic techniques. These workers showed that transformation of NIH3T3 fibroblasts by *v-mos* caused a decrease in the levels of type I collagen RNA. They linked the mouse α2(I) collagen promoter to the *neo* gene and transfected mouse cells to G418^R. Subsequent transformation with *v-mos*

GENE ISOLATION AND IDENTIFICATION

caused loss of G418^R due to the 'repression' of the collagen promoter. The cells were mutagenised and then selection for G418^R was imposed and G418^R colonies were selected. Two of these colonies showed increased expression of the *neo* gene and the cellular type I collagen gene and no reduction in *v-mos* RNA. This indicated that the acquisition of G418^R was due to a specific effect on the collagen promoter and suggests that the mutation affects the *trans*-acting factors that control the $\alpha2(I)$ collagen promoter.

Rapid identification and cloning of new mutations

One of the fastest approaches to identifying the regions of a protein that are critical for function is to look at the effect of mutation in putative key regions. If there is considerable structural information already available then this is done by site-directed mutagenesis (Chapter 8). If however there is no information on the structure of the protein and/or it is a large protein then a large number of sites must be mutated and there is then a problem in screening all the mutations for biological activity. The best approach is therefore to prepare a collection of clones mutated randomly *in vitro* and then select the key mutations on the basis of *in vivo* function by gene transfer into the appropriate cell. For example, the HLA antigens are highly polymorphic and this forms the basis for specific recognition of allogenic cells and for restriction of cytotoxic T lymphocytes (see Chapter 10). There are many amino acid variations between allelic variants but it is not yet known which are the key changes that affect cellular recognition. To answer this question HLA genes can be randomly mutated *in vitro* and then transferred into recipient cells where the antigen will be expressed at the cell surface.

Cells with mutant HLA antigens can be isolated by positive immuno selection or by selection against cells expressing wild type antigens (e.g. Pious *et al.*, 1982). The mutant gene can be characterised by rescuing the gene into *E. coli*. Irrelevant mutations, i.e. in non-functional parts of the gene, will not be picked up in this type of screen. Di Maio *et al.* (1984) have begun such an analysis using a BPV shuttle vector (Chapter 4). Substantial amounts of a human HLA antigen were expressed on the surface of mouse cells following transfer of a BPV-HLA recombinant vector. The advantage of using BPV vectors are the high levels of expression that result from multiple copies of the vector and autonomous replication. This means that transferred genes can be readily rescued by simply using a plasmid-enriched Hirt supernatent (Chapter 2) to transform *E. coli* without having to reclone DNA. Unfortunately, in this study, the BPV-HLA recombinant integrated into the genome and so the simple rescue step was not possible. This highlights a major problem with BPV vectors in that their behaviour is influenced by the inserted DNA and the configuration of the plasmid. There is currently considerable interest in modifying the vectors to ensure autonomous replication.

Allele rescue by gap repair in *Saccharomyces cerevisiae*

A standard approach to the cloning of mutant alleles is to isolate mutant organisms with a particular phenotype by conventional genetic methods, construct a genomic library from each of these mutants and then screen that library with the cloned wild-type gene. This is time consuming but necessary for most eukaryotes. A rapid method for isolating mutant alleles has been developed in yeast (*S. cerevisiae*) (Orr-Weaver and Szostak, 1983; Orr-Weaver *et al.*, 1982). The technique exploits the ability of yeast cells to repair a gap in a plasmid using the chromosomal allele as a template (Fig. 7.11). A deletion that spans the predicted mutation site is created in the cloned wild-type gene by appropriate restriction enzyme and/or nuclease digestion. If the gapped gene is on a yeast plasmid the plasmid will only survive if the gap is repaired. This can occur either by integration (Chapter 3) or by a direct repair of the gap. In this latter case the plasmid carrying the repair which in turn carries the mutant allele can be rescued directly by transforming *E. coli*.

Gene mapping

Mammalian genomes typically contain about 1.5×10^8 bp of DNA. This huge size makes any genetic analysis difficult. Cosmid clones can be isolated but these only contain 40 000 bp at the most. Many complex genetic loci contain many different genes and span, several hundred kilobases for example, the histocompatibility loci. There are also many situations where classical genetic analyses, e.g. pedigree and cytogenetic analysis in man (reviewed by Gusella, 1986), has located an important region of a chromosome where a gene is presumed to be present but the target region is many millions of base pairs. A number of techniques are being developed to analyse these larger chromosome segments.

In most of these analyses it is important to generate a genetic map so that the order of the genes is known. One procedure for obtaining a genetic map of a large region of DNA, i.e. of the order of 10^6 bp has been described by Weis *et al.* (1986) (Figure 7.12). They were interested in studying the mouse MHC locus as a model system to develop a general mapping procedure. To do this they first infected mouse cells with a retrovirus vector that contained the *neo* gene. Retroviruses integrate randomly and so they selected many G418 resistance transformants and analysed them to find a cell line in which the retrovirus had integrated into the MHC locus.

Metaphase chromosomes from this line were used to re-transform cells of a different species (hamster) selecting for G418 resistance. This produced a panel of G418^R cell line contained different sized fragments. The advantage of using metaphase chromosomes for gene transfer is that the nucleoprotein packaging protects the DNA from degradation and recombination so that large, contiguous stretches of DNA up to 10^7 bp can be transferred. It is also possible to isolate separate chromosomes by an electrophoretic procedure known as pulsed field gel electrophoresis (Schwartz and Cantor, 1984). This means that specific chromosomes can be selected for transfer.

GENE ISOLATION AND IDENTIFICATION

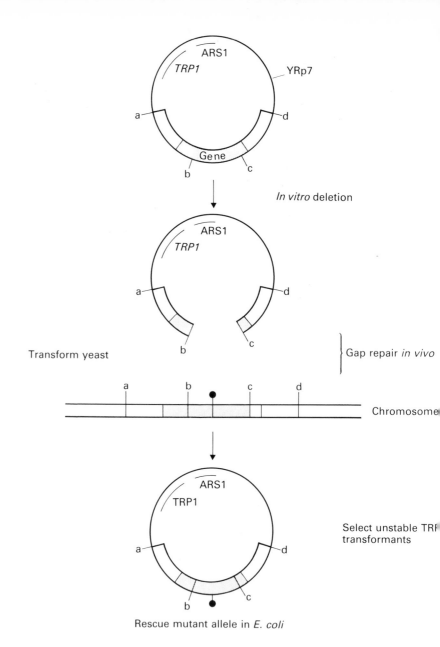

Fig. 7.11. Allele rescue in *S. cerevisiae* by gap repair.

A cosmid library was prepared from the G418 resistant hamster ce▐ and screened for the presence of mouse DNA. This uses the technique th▐ we have discussed of probing the clones with species-specific repetiti▐ DNA. Clones containing mouse DNA were then analysed by Souther▐ blotting using specific segments of the MHC locus that were already ava▐ able. The linear order of the genes can be rapidly deduced, because t▐ frequency with which a particular MHC gene is transferred with G41▐ resistance will depend upon how close it is to the *neo* gene, i.e. the *genet* *linkage*. In our example (Figure 7.12) the H-2K gene is always detecte▐

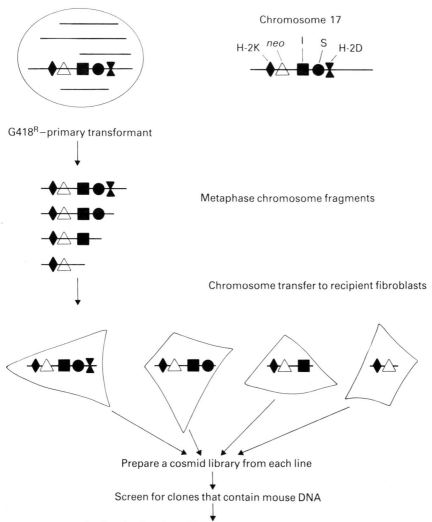

Chromosome 17

H-2K *neo* I S H-2D

G418^R—primary transformant

Metaphase chromosome fragments

Chromosome transfer to recipient fibroblasts

Prepare a cosmid library from each line

Screen for clones that contain mouse DNA

Analyse by Southern blotting using MHC specific probes

Fig. 7.12. Linkage analysis with *neo*-tagged chromosomal fragments.

with the *neo* gene so it must be very close. Whenever the S gene is transferred the I gene is always present but the I gene can be transferred without the S gene. This means that I is closer to *neo* than S. The gene order is therefore H2K-*neo*-I-S. This map has therefore been obtained without extensive sequence and chromosome walking.

This retrovirus insertion followed by selecting cosmid clones by species specific screening can be used to isolate DNA clones from any small defined region of the target chromosome.

Summary

Gene transfer techniques have simplified gene isolation in many cases, particularly from the larger genomes. In situations where the gene product

was unknown, e.g. the *per* locus in *Drosophila*, cloning is only possible b
gene transfer and complementation of the defect in the recipient organism
In the case of the oncogenes gene transfer not only facilitated cloning bu
gave the first clue that single chromosomal genes could produce a tumour
This was stronger evidence for the genetic basis of human cancer. Th
general procedure of tagging pieces of chromosomes with a readily selecte
and detected marker, e.g. a retrovirus, a transposon or a resistance gene suc
as *neo* is particularly important. Because these markers integrate randomly
potentially any region of a genome can be cloned. For the analysis o
mammalian genomes the ability to transfer DNA between species allows th
experimenter to focus on a small region of a chromosome without the adde
complexities of the rest of the genotype because species-specific probes ca
be used. Direct cloning in mammalian cells is at present only possible for
few genes where there is a strong detection system, e.g. surface proteins
Improvements in transformation efficiency may, however, extend thi
approach.

There is great interest in the genetic analysis of the human genome, i
particular to isolate genes that are involved in diseases such as cystic fibrosi
and muscular dystrophy. There is currently a proposal to attempt to se
quence the entire human genome and a preliminary map has been prepare
(Doris−Keller *et al.*, 1987). The types of techniques for gene isolation an
mapping that we have described will be used and developed to achieve thes
aims.

CHAPTER 7

8 Gene Expression

Introduction

It is still not clear how gene expression is controlled in eukaryotic cells. This question presents one of the major challenges in modern biology. The manipulation of cloned genes and their analysis *in vivo* has provided key insights into the nature of DNA sequences that control transcription, into the mechanisms of transcription termination and splicing and into translational control. Such studies are now making these processes more amenable to defined biochemical analyses *in vitro* and the combined approaches of gene transfer and biochemical analyses should lead to a mechanistic understanding of the control of gene expression analogous to that achieved in prokaryotes (see Ptashne, 1986a). In addition, gene transfer provides new approaches for studying the complex process of development and differentiation in whole organisms.

Before the advent of gene cloning and DNA manipulation, eukaryotic viruses were used as models for cellular processes because of their small genome size and the ability to isolate and characterise mutants by classical techniques. The use of recombinant DNA techniques has accelerated the rate of these analyses and many of the key features of eukaryotic gene expression are being elucidated in these viral systems.

The first mammalian gene to be cloned was the human β-globin gene (Maniatis *et al.*, 1978). Since then many hundreds of genes from a whole range of eukaryotes have been cloned and are at various stages in their analyses. In this brief chapter it is impossible to cover the breadth of knowledge that is accumulating. We have tried, therefore, to choose examples of viral and cellular genes that illustrate key technical approaches and provide insight into features of gene expression that are common to many genes.

Identifying transcriptional control signals

The basic elements of transcription control were identified using a range of simple *in vitro* mutagenic techniques followed by gene transfer into appropriate cells and then functional analyses. The mutagenic techniques can be classified as one-sided deletions series, internal deletions, linker-scanning mutagenesis and site-directed mutagenesis.

The production of one-sided deletions and internal deletions

One of the key techniques that has been used in the preliminary identification of *cis* acting regulatory regions has been the construction of deletion mutants *in vitro* using exonucleases such as Ba131. This is an exonuclease which uses both double-stranded and single-stranded DNA as a substrate and which works in both 5′ to 3′ and 3′ to 5′ directions.

A typical, simple Bal31 deletion analysis is shown in Figure 8.1 to identify transcription control elements in the SV40 promoter (Benoist and Chambon, 1980, 1981). Sequence analysis and transcript mapping had already localised the promoter region and indicated some potentially interesting features, in particular, two direct repeats of 72 bp, three repeats of a 21 bp GC rich region and an A + T rich region called the TATA box with the consensus 5′TATAT/AAT/A. Two sets of one-sided deletion mutants, PS and HS, were constructed extending upstream and downstream from the cap site. These are referred to as the 5′ deletion and 3′ deletion series respectively. To construct the PS series the plasmid was cleaved at the EcoRI site and treated with Ba131. This produced fragments which had deletions in one or both of the 72 bp repeats. The HS deletion series was constructed by cleavage at a BamHI site that had been artificially created in the SV40 genome. The Ba131 treatment produced molecules which had lost the TATA box and GC-rich regions and the cap-site proximal 72 bp repeat. One of the problems of this type of analysis is that because the enzyme digests DNA in both directions other possibly essential sequences may be removed, e.g. part of the vector DNA is removed in the PS series and the T antigen coding region is progressively lost in the HS series. This results in the loss of key sequences that are required for vector function and in the juxtaposition of different sequences next to the deletion end points which could affect the properties of the mutants. To avoid these complications the deleted promoter fragments are usually recloned into a new vector such that the only change in the molecule is the loss of specific promoter DNA.

Gene expression directed by the SV40 promoter mutants was analysed by determining T antigen production in transient transfections of monkey CV-1 cells. In the HS deletion series, when the TATA box was removed, there was no drop in the efficiency of transcription but the mRNA start sites were variable. Some of the larger deletions in the HS series showed reduced levels of transcription, particularly those that disrupted the 21 bp repeats. This result hints at later discoveries of the importance of a GC-rich sequence (see below). The region of major importance in transcription was, however, defined by the PS series. Deletions such as ps366 that disrupted both of the 72 bp repeats showed no transcriptional activity. This gave one of the first indications that in eukaryotes sequences far upstream from the cap site were critical transcription control elements. In this case, these sequences were located about 250 bp upstream. These far upstream sequences have been called *enhancers* and they are essential for transcription in many genes. Analogous sequences are rarely present in prokaryotic genes and therefore, the requirement for an enhancer indicates a major difference in the structure of eukaryotic and prokaryotic promoters.

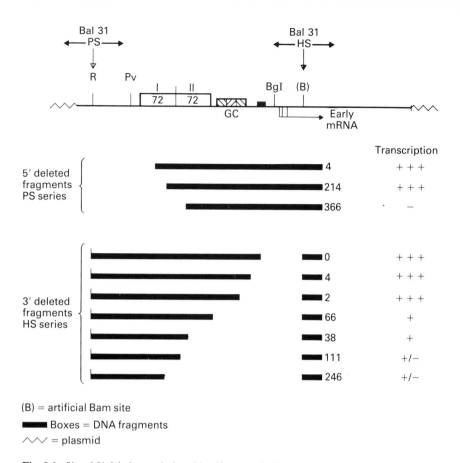

Fig. 8.1. 5' and 3' deletion analysis to identify transcription elements in the SV40 promoter.

One of the problems of this simple type of Bal31 analysis is that it does not allow for the detection of multiple control elements. The first sequence to be deleted may be necessary for transcription but not sufficient. However, as deletion abolishes transcription no other sequences can be studied. This can be solved by creating internal deletions by combining molecules from the 5' series with molecules from the 3' series. In this way a series of small or large internal deletions can be made throughout the promoter. The early analyses of the β-globin gene promoters used this technique (Grosveld *et al.*, 1982; Dierks *et al.*, 1983) and some of the results (Dierks *et al.*, 1983) are shown in Figure 8.2. Nucleotide sequence analysis had revealed an A+T-rich region that was related to the TATA box and was called the ATA box. A second element that was highly conserved between a number of eukaryotic genes was found at -75 and had the consensus sequence 5′GGC/TCAAT/ACT −3′ (Benoist and Chambon, 1980); this was called the *CCAAT* box. A 5' deletion analysis indicated that deletion of sequences around -100 reduced transcription by 5-fold and this region was called the *-100 region*. The internal deletion series showed that even when the -100 region was present a deletion of the CCAAT box, e.g. deletion 631 in Figure 8.2, caused a 5-fold reduction in transcription. This indicated that there were at least two

GENE EXPRESSION

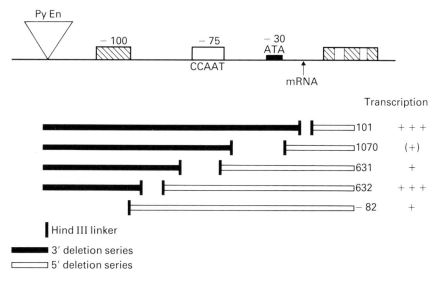

Py En

−100

−75
CCAAT

−30
ATA

mRNA

Transcription

101   + + +

1070   (+)

631   +

632   + + +

− 82   +

| Hind III linker

◼ 3′ deletion series

▭ 5′ deletion series

Fig. 8.2. Internal deletion analysis to identify control elements in the rabbit β-globin gene.

domains in the promoter that were essential for transcription, the -100 box and the CCAAT box. Some internal deletions, e.g. deletions such as 632, Figure 8.2 between the -100 box and the CCAAT box had no effect on transcription, indicating that key transcription control sequences consisted of discrete elements with non-essential regions between them.

In contrast to the results with the SV40 promoter, internal deletions that removed the ATA box from the β-globin gene caused a significant drop in the level of transcription. Some of the residual transcripts in these deletions, however, initiated correctly, implying that the cap site itself may contain some information for initiation. The ATA/TATA box region in a number of other promoters has also been shown to affect efficiency of transcription and therefore this region may have both initiation and efficiency determinants (reviewed by Wasylyk, 1986). In order to analyse β-globin expression, a plasmid with a viral enhancer, in this case polyomavirus, must be used to ensure a sufficient level of transcription in a transient assay. The -100 region, is therefore, not considered to be an enhancer because of its weak activity and because its deletion does not abolish expression. It is one of a class of *upstream elements* that contributes to the overall level of transcription and in some cases to regulation of expression (see later).

Linker scanning mutagenesis

The one-sided and internal deletions that we have discussed so far are extremely useful but rather crude as they alter the spatial relationships between promoter elements. A more sophisticated means of analysis was developed by McKnight and Kingsbury (1982). This is *linker scanning mutagenesis*. A series of one-sided 3′ and 5′ deletion mutants are produced and then joined by a synthetic linker such that the correct spacing between the deletion end points is preserved. For example, if the direct joining of the 5' and 3' ends creates a 10 bp deletion then a 10 bp oligonucleotide is inserted. By combining different end points a linker can be positioned at any point in the promoter and gives a rapid way of creating single-site small insertion/deletion mutations (Figure 8.3). When this technique was applied to the promoter of the HSV *tk* gene, three distinct control signals were identified. One of these was the TATA box region which, when disrupted by a linker at two sites, resulted in dramatic drop in transcription paralleling the β-globin data. Two other important sites were also identified. These were called *distal sequences* (DSI and DSII) and were partially homologous inverted repeats that contained a *GC-rich box* analogous to the GC boxes in the 21 bp repeats in SV40. These GC boxes have now been found in a number of promoters and are key elements in transcription. The consensus site that has been derived is 5'GTGGGCGGG/AG/AC/T−3' (reviewed by Kadonaga *et al.*, 1986). The element DSII also overlaps a region that has homology to the CCAAT box and subsequent analyses have shown that this is a functionally significant site (Jones *et al.*, 1985). The CCAAT box is therefore, a key element in many promoters. The CCAAT box and DS elements or GC boxes are members of a class of *general upstream elements*

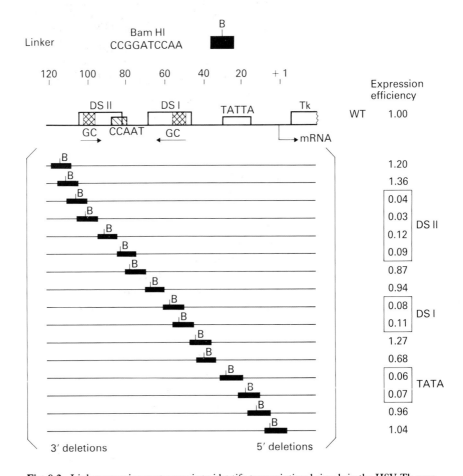

Fig. 8.3. Linker scanning mutagenesis to identify transcriptional signals in the HSV-Tk gene.

GENE EXPRESSION

that are found in many otherwise unrelated genes. The HSV *tk* promoter, therefore, requires two GC-rich boxes, a CCAAT box and a TATA box for efficient and accurate transcription.

Fine structure analysis by site-directed mutagenesis

Once a transcriptional control element has been identified it is important to locate the key nucleotides in any consensus sequence. The linker scanning approach has the drawback that the oligonucleotides may influence the function of neighbouring sequences and might even have some fortuitous 'promoter' activity. The most precise way of confirming the significance of control elements and identifying key nucleotides is to make small changes, down to single base changes, in the sequence of interest. This is done by a procedure known as *site-directed mutagenesis*. There are a variety of techniques for site-directed mutagenesis (reviewed by Shortle *et al.*, 1981; Smith, 1985; Botstein and Shortle, 1985) and three widely used procedures are outlined in Figures 8.4, 8.5 and 8.6, additional examples can be found in Chapters 10 and 11.

Oligonucleotide-mediated site-directed mutagenesis

In this procedure (Figure 8.4) the region of interest is cloned into the phage M13 and single-stranded templates are prepared exactly as in the 'Sanger' − dideoxy sequencing technique (Messing *et al.*, 1977; Sanger *et al.*, 1977). A synthetic oligonucleotide is constructed that is homologous to the sequence of interest except, for example, for a single base change. This oligonucleotide anneals to the template with a single base mismatch and is used to prime second-strand synthesis. The double-stranded DNA is then used to transfect *E. coli*, whereupon it replicates to produce single stranded M13 phage, half of which contain the wild-type DNA and half of which are mutated. The mutants can be screened by hybridising plaques with the original oligonucleotide which will give a stronger signal with the mutant than the wild-type DNA because of the mismatch with wild type. The restriction fragment A−B (Figure 8.4) containing the mutation can then be isolated from the replicative double-stranded form of the mutant phage and reintroduced into the promoter being studied. Oligonucleotide-mediated site directed mutagenesis is the most efficient procedure to use when the target area is small. More sophisticated developments of this procedure can increase the recovery of mutated DNA by selection against non-mutant DNA (e.g. Kunkel, 1985).

Random chemical mutagenesis

This procedure relies upon the fact that specific chemicals will alter bases in DNA *in situ*. Some chemical mutagens such as sodium bisulphite act preferentially on single-stranded DNA and therefore their site of action can be targeted by using a molecule that is single stranded only over the region of interest. Other chemicals attack both single- and double-stranded DNA and therefore the mutations cannot be directed to the target. In this case, the region of interest must be sub-cloned after mutagenesis to isolate mutations

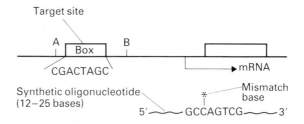

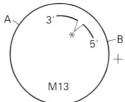

1) Anneal oligonucleotide to target site subcloned into M13

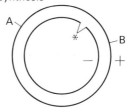

2) Add DNA polymelase, ligase and the 4 dNTPs for − ve strand synthesis

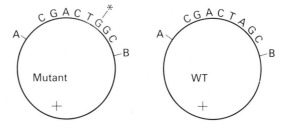

3) Transform *E. coli* and isolate phages containing single + ve strands

4) Screen plaques by hybridisation with mutant oligonucleotide

5) Introduce mutant fragment into the promoter

Fig. 8.4. Oligonucleotide-mediated site-directed mutagenesis.

that are localised only to the region of interest. Chemicals that are commonly used are listed in Table 8.1.)

An outline of a random chemical mutagenesis procedure devised by

1) Treat single stranded DNA with low concentrations of chemical mutagen

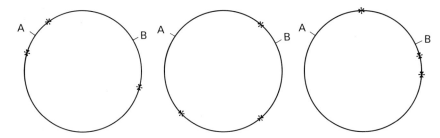

2) Synthesise second strand with reverse transcriptase

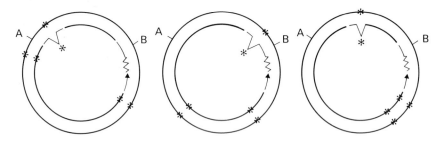

3) Subclone the A-B fragments into specialised vectors to allow electrophoretic separation of fragments

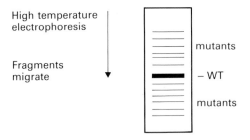

4) Isolate fragments from the gel and reclone into the promoter

Fig. 8.5. Random chemical mutagenesis. Asterisks indicate mutations.

Table 8.1. Chemicals used for site-directed mutagenesis.

Chemical	Target
Sodium bisulphite (deamination)	dC → dT (only in single stranded DNA
Nitrous acid (deamination)	dC → d Unridine dA → deoxyhypoxanthine dG → deoxyxanthine
Formic acid	dA dG glycosyl bonds break
Hydrazine	dC dT pyrimidine rings break

CHAPTER 8

1) Create a gap in the target

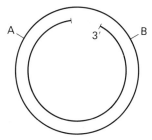

2) Repair gap with DNA polymerase + 3 dNTPs and 1
   α thiophosphate NTP

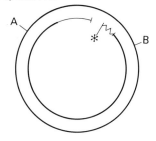

3) Finish repair with DNA pol + 4 dNTPs

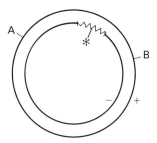

4) α thiophosphate base induces misincorporation
   at replication

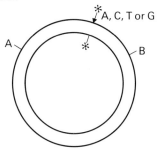

Fig. 8.6. Gap-misrepair mutagenesis.

Myers *et al.* (1985) is shown in Figure 8.5. The mutations are generated by treating single-stranded DNA with limiting concentrations of chemical mutagens. The mutated DNA is then used as a template for copying by reverse transcriptase which, when copying damaged DNA, e.g. at the site

GENE EXPRESSION

of a depurination, inserts any of the four bases at the site. The double-stranded DNA is then digested with restriction enzymes A and B to excise the target DNA, which is sub-cloned into specialised vectors for screening or selecting for mutants. It is possible to isolate the A–B fragments from a pool of these sub-clones and separate the different mutant fragments from each other and from wild-type fragments. This involves a sophisticated electrophoresis at high temperature through a gradient of the denaturing agent urea (Myers *et al.*, 1985). The mutant fragments can be purified directly from the gel and re-cloned into the promoter for analysis.

Gap-misrepair mutagenesis

The third approach is to use DNA polymerase to repair a single-stranded gap in the presence of nucleotide analogues which will be incorporated at a high rate.

Nucleotide analogues such as α-thiophosphate nucleoside triphosphates block the normal error-correcting 3′ to 5′ nuclease activity of the polymerase but they can still be extended in a second reaction to complete the repair. When the strand is copied during replication there is a high rate of mis-incorporation opposite the analogue to produce single base changes (Figure 8.6).

It is possible to use these techniques of site-directed mutagenesis to change any base into any of the other three bases at any position in a stretch of DNA. The same techniques will be encountered in other areas of gene manipulation, e.g. to alter replication regions (Chapter 9) and coding regions (Chapters 10 and 11).

Many control elements are composed of multiple smaller elements that often act in combination and therefore changes in individual elements may not produce a marked effect on transcription (see later). In practice, transcriptional control elements are identified using a number of approaches. A rough plan of the promoter is obtained using large internal deletions. Smaller deletions or linker-scanning mutations are then studied in more defined areas, then single base changes are made in these areas and finally multiple base changes are made.

A general scheme of a pol II promoter

The results that we have outlined from these early studies have highlighted the importance of some key transcription control elements that have now been repeatedly confirmed in many other systems. They allow us to draw a generalised scheme of a simple eukaryotic pol II promoter (Figure 8.7). Most promoters contain an A+T-rich region, the *TATA box*, which is involved in specifying transcription initiation and many also influence the levels of transcription. The TATA box is located 20–30 bp upstream of the cap site in most eukaryotic cells including mammalian, insect and plant cells but the spacing is more variable in lower eukaryotes such as *S. cerevisiae* when it can be 80–100 bp upstream (Dobson *et al.*, 1982; Waslyk, 1986; Kuhlemeier *et al.*, 1987a). Many promoters contain *upstream elements* located

Fig. 8.7. Basic requirements for pol II gene transcription. The coding exons are hatched.

at about 80–100 bp upstream of the cap site that are involved in determining the efficiency of transcription. Two such elements have been identified in a number of different genes. These are the *CCAAT box* and the *GC box* and either or both may be present, sometimes in multiple copies. These are *general upstream elements*. They are not found in all species, e.g. very few yeast or plant genes contain a CCAAT box (Dobson *et al.*, 1982; Kuhlemeir *et al.*, 1987b). Other upstream elements may be promoter-specific or restricted to gene families such as the -100 box in the β-globin gene. Many promoters also have far upstream sequences which, if present, are usually essential for transcription; these are the *enhancers*.

Enhancers

The first enhancer was identified in the SV40 promoter but they have now been found in the majority of cellular pol II genes in both animals and plants (Serfling *et al.*, 1985; Kuhlemeier *et al.*, 1987b). Analogous elements are also found in pol I genes (Sollner-Webb and Tower, 1986). In *Saccharomyces cerevisiae* a similar element called the *upstream activating sequence (UAS)* has been found in most genes (Guarente, 1984, 1988). Enhancers function as positive activators of transcription but their precise mechanism of action is not clear. A variety of different molecular approaches have been used to describe their properties.

General properties of enhancers

The first property of enhancers to be discovered was that *they are not promoter-specific*. When the rabbit β-globin gene is introduced into HeLa cells, the levels of mRNA produced in a transient assay are low. If the SV40 enhancer region is inserted upstream from the β-globin promoter the level of transcription is increased 200-fold (Banerji *et al.*, 1981). This effect is only achieved when the enhancer is on the same plasmid as the β-globin gene, i.e. it is *cis active*. Another discovery was that enhancers can be inverted and still function, showing that enhancers are *orientation-independent*. Enhancers also function at various positions upstream or downstream from the promoter, i.e. they are *position-independent*. The enhancer was still active when placed over 1000 bp away from the cap site illustrating another property of enhancers that they can *act over long distances*. In most chromosomal genes the enhancer is located 100 to 500 bp upstream from the cap site but some very long range enhancers have been identified, e.g. the albumin gene enhancer is 10 kb upstream (Pinkert *et al.*, 1987). Even though the enhancer was positioned at a variety of sites relative to the β-globin gene, transcription continued to initiate at the correct cap site indicating that enhancers are not involved in determining accurate transcription initiation. These features have proved to be general properties shared by most enhancers but there are exceptions and differences in detail. A detailed analysis of the functions of the SV40 enhancer has for example shown that there is an optimal spatial relationship between the enhancer and RNA start site to ensure maximum transcription.

Enhancers can function within a transcriptional unit

The analysis of the first cellular (i.e non-viral) enhancer to be identified showed that enhancers can also function within the transcribed region. This enhancer was found in the immunoglobulin heavy chain gene. When a rearranged mouse heavy chain gene was introduced into myeloma cells high levels of immunoglobulin were synthesised. If, however, a large deletion was made in the major intron that separates the VDJ exons and the Cp exon (Figure 1.11) then no expression of immunoglobulin was directed by the transferred gene (Queen and Baltimore, 1983). This experiment showed that there was a key transcriptional control element located in the major intron. This element was then shown to have the properties of an enhancer by subcloning the fragment that contained the enhancer into an enhancerless SV40 promoter. The immunoglobin fragment activated transcription and functioned in both orientations showing that it is a classical enhancer.

The discovery of this essential enhancer in immunoglobulin genes partly solves one of the puzzles of immunoglobulin gene expression. The variable genes have a 5′ flanking region that looks like a typical promoter, yet there is no significant transcription in germ-line cells or non-antibody-producing somatic cells. In the B cell lineage, however, the antibody genes rearrange to construct the mature Ig heavy chain gene with the variable and constant regions joined to produce a single transcriptional unit (Chapter 1). This rearranged gene can now be transcribed from the V gene promoter which has been activated by the enhancer in the C region intron. An enhancer has also been found in the light-chain genes in the intron between the J and the C region. In addition to the control of immunoglobulin gene expression by DNA rearrangement, there are specific protein factors that are only active in B cells (see later).

Some enhancers are cell-specific

In addition to activating genes many enhancers confer host and tissue specificity, e.g. they are cell-specific (reviewed by Maniatis et al., 1987). To study the cell-specific properties of enhancers it was necessary to develop rapid, quantitative, transient transfection assays. To identify the basis of cell-specific expression of the insulin and chymotrypsin genes the 5′ flanking regions of these genes were inserted into pSVOCAT (Figure 4) so that they would drive the expression of the CAT reporter gene (Walker et al., 1983). These plasmids were then transfected into three different cell types. These were CHO cells, a fibroblast line derived from chinese hamster ovary, HIT cells which are an insulin-producing tumour (insulinoma) cell line from hamster pancreatic endocrine cells and AR4-2J cells which are a chymotrypsi-producing cell line from a rat exocrine pancreatic tumour. The control plasmid was pRSV-CAT (Figures 4.11 and 4.12). About 44 hour after transfection, cell extracts were assayed for CAT activity. High levels of activity were detected in all three cell lines with pRSV-CAT but the plasmid driving CAT with the insulin promoter only expressed CAT in the insulinoma cells and the plasmid driving CAT with the chymotrypsin promoter only expressed CAT in the exocrine tumour cells. This suggested that these two genes contain tissue-specific enhancers.

The ability of enhancers to determine tissue specificity has also been studied in transgenic mice. This has been shown, for example, with an immunoglobulin κ light chain gene (Brinster et al., 1983), the elastase I gene (Swift et al., 1984), a myosin gene (Shani, 1985) and many others (reviewed in Palmiter and Brinster, 1986). These tissue-specific enhancers seem to have been conserved during evolution because genes from a wide range of animals, e.g. chickens, humans and rats, still function correctly in mice.

Some enhancers show a partial tissue-specificity in that their activity is restricted to several different cell types rather than to a single cell type e.g. alpha-foetoprotein gene is expressed in yolk sac, foetal liver and gastrointestinal cells (Krumlauf et al., 1985) and the transferrin gene is expressed in liver, brain and testis (McKnight et al., 1983). This broader cell specificity appears to be determined by multiple but functionally different enhancers that activate expression in each cell type. The alpha-foetoprotein gene has three separate enhancers that contribute differently to the efficiency of expression in each tissue (Hammer et al., 1987). In *Drosophila* the two yolk protein genes *yp1* and *yp2* are only expressed in two tissues of the adult female fly. Using P-element transformation two *cis* elements have been found in both genes. One is required for expression in the ovaries and the second is required for expression in the fat body. These enhancers are not only tissue-specific but adult-specific and sex-specific (Garabedian et al., 1986). Some promoters show no obvious specificity and function in a wide range of cells, for example, the promoter of the metallothionein gene.

Although enhancers are clearly important in determining tissue specificity in many genes they are not the only *cis*-acting elements that are required (see below).

Enhancers interact with cellular proteins

The mechanism of enhancer action is being studied in many different systems. In simple regulated genes in yeast it has been possible to exploit classical genetic analyses to isolate genes that code for regulatory proteins. The *GAL4* gene, for example, is required for regulation of the *GAL1* and *GAL10* genes. The *GAL4* gene has been cloned and the GAL4 protein has been produced in sufficient quantities *in vivo* and *in vitro* to demonstrate that it binds to a 17 bp sequence in the UAS. This system is discussed in more detail later. In mammalian cells it is not possible, in general, to isolate regulatory mutants and as a result there is no access to the genes by conventional routes. It has, therefore, been necessary to develop techniques to prove that proteins interact with the enhancers and then to use biochemical methods to purify these proteins.

Enhancer competition assays

Protein binding *in vivo* can be demonstrated indirectly by using a technique called *enhancer competition* (Scholer and Gruss, 1984). A typical competition experiment (Mercola *et al.*, 1985) using the immunoglobulin enhancer is shown in Figure 8.8. The principle of a competition assay is that if there is a protein or proteins binding to the enhancer then the addition of excess enhancer DNA should compete for these proteins. An enhancer assay plasmid was constructed which contained the immunoglobulin heavy-chain enhancer linked to the CAT reporter gene. The activity of the enhancer is measured by quantitating CAT enzyme activity (Chapter 4). The assay plasmid is then mixed with increasing concentrations of a competitor plasmid that carries only the enhancer and no CAT sequences. Plasmid DNA devoid of any eukaryotic sequences is used to maintain the overall DNA concentration. In this study the mixtures of plasmids were used to transfect plasmacytoma cells. As the ratio of competitor to assay plasmid increases the CAT activity decreases. These data are interpreted as showing that proteins that are required for enhancer-dependent CAT activity are being 'mopped up' by the competitor plasmid. This type of competition assay can also be used to show that different enhancers interact with shared cellular factors. The number and location of different protein binding sites can be established by using smaller sub-fragments or even cloned oligonucleotides as the competitor DNA.

In vivo demonstration of protein–nucleic acid interactions

Direct evidence that proteins bind to enhancer sequences *in vivo* can be obtained using the techniques of *dimethyl sulphate (DMS) protection* and *genomic sequencing* (Church and Gilbert, 1984). This is called *in vivo* footprinting and is illustrated in Figure 8.9 which describes a simplified version of an experiment to identify proteins binding at the immunoglobulin heavy-chain enhancer (Ephrussi *et al.*, 1985). Whole cells are treated with dimethyl sulphate at a concentration that will partially methylate purines. This is the reaction that is used in the Maxam and Gilbert method for sequencing DNA *in vitro* (Maxam and Gilbert, 1980). The DNA is then extracted from the cells and digested with a restriction enzyme that cuts about 100 bp upstream from the immunoglobulin heavy-chain enhancer. The digested DNA is then treated with piperidine which chemically cleaves all the methylated guanine residues. The resulting fragments are separated by gel electrophoresis and transferred to a filter by standard blotting procedures. The filter is then probed with a radioactively labelled single-stranded DNA that is homologous to about 100 nucleotides adjacent to the restriction enzyme site. This allows the detection of all the fragments on the filter that end at this restriction enzyme cut. There will be a range of

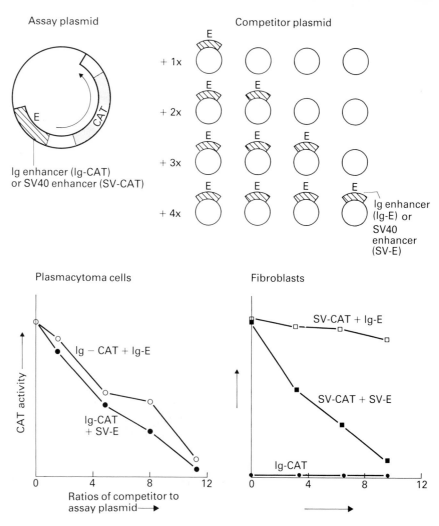

Fig. 8.8. *In vivo* competition experiment to show the interaction of *trans*-acting factors with the Ig and SV40 enhancers.

fragment sizes because of the random cleavage at guanine residues, and so a typical sequencing-type ladder of fragments is obtained. This procedure of labelling fragments with shared ends by using a short hybridisation probe is called *indirect end labelling* and is described in detail by Wu (1980).

If there is a protein that normally binds to the enhancer region *in vivo* then it will 'hide' the DNA from the dimethyl sulphate treatment. These guanine residues that are hidden beneath the protein will not become methylated and so will not be cleaved by piperidine. This will show up in the sequencing ladder as a lack of fragments that end at these G residues. There will be a gap in the ladder. This gap is usually called a *'footprint'*. The analysis is usually made more sophisticated than shown in Figure 8.9 by studying the methylation protection pattern on the two strands of the DNA. This is done by simply using different single-stranded probes that are homologous to each of the strands. Using this *in vivo* footprinting technique

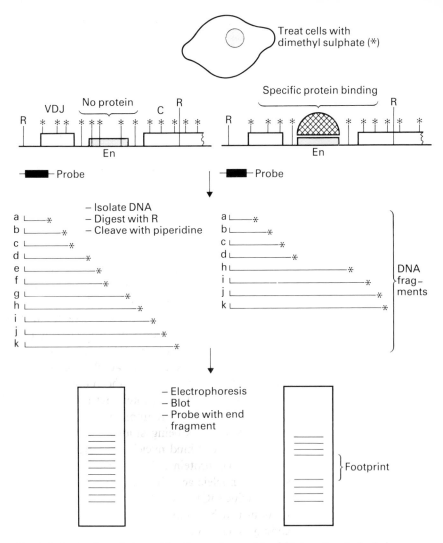

Fig. 8.9. *In vivo* footprinting to identify protein–nucleic acid interactions in the Ig heavy chain gene enhancer. The status of the immunoglobulin gene DNA that has been methylated *in vivo* by treating cells with di methyl sulphate is shown when a protein is bound to the enhancer (right) or when the enhancer is free from protein (left). The VDJ and C exons are shown as open boxes and the enhancer is shaded.

several G residues were found to be protected on both strands in the immunoglobulin heavy-chain enhancer region. Many of these residues were clustered around four regions called βE1, βE2, βE3, and βE4 that showed partial homology with the sequence 5′CAGGTGGC–3′. Another significant protected region was outside these four regions and contained the sequence 5′ATTTGCAT–3′ that is also found in the 5′ flanking region of immunoglobulin heavy and light-chain genes and is called the *immunoglobulin 'octomer'*. The methylation protection was, however, only seen when the experiment was done with DNA from cells in the B lineage. This indicated that tissue specificity was related to the specific binding of cellular factors to the enhancer. The finding that the enhancer region contains multiple and different binding sites suggests that enhancers have complex structures (see later).

In vitro demonstration of protein–nucleic acid interactions

The interaction of proteins with specific DNA fragments can also be studied *in vitro* using very similar techniques. The DNA is mixed with a protein extract and incubated to allow any binding to occur and then treated with DMS or more usually with limiting concentrations of DNaseI. This enzyme will cleave the DNA completely randomly, generating a more closely spaced ladder of fragments than the G cleavage in a DMS protection assay. If there is a protein binding to the DNA then it protects the DNA from DNaseI cleavage and a footprint will be obtained. The ladder is usually visualised directly by end-labelling the DNA fragment with ^{32}P rather than indirectly by hybridisation.

Another important technique for studying protein–DNA interactions *in vitro* is 'band retardation' (Fried and Crothers, 1981; Strauss and Varshavsky, 1984; Singh *et al.*, 1986; Crothers, 1987). When naked DNA is subjected to an electric field as in gel electrophoresis the distance that the DNA travels towards the anode is approximately inversely proportional to the logarithm of molecular weight of the DNA. If, however, the DNA is bound to a protein then the migration in an electric field is retarded and the fragment travels a shorter distance. If the mobility of a DNA fragment is reduced when it is mixed with a protein extract this is an indication that a protein is binding to the DNA. A number of controls must be used to ensure that a specific reaction is being studied because, of course, a cell contains many proteins that will bind nucleic acids ranging from histones, through nucleases to ribosomal proteins. This non-specific binding is reduced by adding a synthetic nucleic acid duplex which is a polymer of deoxyy inosine (dI) and deoxycytine (dC) poly (dI.dC) (Singh *et al.*, 1986). The specificity of binding is confirmed by competing out the binding to a labelled DNA fragment by adding excess amounts of the same DNA that is not labelled (Figure 8.10). The retardation is lost at a relatively low concentration of the specific DNA because there is competition for the bound protein.

Regulated enhancers

We have seen that enhancers are positive activators of transcription but that in some cases they only function in specific cell types. There are also enhancers that only function in response to specific stimuli. That is, they are regulated enhancers. A good example of such regulation is the induction of gene expression by steroid hormones. Many cellular genes such as the ovalbumin, lysozyme and growth hormone genes and some viral genes, e.g. mouse mammary tumour virus (MMTV) are induced by steroids (reviewed by Yamamoto, 1985). Steroids are internalised and bind to specific receptor proteins which interact with specific DNA sequences to stimulate transcription (Becker *et al.*, 1986; Willman and Beato, 1986). Nucleotide sequence comparisons and mutagenesis studies have identified a consensus

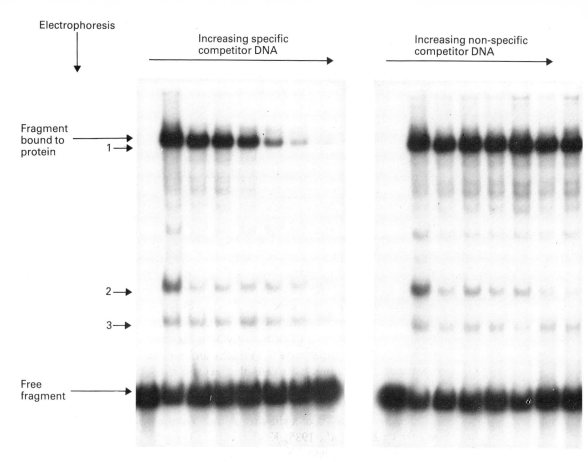

Fig. 8.10. Analyses of protein binding by band retardation. This experiment shows that a protein binds to the UAS of the yeast PGK gene. The free fragment is the UAS and a major retarded band (1) is seen in the presence of a yeast protein extract. This is specifically competed with UAS competitor DNA (left) but not with an unrelated non-specific DNA (right). Bands 2 and 3 are DNA binding proteins that are not specific for the PGK gene because they are competed equally with specific and non-specific DNA. The additional minor bands may represent other proteins that interact with the UAS. (Data from Stanway *et al.*, 1986.)

sequence that binds to the glucocorticoid receptor protein. This is the *glucocorticoid response element (GRE)* which has the consensus 5′−TG/CGT−A/TCAA/T−TGTT/CCT−3′ (Karin *et al.*, 1984). The GRE functions as a classical enhancer; it brings heterologous non-hormone responding genes under steroid hormone control. It functions in both orientations and it is distance-independent. The GRE also shows considerable conservation between species.

Another highly specific regulatory response is the induction of specific genes by physiological shock, in particular, heat shock. A specific subset of heat-shock proteins (HSP) are synthesised in yeast, *Drosophila*, *Xenopus*, plant and mammalian cells and the promoters of the relevant genes share a sequence called the *heat-shock response element (HSRE)* (Lindquist, 1986; Baumann *et al.*, 1987; Wiederrecht *et al.*, 1987). This binds to a protein called *heat-shock transcription factor (HSTF)*. Many of the UASs in yeast also confer specific regulation, e.g. the UAS from the *GALI* gene mediates activation of transcription in response to galactose.

In most cases induction of transcription is mediated by a relatively short sequence of 10 to 20 bp which we will call an '*activator site*'. It is usually the case, however, that this sequence is repeated several times, e.g. the 17 bp recognition site for *GAL4* protein that mediates induction by galactose is repeated four times (Giniger *et al.*, 1985) and the HSRE is repeated four times in the *Drosophila hsp*70 gene (Topol *et al.*, 1985). It appears that activation is more efficient if there are multiple proteins binding to the regulated enhancers. In the case of heat-shock-dependent transcription, multiple binding must occur before there is any significant activation (Shuey and Parker, 1986a, b). Similarly in the mouse metallothionein gene promoter there are multiple elements that can respond to heavy metals. These are the *metal response elements (MRE)* (see later). A single element, when transferred to a heterologous promoter, cannot confer metal responsiveness: two or more are required (Searle *et al.*, 1985). Several plant genes that show regulated expression also contain multiple short sequences, e.g. the light induction sequences in the pea *rbc*-S gene (Fluhr *et al.*, 1986).

So far we have considered enhancers only as positive activators of transcription. However, it has recently been shown that some enhancers can activate the gene under some conditions and repress the gene under other conditions. For example, in cells that express the adenovirus ElA protein the polyoma, SV40 and immunoglobulin enhancers are repressed. This is due either to binding directly to ElA or to a cellular factor that is induced by ElA via recognition of a site that we will call the *repression site* (Borrelli *et al.*, 1984); Hen *et al.*, 1985; Kovesdi *et al.*, 1986; the properties of ElA are reviewed by Berk, 1986). There are also sequences that behave like enhancers in that they are orientation- and distance-independent and act on heterologous genes but they always repress transcription. These have been called '*silencers*'. They have been found in yeast (Brand *et al.*, 1985), mammalian (Laimins *et al.*, 1986) and plant (Simpson *et al.*, 1986) genes.

Enhancers may therefore contain regions that activate transcription in specific cells or in response to environmental signals or constitutionally. In some cases they may mediate repression, and an enhancer may activate or repress depending upon conditions. Enhancers may contain multiple copies of the activation sites and different activation sites. The 'enhancer' is therefore not a single discrete entity and it is perhaps more useful to consider it as an 'enhancer domain'. The complexity of an 'enhancer' is illustrated further when we look at the SV40 enhancer in more detail.

Fine structure analysis of the SV40 enhancer

Preliminary analysis of the SV40 promoter indicated that the region containing the 72 bp repeats was essential for enhancer activity (Figure 8.1). When nucleotide sequence comparisons were made between a number of enhancers a shared region was identified. This had the sequence 5′GTGGA/TA/TA/TG–3′ and is called the *enhancer core*. The functional significance of this sequence was established by Weiher *et al.* (1983) by making point mutations in the

72 bp repeat. Mutations in the core sequence significantly reduced transcription.

Recently a number of groups have undertaken a more detailed analysis of the enhancer and revealed a much more complicated structure (Ondek *et al.*, 1987; Herr and Clarke, 1986; Zenke *et al.*, 1986; Wildeman *et al.*, 1986; Schirm *et al.*, 1987). The results of one of these studies is shown in Figures 8.11 and 8.12. Zenke *et al.* (1986) constructed a plasmid, pAO, that contained a promoterless β-globin gene into which they inserted various derivatives of the SV40 promoter that had mutations in the enhancer region. In order to facilitate the analysis they first constructed a derivative of the SV40 promoter that precisely lacked one of the 72 bp repeats. This had little effect on transcription levels but ensured that significant mutations would not be missed because the functional region is duplicated (Figure 8.11). A set of linker-scanning deletion mutants and a collection of single and multiple-point mutations were constructed. The effect of the mutations was monitored in a transient transfection of HeLa cells by quantitating β-globin-specific mRNA. The results are summarised in Figure 8.11.

The essential enhancer is divisible into two domains called A and B. Deletion of either domain abolishes enhancer activity, therefore cooperation is required between A and B. These domains, however, appear to function

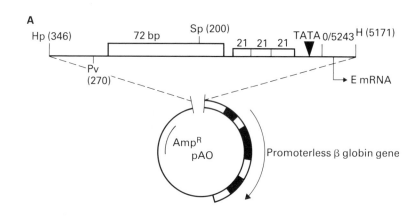

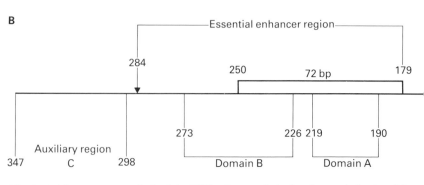

Fig. 8.11. Fine structure analysis of the SV40 enhancer. A single enhancer derivative of the promoter region is shown in A and details of the enhancer are shown in B. The numbers are the nucleotide coordinates of the SV40 genome.

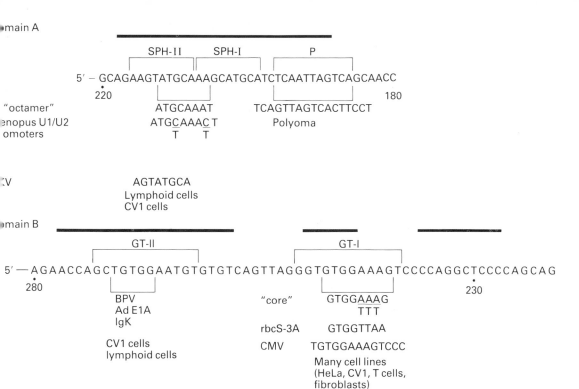

Fig. 8.12. Functional sequence motifs in the SV40 and other enhancers. The nucleotide sequence of the essential enhancer region is shown, with motifs found in other promoters indicated.

as independent entities. They can be present in any orientation with respect to each other and still function when they are separated by variable spacing up to 58 nucleotides. A partial restoration of enhancer function can be obtained by duplicating either domain but this is not as effective as using the two different domains. The mutations that abolish enhancer activity are scattered throughout the enhancer. None of the scanning mutations or single point mutations reduced transcription by more than 6 to 8 fold. This indicated that multiple sites were involved in full enhancer function. At least five separate sites were identified (Figure 8.12). In domain A there are two repeat sequences that contain an SphI site and are consequently called SPH-I and SPH-II boxes. The SPH boxes share homology with several other papovirus enhancer regions (e.g. BKV) and with cellular enhancers (e.g. the immunoglobulin octamer (opposite strand) and the *Xenopus* U1/U2 promoters). A third sequence called P has significant homology with the polyomavirus enhancer. Domain B contains two sequences called the 'GT motifs'. The downstream of these, GTI, contains the previously identified 'core' sequence. Similar GT motifs are found in bovine papillomavirus, adenovirus E1A, immunoglobulin light-chain genes, cytomegalovirus (CMV) and some plant promoters (Schirm *et al.*, 1987; Johnson *et al.*, 1987; Kuhlemeier *et al.*, 1987b). DNaseI protection studies have now shown that proteins bind to these five regions, indicated by the solid bar in Figure 8.12

(Wildeman *et al.*, 1986; Johnson *et al.*, 1987). These data show that the SV40 enhancer is composed of several sequence motifs, which can be found in other promoters. The differences between different viral enhancers may simply reflect different combinations and spatial arrangements of small sequence elements that have occurred during evolution. The different elements may contribute differently to promoter efficiency and/or to tissue specificities. For example, the CMV enhancer appears to be very active and contains five copies of the core sequences (Boshart *et al.*, 1985). The different motifs in the SV40 enhancer possess different cell specificities (Ondek *et al.*, 1987; Schirm *et al.*, 1987).

Control of transcription by multiple elements

Enhancers play a key role in determining tissue specificity and regulation of transcription but they are not the only elements that confer these properties (reviewed by Kelly and Darlington, 1985; Maniatis *et al.*, 1987). This is illustrated by a detailed look at the immunoglobulin gene where sequences within the transcribed region are also important. In addition, we have discussed the complexity of enhancers but even though we have extended the enhancer to an 'enhancer domain' it has still appeared as a defined element that is distinct from upstream elements and the initiation site. This is not, in fact, always, the case. Sequences that are found in the enhancer can also be identified as upstream elements in some genes and vice versa. In fact, the 'enhancer domain' can spread through the entire 5' flanking region. This is illustrated by a detailed look at the mouse metallothionein gene.

Immunoglobulin gene tissue-specific expression requires three control regions

Grosschedl and Baltimore (1985) showed that the expression of immunoglobulin genes in myeloid cells involved three separate elements. They constructed various hybrid genes containing immunoglobulin heavy-chain sequences combined with portions of tissue non-specific transcription units. These were derived from Moloney murine leukaemia virus (Mo-MLV) which is expressed in a wide range of cell types, and the *E. coli gpt* gene which acts as a reporter gene for any cell type. The hybrid genes were inserted into a transient polyomavirus-based expression vector and transcription was quantitated 48 hr after transfection. When the immunoglobulin enhancer (IgEn) was deleted, transcription, as expected, was abolished but it was restored by inserting the enhancer from MoVLV (MLV-En). This hybrid gene was expressed efficiently in myeloma cells but despite the broad cell specificity of MLV-En there was no expression in fibroblasts. This shows that a tissue non-specific enhancer can substitute for IgEn but because tissue specificity is still retained, other regions of the gene must be important in imposing this specificity. To identify the additional tissue specificity determinants the immunoglobulin 5' flanking region was deleted and replaced with the entire promoter and enhancer of MLV (MLV-P). The level of expression of this hybrid gene was about 8 fold higher in myeloma cells than in fibroblast. This showed that the immunoglobulin transcriptional unit conferred increased expression in myeloma cells and must therefore, also contain a tissue-specific element. The enhancerless immunoglobulin promoter activated by the non-tissue-specific MLV enhancer was then fused to a *gpt* reporter gene. This hybrid gene was expressed at 10-fold higher levels in myeloma cells than in fibroblast indicating that VH-P also contains tissue specific elements. The full expression of the immunoglobulin heavy-chain gene in myeloid cells, therefore, requires an enhancer which is only active in these cells and it is modulated or augmented by sequences in the rest of the promoter and the transcribed region.

Several other genes are now known to require sequences within the transcribed region for the correct control of transcription. For example, during mammalian development distinct haemoglobins are synthesised at specific stages. This is referred to as *haemoglobin switching*. The sequences that are required for the switch from foetal β-globin synthesis to adult β-globin synthesis are located in the 3′ half of the adult β-globin gene (Kollias *et al.*, 1986). Similarly myoblast specific repression of cellular thymidine kinase (Merrill *et al.*, 1984), brain specific expression of HPRT (Stout *et al.*, 1985), and expression of *Drosophila Adh* (Fischer and Maniatis, 1986) require downstream sequences.

Multiple interspersed control elements in metallothionein genes

The enhancer is often not a single discrete element but can be composed of multiple motifs that may determine tissue specificity, regulated expression and may influence levels of activation. We have also seen that other regions of the promoter can contain tissue-specific and regulatory elements. This complexity of gene control elements can be illustrated further by considering the expression of the metallothionein gene.

Metallothioneins are small metal-binding proteins whose function appears to be to inhibit the toxicity of heavy metal ions. They are synthesised in many different cell types at a basal level but their synthesis is further induced at the transcriptional level by a variety of agents including heavy metals such as zinc and cadmium, glucocorticoid steroids and bacterial lipopolysaccharides (reviewed by Hamer, 1986). The 5′ flanking regions of these genes have been studied by the procedures that we have already discussed, i.e., various *in vitro* mutants have been analysed by transient transfections and protein–nucleic acid interactions are being studied by various footprinting techniques, in addition an *in vitro* transcription system has been established that responds to metal induction. The discoveries about the structure of the human metallothionein gene, h*Mt*II$_A$, promoter are summarised in Figure 8.13. There are at least five distinct control elements but these can be divided into two categories. These are *basal* and *induced* elements. The basal sequences include a TATA box, a GC box and at least two basal level enhancers (BLE). There are two types of induced elements, those that respond to heavy metals, i.e. metal regulatory elements (MRE) and those that respond to glucocorticoids (GRE). The different elements are interspersed throughout about 300 bp of the 5′ flanking region. In the mouse *Mt*-1 gene there are five MREs which are related by a

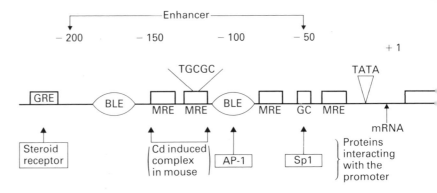

Fig. 8.13. The structure of the human metallothionein gene promoter. The key elements in the promoter are GRE = glucocorticoid response element, BLE = basal level enhancer, MRE = metal response element and GC = the Sp1 site; Cd = cadmium.

core sequence homology but which have different activation efficiencies. The efficiency of weaker elements can be improved by increasing the number of copies (Searle *et al.*, 1985).

There are a number of proteins that bind to the *Mt*-1 promoter. The general transcription factor Sp1 (Briggs *et al.*, 1986) binds to the GC box and another factor AP-1 binds to the basal level enhancer (BLE) sequence (Lee *et al.*, 1987). A specific protein complex is formed at the MREs in the presence of cadmium (Seguin and Hamer, 1987). The *Mt*-1 promoter is transcribed at a constitutive level through the action of the BLE sequences. This basal level is then increased by proteins that bind to the MRE and GRE sequences. The basal activation of the gene and its regulation are therefore separable.

Although the *Mt*-1 entire 5′ flanking region consists of multiple constitutive and regulated elements, when the entire region is inserted into an enhancer assay plasmid such as pA10 CAT, it functions as a simple enhancer (Weber *et al.*, 1983). Individual elements, e.g. a single MRE, will not function as an enhancer and could, therefore, have been defined as an upstream promoter element but multiple MREs act as enhancers (Searle *et al.*, 1985). The distinction between upstream promoter elements and enhancers is therefore no longer clear.

Communication between control elements

The multiple elements in a pol II transcribed gene may be separated by hundreds of thousands of nucleotides. Transcription is regulated by DNA binding proteins but these proteins must be able to influence the activity of RNA polymerase at some distance downstream at the transcription start site. A key question in the control of transcription in eukaryotes is how do the different elements communicate with each other to ensure accurate and regulated initiation of transcription?

General models for communication between control elements

A number of general models have been proposed to explain these interactions (Echols, 1986; Ptashne, 1986b) and these are outlined in Figure 8.14. The models have been called, *twisting, oozing, sliding* and *looping*. The twisting model proposes that regulatory proteins bind to an altered form of DNA, e.g. left handed DNA, or that they alter the DNA conformation upon binding, e.g. by unwinding it. The conformational changes would then be propagated through the DNA allowing new proteins and ultimately RNA polymerase to interact with the DNA. The sliding model proposes that a protein(s) binds to the upstream site and then tracks, or slides along the DNA to another specific control element until the start site is reached and

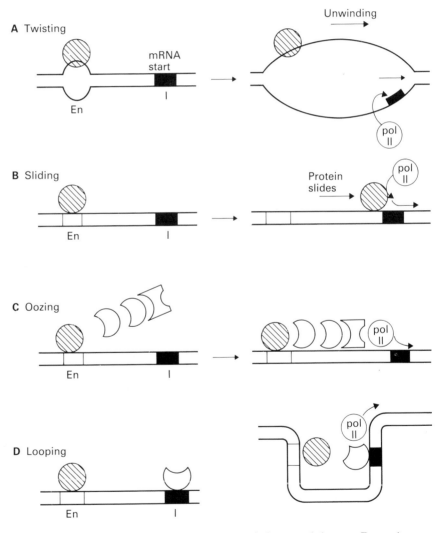

Fig. 8.14. Models for communication between transcription control elements. En = enhancer and I = mRNA initiation region; the hatched and open shapes represent proteins that interact with the promoter. The initial events are shown on the left and they lead to polymerase II interacting with the promoter to allow transcription (right).

GENE EXPRESSION

transcription initiates. The oozing model proposes that a protein(s) binds to the upstream site and facilitates the binding of other proteins to adjacent sequences: this second binding then facilitates a further binding to adjacent sites to produce a procession of proteins 'oozing' along the DNA. The looping model proposes that proteins bind to widely separated sites and contact each other via a protein—protein interaction with the intervening DNA looping to allow the proteins to contact each other. These models for transcription initiation are being tested by making specific mutations in promoters and in the genes that encode regulatory proteins. The most detailed analyses have been done in yeast using the *GAL1—GAL10* gene and in mammalian cells using the SV40 early promoter region.

Interactions at the yeast GAL1—GAL10 promoter

The expression of the *GAL1—GAL10* gene in yeast is controlled by the interaction of a positive regulator protein *GAL4*, which binds, probably as a dimer, to the upstream activation sequence (UAS$_G$). The UAS$_G$ was identified by standard deletion analysis. It is located about 250 bp upstream from the mRNA start site and is still functional when moved 600 bp upstream (Guarente *et al.*, 1982; West *et al.*, 1984). There are four 17 bp sequences within UAS$_G$ that bind *GAL4* protein. This has been shown by *in vivo* and *in vitro* footprinting (Giniger *et al.*, 1985; Bram and Kornberg, 1985). When *GAL4* protein binds to UAS$_G$ in the presence of galactose, transcription is activated by about 1000 fold. To establish how binding of the *GAL4* protein to UAS$_G$ causes transcriptional activation the *GAL4* gene has been manipulated to produce a series of mutant and fusion proteins (Figure 8.15, Brent and Ptashne, 1985; Keegan *et al.*, 1986; Ma and Ptashne, 1987).

A Wild-type GAL4 protein : wild type *GAL1* target gene

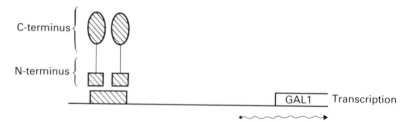

B Wild type GAL4 protein : *lex*A-*GAL1* target gene

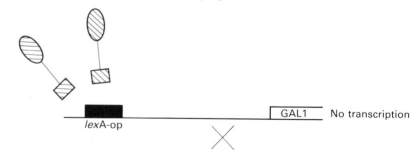

CHAPTER 8

The aim of the study was to distinguish between two possible general mechanisms for transcriptional activation. In the first the *GAL4* protein would bind to DNA and stabilise an unusual DNA structure which would facilitate the binding of other proteins at the transcription start. In the second, *GAL4* would bind DNA but without perturbing the structure

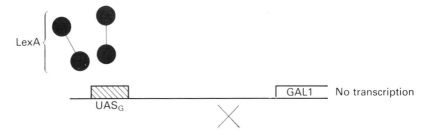

C LexA protein : wild type *GAL*1 target

D LexA protein : *lex*A-*GAL*1 target gene

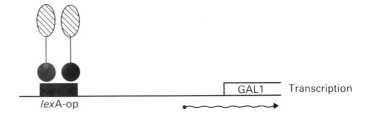

E LexA-GAL4 fusion protein : lexA-GAL1 target gene

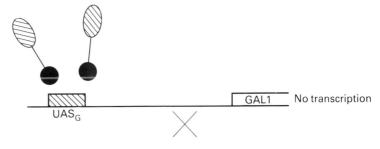

F LexA-GAL4 fusion protein : wild type *GAL*1 gene

Fig. 8.15. Transcriptional activation by GAL4 and LexA-GAL4 proteins.

GENE EXPRESSION

would then activate transcription by contacting other proteins. A simplified account of this analysis is shown in Figure 8.15.

The bacterial protein, LexA, is a repressor protein that regulates its own synthesis by binding to an operator site, the *lexA*-op. This protein can be expressed in yeast and if the *lexA*-op DNA is also present then LexA protein will bind to the specific recognition site in *lexA*-op (Brent and Ptashne, 1985). A modified *GAL4* protein was constructed by making a hybrid gene between the bacterial *lexA* gene coding sequence and the *GAL4* gene coding sequence. This produces a fusion protein with LexA at the N-terminus and GAL4 at the C-terminus. Yeast strains were constructed that were expressing either wild type GAL4 protein or protein or the LexA–GAL4 fusion protein.

Two types of target gene were constructed to test the properties of these proteins. These genes both had the promoter from the *GAL1* gene fused to the β-galactosidase coding region (*lacZ*) as a reporter gene. In one case, 'wild-type target', the wild-type promoter containing UAS_G was used and in the second '*lexA* target', UAS_G was deleted and replaced with a fragment of bacterial DNA that carried the *lexA* gene operator sequence (*lexA*-op). These target genes were transferred to the host strains that were expressing the different regulatory proteins and transcription from the target genes was quantitated by assaying β-galactosidase levels (Figure 8.15).

Normal GAL4 protein activates the wild-type target gene (Figure 8.15A) but it does not activate the *lexA* target gene (Figure 8.15B). These results were as expected and show that there is no interaction between the GAL4 protein and *lexA*-op. The normal LexA protein did not activate the wild-type target gene (Figure 8.15C) and this was expected because there was no recognition site in the DNA. The normal LexA protein did not, however, activate the *lexA* target gene (Figure 8.15D). This showed that even though the LexA protein binds to the *lexA* operator, this simple binding is not sufficient to cause activation. When the LexA–GAL4 fusion protein was used, a different result was obtained. The fusion protein now activated transcription from the *lexA* target gene (Figure 8.15E) but not from the wild-type target gene (Figure 8.15F). This showed that activation required both a specific binding mediated by lexA and the presence of GAL4 protein. Binding alone was insufficient and the most likely interpretation is that following binding to the DNA the GAL4 protein makes contacts with other proteins to activate transcription. These contacts must have a degree of specificity because LexA protein cannot make the contacts. The contacts do not appear, however, to be gene-specific because LexA–GAL4 protein activates transcription of *CYC1*–*lacZ* fusion gene provided that the *CYC1* UAS is replaced with *lexA*-op.

The simplest explanation of the properties of GAL4 protein is that it has two functional domains, a DNA-binding domain and a gene activation domain. This has now been confirmed by a detailed *in vitro* mutagenesis of the *GAL4* gene to produce proteins that have N-terminal and C-terminal deletions (Keegan *et al.*, 1986; Ma and Ptashne, 1987, Figure 8.16). A 98 amino acid fragment from the N-terminus binds UAS_G but fails to activate and in the converse experiment deletion of the N-terminal amino

A GAL4

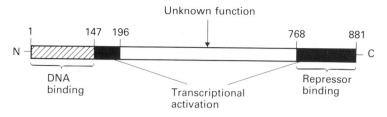

B Glucocorticoid receptor

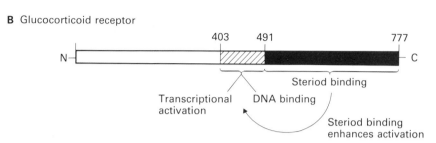

Fig. 8.16. Functional domains in regulatory proteins.

acids fails to activate and does not bind DNA. The N-terminus must contain a DNA-binding domain and the C-terminus contains the activating function which is in fact localised in two separate regions of the C-terminus. The favoured hypothesis is that the activator domain functions by making contact with proteins bound at other control elements. The study argues against the twisting model in its simplest form because this would only require a DNA-binding domain to stabilise a topological change or alter chromatin phasing.

A number of other studies have shown that transcriptional regulatory proteins have separable DNA binding and activation domains, e.g. the ADR1 protein that regulates *ADH1* expression (Hartshorne *et al.*, 1986) and the GCN4 protein that regulates *HIS4* expression (Hope and Struhl, 1986) in yeast and the glucocorticoid receptor that regulates expression of hormone responsive genes in the rat (Giguere *et al.*, 1986; Miesfeld *et al.*, 1987; Green and Chambon, 1987; Godowski *et al.*, 1987; Figure 8.8). Many of these proteins appear to be related by the fact that the activating domains contain a high proportion of acidic amino acids (Struhl, 1987). It is likely that they interact with general transcription factors to activate transcription and that their binding to the recognition site on the DNA confers gene specificity.

Optimal alignment of regulatory proteins is required for SV40 enhancer function

Evidence in favour of the looping model of control element communication has been obtained by analysing the effect of varying the spacing between the control elements in the SV40 promoter (Takahashi *et al.*, 1986). Efficient transcription requires that proteins bind to the upstream enhancer in the 72 bp repeat, to GC boxes in the 21 bp repeats, and to the TATA box region (Figure 8.18). A series of synthetic oligonucleotides were inserted into the promoter at a unique site between the enhancer and the 21 bp repeats. These oligonucleotides increased the spacing between these elements by multiples of a half turn or a full turn of the DNA in a B-helix form (a full turn is about 10.5 bp). Multiples of half a turn, e.g. insertions of 5 and 15 bp, reduced transcription by about 90% but multiples of a full turn, e.g. 10 ·and 21 bp, had little effect on transcription. The insertion of 'half-turn' multiples serves to rotate the protein-binding sites around the helix (Figure 8.17). This causes the protein-binding sites to alter their relative positions.

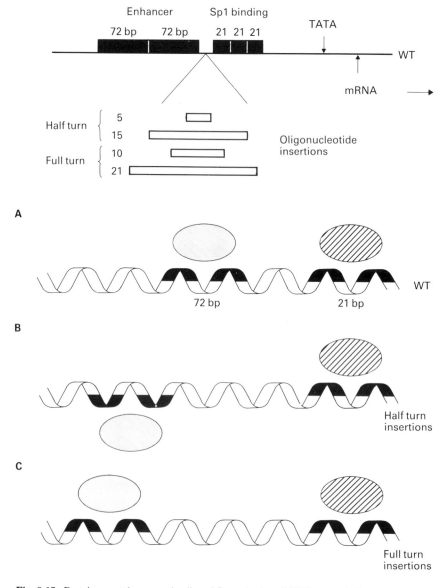

Fig. 8.17. Proteins must be correctly aligned for activation of SV40 transcription.

CHAPTER 8

In the wild-type configuration they lie on the same face (Figure 8.18A) and with a half-turn insertion they lie on opposite faces (Figure 8.18B). An insertion of 10 or 21 bp moves the binding sites apart but they still end up on the same side of the helix (Figure 8.18C). Clearly this study suggests that there is a requirement for proteins that control transcription to be optimally aligned. The model that is most likely to be affected by misalignment is the looping model. In Figure 8.14D you can see that it is easier to form a loop when the proteins are on the same side of the DNA. This study therefore favours, but does not prove, the looping model.

Research into transcription mechanisms is one of the most active areas of molecular biology. We have described some of the types of gene transfer experiments that are being used to provide the answers. To date it seems that DNA binding followed by protein-to-protein contacts is important, that simple unwinding is not required and that communication could involve looping. It is, however, far too soon to determine whether there will be a universal model for all eukaryotic genes and indeed whether any of the simple models outlined in Figure 8.14 are adequate.

A summary of the control of pol II gene transcription

We presented a simple plan of basic pol II gene promoter in Figure 8.7 but as we have now seen, the control of pol II gene transcription can be more complex. There are four groups of elements in a pol II gene that can control transcription, although not all the elements may be present in all genes (Figure 8.19). These elements are the *initiation elements* the *upstream promoter elements*, the *enhancer* and the *downstream elements*. The initiation elements consist of the TATA box and the cap site (the I site in yeast) which ensure

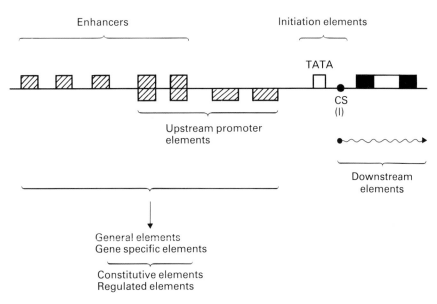

Fig. 8.18. Multiple elements control pol II gene transcription. The hatched boxes are control elements, the exons are shaded, the TATA box and cap site (CS or I) are shown.

GENE EXPRESSION

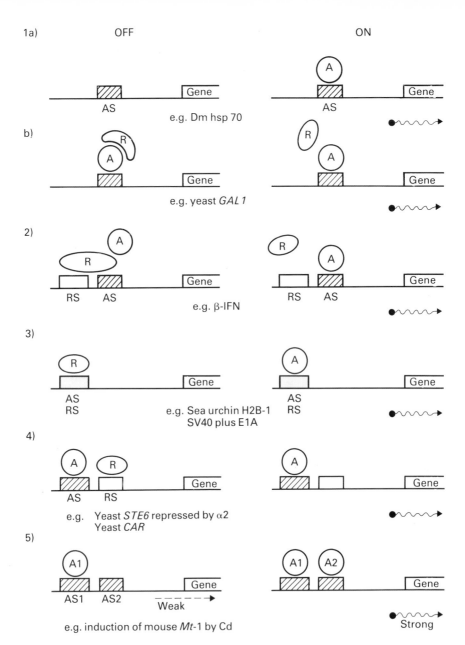

Fig. 8.19. Regulation of pol II gene expression. AS = activator sequence; RS = repressor sequence. The shapes marked A and R are activator and repressor proteins respectively. The stippled boxes indicate where the same or an overlapping sequence is recognized by activator and repressor proteins.

accurate initiation of transcription. These are found in some form in all genes.

The distinction between the other elements, i.e. enhancers, upstream promoter elements and downstream elements, is becoming rather blurred and it is probably only useful to use these descriptions as operational definitions to describe the location in the gene and the behaviour of the sequence in enhancer assay plasmids such as pA10CAT. Even this is less than satisfactory because in several genes, sequences which behave as classical enhancers are located extremely close to the cap site. For example, the β-interferon gene enhancer displays all the functions of an enhancer in a heterologous promoter but it is found only 55 nucleotides upstream from the cap site (Fujita *et al.*, 1985). Also some sequences fail to act as enhancers in a single copy but are typical enhancers in multiple copies. This is true for heat-shock response and metal response elements (Bienz and Pelham, 1986; Hamer, 1986). Some of the elements that have been found in several genes are in different locations, e.g. the 'octamer' is found in the SV40 enhancer, in the immunoglobulin heavy-chain upstream region and in the immunoglobulin heavy-chain enhancer which is in the intron and could, therefore, be considered a downstream element. In Figure 8.19 we have shown the enhancer region overlapping the upstream promoter elements to reflect the difficulty of distinguishing between them in some cases. There are multiple functional elements within these regions. Some of the elements may be found in a number of otherwise unrelated genes and are conserved between organisms as diverse as peas and frogs; these are *general elements*, e.g. the CCAAT box and the 'octamer' site. Others are gene or gene-family-specific elements such as the heat-shock response elements. The elements can be functional under all conditions, i.e., they are *constitutive elements* or they can function only in specific cells or under specific conditions, i.e. they are *regulated elements*.

All the elements seem to function by binding to proteins and some of these proteins have now been purified and are being characterised (reviewed by Dynan and Tjian, 1985; Maniatis *et al.*, 1987; Short, 1987). It appears that a protein binds to the TATA box (Parker and Topol, 1984; Sawadogo and Roeder, 1985). The CCAAT box binds CAAT transcription factor (CTF) also called NF-1 (nuclear factor-1) (Jones *et al.*, 1987). Upstream promoter elements that are found in a number of genes bind the proteins Sp1 (Gidoni *et al.*, 1984; Briggs *et al.*, 1986) and Ap1 (Lee *et al.*, 1987). Proteins that bind to regulatory elements have been identified, e.g. *heat-shock transcription factor (HSTF)* (Wiederrecht *et al.*, 1987), steroid receptors that interact with *glucocorticoid steroid receptor elements (GRE)* (Becker *et al.*, 1986; Willman and Beato, 1986); *GAL4* that interacts with the *GAL1* UAS in yeast (Giniger *et al.*, 1985). Tissue-specific regulation of immunoglobulin genes is mediated by nuclear factors NF-βE1 and NFβE3 that bind to the E1 and E3 elements in the enhancer and nuclear factor NF-A1 and NF-A3 that bind to the octamer sequence (Bohmann *et al.*, 1987). In addition regulation of the light-chain gene expression is mediated by NF-κB and NF-κ (Sen and Baltimore, 1986a; Weinberger *et al.*, 1986; Gerster *et al.*, 1987). One complexity is that some proteins have both activator and repressor functions. This has been shown for the BPV-1 transactivating protein E2 (Lambert *et al.*, 1987), for the cellular oncogene c-*myc* (Kaddurah-Daouk *et al.*, 1987) and the adenovirus E1A protein (reviewed by Berk, 1986). Also in yeast a regulatory protein, HAP1 that controls the expression of cyto-chrome genes binds to different DNA sequences in the *CYC1* and *CYC7* genes (Pfeifer *et al.*, 1987).

Many proteins that regulate pol II gene expression cannot be newly synthesised in response to the regulatory stimulus. This is because responses can be rapid and occur without protein synthesis. Tissue-specific and regulatory proteins are probably present in many cells but they are either inactive or compartmentalised. Alternatively the target site on the gene may be inaccessible. For example, the NF-κB protein that is required for the expression of immunoglobulin light chain genes is found in many different cell types but must be activated. This can be achieved by treating cells with lipopolysaccharides, phorbol esters or cyclohexamide. Activation of NFκB is not the whole story however, because it can be activated in cells such as HeLa and T cells which do not naturally express immunoglobulins (Sen and Baltimore, 1986b; Nabel and Baltimore, 1987). Steroid hormone receptors are present in cells and bind cognate sites on DNA *in vitro* in the absence of hormone (Willman and Beato, 1986). *In vivo* binding is highly hormone specific and it is possible that the hormone influences nuclear localisation of the receptor (Becker *et al.*, 1986). Proteins that interact with the rat albumin promoter are also present in a number of cell types that do not express the gene (Babiss *et al.*, 1987).

Analyses of *in vitro* transcription, which is usually independent of the presences of enhancers, have indicated that the first stage in transcription initiation is the formation of a protein complex that assembles in the TATA/cap-site region (Sawadogo and Roeder, 1985; Montcollin *et al.*, 1986). This is called the *pre-initiation complex*. Additional proteins then interact with polymerase II to activate it for transcription (Zheng *et al.*, 1987). The availability of the initiation elements for forming a pre-initiation complex *in vivo* may be determined by other protein interactions in the promoter, e.g. in the upstream and enhancer regions which regulate transcription.

Several types of regulation of pol II gene expression have been found and regulation is mediated by specific sequences which we have called *activation sites (AS)* or *repressor sites (RS)*. Although the precise details of mechanism have yet to be established we have drawn some general models in Figure 8.20. Type I regulation is induction of transcription that is caused by an activator protein (A) binding to an activation site. An example of this type of regulation is the heat-shock response. A gene such as *hsp*70 is not transcribed until the activator protein, in this case HSTF, is activated by heat shock (Kingston *et al.*, 1987). In some cases, e.g. in the yeast *GAL1* gene, the activator protein (GAL4) appears to be bound to the DNA but masked by a repressor protein (GAL80) which is only removed in the presence of the inducer (galactose) to allow transcription (West *et al.*, 1987).

Type II regulation is also an induction but in this case the gene is not transcribed because the binding of a repressor protein (R) to a repressor site (RS) blocks the access to the adjacent activator site (AS). When the repressor is inactivated the activator protein binds and transcription occurs. An example of this type of negative regulation of induction is the induction of the human β-interferon gene by virus infection (Goodbourn *et al.*, 1986; Zinn and Maniatis, 1986). In uninfected cells a sequence in the promoter can be deleted and transcription occurs. DNaseI footprinting showed that a protein is bound to this sequence but only before induction. After induction a footprint was obtained over a different site very close to the first site. The first site is a repressor site that binds a repressor protein and the second site is an activation site that binds an activator protein.

Type III regulation is a variation of type II where the gene is switched off by binding a repressor and activated by binding an activator but this time the same sequence mediates the repression and activation. This may be the case with the EIA-mediated repression of certain enhancers although the site of action has not been identified (Hen *et al.*, 1985). Recently the regulation of sea urchin histone H2B-1 gene has been shown to be repressed by a protein (CDP) that displaces CAAT transcription factor (CTF), i.e. the CDP (CCAAT displacement protein) and the CTF bind to the same sequence of DNA (Barberis *et al.*, 1987). The silencer sequences in the yeast *HMR* and *HML* genes (Brand *et al.*, 1985), rat insulin-1 gene (Laimins *et al.*, 1986), the c-*myc* gene (Remmers *et al.*, 1986) and the UASs in yeast *CYC1* and HO genes may also function in this way (Pfeiffer *et al.*, 1987; Nasmyth *et al.*, 1987).

Type IV regulation is where the 'ground state' of the gene is to be active through the action of an activator site but an additional repressor site can block activity when the repressor protein binds. An example of this is the repression of the yeast *STE6* gene by the mating type protein (Wilson and Herskowitz, 1986). The repressor site can also function upstream of the activation site. Such *upstream repressor sites (URS)* have been found in the yeast *CAR* genes (Sumrada and Cooper, 1987).

Type V regulation is also where the 'ground state' of the gene is to be active through the action of an activator site but the regulated binding of proteins to additional different activator sites increases transcription. This is the situation with metallothionein gene expression which is increased in response to steroids or heavy metals.

While these models are clearly over-simplifications they may be useful for structuring the mass of information about gene expression in many organisms that is pouring into the literature.

Many of these regulatory interactions are not 'all or none'. The level of activation or repression may depend upon the number of proteins binding and interactions with other elements in the promoter. For example, in yeast there are two types of TATA box elements, those that are used for constitutive expression and those that are used for regulated expression (Struhl, 1987). In yeast there are also *negative elements* which are DNA sequences which when deleted allow the levels of transcription to increase several fold, e.g. in the *CYC1* gene (Wright and Zitomer, 1985) and *TRP1* gene (Kim *et al.*, 1986). These may bind repressors that serve to 'dampen' expression or they may alter the structure of the DNA to reduce the efficiency of interactions between elements.

It is a common feature of pol II promoters that control elements are present in multiple copies. The copies often have slight sequence variations and different efficiencies. This allows subtle modulations of expression if the concentration or the binding affinity of the protein factors is varied under different conditions. It also implies that multiple protein contacts are

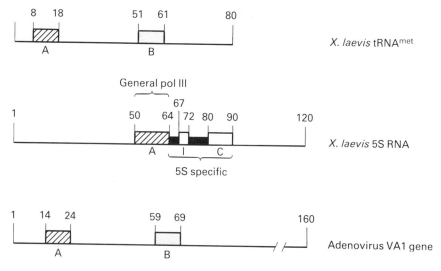

Fig. 8.20. The internal control regions of pol III genes.

required for efficient transcription and this has been shown for example, with the heat shock genes.

Individual sequence elements are often functionally independent, i.e. they have some activity when 'transplanted' into new promoter environments and it is possible to build new promoters by assembling these different elements. It is often true, however, that the assembly of heterologous elements or modules does not produce a promoter that is as efficient as the homologous promoter (e.g. Garcia *et al.*, 1986). This implies that precise interactions between different elements in the promoter might be important for optimum function. It is possible that all pol II promoters have evolved from the assembly of a small number of primordial blocks of sequences which then produce different promoter efficiencies and specificities depending upon the number and arrangement of these sequences (Cochran and Weissman, 1984; Serfling *et al.*, 1985; Wasylyk, 1986).

The role of stable transcription complexes in gene expression

The control of transcription in response to stimuli such as steroids, heavy metals or galactose induction clearly involves the binding of specific protein factors. The tissue-specific expression of genes such as immunoglobulin gene expression in B cells, also requires specific nucleic acid binding proteins. There are, however, at least 10 000 tissue-specific genes in typical vertebrate but in a given cell type most of these will not be expressed. Cell lineages are established during development and cells become committed to differentiate into a restricted number of cell types. During this commitment stage many genes must be completely repressed whereas others must retain the potential for expression at subsequent stages of differentiation even if they are inactive in the stem cell (see Chapters 1, 4 and 5 for a discussion of cell lineages). It is still not clear how a repressed or potentially activatable state of gene activity can be maintained and how the status can be reprogrammed to release genes from repression during differentiation. Studies with *Xenopus* 5S RNA genes are beginning to provide some clues. The studies are extremely important because they provide a basis for considering a higher order control of gene expression at the level of chromosome organisation.

5S RNA gene transcription

There are two 5S RNA gene families in *Xenopus*. These are the somatic-type gene family that has about 800 copies and the oocyte type gene family that has about 40 000 copies. The two gene families are differentially regulated during development. In oocytes only the oocyte-type genes are expressed. By the time of gastrulation these oocyte genes are repressed and the somatic-type genes are transcribed (Brown, 1984). Genes such as 5S RNA genes, tRNA and VA RNA, that are transcribed by pol III all contain the major promoter elements within the transcribed region at a site called the *internal control region ICR* (Chapter 1). The characteristic features of ICRs and the proteins that they interact with have been established by the mutagenesis and protein-binding procedures that we have described for pol II genes. The ICR is composed of multiple functional domains, which in the 5S RNA genes are almost contiguous and in the tRNA genes are separated into two blocks called the A block and the B block. All pol III genes share homology in the A block but 5S genes lack the B block but contain 5S specific elements. These are the intermediate and C blocks (Figure 8.20). Three classes of transcription factor (TF) bind to the ICRs. These are TFIIIA which is specific for 5S RNA genes and TFIIIC and TFIIIB which interact with all pol III genes. These transcription factors were identified by chromatographic fractionation and TFIIIA has been purified and the gene has been cloned. The TFIIIB and TFIIIC factors may contain several protein species (for background see Pelham *et al.*, 1981; Sakonju and Brown, 1982; Ciliberto *et al.*, 1983; Ginsberg *et al.*, 1984; Camier *et al.*, 1985; Bogenhagen, 1985; Pieler *et al.*, 1985, 1987; Setzer and Brown, 1985; Cannon *et al.*, 1986).

When cloned oocyte 5S RNA and somatic 5S RNA genes are mixed with crude extracts from amphibian or human somatic cells they are transcribed equally. If, however, chromatin is used as the template only somatic cell chromatin is transcribed to produce somatic-type 5S RNA whereas the oocyte-type genes are repressed. The differential transcription of 5S RNA genes, therefore, requires that the genes are packaged into chromatin (Schlissel and Brown, 1984). The somatic 5S RNA genes which are active are packaged as stable active transcription complexes that contain stoichiometric amounts of TFIIIA, TFIIIB and TFIIIC and 5S DNA. The addition of the pol III is then sufficient to ensure transcription. The oocyte-type 5S rRNA genes do not form a transcription complex because the factors are prevented from binding to the DNA by histone H1. The active transcription complex is assembled in multiple stages (Figure 8.21). First, TFIIIA interacts with the C block and the I block. This interaction then facilitates the binding of TFIIIC to the A block. In the absence of TFIIIA there is only

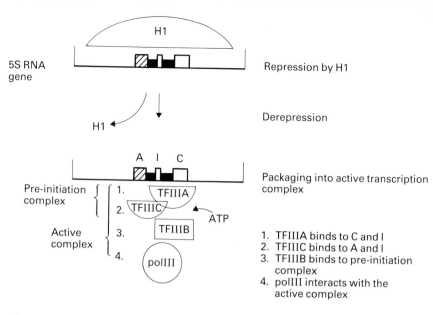

Fig. 8.21. The formation of 5S RNA gene transcription complexes.

very weak binding of TFIIIC. TFIIIA binding to the ICR is not a highly stable interaction but once TFIIIC has bound the complex is stable and the gene has been assembled into a pre-initiation complex and has the potential to be transcribed, i.e. the gene is committed. A fully active complex is only formed by the subsequent binding of TFIIIB which appears to contact the proteins in the preinitiation complex rather than the DNA. There is a requirement for ATP at this stage and this is the rate limiting step. Pol III will then interact with this stable transcription complex, the complex is maintained during repeated rounds of transcription *in vivo* and *in vitro* (Bieker *et al.*, 1985; Wolffe *et al.*, 1986). The developmental switch from oocyte to somatic gene expression appears to be controlled by the concentration of TFIIIA and its competition with H1 and by the affinity of the ICRs for TFIIIA although the details are not yet clear (Brown, 1984; Kmiec *et al.*, 1986). The concentration of TFIIIA may be regulated in part by its binding to the 5S RNA itself (Pelham and Brown, 1980).

A general role for stable transcription complexes in development and differentiation

The observations with 5S genes can be extended to all eukaryotic genes and a general scheme for development and differentiation based on the formation of stable transcription complexes has been proposed (Brown, 1984; Weintraub, 1985). A simplified version is shown in Figure 8.22. During early development genes become generally repressed by packaging into chromatin. At specific stages transcription factors are synthesised that form pre-initiation complexes on some genes. These genes are therefore determined; the complex is stably inherited but not complete. A complete, active complex is assembled

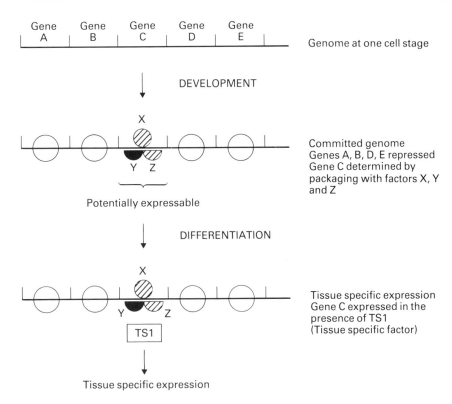

Fig. 8.22. The control of gene activity during development and differentiation.

when the concentration of a tissue-specific factor or other regulatory factor is increased and transcription will then occur. Differentiation is, therefore, seen as the modulation of the activity of determined genes. The transcription complexes form and the polymerases then interact with them. This is significantly different from the general process of transcription is prokaryotes where the RNA polymerase binds directly to the DNA. It is possible that none of the eukaryotic RNA polymerases recognise specific sequences in the DNA.

It seems likely that histones and probably largely H1 are responsible for a general repression of all genes and that activation of a gene involves peturbation of the chromatin structure to displace H1 with a transcription complex. One easily envisaged mechanism of chromatin peturbation is DNA replication. If the concentration of transcription factors is increased at DNA replication then there will be effective competition with histones as the nucleosomes reassemble and a transcription complex will form. Conversely, if the concentration of transcription factors falls then the genes are more likely to assemble into repressed chromatin. It is clear that not all repression is mediated by histones as we have already seen that there are specific repressors of pol II gene expression. It is also clear that factors other than DNA replication can perturb chromatin structure, e.g. globin gene expression is activated by *trans*-acting factors (Baron and Maniatis, 1986). One possibility is that the enhancer may play a role in mediating chromatin reorganisation. This could be by the simple interaction with proteins that then prevent full nucleosome assembly or the interaction may influence the physical structure of the DNA. It is also possible that DNA must be localised at the nuclear matrix before transcription can occur. A number of genes such as the *Drosophila Adh* and the immunoglobulin kappa light-chain gene have specific sequences in their flanking regions that allow them to associate with the nuclear matrix: these are called *scaffold association regions (SAR)* or *matrix association regions (MAR)* (Gasser and Laemmli, 1986; Cockerill and Garrard, 1986). Enhancers may only be required for the initial assembly of the gene into an active transcription complex but they may then be dispensable for maintenance of transcription. This idea is supported by the finding that endogenous immunoglobulin genes which have deleted the enhancer during B-cell maturation are still transcribed whereas similary deleted transfected immunoglobulin genes are inactive (e.g. Zaller and Eckhardt, 1985). Similarly, in a delayed enhancer competition experiment, once gene expression directed by the SV40 promoter is established in a transient transfection of temperature-sensitive COS cells, the subsequent amplification of a co-transfected SV40 mini-replicon by raising the temperature has no effect on the ongoing transcription, i.e. the enhancer on the mini-replicon is not competing for transacting proteins (Wang and Calame, 1986).

Mechanisms for maintaining stable transcription complexes

It is still not clear how transcription complexes are stably maintained during repeated transcription and during DNA replication. In both these processes polymerases must pass by the transcription complex without destabilising it. A possible mechanism is suggested by the structure of the TFIIIA protein (Ginsberg *et al.*, 1984; Miller *et al.*, 1985c). This is a 40 kD protein that contains nine tandem repeats of a 30 amino acid sequence at its N-terminus. Each of these repeats is folded into an independent domain which is stabilised by interactions between cysteine and histidine residues and zinc ions. There are 7 to 11 zinc atoms in the protein. The ends of each domain are 'anchored' by the formation of a tetrahedral zinc ligand. This produces a protein that is extended so that it can bind to the rather long I-box and C-region and which has a series of loops or 'fingers' that can contact the DNA at multiple points. The protein is asymmetric and aligns on the DNA with the C-terminus proximal to the 5′ end of the gene. Such asymmetry is an anticipated feature of a protein that must direct the RNA polymerase to move in one direction. The C-terminus also contains basic amino acids which may facilitate an interaction with other DNA sequences. This could, for example, be involved in interacting with upstream regions that are known to be important for tRNA transcription. Alternatively, this region may interact with other proteins in the transcription complex. The part of the protein containing the finger domains can interact with about 50 bp of DNA and each finger tip could, therefore, bind to about 5-6 bp. The fact that transcription factors can make multiple contacts over an extended stretch of DNA will allow the passage of enzymes through the complex

(Figure 8.23). One can imagine the fingers transiently and sequentially dissassociating from DNA as the polymerase passes but at all times there will be a few fingers binding to retain the complex. More complex structures can be built up by inter-digitating DNA binding sites between two proteins and then stabilising the complex by protein−protein interactions or by placing a protein bridge across the DNA binding proteins. This probably occurs in the transcription of tRNA genes where the ICR spans a much larger region. Proteins can dissociate from box A but be held in place by interactions with proteins binding at box B so that the complex is rapidly reformed (e.g. Camier *et al.*, 1985).

A number of regulatory proteins have now been found to contain zinc-finger DNA-binding domains (Vincent and Monod, 1986; Berg, 1986) and

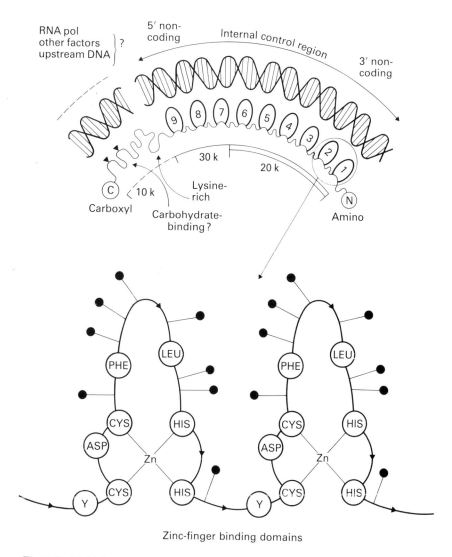

Zinc-finger binding domains

Fig. 8.23. Multiple contacts between TFIII A and the 5S RNA ICR. The dark circles indicate the probable DNA binding side chains. (Adapted from Miller *et al.*, 1985c.)

GENE EXPRESSION

footprinting analyses indicate that it is common for there to be multiple contact points between protein and DNA (e.g. Gidoni et al., 1984). It has also been shown that there is an underlying structural periodicity in the DNA sequences of transcriptional control regions that involves a short run of guanosine every five nucleotides (Rhodes and Klug, 1986). These observations suggest that many aspects of the control of eukaryotic gene transcription may involve the formation of stable complexes via multiple interaction sites and that the zinc-finger structure will be a common feature of eukaryotic DNA-binding proteins. Of particular interest is the discovery that one of the *Drosophila* segmentation genes, *Kruppel*, is a zinc-finger DNA-binding protein (Rosenberg et al., 1986). As these genes are required to establish the body plan of the fly during development it seems reasonable that these proteins should have the capacity to form stable transcription complexes. Not all the homeotic genes, however, have this structure, many of them contain a conserved region called the *homeo box* which could potentially form a α-helix−turn−α-helix structure which is typical of DNA-binding proteins described in prokaryotic systems (Ptashne, 1986a; Ruddle et al., 1985). There are likely to be several types of protein−DNA interactions and it is not yet clear if there will be differences between proteins that form pre-initiation complexes and those that mediate transient responses.

It is also interesting that the gene for TFIIIA is composed of multiple exons which each correspond to a finger-binding domain (Tso et al., 1986). By varying the number of finger-binding domains and the spacing between them and modifying the strength of individual interactions, a whole variety of regulatory proteins could have evolved from a single modular repeat.

This brief account of the control of transcription indicates a level of complexity that is not found in prokaryotic systems. This is necessary to achieve the situation where the majority of genes in a cell are inactive yet a few genes can respond to highly specific developmental and environmental stimuli. Despite this high degree of control over gene expression at the level of transcription further controls can be imposed at later stages as we shall see in the next sections.

The 5' untranslated region

Although some transcriptional controls are located around the cap site and within the region corresponding to the 5' untranslated leader a major function of the 5' untranslated region is to direct the ribosomes to the initiation site and in some cases to regulate translation.

The scanning model for translation initiation

Most of the evidence points to a *scanning model* of translation initiation where the small (40S) ribosomal subunit binds to the 5' cap and moves along the 5' leader until the first AUG is encountered. This is where the 80S ribosome is assembled and translation begins (Kozak, 1983a). The consequence of this mechanism is that the first AUG in any mRNA will usually be used to initiate translation. An elegant demonstration of this model is shown in Figure 8.24 based on experiments described by Kozak (1983b). Plasmids were constructed that contained the rat insulin II gene with additional copies of the initiating AUG and 5′ leader sequence inserted upstream from the authentic AUG. All the upstream AUG (ATG in the DNA) codons were in frame with each other and with the authentic AUG. Two of these plasmids are shown in Figure 8.24. The expression and replication of these genes was driven by the SV40 early promoter and replication region and was analysed 40 hr after transient transfection of COS cells (see Chapter 4). If the upstream AUGs are active in translation initiation then the plasmids should direct the synthesis of elongated insulin polypeptides. The polypeptides were synthesised in the presence of β-hydroxyleucine which inhibits signal peptide cleavage and the primary translation products were analysed by polyacrylamide gel electrophoresis after immunoprecipitation with insulin-specific antiserum. In all cases the size of the major polypeptide product showed that the first AUG in each of the plasmids was used. This suggests that ribosomes do not bind directly to internal AUGs at a ribosome binding site, as is the case in prokaryotes (Shine and Dalgarno, 1975), but that they scan the leader from the 5′ end.

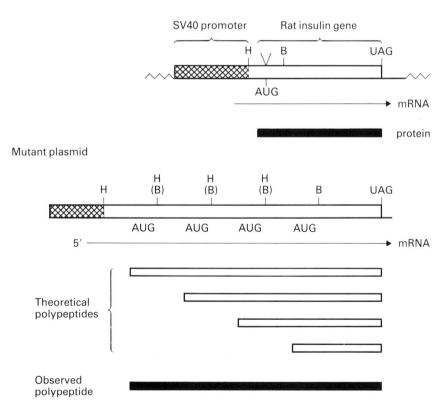

Fig. 8.24. Demonstration of the scanning model of translation initiation.

Optimum sequences for initiation

The simplest version of the scanning model, i.e. that the leader is scanned and the first AUG is used, does not hold true in all cases. There are about 5–10% of genes where initiation does not occur at the first AUG (Kozak 1984) particularly in some viruses (Mardon and Varmus, 1983). This appears to be due, in part, to requirements for a particular nucleotide sequence around the AUG for efficient initiation to occur. A survey of higher eukaryotic mRNA leader sequences has identified a preferred sequence for initiation (Kozak, 1984). Kozak (1986a) has made a series of single and multiple point mutations in this sequence that reduce the efficiency of translation. This has allowed the identification of an optimal sequence for translation initiation by mammalian cell ribosomes. Mutations at position −3 with respect to the A of the initiating ATG had the most inhibitory effect whereas mutations at −2 stimulated translation, particularly if the rest of the sequence was sub-optimal. The importance of the AUG initiation sequence environment is further emphasised by the discovery that a disease in man, α-thalassemia is caused by a deficiency in α-globin synthesis due to a mutation within this sequence at the -3 position (Figure 8.25).

The consensus sequence for translation initiation of *S. cerevisiae* is somewhat different (Figure 8.25; Hamilton *et al.*, 1987). There is, however, little evidence that this precise sequence is important for efficient translation and in one study a large number of single-point mutations around the initiating AUG of the *CYC1* gene had little or no effect on translation (Sherman and Stewart, 1982). There does appear, however, to be a general requirement for a high A/U context in the immediate 5' leader sequence as GC-rich regions show a 3- to 5-fold lower level of translation (Kingsman and Kingsman, 1983) and extreme sequences such as long stretches of C residues can reduce translation by 100-fold (Kniskern *et al.*, 1986). The 5' untranslated region of plant genes also shows differences from the mammalian consensus sequence. There is still a marked preference for A at −3 but this is not flanked by C residues. As with yeast, the 5' leader in plants is A/U rich (Heidecker and Messing, 1986; Lütcke *et al.*, 1987).

The role of secondary structure in the leader

The efficiency of translation can also be influenced by the secondary structure of the 5' leader sequence. This appears, however, to be significant only when the secondary structure is extreme. Kozak (1986b) constructed a series of plasmids in which the initiating ATG of the pro-insulin transcript was contained within synthetically created hairpin loops of increasing stability or where highly stable hairpins were inserted between the cap and the AUG. A hairpin loop with a stability of $\Delta G = -30$ kcal/mol^{-1} had little effect on translation. The ribosome could readily melt these structures. Hairpin loops of even greater stability $\Delta G = -50$ kcal/mol^{-1}, inhibited translation by 85–95% when they were in the leader or surrounding the AUG. This contrasts markedly with the situation in prokaryotes where secondary structures of only $\Delta G = -12$ kcal/mol^{-1} can inhibit ribosome

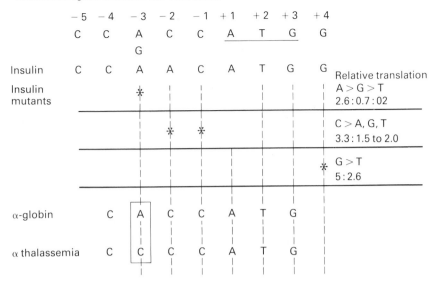

Mammalian gene initiation site consensus

	−5	−4	−3	−2	−1	+1	+2	+3	+4	
	C	C	A	C	C	A	T	G	G	
			G							
Insulin	C	C	A	A	C	A	T	G	G	Relative translation
Insulin mutants			*							A > G > T
										2.6 : 0.7 : 02
				*	*					C > A, G, T
										3.3 : 1.5 to 2.0
									*	G > T
										5 : 2.6
α-globin		C	A	C	C	A	T	G		
α thalassemia		C	C	C	C	A	T	G		

Optimum higher eukaryotic initiation sequence

	−3	−2	−1	+1	+2	+3	+4
	A	C	C	A	T	G	G

Yeast consensus sequence

$$\begin{array}{ccccccccccc} & & & & & +1 & & & & & \\ \frac{A}{T} & A & \frac{A}{C} & A & \frac{A}{C} & A & A & T & G & T & C & \frac{T}{C} \end{array}$$

Plant consensus sequence

$$\begin{array}{ccccccccccc} C & A & A & C & A & A & T & G & G & C & N \end{array}$$

Fig. 8.25. Optimal translation initiation environments. The asterisks indicate nucleotide positions that were mutated in the study of Kozak (1986a).

binding (Munson *et al.*, 1984). Although secondary structure alone may not affect translation in eukaryotes it is conceivable that translation could be regulated by other interactions that stabilise weak secondary structures.

A secondary structure that does influence translation is hybrid formation between the 5′ and 3′ untranslated regions. This is important in a number of plant genes including zein. The zein mRNA has imperfect inverted repeat sequences in the 5′ and 3′ untranslated leader which hybridise to form a duplex with 20 bp matches and a ΔG of −20 kcal/mol[−1]. When the inverted repeats are deleted translation is increased by 50-fold (Spena *et al.*, 1985). The 3′ untranslated region has also been shown to influence translational efficiency in yeast (Zaret and Sherman, 1984). Intermolecular hybridisation between the 5′ leader and other RNA molecules may also influence translation. This has been shown experimentally by introducing

RNA that is complementary to the 5' untranslated leader, so called *antisense RNA*, into cells and showing a reduction in translation (Weintraub *et al.*, 1985). The formation of double-stranded RNA can also inhibit translation by activating a double-stranded RNA-dependent kinase (DAI) which phosphorylates eIF-2α and prevents translation. This is part of the cellular response to inhibit viral replication (Ochoa, 1983; Pestka *et al.*, 1987).

Regulation of translation

There are a number of clear cases where gene expression is controlled by regulating translation. When *Drosophila* cells are heat shocked, the mRNA encoding the heat-shock proteins (HSP) are translated at high rates whereas the translation of cellular mRNAs is repressed (Lindquist, 1986). The *hsp70* leader sequence contains sequences that are essential for efficient translation at high temperature (Klemenz *et al.*, 1985). Heat shock message leaders have an unusual structure. They are between 180 to 250 bases long and have a high proportion of A residues. The extreme 5' end of the leader appears to be important as the addition of extra bases at the 5' end abolishes the high translation (McGarry and Lindquist, 1985). Also the deletion of the most 5' 26 nucleotides abolishes high translation after heat shock whereas deletion of the rest of the leader has no effect (Hultmark *et al.*, 1986).

A well characterised example of translational control is the regulation of amino acid biosynthesis in *S. cerevisiae*. When yeast is starved of any single amino acid there is an increase in transcription in many of the amino acid biosynthetic genes which are not genetically linked. This increase in transcription is mediated by the binding of a regulatory protein to the UASs in the promoters of responsive genes (Hope and Struhl, 1986). This protein is encoded by the *GCN4* gene and the expression of this gene is in turn regulated by other *trans*-acting factors in a complex cascade of interactions that involves a number of different genes. Part of the regulation of *GCN4* expression occurs at the translational level (Figure 8.26). Translation is repressed under non-starvation conditions and stimulated under starvation conditions. The 5' leader of *GCN4* mRNA is 600 nucleotides, which is unusually long. It also contains four AUG codons upstream from the authentic AUG used for synthesising *GCN4* protein. Each of these upstream AUG codons is followed some way downstream by an in-frame termination codon so that there are four very short open reading frames (ORFs) that can be translated before the *GCN4* coding region (Hinnebusch, 1984; Thireos *al.*, 1984). The upstream ORFs are essential for translational repression. This was shown by making point mutations in the four upstream AUG codons and measuring the effect on expression of *lacZ* which was fused to the *GCN4* leader (Mueller and Hinnebusch, 1986). With the wild type leader there is 8-fold de-repression of *GCN4− lacZ* activity in response to amino acid starvation. Progressive removal of AUG codons from the 3' end of the leader results in a progressive increase in β-galactosidase activity under repressed conditions. When all four AUGs are removed expression is fully de-repressed. When the AUGs are progressively removed from the 5'

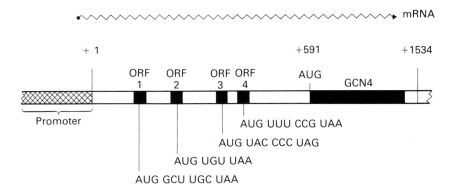

A GCN4 gene

B Expression of GCN4-lac Z fusions

1	2	3	4	lacZ		β-galactosidase activity	
						Repressed	Derepressed
■	■	■	■		WT	10	80
■	■	■	×			20	80
■	■	×	×			110	290
■	×	×	×			240	390
×	×	×	×			740	470
×	■	■	■			3	20
×	×	■	■			1	10
×	×	×	■			5	12
×	×	×	×			740	470

X = mutation

Fig. 8.26. Translational regulations of GCN4 synthesis in yeast.

end a somewhat different result is obtained. Removal of the two 5′ proximal AUGs leads to decreased expression and only removal of all four AUGs allows de-repression. The most 3′ ORF appears to be the dominant repressing element. This study shows that eukaryotic ribosomes can re-initiate, and it is possible that *trans*-acting factors influence the efficiency of re-initiation possibly by inhibiting ribosome disassembly. The study also shows that there are more complex possibilities for regulation because the different AUGs contribute differently to translational repression and the 5′ AUGs influence the activity of the 3′ AUGs, implying some translational coupling.

RNA splicing

Many of the details of the biochemical reactions that are involved in splicing have been established using *in vitro* splicing systems. The minimal sequence requirements and factors affecting the regulation and efficiency of splicing

GENE EXPRESSION

have largely been established by studying experimentally produced and naturally occurring mutations (e.g. Padgett *et al.*, 1986).

General features of splicing

The general features of the splicing reaction are outlined in Figure 8.27. The steps are separated for clarity but it is likely that they occur in coupled reactions. The splicing intermediates are present in a multicomponent complex containing a number of proteins and snRNPs. This has been called the

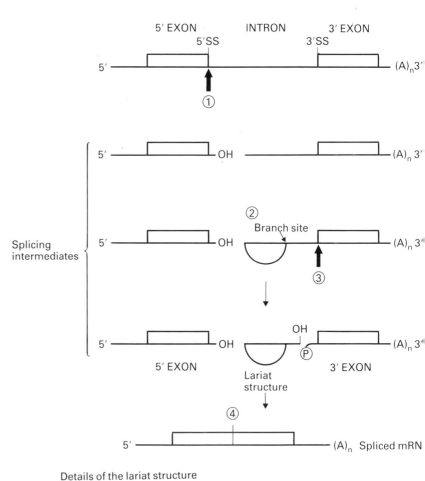

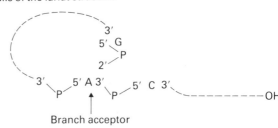

Fig. 8.27. General features of mRNA splicing.

splicesome (Brody and Abelson, 1985; Choi *et al.*, 1986; Osheim *et al.*, 1985). It is formed on the nascent pre-mRNA and the entire splicing reaction occurs in the nucleus. The first stage is precise cleavage at the 5' splice site to release the 5' exon. This is rapidly followed by or coupled to the second reaction in which the 5' end of the intron folds round to interact with a site within the intron called the *branch site*. The branch site is found 20-50 nucleotides upstream from the 3' splice site. the 5' splice site and branch site are covalently linked through an adenine in the branch site via a 2' to 5' phosphodiester bond (Ruskin *et al.*, 1984). The third stage involves a precise cleavage at the 3' splice site. This releases the intron as a circular structure with a tail, and is described as a *lariat structure*. The fourth stage is a phosphodiester linkage between the 5' and 3' exons to form the mature spliced mRNA (reviewed by Padgett *et al.*, 1986).

Sequence requirements for splicing

When many intron sequences are compared there is very little homology between them except at the splice donor and acceptor junctions. A consensus sequence for these junctions in vertebrate genes has been deduced Padgett *et al.*, 1986; Figure 8.28). The key features are the conserved dinucleotides. GT and AG within the intron at the 5' and 3' splice sites respectively and a pyrimidine-rich tract near the 3' splice site. The majority of introns have these sequences. The consensus sequences in yeast are somewhat different. The 5' splice site is highly conserved but the 3' site is variable and there is no pyrimidine-rich tract. The branch site in yeast is, however, highly conserved. This is in contrast to the situation in vertebrate genes where only very loose homology is present which cannot be readily detected by a sequence comparison without functional data.

The importance of these sequences has been established by mutating them and following the splicing reactions *in vivo* and *in vitro*. The junction sequences and a minimal size of about 80 bp is required for splicing but essentially random sequences can be placed within the intron. The only mutations that reduce the amount of normally spliced mRNA are in the first six bases of the 5' splice site and the last two bases of the 3' splice site. Deletion of the pyrimidine tract in the 3' splice site prevents lariat formation but single base changes in the 3' splice site AG dinucleotides or the five 5' splice site nucleotides do not affect lariat formation (Reed and Maniatis, 1985). Although there is considerable flexibility in the sequences at the branch sites these sequences enhance the efficiency of splicing and mutations at the acceptor A nucleotide that is linked to the 5' end of the intron reduce the efficiency (Rautmann and Breathnach, 1985).

In *S. cerevisiae* the branch site, called the TACTAAC box is highly conserved (Pikielny *et al.*, 1983) and deletions or mutations in this sequence abolish splicing (Langford *et al.*, 1984; Jacquier *et al.*, 1985; Parker and Guthrie, 1985). Studies of a number of introns have also suggested that some non-specific features of the introns can affect the efficiency of splicing (e.g. Pikielny and Rosbash, 1985).

A model for the assembly of the splicesome has been developed (Maniatis

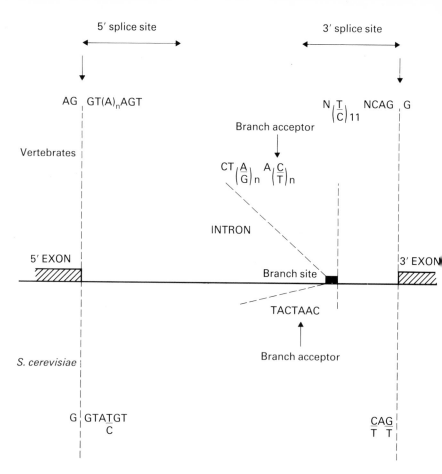

Fig. 8.28. Consensus DNA sequences in introns.

and Reed, 1987; Berget and Robberson, 1986). It is proposed that the 5′ and 3′ splice sites are independently recognised by U1 and U5 snRNPs respectively to form a 35–40S complex. The formation of the branch site and lariat intermediate occurs and splicing then involves U2, U4 and U6 snRNPs in a 50–60S splicing complex. This stage may be influenced by additional regulatory proteins. The splicing complex also contains heterogeneous nuclear RNPs which are essential for the splicing reaction (Choi *et al.*, 1986).

Splice site selection

Most genes contain multiple introns and, therefore, there must be mechanisms to ensure that the correct exons are ligated together. the mechanism of splice site selection must, however, have some flexibility because there are a number of examples where differential splicing is used to generate multiple proteins from a single transcriptional unit (see Chapter 1 and Leff *et al.*, 1986; Breitbart *et al.*, 1987). Exon-swapping experiments have shown that any 5′ splice site can be joined with any 3′ splice site. For example, a hybrid gene was constructed where the first exon coding for the SV40 T antigen was placed upstream from the third exon of the mouse β-globin gene. When the gene was introduced into monkey cells it was transcribed and spliced using the 5' splice site from the SV40 exon and the 3' splice site of the β -globin gene (Chu and Sharp, 1981). Also *in vitro* studies have shown that *trans* splicing is possible where the 5' splice site from one gene can be joined to the 3' splice site of a second unlinked gene (Konarska *et al.*, 1985). These studies indicate that there is no barrier to any interaction between potential 5' and 3' splice sites. This has to be reconciled with the fact that many genes contain multiple introns yet the correct 5' and 3' splice sites are used to produce the mature mRNA. It has been suggested that there is a scanning mechanism whereby the introns are sequentially removed in a 5' and 3' direction. Many observations do not fit such a model. Intermediate partially spliced RNAs can be found in the nucleus where introns in the middle of a gene have been removed before 5' or 3' proximal introns. Each intron appears to be processed independently and at different rates (reviewed in Padgett *et al.*, 1986).

Sequence environment and proximity of exons are important factors influencing splice site selection (Reed and Maniatis, 1986) but as identical precursor RNAs may be spliced differently in different cells at different developmental stages it cannot only be *cis*-acting sequences that are important. Recent studies of the calcitonin/calciton gene related peptide (CGRP) gene and muscle troponin genes have suggested that differential splicing is achieved by the modulation of pre mRNA secondary structure by *splice commitment regulatory factors* (Leff *et al.*, 1987; Breitbart and Nadel-Ginard, 1987). This is likely to be a protein or snRNP that interacts with the splice junctions and influences the assembly of the splicesome.

The end of the message

Transcription termination in many transcriptional units seems to be simply a necessary 'mechanical' process but there are several examples of complex transcriptional units where the production of different proteins is determined in part by differential 3' end formation, e.g. the production of secretory versus membrane bound IgM (see Chapter 1, Platt, 1986; Leff *et al.*, 1986). The 3' untranslated region may also be important for determining the stability of the mRNA and in some cases can influence translation.

A number of nucleotide sequences have been identified that are important for specifying the correct 3' end of eukaryotic RNAs. In pol III genes transcription termination usually occurs in simple DNA sequences that contain clusters of four or more T residues. The 3' end of the RNA generated by transcription termination is not further modified or processed, indicating that the end of mature pol III transcripts is specified by the site of termination of transcription. In the case of pol I transcription the end of the pre-rRNA is not coincident with the end of the mature 28S rRNA. Transcription usually terminates several hundred nucleotides further downstream and this is followed by rapid processing to produce the mature 3' end. The site of transcription termination is closely associated with transcription initiation at the next tandem repeat (see later). In pol II genes the production of the 3′ end of the mature mRNA is also a two stage process of transcription termination followed by mature 3′ end formation by post-transcriptional processing to produce mRNA with 3′ ends. The majority of pol II transcripts are polyadenylated and these genes are characterised by the presence of the sequence 5′AAUAAA−3′ 10 to 30 nucleotides upstream from the end of the 3′ untranslated region (Proudfoot and Brownlee, 1976). Although this site is usually called the '*poly A site*' it is more generally involved in the whole process of 3′ end formation (see later). The precursor mRNA molecules can be extremely long, e.g. the mouse β-globin pre-mRNA terminates between 600 to 1500 nucleotides downstream from the poly A site (Hofer *et al.*, 1982). It is difficult to identify the termination sites because nascent RNA must be analysed and as the processing reactions may be extremely rapid the primary end point can easily be missed. It is, however, becoming clear that there are specific termination sites for pol II transcription rather than a simple random loss of processivity of the polymerase.

The mechanism of pol II transcription termination

The mechanism of pol II transcription termination has been studied in most detail for the non-polyadenylated histone gene transcripts in sea urchin (Birnstiel *et al.*, 1985). The histone genes lack the AAUAAA sequence but the 3′ untranslated region contains a 23 nucleotide sequence that is conserved across all eukaryotes that have been studied except *S. cerevisiae*. The sequence contains a hyphenated inverted repeat that can form a stable hairpin and loop structure (Figure 8.29). The sequence requirements for termination have been analysed by injecting wild-type and mutated H2A genes into *Xenopus* oocytes which correctly terminate heterologous histone H2A transcripts. Deletion of the inverted repeat caused the polymerase to read through into the spacer region indicating that this is a key element for transcription termination (Birchmeier *et al.*, 1982). The inverted repeat is not, however, sufficient because deletion of the spacer DNA significantly reduced the level of authentically terminated transcripts. Efficient and accurate termination depends upon the formation of the stem loop structure as single-point mutations that destabilise the loop reduce termination. The important spacer region is 80 bp immediately adjacent to the termination site. This contains another conserved sequence 5′−CAAGAAAGA−3′ (Birchmeier *et al.*, 1983). To distinguish between a requirement for this sequence as a termination site for pol II or as a stop site for 3′ end processing, a series of histone mRNAs with 3′ extensions were constructed by using *E. coli* RNA polymerase to transcribe histone genes *in vitro* (Birchmeier *et al.*, 1984). When these long transcripts were injected into *Xenopus* oocytes they were rapidly processed to H2A mRNA with the correct 3′ ends. This indicated that the processing events could be separated from the termination events.

An insight into the mechanism of transcription termination came from

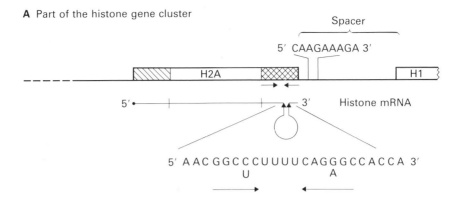

A Part of the histone gene cluster

Spacer

5' CAAGAAAGA 3'

H2A

H1

5' • ———|—————————————|——▲▲— 3'   Histone mRNA

5' A AC G G C C C U U U U C A G G G C C A C C A 3'
          U   A

B Mechanism of termination

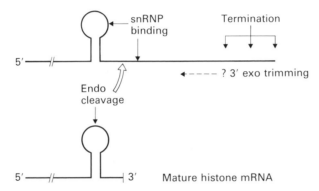

snRNP binding

Termination

5' ———//————

? 3' exo trimming

Endo cleavage

5' ———//———— 3'   Mature histone mRNA

Fig. 8.29. Requirements for transcription termination of sea urchin histone H2A mRNA.

the observation that when artificially extended histone H3 transcripts were injected into oocytes they were not processed unless they were coinjected with a specific fraction of sea urchin chromatin. The essential component was an snRNP containing U7 snRNA and it is proposed that snRNPs are required to form a 3' end processing complex. One model is that the U7 snRNP interacts with the hairpin loop and directs an endonuclease to a specific cleavage site. An alternative is that it recognises the spacer sequences and forms a specific block to protect the 3' end from processive exonucleolytic processing (Platt, 1986).

Termination of polyadenylated transcripts has also been studied in oocytes and in transient transfections of animal cells. Deletion of the AAUAAA signal in the SV40 late transcription unit prevents the formation of stable mRNA (Fitzgerald and Shenk, 1981), but this is not simply due to a lack of polyadenylation. Point mutations in the sequence result in the accumulation of long transcripts implicating the AAUAAA signal in transcript processing (Wickens and Stephenson, 1984). One mutation to AAGAAA reduced the efficiency of processing but still allowed polyadenylation of the normally processed and the longer transcripts (Montell *et al.*, 1983). This suggests that polyadenylation and cleavage are separate reactions.

A functional sequence for processing called the *T-element* has been located by deletion analysis 59 nucleotides downstream from the AAUAAA site of the SV40 late transcription (Sadofsky *et al.*, 1985; McDevitt *et al.*, 1984). Sequence comparisons between mammalian mRNA 3′ termini have revealed another signal located approximately 30 bp downstream from the AAUAAA sequence in 67% of genes. This is called the *GT-element* (McLauchlan *et al.*, 1985). Either of these downstream elements in conjunction with the AAUAAA sequence can mediate 3′ end formation in many genes (McDevitt *et al.*, 1986). In the rabbit β-globin gene, however, both downstream elements are required and appear to act synergistically to allow 3′ end formation (Gil and Proudfoot, 1987). In *S. cerevisiae* a number of different sequences have also been implicated in 3′ end formation, e.g. deletion of a 38 bp region downstream of the *CYC1* gene results in read-through transcription (Zaret and Sherman, 1982). This region contains a loose homology with several other yeast genes to give the consensus 5′−TAG−−−TAGT−−(AT)n−−TTT−3′ (Zaret and Sherman, 1984). The essential functional sequences in the *Drosophila ADE8* terminator which is functional in yeast is, however, 5′−TTTTTATA−3′ (Henikoff and Cohen, 1984). All transcripts are polyadenylated in yeast but no polyadenylation site related to the AAUAAA site is present although a role for the sequence ATAAATAAA/G which is found in a number of genes has been proposed (Bennetzen and Hall, 1982b). Either cleavage and polyadenylation are tightly coupled in yeast or any free 3′ end is sufficient for polyadenylation.

Secondary structure features may also be important; several genes have the potential to form stem loop structures. In the pre-mRNA this usually flanks the AAUAAA site, e.g. IgM (Danner and Leder, 1985) and bovine growth hormone (Woychik *et al.*, 1985 and either involves the AAUAAA in the loop (IgM and bGH) or in the stem, as seen with adenovirus E2A mRNA (McDevitt *et al.*, 1984).

In vitro systems have been established for studying 3′ end formation and these have confirmed that processing and polyadenylation are not coupled and that processing is via an endonucleolytic cleavage (Moore and Sharp, 1985). There is increasing evidence that snRNPs are involved in cleavage and polyadenylation and U1 and U4 snRNPs have been implicated (Moore and Sharp, 1985; Hashimoto and Steitz, 1986). These may function by base pairing with the 3′ ends to which they show significant homology.

Regulation of transcription termination

Some genes contain multiple polyadenylation sites and the use of these sites is regulated by cell type or developmental stage, e.g. growth dependent variations in DHFR expression involve the use of alternative poly-A sites (Kaufman and Sharp, 1983). One of the best characterised examples is the switch from membrane-bound to the secreted form of IgM during B-cell maturation (see Chapter 1). Membrane-bound antibody is synthesised from a long mRNA that terminates at a second poly-A site to reveal an additional exon. Secreted IgM is synthesised from a shorter mRNA that terminates at the first polyadenylation site. The precursor RNAs to both forms, however, extend beyond the second poly-A site. Evidence that the polyadenylation site is important has been obtained from studying deletion mutants in transient transfections of early- and late-stage murine B cells (Danner and Leder, 1985). A deletion which removed 35 bp spanning the AAUAAA signal for secreted IgM resulted in a switch to membrane-bound IgM. The model is that termination of transcription occurs similarly in both cell types but that the subsequent cleavage and polyadenylation reaction is regulated differentially. Regulation of polyadenylation and cleavage appears to be a key feature of adenovirus gene expression (Ziff, 1980; Nevins, 1983). Regulation of differential polyadenylation/cleavage may involve the differential association of snRNPs with termination signals (Platt, 1986).

Recently an unusual feature of transcription termination has been identified in rRNA gene expression. This occurs in the nucleolus using an rRNA dedicated enzyme: pol I and the rRNA genes are arranged as large clusters of tandem repeats (see Chapter 1). Deletion of a conserved 7 bp element in the spacer DNA abolishes both transcription termination and transcription initiation of the adjacent downstream repeat. This arrangement may facilitate initiation by ensuring that the promoter is 'loaded' with polymerase from the preceding gene. Such polymerase recycling could ensure that rDNA promoters achieve a high level of transcription (Baker and Platt, 1986).

Transcript stability

The 3' untranslated region of pol II mRNA may also play a major role in determining transcript stability. The mRNAs of genes that are only expressed transiently, for example mRNAs for lymphokines, cytokines and proto-oncogenes, contain an AU-rich sequence in the 3' untranslated regions with the consensus 5'AUUUA3' (called the AT sequence in the DNA) often in multiple copies. These RNAs have half-lives of less than 30 min whereas long-lived mRNAs, e.g. β-globin, that have half-lives of greater than 17 hr do not contain these sequences. Shaw and Kamen (1986) introduced a 51 nucleotide AT sequence from the human lymphokine gene GM-CSF (granulocyte—monocyte colony stimulating factor) into the 3' untranslated region of the rabbit β-globin gene. This caused the β-globin transcript to become highly unstable. This instability was partially alleviated by cyclohexamide suggesting that the AT sequence may be a recognition site for a transiently expressed enzyme that degrades key cell-regulatory mRNAs. Recently Clemens (1987) has noted that these AT sequences could potentially base pair with the 3' ends of the repetitive B2 RNAs which may, therefore, have a role in regulating the levels of these RNAs. Excessively long transcripts in *S. cerevisiae* are also unstable (Zaret and Sherman, 1984). Regulating the use of different termination sites to generate longer or shorter mRNAs could provide a means of regulating gene expression.

Protein stability

There is a huge range of protein stabilities in eukaryotic cells ranging from proteins whose half-lives exceed a cell generation time to proteins whose half-lives are extremely short. Many regulatory proteins fall into this latter category with half-lives of a few minutes. The selective degradation of proteins in eukaryotic cells is mediated by the covalent conjugation of a small protein called ubiquitin. Linkage is usually to the side-chain amino groups within proteins but a similar process occurs when ubiquitin is linked to the α-amino group, i.e. at the N-terminus either chemically or by expressing an appropriate gene fusion created *in vitro*. Bachmair *et al.* (1986) have now shown that the rate at which a ubiquitinated protein is degraded in yeast is critically dependent upon the N-terminal amino acid residue. They constructed fusion genes that had the ubiquitin coding sequence linked to the β-galactosidase N-terminus. The expression of the fusion gene was driven by the *GAL* UAS (Figure 8.30). The removal of ubiquitin occurred equally in all the constructions except those in which proline was the adjacent amino acid but the released β-galactose proteins had very different stabilities ranging from 2 min to over 20 hr. The most 'stabilising' amino acids are also those that are never found blocked by acylation. Proteins that are found in compartments tend to have destabilising

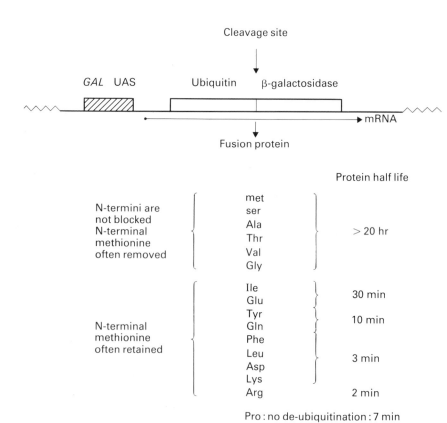

Fig. 8.30. Protein stability is a function of the N-terminal amino acid.

amino acids at their N-terminus. This would ensure that failure to compartmentalise correctly would result in rapid intracellular degradation. Stability of proteins may be regulated, therefore, both at the level of degradative enzymes that recognise the different N-termini and by a class of enzymes, the amino acyl tRNA-protein transferases which conjugate specific modified amino acids post-translationally to N-termini.

New ways of studying development and differentiation

We have described the approaches that are being used to analyse the expression of individual genes whose function is well known. In the process of development and differentiation, however, a whole battery of different genes is involved and in some cases the functions are not known. The precise time of expression as well as the levels of gene product are also important. Techniques are needed to be able to interfere with or analyse gene expression at precise stages during development and to implicate cloned genes in specific developmental processes. Three such techniques involve using firefly luciferase gene as a reporter, pulsing expression by heat shock and inhibiting expression by antisense RNA.

Plants that 'glow in the dark'

Firefly (*Photinus pyralis*) luciferase catalyses the ATP-dependent oxidation of luciferin which results in the emission of light. Ow *et al.* (1986) have placed the coding sequence of the luciferase gene under the control of the CaMV 35S RNA promoter and produced transgenic tobacco plants by leaf disc transformation using the *A. tumefaciens* Ti plasmid binary vector Bin 19 (Chapter 6). The primary transformants were screened by analysing pieces of the disc for luciferase activity. Transgenic plants were then propagated and analysed for the expression of luciferase in the whole plant without destroying any tissue. To do this, luciferin was introduced into the plants by watering small plantlets with a 1mM luciferin solution and then the plants were exposed by contact to photographic film for 24 hr. Light emission was observed in the plant reflecting the expression of the gene in different tissues. If the CaMV promoter were to be replaced by the promoter from an organ-specific gene such as a *cab* gene or from developmentally regulated genes such as storage protein genes, then expression could be monitored during the whole process of development and growth from callus to mature fertile plant. Monitoring of expression could be continuous by using image-intensifying video equipment. This will allow very subtle effects on gene expression to be analysed. The luciferase reporter gene can also be used in cultured animal cells (de Wet *et al.*, 1986) and could, for example, be used to follow cell-cycle-regulated gene expression or altered gene expression during malignant transformation. Also as the activity of luciferase is dependent on the pH and ATP levels in the cell it will be a very sensitive indicator of the cell's metabolism. This could be useful both in optimising production systems for biotechnology or for generally monitoring the growth of cell cultures.

Redesigning the body plan of *Drosophila*

There are a number of genes in *Drosophila* that play a major role in specifying correct development. These are referred to as homoeotic genes and they were originally recognised by mutations that caused catastrophic changes in morphology, e.g. the appearance of legs where wings should be or the transformation of antennae into legs (see Chapter 5). This latter effect is caused by mutations in the gene *Antennapoedia* (*Antp*). It has been suggested that this transformation is caused by the over-production of the *Antp* protein at a particular stage in development. To test this idea Schnewly *et al.* (1987) constructed a fusion gene that had the heat inducible *hsp*70 promoter driving the expression of the *Antp* cDNA. This was introduced into *Drosophila* embryos using a derivative of the Carnegie 20 P element vector (Chapter 5). Several homozygous transgenic lines were established that had P-element insertions at different chromosomal locations. The developmental effects of *Antp* protein were then tested by subjecting the embryos from these flies to a heat shock at different times.

At the normal temperature (25°C) the transgenic embryos showed no abnormalities. Heat shock during the first and second larval instar also produced no effect but heat shock during the third larval stage produced antennae to leg transformations in most of the flies. Later heat shock at the prepuparation stage had little or no effect. This experiment confirms the role of the antennapoedia protein in determining the second thoracic segment structures and pinpoints the time of development when the level of expression is critical. Clearly if this technique is applied to the other homeotic genes singly and in combinations the competitive regulatory interactions between different homeotic proteins can be established and the hierarchy of gene expression that results in a precise developmental programme can be deduced.

Inhibiting gene expression by antisense RNA

The observation that *antisense* RNA results in the inhibition of translation of its complementary mRNA (Weintraub *et al.*, 1985; Melton, 1985) allows the specific inhibition of individual genes *in vivo* by simply introducing the 'antisense' RNA. A typical experiment is shown in Figure 8.30 (Rosenberg *et al.*, 1985). RNA can be produced in large amounts *in vitro* using specialised vectors which have high-efficiency bacterial promoters to drive transcription, e.g. the SP6 vectors use the *Salmonella* phage p22 polymerase to produce RNA. The *Kruppel* gene of *Drosophila* is required in the early embryos to specify the correct segmentation pattern. When the *Kruppel* coding sequence is inserted into an SP6 vector in the wrong orientation such that the 3′ end of the coding DNA strand is nearest to the promoter, the non-coding strand is copied into RNA. This RNA is exactly complementary to the normal mRNA produced in *Drosophila* cells and therefore it can form a hybrid with the mRNA. When the antisense RNA was injected into wild-type embryos they failed to develop normally and showed segmentation defects typical of *Kruppel* mutations, i.e. they were phenocopies (phenotypically mutant but with a normal chromosomal gene). It may be possible to produce transgenic

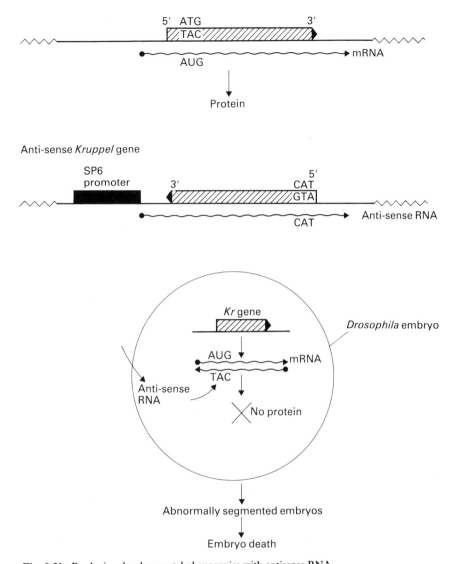

WT *Kruppel* (Kr) gene

Anti-sense *Kruppel* gene

Drosophila embryo

Fig. 8.31. Producing developmental phenocopies with antisense RNA.

flies carrying antisense plasmids that can be expressed at different stages, e.g. by heat-shocking. This will allow the effect of inhibiting gene expression on development to be tested. Similar approaches have been used to block the gene expression in plant cells (Ecker and Davis, 1986).

In some cases genes have been cloned but their biological role is not clear. By introducing antisense plasmids into cells and observing the phenotypic effects clues as to the biochemical properties and physiological significance of the gene and its products can be obtained. Also as regions of complementarity of only 50 bp can inhibit gene expression it may be possible to design antisense plasmids that selectively inhibit closely related

GENE EXPRESSION

members of gene families so that the individual contributions of genes to a process can be evaluated.

Summary

A eukaryotic gene contains an amazing number of specific sites in the primary nucleotide sequence that control expression. Most of the significant control occurs at the level of transcription but post-transcriptional controls, e.g. control of splicing and translation, are also important.

The eukaryotic promoter appears to be assembled from many short sequence elements some of which can be found in many different genes and even in different organisms. Some elements are, however, gene-specific. There are elements that are required for any transcription to occur at all, that is, they are constitutive. Other elements are regulated or function only in certain cells. A simple explanation for eukaryotic promoter structure is that a small number of primordial elements have been 'mix and matched'. They have been repeated in some genes and interspersed with different elements. Many of the elements are related but not identical. These small nucleotide changes in elements in different genes then allows subtle differences in gene expression. In some cases the elements are dispersed throughout the gene into 5' leader sequences and introns and possibly even coding sequences. The number of different elements, their spatial arrangement and small changes in the sequence confer a unique expression profile on individual genes or gene families.

The rapid response of genes to environmental stimuli such as heat shock, or steroid hormones appears to be mediated by interactions with pre-synthesised proteins that are activated by the inducer. This could occur by binding the inducer as in hormone induction or by post-translational modification such as phosphorylation, or by changes in the compartmentalisation of the protein. During development and differentiation the activity of the gene may have to change. This may be dictated by the ability of the gene to be assembled into a pre-initiation complex. This may be mediated by the enhancers which ensure that the gene is 'prepared' for transcription. Transcription then occurs when specific regulatory proteins are either synthesised or activated. Tissue specificity is probably determined by the lack of availability of the promoter elements because many of the activator proteins are usually present in cells that do not naturally express the gene. It is likely, however, that not all activator proteins are present in all cells.

The control of translation often occurs when a very rapid response is required as in heat shock, or when very large amounts of protein are required and therefore pre-synthesised mRNA must be used, as for example during early embryonic development. It also allows subtle regulation to occur as in the response of yeast amino acid biosynthetic genes to the constantly varying nutritional status of a culture medium.

Control of splicing allows diversity of gene expression. This is an 'economy measure' because a single transcriptional unit with all its heirarchy of controls can be used to generate multiple products simply by the regulation of splicing factors.

The models for the control of gene expression and the mechanistic basis of the protein−nucleic acid interactions that produce this control are far from clear. The combination of recombinant DNA techniques and biochemical analyses of *in vitro* reactions is building up the picture. It seems, however, that much of the analysis must be done *in vivo* because the specificity of the protein−nucleic acid interactions in terms of regulation and cell specificity may be lost *in vitro*. Also the role of higher order chromosome structure and the formation of stable transcription complex is clearly important *in vivo*. In particular the enhancers are often not required for transcription *in vitro* (e.g. Mathias and Chambon, 1981) but they are essential *in vivo* and this may reflect their importance in 'preparing' the gene for transcription. New methods for studying gene activity *in vivo* are being devised that allow genes to be switched on and off at precise times by the experimenter or which use reporter genes such as luciferase which can be monitored continuously and in some cases, e.g. in plants, non-invasively. Current research is also aimed at building up heirarchies of transcriptional controls by introducing target genes and genes for regulatory proteins into cultured cells in precise combinations.

9 DNA Components of Chromosomes

Introduction

Gene transfer technology is being used to address fundamental questions about the organisation and behaviour of chromosomes. The most significant advances have been in *Saccharomyces cerevisiae* where the key chromosome elements namely replicators, centromeres, and telomeres have been isolated. This has allowed the construction of entire chromosomes *in vitro*. The nature of other genetic elements such as transposons and the mechanisms of recombination and mutation, are also being analysed.

Replication

The replication of some pieces of DNA is initiated at a precise nucleotide sequence called the *replication origin*. This has been amply demonstrated for bacterial and bacteriophage genomes and for many eukaryotic viral genomes. There is, however, still considerable uncertainty about the identity and significance of specific replication origins in eukayrotic chromosomes. There are two key features of eukaryotic chromosome replication. Firstly there are multiple replication units called *replicons* and secondly chromosomes normally only replicate once per cell cycle and therefore initiation of replication must be subject to some strict controls. Two gene transfer systems have been used extensively to investigate the specificity and control of eukaryotic chromosomal DNA replication. These use either *S. cerevisiae* or *Xenopus* eggs. Recently papillomaviruses and *Drosophila* have also been used to study the control of DNA replication. We will describe some of the key experiments that have used gene manipulation and gene transfer techniques to enhance an understanding of DNA replication. More details can be found in reviews by Hand (1978), Williamson (1985) and Campbell (1986).

Characterisation of *S. cerevisiae* ARSs

In Chapter 3 we discussed the development of gene transfer vectors for *S. cerevisiae* and showed that integrative vectors could be converted to autonomous vectors by the addition of certain fragments of chromosomal DNA. These fragments, referred to as *autonomously replicating sequences (ARSs)*, are good candidates for authentic chromosomal replication origins. The frequency of isolation of ARSs from chromosomal DNA is about one in 40 kb which is similar to the average spacing between replicons that is observed by electron microscopy. The first ARS to be characterised was linked to the *TRP1* gene; it is referred to as *ARS1* and is present on the vector YRp7 (Struhl *et al.*, 1979; Stinchcomb *et al.*, 1979; Chapter 3). Subsequently many different ARSs have been isolated by simply 'shotgun' cloning chromosomal DNA into an integrative vector and then selecting transformants. The integrative vector produces only 1–10 transformants per µg of DNA whereas plasmids containing an ARS produce about 50 000 transformants per µg. In a typical experiment a chromosomal shotgun collection yielded 45 transformants and two thirds of these were unstable indicating that they carried autonomous plasmids (Chan and Tye, 1980).

The characteristic property of an ARS is the ability to confer high-frequency transformation on a plasmid. The plasmids are not, however, efficiently maintained in transformants which are, therefore, unstable for the transferred phenotype. Different ARSs have different stabilities and this can be monitored by determining the percentage of cells in a culture that display the transformed phenotype. Another measure of the replication competence of an ARS plasmid is to measure the doubling time of the culture. Transformants which lose the plasmid at a high rate grow more slowly under selective conditions than transformants containing more stable plasmids.

The nucleotide sequences of several ARSs have been compared and an 11 bp consensus sequence has been identified (Stinchcomb *et al.*, 1981; Broach *et al.*, 1983b). The minimal sequences required for replication and the functional significance of the consensus sequence are being studied by using *in vitro* mutagenesis and then assaying the mutant plasmids for their transformation and replication competence. For example, Kearsey (1984) has analysed an ARS that is linked to the *HO* gene (this gene is involved in mating type switching). There is a region that shows a 10/11 bp match with the consensus sequence and contains seven nucleotides that are highly conserved between different *S. cerevisiae* ARSs (Figure 9.1). The consensus is shown in bold type and the conserved nucleotides are marked with an asterisk. A series of deletions into this region abolished the ARS activity of the fragment as measured by high-frequency transformation. This defined a 14 bp core sequence that was essential for ARS activity. Nucleotide changes within the consensus sequence also abolished ARS activity and in two cases a single change, e.g. a T-to-C at position 9 and a T-to-A at position 3 were sufficient to abolish ARS function. Some changes outside the consensus sequence reduced the efficiency of the ARS but never abolished function. This study showed that the consensus sequence was critical for function, that the specific sequence of bases was important and not simply the A+T richness and that flanking sequences could influence the function of the consensus sequence. This study has been confirmed and extended by a detailed *in vitro* mutagenesis and *in vivo* analysis of ARS1 (Celniker *et al.*, 1984; Srienc *et al.*, 1985). The EcoRI fragment in YRp7 that carries the *TRP1* gene and ARS1 is shown in Figure 9.2. The ARS consensus sequence lies adjacent to a unique BglII site in this fragment and has been called the

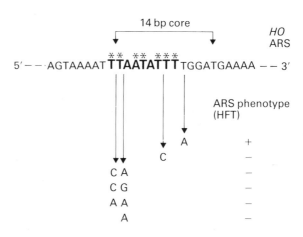

Fig. 9.1. The functional significance of the ARS consensus sequence. HFT = high frequency transformation.

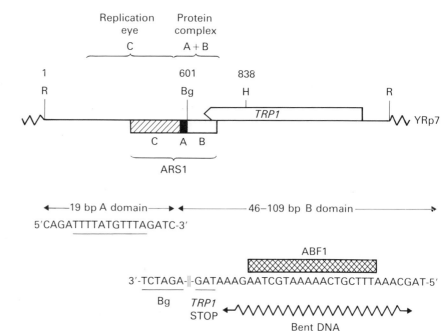

Fig. 9.2. The structure of ARS1.

A domain. When deletions were made in DNA to the 3' side of the A domain, the fragment was still capable of high-frequency transformation but the plasmid stability was reduced. This defined a region called the B domain which was important for efficient stable replication but which could not itself mediate high-frequency transformation. When deletions were made in the DNA to the 5′ side of the A domain a third domain called C was identified; however, the effect of these deletions was only obvious when the B domain was missing. In the absence of the B domain the deletions in C reduced the stability of the plasmids. The significance of the C domain is not clear as heterologous sequences, e.g. *E. coli* DNA, can replace the C domain if B is present.

The minimal functional region of ARS1 was defined by sub-cloning progressively smaller regions containing the 11 bp consensus into an integrative vector and testing for high-frequency transformation. A 45 bp fragment from nucleotide 582 to 627 was functional but a 19 bp fragment spanning 582 to 601 was not functional (Celniker *et al.*, 1984). If, however, a CEN sequence was included on the assay plasmid then the 19 bp sequence allowed high-frequency transformation (Srienc *et al.*, 1985). These studies (Kearsey, 1984; Celniker *et al.*, 1984; Srienc *et al.*, 1985), therefore, demonstrated that a 14−19 bp core that contains the 11 bp consensus sequence is necessary for high frequency transformation but that additional flanking sequences are required to ensure efficient and stable replication. One idea is that the B domain is responsible for attaching the plasmid to nuclear structures. A CEN sequence presumably has a similar function and can therefore replace the B domain although in this case it allows the plasmid to segregate equally to mother and daughter cells whereas ARS plasmids show an asymmetric distribution (Chapter 3).

Recently 'unusual' DNA regions associated with viral replication origins have been shown to be functionally important (e.g. Ryder *et al.*, 1986). This DNA is characterised by the presence of several polyadenosine tracts that are 3−5 nucleotides long and that are approximately 10.5 bp apart, i.e. one turn of a B-form helix. Restriction fragments that contain such sequences show an aberrant mobility during electrophoresis in 10% polyacrylamide gels at 25°C. The fragments migrate 20−40% more slowly than predicted from their molecular weight. DNA of this general structure can bend at the 3′ ends of the adenosine tracts and is therefore called *bent DNA*. It is proposed that bending facilitates the interactions between proteins that are bound to the DNA in that region. A stretch of DNA with the potential to bend has been identified in the 3′ coding region of *TRP1* just before the translation stop and just within the B domain (Figure 8.2; Snyder *et al.*, 1986). Deletions that remove the bent DNA did not affect high-frequency transformation but reduced plasmid stability by 5-fold. A protein present in crude yeast cell extracts binds to this region of bent DNA; this has been called *ABF1 (ARS binding factor 1)* (Snyder *et al.*, 1986). The nucleotide sequence of 13 different yeast ARSs have been compared and a second consensus sequence has been proposed (Palzkill *et al.*, 1986). This is located within domain B at 80 ± 30 nucleotides 3′ to the domain A consensus and has the sequence 5′−CTTTTAGCA/TA/TA/T−3′. Preliminary deletion analyses indicate that these are functionally significant.

Domains A and B may be involved in assembling a replication complex which then promotes replication initiation with the C region. Studies of *in vitro* replicating ARS1 plasmids have shown a protein complex bound to the

DNA in the region of the A domain (Jazwinski *et al.*, 1983). Different ARSs replicate at different times during S phase (Fangman *et al.*, 1983) and respond differently to mutations that are defective in the maintenance of mini-chromosomes (Maine *et al.*, 1984). This suggests that ARS activity may be differentially regulated and it is possible that this regulation is mediated through domain B and possibly domain C.

Heterologous ARSs in *S. cerevisiae*

Specific chromosomal DNA fragments from a number of different eukaryotes have now been shown to function as ARSs in *S. cerevisiae*. These have been isolated by 'shotgun' cloning into an integrative vector and then screening transformants for the presence of autonomous plasmids (e.g. Stinchcomb *et al.*, 1980). They have been found in mammals, plants, amphibians, insects, nematodes, slime moulds, algae, protozoa and other fungi (reviewed by Williamson, 1986). Detailed analyses of these heterologous ARSs has shown that in the case of ARSs from *Xenopus* (Kearsey, 1983) and *Drosophila* (Mills *et al.*, 1986) the functional region contains an identical match to the *S. cerevisiae* ARS consensus sequence. In the case of two human ARSs, however, while both shared an 18 bp homology block this was only partially related to the *S. cerevisiae* consensus sequence (Montiel *et al.*, 1984). The fact that the human sequences could replicate suggests that the sequence requirements for replication in *S. cerevisiae* may not be as stringent as suggested by *in vitro* mutagenesis of *S. cerevisiae* ARSs. The functional organisation of both *Drosophila* and *human* ARSs was found to be similar to that described for *S. cerevisiae ARS1*. There were two spatially distinct domains, the replication sequence (RS) which was essential for high-frequency transformation and the replication enhancer (RE) which was required for maximum replication competence and/or to increase mitotic stability (Montiel *et al.*, 1984; Mills *et al.*, 1986). These domains appear to be analogous to A and B in yeast ARS1. The functional significance of these heterologous ARSs in terms of replication in their cognate chromosomes is not known. ARSs isolated from the mouse genome failed to replicate autonomously in mouse cells (Roth *et al.*, 1983) and sequences from *S. pombe* that have ARS function in *S. cerevisiae* do not replicate in *S. pombe* itself (Maundrell *et al.*, 1985). Given the sequence, organisational and functional similarities between ARSs from different origins it would, however, be surprising if they had no biological function in their cognate chromosomes. It is possible that additional sequences, particularly centromere-like sequences will be required to demonstrate stable autonomous replication of these ARSs in their cognate systems.

Lack of specific sequence requirements for replication in *Xenopus* eggs

DNA replication can be studied by microinjecting DNA into unfertilised *Xenopus* eggs. The DNA replicates and this replication is tightly coupled to the egg's cell cycle (Harland and Laskey, 1980) following activation by pricking (see Chapter 5). The DNA replicates only once per cell cycle, i.e.

re-initiation is completely blocked. In order to study the specificity of initiation and cell cycle control, a range of prokaryotic and eukaryotic viral and chromosomal DNA fragments have been injected into *Xenopus* eggs. The general conclusion from these studies is that there is no specific sequence requirement because bacteriophage DNA replicated as efficiently as *Xenopus* DNA (Harland and Laskey, 1980; McTiernan and Stambrook, 1980). Even sequences that contained known replication origins such as the SV40 origin failed to replicate preferentially (McTiernan and Stambrook, 1982). The amount of synthesis was only correlated with the length of the template and not with the source of the DNA (Mechali and Kearsey, 1984). The replication of all these different DNA fragments was also under strict cell cycle control indicating that the block to re-initiation within a cell cycle was also not sequence-specific. The fact that this was true of viral replication origins such as SV40 which normally show runaway replication suggests that one of the functions of viral encoded replication proteins might be to overcome the host cell cycle control.

These studies with *Xenopus* eggs that indicate no requirement for sequence specificity for replication contrast markedly with the studies in *S. cerevisiae*. It is possible that replication in eggs reflects some special features of early embryos. DNA replication in embryos is extremely rapid (Laskey *et al.*, 1985) and the number of replication eyes, i.e. initiation sites, can be 5-fold higher than in adult cells (e.g. Blumenthal *et al.*, 1974). Alternatively the ARSs may not be true replication origins but rather they may be regulatory sites that determine the time of replication at adjacent non-specific sites. Such regulatory sites may be non-functional in early embryos. It may be relevant that early embryos are transcriptionally quiescent during their phase of maximum replication. The sites of initiation for replication may only be critical when specific regions of the genome must also be transcribed and therefore specific origins of replication may be important. The *Xenopus* egg system for studying replication may allow the identification of the 'global' cellular regulators of cell cycle control of replication whereas studies in *S. cerevisiae* may identify specific regulatory signals.

DNA amplification

DNA replication is usually controlled so that there is no re-initiation within a cell cycle. There are, however, some exceptions. An extreme case is the over-replication of entire choromosomes to produce polytene chromosomes in some terminally differentiated cells in insects. In other cases there is a selective amplification of specific regions of the chromosome. This amplification can confer a selective advantage, for example, by determining drug resistance or it may allow the accumulation of specific gene products during development, as in the amplification of rDNA in amphibian oocytes (reviewed by Stark and Wahl, 1984; Schimke, 1984). A simple model for amplification has been suggested by Botchan *et al.* (1980) and Roberts *et al.* (1983). This model proposes that there is uncontrolled replication at a single initiation site. This generates an 'onion skin' structure containing multiple repeats. Recently a different model has been proposed which involves the production of an inverted duplication followed by recombination across the duplication to generate a molecule with two replication forks moving in the same direction (Passananti *et al.*, 1987). This model is similar to the one proposed for the replication of the 2 μm circle in yeast (see Chapter 3). It is supported by the finding that many amplicons in mammalian cells are flanked by inverted duplications. The amount of amplified DNA is usually large, about 1000 kb and therefore encompasses several replication initiation sites.

Gene transfer is being used to analyse the mechanism of DNA amplification and in one case, that of chorion gene amplification in *Drosophila*, a *cis*-acting control element has been localised. There are two clusters of chorion genes located on chromosome I (the X chromosome) and chromosome III. These encode about 20 different chorion proteins which form the egg shell. These proteins are only synthesised at the terminal stages of oogenesis in the ovarian follicle cells (Spradling and Mahwold, 1980). At this stage the chromosomes of the follicle cells are already polytene so amplification can be viewed as a localised increase in polyteny. At about 18 hrs before eggshell formation the chorion proteins over replicate in relation to the rest of the DNA. They increase to produce 16 copies of the cluster on chromosome I and 60 copies on chromosome III. The region of DNA that is amplified is about 80–100 kb but the chorion genes themselves are located in a 12–20 kb region at the centre of the amplified DNA (Spradling, 1981). To study the mechanism of amplification the 66D locus of chromosome III that contains the chorion genes has been cloned and segments have been introduced into new chromosome locations by P-element-mediated gene transfer (De Ciccio and Spradling, 1984; Figure 9.3).

Initially, a 7.7 kb EcoRI fragment was analysed. A transgenic line containing a single P element transposon containing this fragment was constructed. The transposon was integrated at site 68A, that is, in a region that is not normally amplified. DNA was prepared from egg chambers and non-ovarian female tissues from these flies at various developmental stages and analysed by Southern blotting. The transposon DNA was amplified at precisely the same time as the endogenous chorion genes. Furthermore, the amplification was tissue-specific occurring only in ovarian tissue but the level of amplification was only 20-fold. This study showed that the 7.7 kb EcoRI must contain specific sequences that determine tissue specific and stage specific DNA amplification. Similar results were obtained with a smaller 3.8 kb fragment that contains the two major chorion genes. This fragment induced local amplification at the correct stage and in the correct tissues. Amplification with this fragment was sensitive to the position of integration: for it was only shown in 53% of the sites and the levels of amplification varied between sites. Despite this position effect further analysis showed that amplication was still directed by a 510 bp fragment (Orr-Weaver and Spradling, 1986). This has been called an *amplification control element (ACE3)*. The chorion genes on the X chromosome have been subjected to a similar analysis and a 467 bp essential has been identified called ACE1 (Spradling *et al.*, 1987). This contained a 32 bp region of partial sequence homology with the ACE3 *cis* active fragment. It seems likely that this region constitutes, at least in part, the replication origin used to amplify the region. It contains several repeats containing the motif

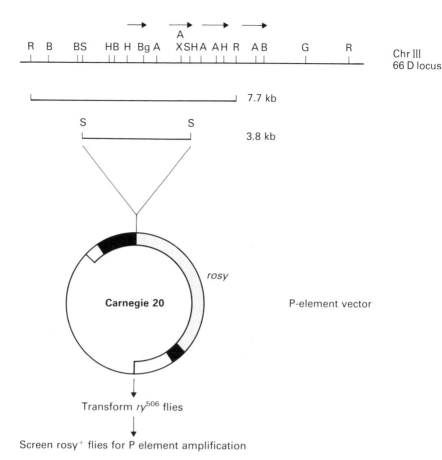

Fig. 9.3. Chorion gene amplification in *Drosophila*.

5'−AATAC−3'. The requirement for additional flanking sequences for maximum amplification and to overcome position effects is thought to be to protect the ACE from inhibitory sequences. The possibility that eukaryotic chromosomes contain elements that function to restrict amplification is also suggested by the discovery of an unusual sequence element at the end of a carcinogen-induced mouse amplicon (Baron *et al.*, 1987). A number of mutants that effect chorion gene amplification have also been identified and these may define *trans*-acting factors that interact with ACE (Orr *et al.*, 1984).

Regions of the mammalian genome that amplify, e.g. the *dhfr* gene and CAD complex genes, comprise several hundreds of kilobases. In the case of a *dhfr* amplicon, replication initiates in the same place (Burhans *et al.*, 1986 and a 260 kb cosmid clone from a CAD amplicon replicates autonomously in mouse cells (Carroll *et al.*, 1987) suggesting that mammalian amplicons contain functional replication origins.

Centromeres

A major achievement of gene transfer techniques has been the isolation and characterisation of centromeres. This provides a route to understanding the

fundamental process of chromosome pairing and segregation at mitosis a█
meiosis. Most of the information to date has been obtained in *S. cerevis*█
but putative centromeres from other eukaryotes namely *Schizosaccharomy*█
pombe (Clarke *et al.*, 1986) *Caenorhabditis elegans* (Stinchcomb *et al.*, 198█
and mouse (Rassoulzadegan *et al.*, 1986) have recently been isolated. T█
structure and function of centromeres has been reviewed by Clarke a█
Carbon (1985), Blackburn and Szostak (1984) and Murray and Szost█
(1985).

Isolating centromeres

The initial isolation of a yeast centromere, CEN3, has been described █
Chapter 3. The key property of a centromere sequence is the ability █
confer stability on an ARS plasmid during both mitotic and meiotic c█
division (Clarke and Carbon, 1980). This property has been used to sel█
other centromere sequences from *S. cerevisiae* (Hsaio and Carbon, 1981)█
yeast library of Sau3a fragments in YRp7 was transformed into *S. cerevis*█
selecting for *TRP*$^+$. Transformants were then cultured for extended peri█
without selection. All unstable plasmids would be completely lost by t█
procedure but when the cultures were replated on to selective media █
clones were still *TRP*$^+$. About 90% of these contained autonomous plasmi█
half contained fragments of endogenous 2 μm circle and were not analy█
further but 13 plasmids contained chromosomal DNA and five of th█
chosen randomly showed the ordered meiotic segregation expected █
centromere plasmids. A similar procedure has been used to isol█
centromere-like sequences called *segregators* from *Caenorhabditis elegans*.█
this case repeated rounds of growth in selective and non-selective me█
were used to enrich for segregators (Stinchcomb *et al.*, 1985a). Hieter *et*█
(1985a) have now devised a rapid direct selection procedure for isolat█
stabilising sequences. They have shown that a cloned ochre suppressi█
tRNA gene, *SUP11*, is lethal at high copy number and causes cell de█
when it is present on an ARS plasmid. The provision of a CEN sequen█
however, reduces the copy number to one (Chapter 3) and therefore, t█
cells survive. When random chromosome fragments were cloned into█
SUP11-ARS plasmid ten distinct segments were isolated which c█
responded to chromosomal centromeres.

Genomic substitutions of centromeres

In order to confirm that CENs are, in fact, functional centromeres and █
gain further insight into their properties, Clarke and Carbon (1983) design█
a series of *in vivo* gene disruption experiments. A region of chromosome █
containing CEN3 was manipulated *in vitro* so that the core 627 bp CE█
region was replaced by the *URA3* gene. A unique BamHI cloning site w█
retained so that different CEN fragments could be re-inserted. Lin█
fragments were targeted to chromosome III using the homology of t█
DNA regions that flank the core CEN sequence (see Chapter 3 for det█
about gene targeting). Three types of genomic substitutions are shown█

Figure 9.4. If CEN3 is indeed a genuine centromere then fragment I should target to the centromere and cause a deletion of the chromosome III centromere. The fragment was introduced into a diploid host background to be certain that viable transformants could be obtained which had one functional chromosome III and one altered chromosome III. The two copies of chromosome III carried different genetic markers so that their inheritence during mitotic and meiotic cell division could be followed. The result of this genomic substitution was as expected. There was extreme instability of one of either of the copies of chromosome III in a large proportion of the transformants. In addition, it was not possible to obtain haploid strains of both chromosome III genotypes by sporulating the transformed diploids. The targeting fragment 1 could therefore produce a chromosome III that showed the high instability that is characteristic of an acentric chromosome. This confirmed that the cloned CEN3 was a functional centromere. When targeting fragment 2 was used, there was a high frequency of transformation and Southern hybridisation experiments showed that the centromere of chromosome III had been inverted relative to flanking markers. All the transformants, however, showed normal chromosome III behaviour. Similar results were obtained when targeting fragment 3 was used. This showed that centromere function is not orientation-dependent and that centromeres are not chromosome specific because clearly CEN11 can function in a chromosome III background. This suggests that chromosome pairing during meiosis must depend upon homology between chromosome arms and not on specific centromere interactions.

Although *Saccharomyces* centromeres are interchangeable between chromosomes they appear to be highly species specific. They do not function as centromeres in *Klyveromyces lactis*, *S. pombe*, *Neurospora crassa* or cultured animal cells (Carbon, 1984).

S. cerevisiae is clearly a good system for studying chromosome behaviour because of the ease of genomic substitution and the sensitive genetic methods of following chromosome behaviour during meiosis and mitosis. It is now possible to make single base changes in a centromere *in vivo* by *in vitro* mutagenesis and gene targeting and to assess the effect on chromosome function.

The structure of CEN sequences

Nucleotide sequence comparisons between different centromeres have indicated some highly conserved features. There are three major domains called *centromere DNA sequence elements (CDE)* I, II and III. These were first described by Fitzgerald-Hayes *et al.* (1982) by comparing CEN11 and CEN3 but a more complete description has now been obtained by comparing additional CEN sequences (Panzeri and Philippsen, 1982; Hieter *et al.*, 1985a; Nietz and Carbon, 1985). The complete sequence of CEN3 and a consensus derived from comparing ten different CEN sequences (Hieter *et al.*, 1985a) is shown in Figure 9.5. CDEI comprises an 8 bp block of perfect homology provided that a purine ambiguity is allowed at positions 1 and 6. CDEII comprises a region that is roughly conserved in length: it

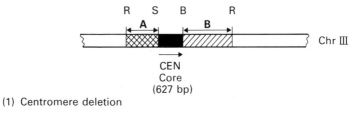

CEN Core (627 bp) — Chr III

(1) Centromere deletion

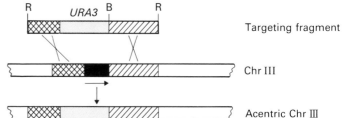

Targeting fragment

Chr III

Acentric Chr III

(2) Centromere inversion

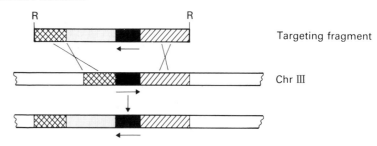

Targeting fragment

Chr III

(3) Centromere replacement

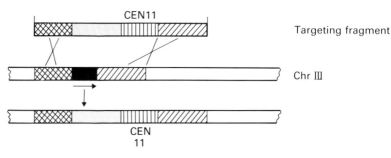

Targeting fragment

Chr III

CEN 11

Fig. 9.4. Genomic substitutions of centromeres.

is between 78 to 86 bp long. There is no sequence homology but the region is greater than 90% A+T. CDEIII is a 25 bp domain in which eight positions are precisely conserved between all the CENs. This domain displays dyad symmetry around a central cytosine (indicated by the arrows in Figure 9.5B). The functional significance of these conserved blocks is being established by *in vitro* mutagenesis and gene transfer.

Centromere function can be assessed either by analysing chromosome behaviour following genomic substitutions or, for example, by using the fact that mutant centromeres will not block the lethality of *SUP11* plasmids. Some of the mutations that disrupt centromere function are shown in Figure 9.5. McGrew *et al.* (1986) have shown that single base changes a

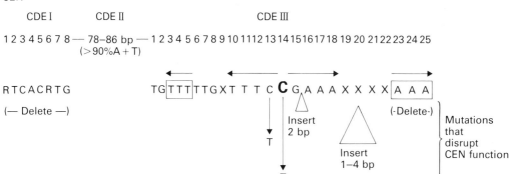

A

		II			
I		87 bp (93% A + T)		III	IV

EN3 | ATAAGTCACATGAT | GATATTTGATT TT ATTATAT TTT T AAAAAAAGTAAAAAATAAAAAGTAGT TTATT TTT T AAAAAATAAAATTTAAAATATTAGTGTATT | TGATTTCCGAA | ← 248 bp → | TTTAGAGCAA |
TATTCA GTGTACTA | CTATAAACTAAAATAATATAAAA A TTTT TTT T CATTTTT TAT TTT T TCATCAAATAAAAA T TTTT TATT TTAAATTTTATAATCACATAA | ACTAAAGGCTT | | AAATCTCGTT |

B

Consensus CEN

         CDE I   CDE II   CDE III

  1 2 3 4 5 6 7 8 — 78–86 bp — 1 2 3 4 5 6 7 8 9 10 11 12 13 14 15 16 17 18 19 20 21 22 23 24 25
                 (>90%A + T)

    R T C A C R T G   TG TTT TTGX T T T C **C** G A A A X X X X A A A

    (— Delete —)

                                    Insert 2 bp   (-Delete-)

                              T   Insert 1–4 bp   Mutations that disrupt CEN function

                              T

Fig. 9.5. The structure of CEN sequences. A, The nucleotide sequence of CEN3. B, The consensus sequence and the effect of some specific mutations on centromere function. The region of dyad symmetry around the central C in CDE III is indicated by arrows. R = purine and X = any nucleotide.

positions 13 and 14 in CDEIII respectively decrease or abolish centromere activity. It appears that maintenance of the precise organisation of the palindrome in CDEIII is critical as small insertions of 1–4 bp are sufficient to abolish centromere function (McGrew *et al.*, 1986; Hegemann *et al.*, 1986). It may be significant that in CDEIII the triple T and triple A boxes are located 10 base pairs to the left and right of the central cytosine. This places them two helix turns apart on the same face and this might be important in creating a protein–nucleic acid binding site (Hieter *et al.*, 1985a). Deletion of the triple A box abolishes centromere function (Hegemann *et al.*, 1986) and deletion of CDEI dramatically reduces activity (Panzeri *et al.*, 1985). The most important feature of CDEII appears to be the length of the element. Small changes in nucleotide composition can be tolerated and 50% increase in spacing between CDEI and CDEIII has no effect (Panzeri *et al.*, 1985) but doubling the length of CDEII causes a 100-fold increase in chromosome non-disjunction (Gaudet and Fitzgerald-Hayes, 1987). Deletions also interfere with correct function (Clarke and Carbon, 1985; Gaudet and Fitzgerald, 1987) but in many cases function can be restored by inserting the correct size heterologous DNA which does not have to be particularly A+T-rich.

The functional boundaries of a CEN sequence have been delimited by Bal31 deletion and sub-cloning analyses to lie just outside the conserved elements (Panzeri *et al.*, 1985; McGrew *et al.*, 1986). This means that CEN sequences are rather small, that is, only about 120 bp. Centromere regions from other eukaryotes are associated with satellite DNA but in the case of *S. cerevisiae* CENs there is clearly no functional requirement for highly repetitive DNA. Some sequence homologies have, however, been noted between *Drosophila* and bovine satellite DNA and CENs (Fitzgerald-Hayes *et al.*, 1982). The functional significance of this observation has yet to be explored.

Chromatin organisation around CEN sequences

The centromere is responsible for attaching the chromosomes to the spindle apparatus and for ensuring accurate segregation of replicated chromosomes to daughter cells. The precise nature of the interaction is not known but in *S. cerevisiae* a single microtubule is located at each centromere (Peterson and Ris, 1976). In higher eukaryotes the spindle attachment point at the centromere forms a cytologically distinct structure called the *kinetochore*. It is likely that attachment is mediated by specific centromere-binding proteins and the next goal in centromere research is to characterise these protein−nucleic acid and protein−protein interactions. As a first step, Bloom and Carbon (1982) have investigated the chromatin organisation at CEN3. This was established by digesting chromatin with limiting amounts of micrococcal nuclease and probing the digestion products with a cloned CEN3 probe. The CEN3 homologous DNA was present in a more highly ordered ladder than bulk chromatin. This indicated that nucleosomes are uniformly spaced in the centromere region. The position of the nucleosomes was determined precisely by DNaseI digestion and indirect end-labelling. A striking finding was that a 250 bp region of DNA was completely protected from nucleolytic cleavage and this was flanked by fixed nuclease hypersensitive sites. A model for the chromatin organisation around the centromere is shown in Figure 9.6. The formation of the ordered array of nucleosomes does not, however, result from the simple exclusion of nucleosomes from the centromere core. This has been shown by deleting sequences immediately next to the CEN3 core on a plasmid so that pBR322 DNA is moved adjacent to the core. In this case the distribution of nucleosomes over the plasmid DNA is random. Also if the CEN3 core is deleted the ordered nucleosome array still assembles on the flanking DNA (Bloom and Carbon, 1982). This means that the nucleosome phasing depends upon the base sequence and/or composition of the centromere-flanking DNA.

The CDEIII domain lies near the centre of the protected region and it has been proposed that this might act as the microtubule binding site and serve as a primitive kinetochore. The ordered array of nucleosomes extends for 2 to 3.5 kb around the protected core (Carbon, 1984). Cloned centromere DNA has been shown to bind proteins (Bloom *et al.*, 1983): a specific protein binding to a synthetic oligonucleotide corresponding to CDEIII has been demonstrated (Hegemann *et al.*, 1986). A protein that binds to CDEI has also been isolated and it appears to have a counterpart in human cells (Bram and Kornberg, 1987).

The flanking sequences within the ordered nucleosome array are transcribed, but no transcripts traverse the centromere. At least one of the transcripts encodes a gene (SPO15) that is essential for meiosis (Yeh *et al.*, 1986). It is likely that strong transcriptional terminaters flank the CEN core to prevent transcriptional inactivation. This is known to occur because when a CEN core is placed 120 bp downstream from a strong regulated promoter

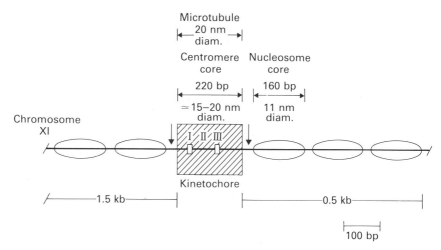

Fig. 9.6. Chromatin organization of CEN. The arrows indicate DNase hypersensitive sites.

the activity of the centromere can be 'turned on or off' by activating or repressing the promoter (Hill and Bloom, 1987).

Telomeres

The telomeres are the structures at the ends of eukaryotic chromosomes. They have two major functions. Firstly, unlike any other DNA end, for example, a double-stranded break caused by enzymes or ionising radiation, they are resistant to exonucleolytic degradation. They therefore preserve the integrity of linear chromosomes. Secondly, they are essential to allow the complete replication of the ends of the chromosome because the activity of DNA polymerase is primer-dependent and polymerisation only occurs in the 5′ to 3′ direction. Gene transfer technology has allowed telomeres to be cloned and models for the terminal replication of chromosomes to be tested.

Cloning telomeres

Because of the special features of the extreme end of a telomere it cannot be simply cloned into a circular plasmid. This problem was circumvented by constructing linear vectors that could be propagated in *S. cerevisiae* (Szostak and Blackburn, 1982). The first source of telomeric DNA was derived from the protozoan *Tetrahymena*. The nucleus of this organism contains multiple copies of autonomously replicating DNA which comprises linear fragments of ribosomal DNA. These linear plasmids were isolated and the ends were liberated by digestion with BamHI (Figure 9.7). These 1.5 kb fragments with a BamHI site at one end and a terminus at the other were then ligated to a yeast ARS plasmid that had been linearised at a unique BglII site and ligated with the terminus fragments. The different ligation products were separated by gel electrophoresis and the required fragment was purified from the gel. This had the ARS plasmid flanked by *Tetrahymena* termini. This fragment was used to transform *S. cerevisiae*, selecting for *LEU2*. The

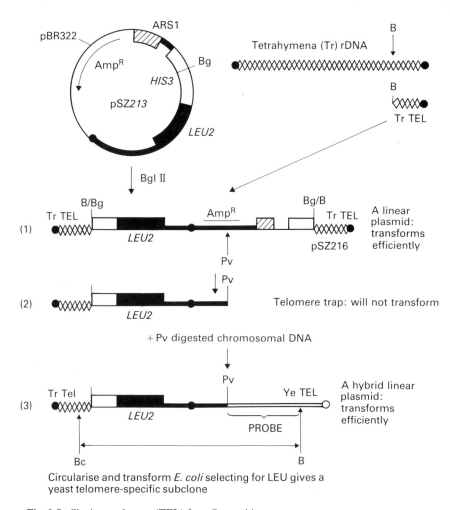

Fig. 9.7. Cloning a telomere (TEL) from *S. cerevisiae*.

fragment transformed efficiently but the transformants were mitotically unstable, indicating that they were autonomous. Restriction mapping showed that the vector had remained linear indicating that the *Tetrahymena* termini retained their special properties and were functioning as telomeres in *S. cerevisiae*.

The discovery that linear plasmids could replicate autonomously in *S. cerevisiae* led to the construction of a '*telomere trap*' vector for isolating *S. cerevisiae* telomeres (Figure 9.7). The linear vector was digested with PvuI to generate a fragment that retained the *LEU2* selectable marker and the pBR322 origin of replication but only one *Tetrahymena* telomere. This plasmid will only transform at a high frequency if telomeric DNA is ligated to the other end. The fragment was ligated with PvuI-digested chromosomal DNA and the ligation mix used directly to transform *S. cerevisiae*. Seventy transformants were obtained and analysis of the autonomous plasmids in these revealed several that had acquired a single PvuI fragment and were therefore hybrid linear plasmids. These plasmids were capable of replicating despite the loss of ARS1 because *Tetrahymena* rDNA has ARS activity (it has subsequently been shown that ARS elements are also present in *Saccharomyces* telomeric DNA). The fragments were restriction-mapped *in situ*, i.e. by digesting total DNA isolated from transformants and then detecting fragments by Southern blotting. The different isolates showed a conserved region of 4 kb at their termini suggesting that all the telomeres in the cell have an identical or very similar structure. Linear fragments cannot be propagated in *E. coli* so the whole of this cloning study had to be undertaken in *Saccharomyces*. In order to prepare a telomere-specific probe to analyse structure and function, the hybrid linear plasmid was digested with BclI and BamHI (these both generate GATC ends) and ligated to produce a circular plasmid that was used to transform a *leuB E. coli*, selecting for *LEU2*.

This probe hybridised to 30−40 fragments in the yeast genome which is consistent with every chromosomal telomere having homology with the probe.

The structure of telomeres

The *Tetrahymena* telomere structure and function is preserved in *S. cerevisiae*. The essential features of telomeres have probably been conserved in evolution. The *Tetrahymena* termini have been characterised and they consist of variable lengths of DNA comprising multiple repeats (20−70) of the simple sequence $5'-C-C-C-C-A-A-3'$ (C_4A_2). At the most terminal side of the cluster there are single-strand breaks after the second A in every two to four repeats. At the more internal side of the cluster there are also breaks on the opposite strand. The DNA terminus is cross-linked and probably forms a hairpin loop (Blackburn and Gall, 1978; Figure 9.8).

The cloned yeast telomeres have similar characteristics but are about 200 nucleotides larger and the repeats are more heterogeneous. The repeats are C_3A, C_2A and CA (Shampay *et al.*, 1984) and there appear to be tracts of poly (dGdT dCdA) (Walmsley *et al.*, 1984).

The function of telomeres

When a linear DNA molecule is replicated by DNA polymerase elongation occurs only in the 5′ to 3′ direction and requires a primer. This produces daughter molecules with protruding 3′ ends which have not been copied (Figure 9.9A).

These ends cannot possibly remain uncopied because chromosomes would become shortened at each round of replication and clearly this is not the case. One model proposed by Cavalier-Smith (1974) and modified by Bateman (1975) suggests that there is a hairpin loop at the chromosome terminus and that replication is continuous around the loop. The loop is then resolved by nicking, denaturation and snapback (Figure 9.9B). A number of experiments have been designed to test this model by analysing the behaviour of cloned telomeres. These are outlined below and have led to the proposal of a new model, the 'repeat addition' model.

335   DNA COMPONENTS OF CHROMOSOMES

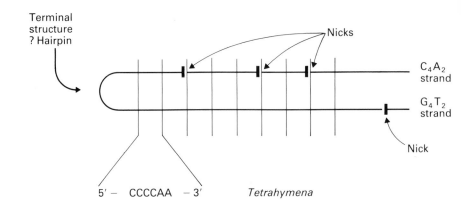

Fig. 9.8. The structure of a telomere.

A hairpin loop is not sufficient for the maintenance of linear plasmids

The idea that the ends of a chromosome can replicate provided that they terminate in a hairpin has been tested using linear plasmids in *S. cerevisiae*. Szostak and Blackburn (1982) constructed a perfect inverted repeat of a 1.9 kb fragment of the *E. coli lacZ* gene.

When this was denatured and renatured, hairpin loops were generated with a BamHI site at the free end. These were ligated to a linearised plasmid similar to pSZ21 (Figure 9.7) to produce a linear duplex flanked by the artificial hairpins. This plasmid was introduced into yeast and transformants were analysed. None of the original linear molecules were maintained as linear plasmids, they were either integrated or 'trapped' as circular replication intermediates. This experiment showed that replication could proceed through a terminal hairpin loop to generate circular molecules but that the hairpin could not be resolved to preserve the linear molecules. Other features of the telomere structure, e.g. the repeats and/or the nicks must, therefore, be important.

Telomeres grow during replication

When *Tetrahymena* termini are propagated in *S. cerevisiae* they undergo a structural change. They increase in size and acquire the ability to hybridise with a synthetic poly (dGT) probe (Shampay *et al.*, 1984). When one sequence was determined there was an additional 228 bp consisting of the $C_{1-3}A$ repeat characteristic of the yeast telomeres. This finding has led to

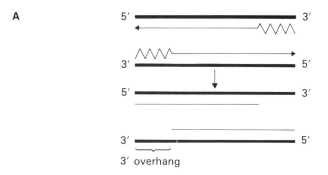

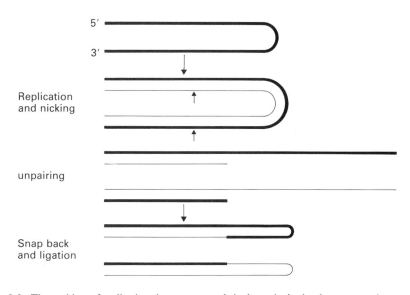

Fig. 9.9. The problem of replicating chromosome ends is shown in A; the chromosomes have an unreplicated 3′ overhang. The problem of replicating linear chomosomes with hair pin loops is shown in B; the chromosome becomes shortened at each round.

the suggestion that the termini are continually modified by the addition and loss of telomeric repeats. The gaps in the G+T strand may be extended from the 3′ end by a terminal transferase (Shampay *et al.*, 1984). Alternatively, repeated sequences may be 'scavenged' from other telomeres or from internally located poly(CA) tracts (Walmsley *et al.*, 1984).

A simplified model for telomere replication called the 'repeat addition' model based on the findings with cloned telomeres is outlined in Figure 9.10 (Murray, 1985). In this model the 3′ DNA strand (B) is extended by the addition of poly (dCA). This then provides a template for RNA-primed DNA synthesis to complete the replication at the terminus. The length of the B strand increases: it may be shortened by exonucleolytic attack specifically during replication. This model does not explain why chromosome ends are stable to degradation and the involvement of a hairpin structure cannot be excluded. It is also likely that the gaps are used in the process of addition and deletion of telomere repeats, possibly by allowing

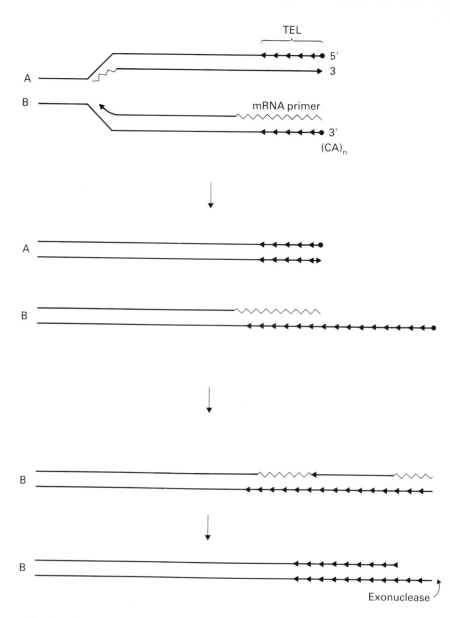

Fig. 9.10. The repeat addition model for telomere replication.

strand invasion across templates to acquire repeats by template copying rather than by terminal addition (Walmsley *et al.*, 1984). Dunn *et al.* (1984) have also suggested that telomeres can grow by recombination between ends of different chromosomes. This is supported by the finding that when linear plasmids are introduced into yeast they acquire chromosomal sequences from 'telomere-adjacent' regions and this does not occur in mutants in the *RAD52* which are deficient in the pathway of homologous recombination.

Further insight into the function of telomeres is promised by the discovery that specific proteins bind to telomeric DNA (Berman *et al.*, 1986) and that a chromosomal gene *CDC17* is essential for telomere elongation (Carson and Hartwell, 1985).

Artificial chromosomes

The availability of cloned centromeres, telomeres and replicator sequences has allowed the construction of artificial chromosomes by simply assembling these elements *in vitro*. Experiments with such artificial chromosomes are now leading to the proposal of new ideas about chromosome behaviour and providing the means to test these ideas (Murray and Szostak, 1985; Murray, 1985). In the first studies, artificial chromosomes of 10–15 kb were constructed but these did not behave as authentic chromosomes when reintroduced into *S. cerevisiae*. These linear molecules segregated randomly and were present at high copy number (Dani and Zakian, 1983; Murray and Szostak, 1983b). When the length of the artificial chromosomes was increased to about 50 kb or when a circular CEN plasmid was used the copy number was maintained at one per cell and the rate of loss reduced to 10^{-2} to 10^{-3} per mitotic cell division (Clarke and Carbon, 1980; Murray and Szostak 1983b; Hieter *et al.*, 1985b). An increased size or circularity, therefore, produces artificial chromosomes that mimic the behaviour of the natural chromosomes. The stability is, however, still several orders of magnitude lower than natural chromosomes, which are lost at a frequency of 10^{-5}. The effect of physical length on the mitotic stability of chromosomes has been analysed by building large artificial chromosomes or by generating sub-chromosome fragments *in vivo*.

Murray and Szostak (1985) have proposed that in order for chromosomes to segragate in an orderly fashion at mitosis, i.e. one sister chromatid moving to each daughter cell, the chromatids must be physically linked at the mitotic spindle. This may be achieved as a result of replication in the absence of topoisomerase II activity (Wang, 1985). In the absence of topoisomerase II the daughter helices will remain intertwined within a topologically closed domain. This catenation is relieved by topoisomerase II which, it is proposed, is regulated during the cell cycle and only acts once the chromosomes are aligned on the mitotic spindle and the kinetochore has been established. The inherent polarity of the kinetochore (Pickett-Heaps *et al.*, 1982) then ensures that the chromatids segregate correctly. Circular molecules, clearly, are topologically closed and the daughter molecules will remain intertwined until resolved by topoisomerase II. Long chromosomes contain topologically closed domains of supercoiled loops and therefore they can become catenated. Supercoiled domains, however, only form in DNA of 30–50 kb long. This means that short chromosomes of less than about 50 kb will not catenate and therefore sister chromatids would interact with the spindle independently and segregation would be random.

It is possible to devise powerful assays for chromosome behaviour in *S. cerevisiae*. Particularly useful colony colour assay systems have been described (Hieter *et al.*, 1985; Koshland *et al.*, 1985). A host strain is constructed that is homozygous for an ochre mutation in *ade2*. These strains accumulate a red pigment and therefore produce red colonies. If a single copy of *SUP11* ochre suppressor is introduced, the partial suppression produces pink colonies. If two or a few copies of *SUP11* are present there is total suppression and white colonies are produced. If cells harbour an unstable plasmid then coloured sectors appear and the degree of sectoring reflects chromosome stability. This assay can be used to monitor ARS and CEN function to study the stability. Future research on chromosome behaviour will involve identifying the precise sequence requirements for replication, and mitotic and meiotic segregation and identifying genes that encode transacting factors that influence these chromosome properties.

Transposons

A variety of eukaryotic transposons have been identified and Figure 1.10 lists the best known examples. Almost all of these genetic elements were first discovered by the genetic events that they mediate, either gross genomic rearrangements, gene disruption or gene activation. Although in some cases, such as the classical controlling elements of maize, these genetic events have been studied since the 1940s it is only recently that gene transfer technology has permitted a detailed analysis of the transposons themselves.

The first eukaryotic transposon to be analysed in depth was the yeast Ty element. Ty is about 5.9 kb in length with a unique internal region, called the epsilon region, flanked by two direct repeats of about 340 bp called delta sequences. There are 30−35 copies of Ty in the haploid genome of most laboratory strains of *S. cerevisiae* and they show some sequence heterogeneity. Their gross structure is similar to that of retroviral proviruses (Chapter 4). This similarity extends further to the production of a terminally redundant 'full-length' transcript that starts in the 5′ delta sequence and ends in the 3′ delta sequence and to the presence at the start of the epsilon region of a sequence that is homologous to the 3′ end of a specific tRNA molecule, tRNAmet. In retroviruses this tRNA binding site is the priming site for first-strand DNA synthesis by reverse transcriptase.

The Ty elements have been shown to mediate all of the genomic events associated with most other transposons. They interact by gene conversion, they act as dispersed regions of homology that can mediate a variety of genomic rearrangements and they can activate and inactivate neighbouring genes (Kingsman *et al.*, 1987b).

The organisation and expression of Ty elements

Insight into the organisation and expression of Ty elements was gained by two approaches. These were the sequencing of entire elements (Clare and Farabaugh, 1985) and the expression of individual elements in high-efficiency expression vectors (Dobson *et al.*, 1984).

In order to determine whether Ty elements encoded any proteins Dobson *et al.* (1984) removed the presumptive 5′ delta promoter from a cloned Ty element and then inserted it into the high-efficiency yeast expression vector pMA91 (Chapter 3). This resulted in a 20-fold increase in the individual Ty-specific mRNA levels and allowed the detection of Ty-specific proteins in cell extracts. This was the first demonstration that a eukaryotic transposon could encode proteins *in vivo*. The major protein, called pl (51 kD) corresponded to a large open reading frame called TYA. A second open reading frame called TYB was also evident from the sequence data but this was in the +1 reading phase with respect to TYA; it did not start with a methionine and the 5' end of TYB was overlapped by the 3' end of TYA.

By marking TYB with an interferon gene and then measuring the size of the interferon fusion protein that was produced when the marked gene was re-introduced into yeast Mellor *et al.* (1985b) showed that TYB is expressed as a TYA:TYB fusion protein. The fusion protein, designated p3, is 190 kD and is the product of the entire coding capacity of the Ty element (Figure 9.11). The fusion of TYA and TYB occurs at the ribosome and involves a specific frameshift event that leads to a shift from TYA to TYB by the ribosome (Wilson *et al.*, 1986). Both Ty primary translation products pl and p3 are subsequently cleaved to produce a series of mature Ty proteins (Figure 9.11). The protease that catalyses these cleavage events is encoded by the TYB gene, towards the 5' end, and shows homology with the proteases that mediate the proteolytic maturation of retroviruses (Adams *et al.*, 1987b). The mode of expression of TYB as a TYA:TYB fusion is also very similar to the way that retroviruses express their *gag* and *pol* genes. Like *TYA*, *gag* is expressed as the primary translation product of the full-length retroviral RNA and *pol* is expressed as a large *gag*−*pol* fusion protein that in retroviruses such as *ALV* requires a frameshift event to fuse the two genes. Retroviral *pol* genes also share limited but significant homologies with TYB. We have already mentioned the protease homologies at the 5' ends of *pol* and TYB but there are also amino acid homologies corresponding to conserved regions of retroviral integrases and reverse transcriptases (Figure 9.11).

These remarkable structural similarities between higher eukaryotic retroviruses and the yeast Ty element were taken to their ultimate conclusion by further analysis of yeast strains containing plasmids massively over-expressing the Ty transcriptional unit. On examination of these strains by electromicroscopy they were found to contain very large numbers of virus-like particles (VLPs) (see Figure 12.5).

These particles, called Ty-VLPs, were shown to contain Ty-encoded proteins related to pl, Ty-encoded reverse transcriptase and Ty 'full-length' RNA. They were directly analogous to retroviral core particles and were capable of synthesising Ty DNA on addition of exogenous nucleotides (Mellor *et al.*, 1985b,c; Garfinkel *et al.*, 1985; Adams *et al.*, 1987a).

Ty elements transpose via an RNA intermediate

The structural similarity of Ty elements to retroviruses strongly suggested that the mechanism of transposition involved an RNA intermediate and reverse transcription. This was proved by a series of gene transfer experiments by Boeke *et al.* (1985). They fused a Ty element to the *GAL1* promoter such that the 5' delta was replaced and so that expression of the element could be regulated by galactose. They also marked the Ty element

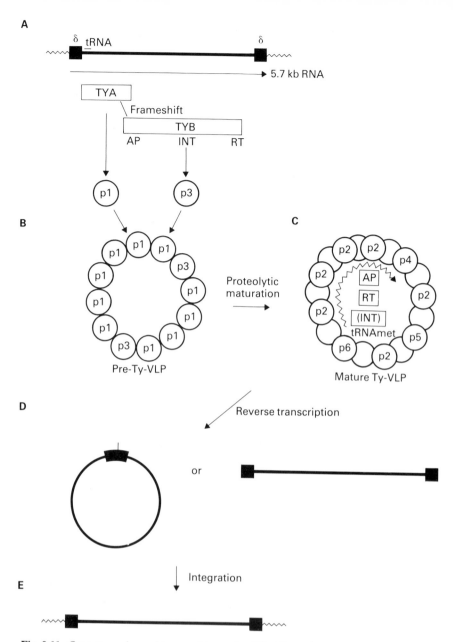

Fig. 9.11. Gene expression and transposition cycle of Ty AP = acid protease; RT= reverse transcriptase; INT = integrase.

with a fragment containing an intron from the yeast ribosomal protein 51 gene (*rp51*). The marked element in the *GAL1* expression vector was introduced into *S. cerevisiae* containing a transposition assay plasmid, pAB100. This is an ARS/CEN plasmid carrying a *HIS3* gene that has a promoter deletion and it therefore cannot complement a *his3* auxotroph. When a Ty element integrates into the defective promoter region, it reactivates the *HIS3* gene producing a HIS⁺ phenotype which can be selected. When the *GAL1* promoter was activated. Ty transposition into pAB100 was increased from 5×10^{-8} to 1×10^{-6}. The majority of Ty insertions into pAB100 under these conditions involved the marked element presumably reflecting the increased concentration of this specific Ty RNA. When the structure of the newly transposed marked Ty element was analysed the following features emerged. Firstly, the 5′ delta sequence which had been truncated in the construction of the expression plasmid was fully recreated in the transposed element. Secondly, a polymorphism in the 5′ delta was transferred to the 3′ delta. This was the loss of an Xho site (X) and a T to A change at position 328. Thirdly, the Ty element acquired point mutations during transposition and finally, the rp51 intron had been precisely removed (Figure 9.12). All these features are characteristic of the formation of retroviral proviruses from viral mRNA. The only way to explain the intron deletion is if the Ty mRNA was spliced and then copied into DNA before integration into the assay plasmid.

These observations, together with the finding that the RNA intermediate in transposition is packaged into Ty-VLPs has led to the hypothesis that the Ty-VLPs act as transposition units or *transposisomes* (Adams *et al.*, 1987a).

The use of expression vectors and gene transfer has therefore, led to a very thorough understanding of the structural and functional organisation and mode of transposition of a piece of DNA that might otherwise have remained labelled as 'junk DNA'. The application of these approaches to other LINEs in the eukaryotic genome and other transposons will increase our understanding of the contribution of such sequences to genome integrity and function.

Recombination and mutation

The integrity of chromosomes may be disrupted by recombination and mutation, and the frequency with which these occur is affected by environmental factors i.e. mutagens such as ultraviolet, by genetic factors as in diseases such as *Xeroderma pigmentosum*, by viral and/or transposon 'invasion' of the chromosome and possibly as part of the cellular ageing process. In the case of immunoglobulin and antigen receptor genes, ordered recombination is essential for the production of functional genes (Tonegawa, 1983; Kronenberg *et al.*, 1986). The processes of recombination and mutation are best studied *in vivo* and gene transfer is providing new experimental approaches to this area, which has previously been difficult to analyse.

A particularly elegant approach for studying recombination in *S. cerevisiae* has been devised by Orr-Weaver *et al.* (1981). This formed the basis for techniques of targeted integration that were discussed in Chapter 3. They showed that if a circular plasmid was linearised and then the ends of the plasmid were deleted to produce a gap, the frequency of recombination was high and the gap was repaired. The gap repair synthesis was dependent upon the *RAD52* gene product. It is essentially a gene conversion process, involving the non-reciprocal transfer of information from a donor to a recipient DNA duplex.

Orr-Weaver and Szostak (1983) have analysed the relationship between gene conversion and crossing over using gapped plasmids (Figure 9.13).

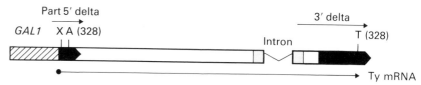

A Ty element before transposition

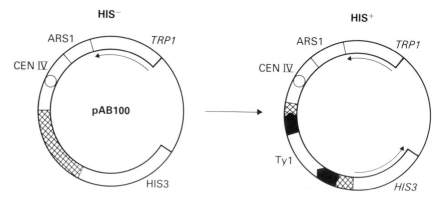

B Transposition assay plasmid

C Ty element after transposition

Fig. 9.12. Ty transposes via an RNA intermediate.

The plasmid carries ARS1 and is, therefore, capable of autonomous replication if it is repaired. The two possible events are gap repair without crossing over, which will generate autonomous plasmids, and gap repair accompanied by a cross-over, which will result in plasmid integration. The finding was that gene conversion was associated with crossing over in about 50% of the cases. This had not been previously established for mitotic recombination. Also by using appropriate genetic markers the size of the converting region can be established by determining what markers are acquired by the plasmid. This system provides the basis for studying the effect of various mutations on gene conversion and recombination and may lead to the identification of shared or separate pathways for the two processes. The gap repair system can also be used to rescue mutant chromosomal alleles without having to prepare and screen a clone bank (Chapter 7).

The specific recombination that is required to generate functional antibody and antigen receptor genes can be studied by transfecting B-and T-cell lines with precursor genes. The recombination occurs in a precise sequence. For example, during the assembly of immunoglobulin genes in pre-B-cells, the D and J segments join first and then the V segment is joined. This appears to be the regulated step and once it occurs no other VJD rearrangements

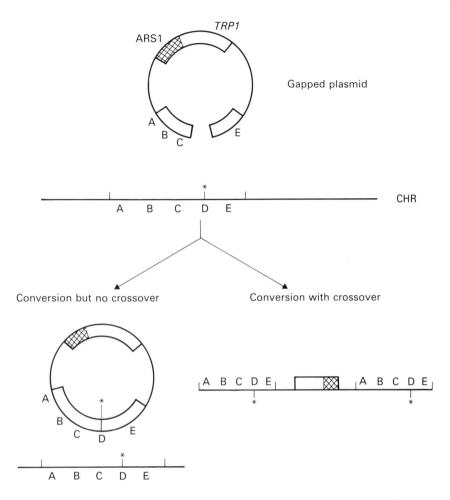

Fig. 9.13. A yeast model system for studying recombination. A mutation is indicated by an asterisk.

can occur ensuring the clonal specificity of the B-cell (i.e. it will only produce one antibody of a single combining specificity). The completed heavy-chain protein is expressed and it 'signals' light-chain gene assembly by V-to-J joining. The recombination is site-specific and is mediated by the heptamer/nonamer DNA segments that flank the functional regions (see Chapter 1). The recombination of immunoglobulin genes only occurs in B-cells and only at precise stages in their differentiation. Similarly, recombination leading to the production of T-cell antigen receptors only occurs in immature thymocytes (e.g. Raulet *et al.*, 1985). There are two hypotheses to explain the specificity of these recombination events. Firstly, the recombination recognition sequences that flank functional regions differ sufficiently to be recognised by different tissue-and stage-specific enzymes and are sufficiently different to distinguish between the sites of recombination between for example, J and D and V and J. Alternatively there may be a universal recombinase which can recognise all the sites on Ig-and T-cell receptor genes but the recognition of recombination targets is controlled by

DNA COMPONENTS OF CHROMOSOMES

other factors such as the 'accessibility' of the DNA (Yancopoulos and Alt, 1986).

Yancopoulos et al. (1986) have tested the idea of a universal recombinase by introducing a segment of the T-cell receptor β1 subunit gene into pre-B-cells. A recombination substrate was constructed which consisted of the D segment then a functional HSV-TK and then two J segments (Figure 9.14). The linear DNA fragment was introduced by electroporation into an A-MLV transformed pre-B-cell, which was tk^-, and HAT^R cell lines were isolated. These contained the linear DNA as shown in Figure 9.14 integrated as a single copy into the genome. The cells were then grown in BUdR which kills cells that express thymidine kinase (Chapter 4). This selects for cell lines that have recombined the D and J regions to delete the HSV-tk gene (Figure 9.13B). Several such cell lines were obtained. They were analysed by Southern blotting and the recombined segments were rescued and sequenced. This showed that the D−J joining was normal, that there was no discrimination between the two J segments and that *de novo* addition of bases at the point of recombination occurred as with normal D−J joining. These cells had previously been shown to rearrange transfected Ig heavy chains and therefore this data shows that there is no discrimination between D and J segments from any of the family of 'immunity genes' supporting the idea of a universal recombinase. It also appears that the recombination mechanism is conserved between species because chicken light-chain genes rearrange in B cells in transgenic mice (Bucchini et al., 1987). The recombination of the transfected DNA was correlated with a higher DNaseI sensitivity of the transfected DNA as compared to the endogenous copies. This may indicate an increased accessibility of the transfected DNA and one idea is that transcriptional enhancers which interact with stage-and cell specific factors and which are present in these sequences may also act as recombination enhancers. This type of study is being extended by introducing mutant recombination substrates to define the minimal sequence requirements for VDJ joining and class switching. For example, Ott et al. (1987) have constructed a retrovirus vector containing the μ to γ switch region from mouse immunoglobulin heavy-chain and deletion derivatives of the switch region. They showed that flanking DNA is not important and that their 49 bp repeat sequence is essential.

A number of gene transfer systems have also been devised to study mutagenesis in mammalian cells, for example, to analyse the effect of carcinogens, to investigate the process of somatic mutagenesis which is a major mechanism for generating variability in immunoglobulin genes and to study DNA damage and mismatch repair mechanisms (Folger et al., 1985a,b; Seidman et al., 1985; Lazo, 1985). A typical technique is that devised by Miller (1982). The E. coli lacI gene codes for the lac repressor. When a wild-type lacI gene is present in a bacterium the expression of β-galactosidase is repressed in the absence of an inducer whereas in the presence of a defective lacI gene the β-galactosidase is expressed. The expression of β-galactosidase is easily monitored by growing the bacteria on the chromogenic substrate 5-bromo-4-chloro-3-indolyl-β-D-galactoside. This gives rise to blue colonies when β-galactosidase is expressed and white colonies when it is repressed.

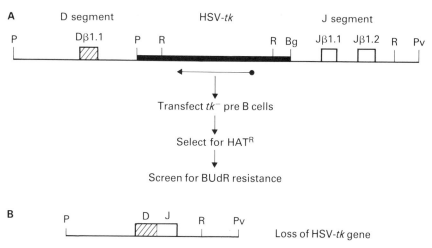

Fig. 9.14. An assay for specific recombination of T cell receptor genes. Recombination between the diversity (D) and joining (J) regions of the gene for the β subunit of the T cell receptor is shown.

The wild-type *lacI* coding sequence is inserted into a mammalian cell expression vector and transfected into different cell types. The vector can then be rescued by transforming *E. coli* and the presence of mutations in *lacI* can be scored by the colour assay (e.g. Miller *et al.*, 1984a). Other mutation markers are *galK* (Lazo, 1985) and *supF* (Seidman *et al.*, 1985). These systems may not, however, accurately reflect the susceptibility of chromosomes to mutagenesis because the mutation frequency of markers in shuttle vectors is about 1%, which is much higher than mutation rates in chromosomes (Razzaque *et al.*, 1984).

Somatic mutation in B cells has been studied in transgenic mice, for example, mice transgenic for a rearranged light-chain gene against phosphoryl choline were hyperimmunised with phosphoryl choline (PC). This imposes 'antigenic selection' which increases the chance of isolating cells which have somatic mutations in the introduced light-chain gene because of clonal amplification. The mice were used to prepare hybridomas, seven anti-PC hybridomas were selected and the light-chain cDNAs were cloned and sequenced. The genes were mutated in some cases at several points, but only in the variable region. The data suggest that there is a specific hyper-mutation mechanism that specifically recognises the V genes as a target (O'Brien *et al.*, 1987).

Summary

The ability to reconstruct an entire yeast chromosome *in vitro* from individual components, i.e. centromeres, telomeres and replication origins is one of the most spectacular successes of gene manipulation. By returning these artificial chromosomes to yeast cells the fundamental processes of chromosome pairing and chromosome segregation are being studied. Gene manipulation techniques have also provided convincing evidence for the involvement of specific sequences in DNA replication in yeast. Future research will show if these are authentic chromosomal replication origins and indicate the essential sequences and their interactions.

The analysis of DNA components of chromosomes has also indicated a large potential for genome fluidity. Some transposons have been shown to have a complex genetic organisation and to produce virus-like particles as intermediates in transposition. The specific recombination that assembles immunoglobulin genes is being shown to be a highly ordered and regulated process. It is also clear from studying immunoglobulin genes that have been transferred to B cells that they are highly mutable but mutation is restricted to hot spots in the variable regions.

One general question posed by these studies is: how is genome integrity maintained? Specifically, what factors localise high-frequency mutagenesis to specific targets? What limits transposition? What prevents massive chromosome disruption by recombination across homologous sequences, e.g. between transposons? What restricts DNA amplification to localised regions? How is chromosome replication generally restricted to once per cell cycle? Presumably it is a breakdown in the controls of genome integrity that result in cell damage and death. The answers to these questions are some of the challenges for the genetic manipulation of chromosomal components in the future.

10 Protein Structure and Function

Introduction

The ability specifically to alter coding sequences and to re-introduce modified genes or proteins into cells or organelles is having a dramatic impact upon our understanding of the properties of eukaryotic proteins.

One of the major advances that has been achieved using the new technology has been in defining the information in protein-coding sequences that is required to direct proteins to different locations in the cell. Specific *protein targeting signals* have now been identified that direct proteins to membranes, to sub-cellular organelles or to be secreted.

Another major area that has benefited from gene manipulation and gene transfer techniques is the analysis of how cells interact with their environment via cell surface receptors and channels. Genes for receptors responsible for recognising hormones, growth factors and neurotransmitters have been cloned and re-introduced into new cells. By specifically altering the receptors and by making hybrid receptors the mechanisms of signal transduction are being established. This is particulary exciting because many oncogenes, in fact, code for defective cell surface receptors. Comparing the expression of defective and normal receptors in transfected cells will therefore provide insight into mechanisms of carcinogenesis. (This is discussed further in Chapter 11.) The ability to manipulate receptors also provides a system for designing new drugs that either interfere with or enhance receptor–ligand interactions. The correct development of an immune response also involves the cell surface proteins which mediate the interactions between different cell types. The most important of these proteins are the histocompatibility antigens (HLA) and these can now be modified and studied in different transfected cell types.

It is also possible to use gene manipulation to test predictions about the relationship between three-dimensional protein structure and function by changing key amino acid residues by site-directed mutagenesis. As the understanding of these structure–function relationships increases, it is possible to specifically alter the protein structure to produce a protein with new functions. This is known as *protein engineering* and it is a strategy that is becoming widely used in the new biotechnology industries.

Each of these properties of proteins encompasses a vast area of research and we cannot possibly cover all aspects in this chapter. We will describe a few key experiments to illustrate the approaches that are being used to study eukaryotic proteins by gene manipulation and gene transfer techniques

Protein targeting

Although proteins are synthesised in the cytoplasm many of them are transported to different sub-cellular compartments or to the plasma membrane or they may be secreted outside the cell. The pathways of intracellular protein transport have largely been described (Walter *et al.*, 1984; Garoff, 1985; Schekman, 1985; Wickner and Lodish, 1985; Colman and Robinson, 1986; Walter and Lingappa, 1986; Singer *et al.*, 1987a, b; Eiler and Schatz, 1988). An outline is given in Figure 10.1. There are two main transport routes, route I is for proteins destined to be secreted outside the cell or to be inserted into the plasma membrane or to enter compartments such as lysosomes and vacuoles. Route II is for proteins that are destined for organelles, i.e. mitochondria and chloroplasts and for peroxisomes and the nucleus. The routes are distinct but they both involve the specific transport of proteins to a membrane-bound compartment; this is usually called *targeting*. The proteins are then either transferred across the membrane or they are

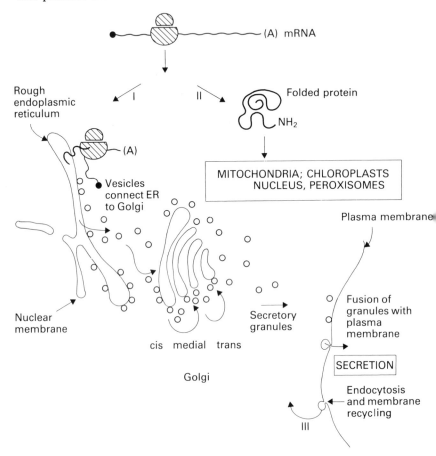

Fig. 10.1. Protein transport in eukaryotic cells.

inserted into the membrane to become *integral membrane proteins*. The process of transfer across membranes is called *translocation* and the process of insertion is usually called *partial translocation* or *intercalation* (Singer *et al*., 1987b).

The route I proteins are targeted to the endoplasmic reticulum (ER). This occurs while the polypeptide is still being synthesised on the ribosome and the ribosome−mRNA−nascent protein complex moves from the cytoplasmic to the cytoplasmic face of the ER. Translocation usually occurs as synthesis is continuing and this process is usually called *co-translational translocation*. The route II proteins are targeted to the organelles once their synthesis has been completed at ribosomes that remain free in the cytoplasm. After targeting they are translocated and this is usually called *post-translational* translocation.

Proteins following route I are subsequently transported unidirectionally from the ER to the Golgi then from the Golgi either to vacuoles and lysosomes or to the plasma membrane or outside of the cell. They are transported in membrane vesicles that are coated with the protein clathrin (Pearse and Crowther, 1987). The Golgi apparatus is a collection of cisternae that have three functional compartments, *cis*, medial and *trans*, through which the proteins are sequentially transported. The ultimate destination of the proteins is determined in the Golgi and this process is called *sorting*. During transfer through the Golgi, proteins are glycosylated by specific glycosyl transferases and mannosidases that are present in each Golgi compartment (Hirschberg and Snider, 1987). This process and the functions of the carbohydrates in many cases is poorly understood. Glycosylation may be important for sorting, e.g. the targeting of proteins to the lysosome is mediated by a receptor protein that recognises a mannose phosphate. It is not essential for secretion because this still occurs in the presence of drugs such as tunicamycin which block glycosylation. For some proteins, however, glycosylation may help to stabilise an optimum conformation that increases the efficiency of secretion (Ferro-Novick *et al*., 1984). Proteins that are secreted are transported in secretory vesicles. They may be constitutively secreted or they may be stored in so-called dense secretory granules and be secreted only in response to trigger stimuli − this is *regulated secretion*. Proteins may also be directed to specific areas of the cell surface for secretion. For example, in some specialised epithelial cells secretion can occur from either the baso−lateral or the apical membrane; this is *polarised transport* (Burgess and Kelly, 1987). Secretion occurs by the fusion of the secretory vesicle with the plasma membrane.

Proteins that follow route II and are destined to enter organelles may have to cross several membranes. In mitochondria, the proteins may partially translocate into the outer membrane, be fully translocated across the outer membrane and remain in the intermembrane space, partially translocate into the inner membrane or be fully translocated across both membranes and enter into the mitochondrial matrix. In the chloroplasts there are three membranes and spaces.

There is one other pathway of protein transport in the cell in the opposite direction where proteins outside the cell are re-incorporated (route

III). This occurs when many cell-surface receptors interact with their ligand and is called *receptor-mediated endocytosis* (Brown *et al.*, 1983).

In order to explain the specific targeting of proteins a number of studies led to the proposal of a '*signal hypothesis*' (Blobel and Dobberstein, 1975). This suggested that amino acid sequences and/or post-translational modifications could specifically interact with target sites. These signals have been called *topogenic sequences* and protein localisation is sometimes referred to as *protein topogenesis* (Blobel, 1980).

Four general classes of topogenic sequences have been proposed. The first is the *signal sequence*. This is essential for the unidirectional transport of proteins from the site of protein synthesis into or across membranes. Most proteins that are targeted to sub-cellular organelles, or that are secreted from the cell, contain a signal sequence. Different signal sequences would, however, be expected to interact with different membranes, specifying, for example, the distinction between pathways I and II (Figure 10.1). Proteins that are destined to remain embedded in membranes must only be partially translocated as compared to secreted proteins that are released into the ER lumen. These integral membrane proteins must therefore, contain a second signal that blocks their complete translocation: this is usually called the *stop transfer sequence* or *membrane anchor sequence*. Many proteins span membranes at multiple points and some proteins insert on to one face of a membrane and do not therefore, necessarily require translocation. It is proposed that there is a third class of topogenic sequence that mediates the insertion of proteins into membranes without translocation: these are *membrane insertion sequences*. Finally, many proteins undergo further localisation after translocation to reach destinations as different as, for example, secretory granules and lysosomes. These signals are referred to as *sorting signals*.

Several mechanisms have been proposed to explain translocation but now seems likely that all translocations, whether they occur at the ER membrane or at organelle membranes, use similar mechanisms (Singer *et al.*, 1987a). It is proposed that this involves specific recognition by the protein of a specialised integral membrane protein called the *translocator protein*. This is an energy-dependent process and the proteins are probably translocated in stages as successive sub-domains until they are fully across the membrane where they fold into their stable conformation in aqueous solution. The difference in post-translational translocation and co-translational translocation are not thought to reflect differences in the translocation mechanism and in fact the translocation of completely synthesised protein across the ER membrane has been demonstrated (Hansen *et al.*, 1986).

Many studies have provided broad support for the existence of targeting and sorting signals but there are a number of questions about the structure and function of these sequences that could not have been readily addressed without the application of recombinant DNA technology. Some of these questions are: (1) what are the specific sequences required to produce a functional topogenic sequence, for example, is there an optimum length, is there an optimum amino acid composition or are specific amino acid residues essential?; (2) are topogenic sequences functionally independent, i.e. can they direct a specific localisation if they are attached to any polypeptide chain?; (3) is the precise location of the topogenic sequence in the polypeptide essential for its correct function? The ability to manipulate topogenic sequences and to produce hybrid proteins has also helped in formulating and testing the current models for translocation. In the next four sections we will describe a few studies that have used gene manipulation to increase our understanding of some of the aspects of protein topogenesis.

Transport to the endoplasmic reticulum

Most proteins that are translocated across or into the ER membrane have a signal sequence at their N-terminus and in many cases this is removed by proteolytic cleavage involving a specific signal peptidase during or after translocation. The signal sequence interacts with a ribonucleoprotein particle that comprises 7SL RNA (Chapter 1) and six different polypeptides and is called the *signal recognition particle* (SRP). This interaction occurs as the polypeptide is being synthesised and causes translation to stop or slow down. The SRP and polysome forms a complex that interacts with a receptor protein on the cytoplasmic face of the ER called the *SRP receptor* (*SRPR*) or *docking protein*. The SRP—polysome complex dissociates, translation continues and the nascent polypeptide is translocated across the ER membrane (Walter *et al.*, 1984; Lauffer *et al.*, 1985; Figure 10.2). The signal sequence interacts with the membrane to facilitate translocation but the precise mechanism is unknown. One idea is that continued protein synthesis in some way provides a 'driving force' to push the protein across the membrane. More recent studies indicate that continued translation is not obligatory for translocation (reviewed by Pfetter and Rothman, 1987).

When signal sequences for route I proteins are compared there is no obvious homology in their amino acid sequence (Watson, 1984). This means that simple analysis of a primary amino acid sequence does not permit the unequivocal identification of a signal sequence. They do however, share some features. They are usually between 20—40 amino acids long and comprise a stretch of basic residues at the extreme N-terminus followed by a block of at least nine hydrophobic residues (von Heijne, 1985). Some homology around the signal peptidase cleavage site has also been noted (Perlman and Halvorsen, 1983). The signal sequence of route II proteins are also extremely variable and tend to be far more hydrophilic than route I protein signal sequences. In particular, they have high numbers of basic and hydroxyl residues but few acidic residues. They also have the potential to form α-helices that have a non-polar face and a hydrophilic face. These are *amphiphilic* α-helices (von Heijne, 1986).

Functional analysis of the invertase signal sequence

Invertase is encoded by the *SUC2* gene in *Saccharomyces cerevisiae*. The protein enters the ER, it is glycosylated and a 19 amino acid N-terminal signal sequence is removed by proteolytic cleavage. Three questions have been asked about the signal peptide. These are: (1) is it functionally independent; (2) is signal peptide cleavage important for secretion outside the

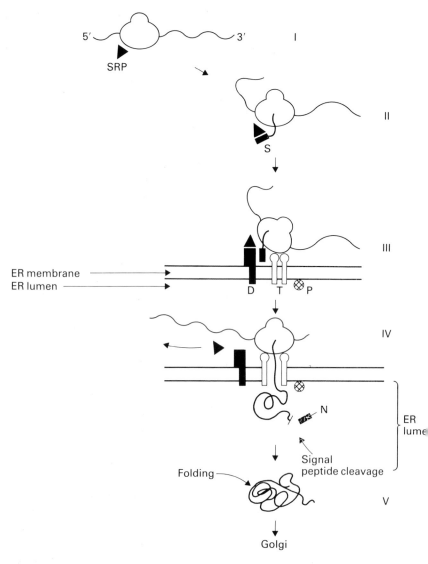

Fig. 10.2. The pathway of secretion into the ER. SRP = signal recognition particle; D = docking protein; P = signal peptidase; S = signal sequence; T = translocation protein.

cell and (3) are specific amino acid residues in the signal functionally significant?

The general approach for determining whether a signal peptide is functionally independent is to produce a hybrid gene which contains the region coding for the signal fused to a heterologous coding sequence. The heterologous coding sequence should encode a protein that has a readily assayable function and which contains no internal topogenic sequences. Initial studies fused the invertase signal sequence to the *E. coli* β-galactosidase coding sequence. The fusion protein was transported to the ER indicating that the signal sequence was functional but the protein was not secreted (Emr *et al.* 1984). It appears that β-galactosidase contains a signal that causes retention

CHAPTER 10

within the ER and other studies have shown that it also contains a nuclear targeting signal (Kalderon *et al.*, 1984b). This enzyme is now rarely used therefore as a marker protein to assay the function of topogenic sequences. The 19 amino acid invertase signal has now been fused to the coding sequence for mature human interferon-α (Chang *et al.*, 1986) and to the coding sequence for calf prochymosin (Smith *et al.*, 1985). Both these proteins were efficiently secreted and the signal peptide was accurately removed, confirming that information for transport into the ER and for signal peptidase recognition was contained within the signal. Both of the marker proteins used in these studies were, however, proteins that are normally secreted and might, therefore, have contained additional topogenic signals that were important for full signal sequence function. In other studies proteins that are normally cytosolic such as chimpanzee α-globin and mouse dihydrofolate reductase have been used as marker proteins (Garoff, 1985).

A number of mutations in the invertase signal sequence have been made in the intact *SUC2* gene by *in vitro* mutagenesis and their effect on secretion of invertase has been studied by re-introducing the mutated genes into yeast cells (Schauer *et al.*, 1985; Kaiser and Botstein, 1986). These studies indicated that deletions in the hydrophobic region of the signal blocked transport. In one case a fully functional invertase enzyme was produced in the cytosol but it retained the signal peptide. Amino acid substitutions in the first five N-terminal amino acids had no effect, suggesting that this part of the signal is not required for targeting. One mutant had a single amino acid change at residue −1. This was defective in signal peptide cleavage and although it reduced the rate of invertase transport to the Golgi by 50-fold some protein was still secreted.

These types of study show that the signal sequence is essential for targeting a protein to the ER and that it is functionally independent, i.e. a signal sequence can be linked to a heterologous mature sequence and the protein will be secreted. Cleavage of the signal sequence does not seem to be an absolute requirement for secretion but if this is impaired, the efficiency of secretion is reduced. It is possible that the retention of the signal sequence after translocation slows the release of the protein from the membrane. These studies also indicated that single amino acid changes in the majority of residues had little effect but disruption of the hydrophobic core blocked targeting. The lack of specific amino acid sequence requirements to produce a functional signal sequence is further emphasised by more recent studies with the invertase signal sequence (Kaiser *et al.*, 1987). A secretion-defective invertase gene was constructed which had a deletion from residue 3 to residue 20 in the signal sequence. Random fragments of human DNA were inserted to replace the deleted signal sequence. About 20% of these random fragments that maintained the correct reading frame with the remainder of the *SUC2* gene functioned as secretion signals. All the DNA sequences were different but the resulting signal peptides were all enriched for hydrophobic residues and had a reduced density of charged residues. The random signal sequences showed a heterogeneous function in both the quantity of invertase secreted and the rate and many showed an intracellular accumulation of protein. A detailed comparison of the properties of these randomly isolated sequences will help to identify the stages of secretion and the functional requirements for a maximally efficient signal sequence.

Integral membrane proteins

Integral membrane proteins are transported to the ER but only partially translocated so that they remain inserted in the membrane. They are held in the membrane by a *transmembrane* or *membrane anchor* domain and display the remainder of the polypeptide at one or both faces of the membrane: these are the *cis* face, which contacts the cytoplasm or the *trans* face, which contacts the ER lumen and is effectively outside the cell. A given polypeptide chain may span the membrane several times. One of the most intensively investigated integral membrane proteins is the haemagglutinin (HA) of influenza virus type A. This was the first integral membrane protein to be cloned and re-introduced into a mammalian cell and a number of studies with *in vitro* mutated HA genes have provided considerable information about the transport process.

The plan of the HA cDNA and a schematic representation of the HA molecule inserted into the viral envelope is shown in Figure 10.3 (Wiley *et al.*, 1981; Wiley and Skehel, 1987). When the viral genome is expressed in infected cells two viral proteins, the HA and neuraminadase (NA), are transported to the ER and then to the plasma membrane. These form the envelope of the virus. The HA polypeptide is cleaved by removal of an arginine residue at position 328 to produce two polypeptides, HA1 and HA2 which are linked by disulphide bridges. Both polypeptides are glycosylated and the molecules assemble into a trimer which projects from the surface of the viral envelope as a 'spike'. The spike has sites that interact with specific cell-surface receptors and contains four antigenic domains, A−D, that are recognised by neutralising antibodies. Amino acid changes at these sites produce variants of the virus which are no longer recognised by protective antibodies. This is known as *antigenic drift* and it is one of the reasons why it is difficult to produce an effective vaccine against influenza (Air and Laver, 1986). The production of anti-influenza vaccines is discussed in more detail in Chapter 12.

The HA displays several features that are shared by integral membrane proteins: a hydrophobic stretch of amino acids at the N-terminus and an uncharged domain near the C-terminus followed by a very hydrophilic 'tail'. The simplest interpretation of the function of these domains is that they represent respectively a signal sequence for transport to the ER and a membrane anchor to prevent full secretion. To determine the importance of these domains a number of *in vitro* mutagenesis and gene transfer experiments have been done. Gething and Sambrook (1981) used an SV40 late replacement vector to transfer HA genes into monkey cells. The HA was produced at about 10^8 molecules per cell, it assembled correctly at the cell surface, it was correctly glycosylated and adsorbed red blood cells, which is a characteristic property of the viral haemagglutinin. The sequences coding for the N-terminal signal sequence were then removed by deleting a restriction

A cDNA

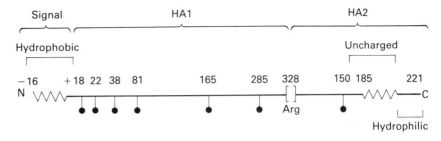

Haemagglutinin trimer
B insertion into the plasma membrane

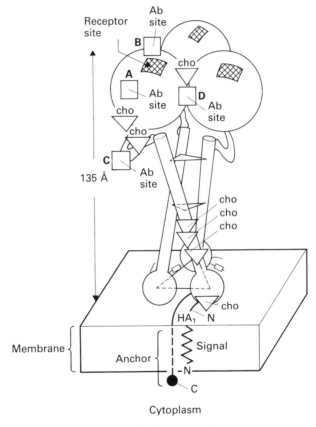

Fig. 10.3. Influenza virus haemagglutinin. Ab sites A−D are regions that are recognized by neutralizing antibodies; cho = glycosylation sites (adapted from Wiley *et al.*, 1981).

fragment that left the initiator ATG but removed all of the presumptive signal up to the first amino acid of the mature polypeptide (Gething and Sambrook, 1982). Cells infected with vectors carrying the signal minus mutant produced only intracellular HA. This was not glycosylated and the yields of protein were very much reduced and cells expressing this mutant did not bind erythrocytes. These data indicated that the signal sequence was

PROTEIN STRUCTURE AND FUNCTION

essential for transport to the cell membrane. A second mutant was constructed which retained the signal sequence but which had 36 amino acids removed from the C-terminus. This resulted in the synthesis of a protein that lacked the presumptive membrane insertion domain and the cytoplasmic tail. This protein was produced in very high yields, about 300 μg per 10^6 cells, but it was secreted into the culture medium rather than remaining as an integral membrane protein. This confirmed that a specific signal could act as a membrane anchor.

It appeared, therefore, that a secreted protein could be converted to an integral membrane protein by the addition of a membrane integration signal following the signal peptide. This idea is confirmed in a separate study by constructing a hybrid gene comprising a rat growth hormone (rGH) cDNA and part of the vesicular stomatitis virus (VSV) glycoprotein G gene (Guan and Rose, 1984). Rat GH has a 26 amino acid N-terminal signal peptide and is secreted usually from secretory vesicles in response to growth hormone releasing factor. G protein is an integral membrane protein which has a 20 amino acid transmembrane domain followed by a 29 amino acid highly charged cytoplasmic tail. The hybrid gene contained the entire rGH coding region except for four C-terminal amino acids which were replaced with the region encoding the 49 amino acid membrane–tail region of the G protein. This was expressed in an SV40 mini-replicon vector in COS cells and rat pituitary GH3 cells which both permit efficient secretion of non-mutant growth hormone. The fusion protein was transported to the Golgi but it was not transported to secretory vesicles. Cell fractionation studies showed that it was integrated into cell membranes. This shows that the G protein membrane anchor was sufficient to block secretion of growth hormone.

A more detailed analysis of the function of the cytoplasmic tail of the influenza HA has been undertaken (Doyle *et al.*, 1985). The tail is 10 amino acids long and the last five residues are absolutely conserved between all influenza virus sub-types. An outline of some of the mutants and their phenotypes is shown in Figure 10.4. The sub-cellular location of wild-type and mutant HA proteins was determined by fluorescent antibody staining of transfected cells to stain specific organelles and HA proteins and then comparing the staining pattern. Different classes of mutant were obtained, all were transported to the ER but some showed markedly reduced transport to the Golgi and one mutant which was slowly transported to the Golgi failed to be transported to the cell surface. This study indicates that the cytoplasmic tail plays an important role in protein sorting after the protein has been directed to the ER. The precise amino acid sequence might, however, not be essential as, for example, the mutant HA71 with three amino acid changes essentially behaved as a wild-type. Large deletions or additions stop transport.

The cytoplasmic tails of some integral membrane proteins may also contain signals for receptor-mediated endocytosis. For example, the low-density lipoprotein (LDL) receptor is important in transporting cholesterol from the plasma into the cells by receptor-mediated endocytosis. About 1 in 500 people inherit a mutant gene for the LDL receptor which predisposes them to premature death from atherosclerosis. In one group of mutations

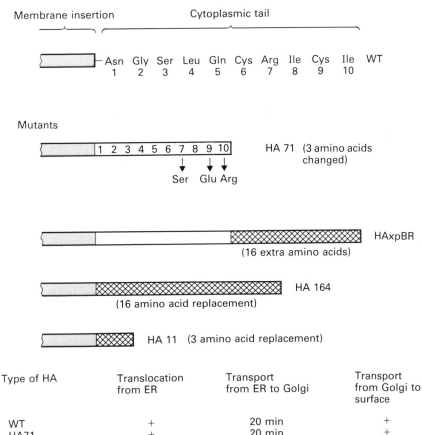

Type of HA	Translocation from ER	Transport from ER to Golgi	Transport from Golgi to surface
WT	+	20 min	+
HA71	+	20 min	+
HAxpBR	+	>8 hr	−
HA164	+	1 hr	+
HA 11	+	1 hr	+

Fig. 10.4. The function of the HA2 cytoplasmic tail.

the receptor is expressed at the cell surface but it is not internalised upon binding to the cholesterol transport protein. These mutant genes have a portion of the membrane-spanning and cytoplasmic domains deleted, indicating that the signals for internalisation are located in these regions (Lehrman *et al.*, 1985a, b).

Influenza virus HA is synthesised as a monomer and it is folded and assembled into a trimeric structure within the ER. Folding takes about 10 min and is completed before the protein is transported to the Golgi. Gething *et al.* (1986) have shown that certain, secretion-defective mutants, e.g. HAXpBR (Figure 10.4) do not fold correctly and do not form trimers. This suggests that efficient transport of HA to the cell surface is also dependent upon the molecule assuming an appropriate native conformation.

Signal sequences can reside within a polypeptide

The majority of proteins that are destined for the ER have a signal sequence at the extreme N-terminus. However, this is not true for chicken ovalbumin,

PROTEIN STRUCTURE AND FUNCTION

which lacks a 'classical' hydrophobic signal at the N-terminus. Ovalbumin secretion has been studied by injecting cDNA under the control of the SV40 promoter into *Xenopus* oocytes (Chapter 5). The protein is correctly targeted to the ER and secreted into the medium. When the cDNA is manipulated to remove the eight extreme N-terminal amino acids, normal secretion still occurs. Deletion of amino acids up to residue 41, however, blocked all transport to the ER. This implicated residues 8–41 as topogenic signals for secretion, although such gross deletions might simply have perturbed the protein structure to effect a non-specific inhibition of secretion. To confirm the location of the signal sequence, gene fusions were constructed in which different regions of the ovalbumin cDNA were fused to the chimpanzee α-globin gene to act as a reporter protein. The addition of amino acids 22 to 41 of ovalbumin were sufficient to direct this normally cytoplasmic protein into the ER. This indicated that the topogenic sequence for transport and translocation into the ER could reside at least 22 amino acids inside a polypeptide.

Integral membrane proteins that span the membrane several times may also possess multiple internal signal sequences. For example, bovine opsin spans the membrane seven times (Figure 10.5). It contains four signal sequences each followed by a membrane domain that acts as a stop-transfer sequence (Friedlander and Blobel, 1985).

Sorting signals

Gene transfer is also being used to identify sequences that determine the sorting of proteins after they are transferred to the Golgi. One particular question is what determines whether a protein is constitutively secreted or whether it is sequestered into dense secretory granules for regulated secretion. The transport mechanism for regulated secretion appears to be highly conserved. This has been demonstrated by simple gene transfer experiments. For example, expression vectors for regulated secretory proteins such as human growth hormone and human insulin have been introduced into pituitary-derived secretory cells such as mouse AtT-20 or rat GH4 which normally package adrenocorticotropic hormone (ACTH) and prepropara-thyroid hormone respectively and regulate their secretion. The foreign proteins are also packaged into dense secretory granules whereas if a normally constitutively secreted protein such as the VSV-G protein lacking the membrane anchor is introduced into these cells it is not packaged into secretory granules but is secreted by the constitutive pathway (Moore *et al.*, 1983; Moore and Kelly, 1985; Burgess and Kelly, 1987). The mechanism for selective packaging into dense secretory granules also appears to be highly conserved between different types of secretory cells. For example, when a human growth hormone expression vector is introduced into a neuronal cell line (PC12), this endocrine hormone is transported to synaptic vesicles. There are a variety of chemicals which stimulate secretion of only those proteins that are packaged in secretory granules; these chemicals are called *secretagogues*. One such chemical, carbachol which stimulates the production of acetyl choline and noradrenalin from PG12 cells also stimulates growth

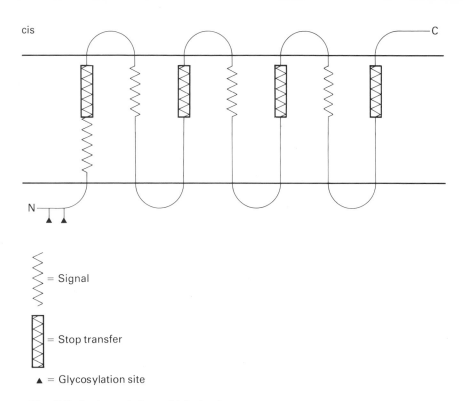

cis

C

N

$\nwarrow\!\!\wedge$ = Signal

$\square$ = Stop transfer

▲ = Glycosylation site

Fig. 10.5. Bovine opsin has multiple signal sequences.

hormone release in transfected cells (Schweitzer and Kelly, 1985). Similarly the exocrine protein trypsinogen is targeted into secretory granules in AtT-20 cells (Burgess *et al.*, 1985). It therefore appears that endocrine hormones, exocrine hormones and neurotransmitters are sorted by similar mechanisms. Not all cells possess the sorting mechanism for regulated secretion. For example, cell lines that are widely used in transfection studies such as COS1, AGMK and L cells will constitutively secrete endocrine hormones. They do not respond to secretagogues and in some cases they fail to correctly process the hormones (e.g. Moore *et al.*, 1983).

In order to define the sorting signals for regulated secretion Moore and Kelly (1986) have constructed a chimaeric plasmid consisting of the RSV-LTR promoter, the hGH-coding region which has the last codon joined directly to the first codon of a VSV-G protein-coding region that contains an intact 433 amino acid luminal region but which lacks the membrane anchor domain. When this plasmid is transfected into AtT-20 cells the fusion protein is targeted to dense secretory granules with the same efficiency as hGH. In addition, the secretion of the fusion protein is regulated by the secretagogue, 8-bromocyclic AMP. It is proposed that proteins such as truncated VSV G protein that are constitutively secreted lack sorting signals and that constitutive secretion occurs by default. Proteins such as hGH, however, have sorting signals which will override constitutive secretion because they contain a positive acting element.

The idea that secretion represents the 'default pathway' is increasingly

PROTEIN STRUCTURE AND FUNCTION

supported by the finding that a number of proteins that are retained in the ER or transported to vacuoles or lysosomes have specific signals (reviewed by Rothman, 1987). For example, three soluble ER proteins found in fibroblasts share a common carboxy terminal tetrapeptide. When this is deleted they are secreted and conversely when this is added to a normally secreted protein, chicken lysosome, they are retained (Munro and Pelham, 1987). Sorting of proteins in yeast for the vacuole as opposed to secretion depends upon an N-terminal sequence in vacuolar enzymes (Bankaitis *et al.*, 1986; Johnson *et al.*, 1987).

Protein import into organelles

Proteins are targeted to mitochondria and chloroplasts after they have been translated and they are thus likely to be folded to some degree before translocation. One major question is, therefore, how does a folded protein which most likely displays a hydrophilic exterior traverse a hydrophobic membrane? The signals which direct transport into organelles may also be more numerous and/or more complex than the ER transport signals because of the organelle structure. This is because both mitochondria and chloroplasts contain a number of discrete compartments. They have double membranes and therefore a protein can be destined for either the inner or outer membrane, the inter-membrane space or the internal matrix (mitochondria) or stroma (chloroplasts). Chloroplasts also contain the thylakoid membranes and the thylakoid lumen. Most of the proteins that are imported into these organelles have an N-terminal sequence that is removed by proteolytic cleavage. This is the *pre-sequence* or *transit peptide*. One idea is that this may contain multiple signals for sub-organellar targeting. The approaches to identifying organelle targeting signals have been similar to those used for identifying ER targeting signals. Mutations have been made in putative signals and their effect upon targeting *in vivo* and in isolated organelles has been studied. Minimal targeting signals have been located by making gene fusions with genes encoding cytoplasmic proteins (reviewed by Colman and Robinson, 1986; Garoff, 1985; Rothman and Kornberg, 1986; Hay *et al.*, 1984; Schatz and Butow, 1983). One of the best studied examples is the targeting of yeast cytochrome C oxidase subunit IV (COXIV) into the mitochondrial matrix.

A signal sequence and an unfolding enzyme are needed for import into the mitochondrial matrix

Most imported mitochondrial proteins are synthesised as precursors with an N-terminal pre-sequence which is removed by proteolytic processing in the mitochondrial matrix. In order to establish that the pre-sequence is a topogenic signal sequence, Hurt *et al.* (1984) constructed a gene fusion using 22 amino acids of the 25 aa COXIV pre-piece and a mouse dihydrofolate reductase (DHFR) gene; DHFR is a cytosolic enzyme and therefore contains no topogenic signals. The fusion gene was expressed using *in vitro* transcription and translation systems and the resulting polypeptide was added to isolated mitochondria in the presence of an ATP-regenerating system. The fusion protein was imported into the mitochondria, indicating that all the information for targeting to the mitochondria and translocation was contained within the N-terminal 22 amino acids of COXIV. In addition, the pre-sequence was proteolytically removed even though the actual cleavage site had not been included in the construction. This indicated that the protease recognition signal was not specific. Progressively shorter pieces of the pre-sequence were fused to the DHFR gene and the first 12 amino acids were shown to direct import, but in this case there was no cleavage. This showed that there may be two domains in the pre-piece, one to direct import and one to direct proteolytic processing and that processing was not essential for import (Hurt et al., 1985).

Eilers and Schatz (1986) have exploited the properties of dihydrofolate reductase to determine whether unfolding is required for transport into mitochondria. The folate antagonist methotrexate inhibits dihydrofolate reductase activity by binding tightly to the active site. X-ray crystallography has shown that this binding does not significantly alter the conformation but that it impedes unfolding of the protein. Methotrexate does not non-specifically affect import into the mitochondria but the specific import of a COXIV−DHFR fusion protein into the mitochondria of methotrexate-treated cells is completely blocked; authentic COXIV protein is however, still imported. These data are interpreted as indicating that the DHFR portion of the protein must normally assume an altered conformation before, or concomitant, with, transport and that when unfolding is blocked by methotrexate binding this can no longer occur.

The mechanism of translocation across the two membranes is still unclear. One proposal is that there are specific 'translocation sites' at which the membranes are close enough to allow simultaneous translocation (Schleyer and Neupert, 1985). There appear to be two stages. An initial interaction with the N-terminal signal which is dependent upon the mitochondrial membrane potential and a second stage which is energy-independent, where the rest of the protein is 'drawn through' the membranes. Similar results have been obtained with other matrix proteins such as human ornithine transcarbamylase (OTC) (Horwich et al., 1985) and yeast alcohol dehydro-genase III (Van Loon et al., 1986) although the key transport sequences are not always located at the extreme N-terminus (Horwich et al., 1986).

Targeting proteins to the inter-membrane space and the outer membrane

There are relatively few studies, to date, of proteins that are targeted to the inter-membrane space in mitochondria. It appears, however, that this may be directed by multiple signals. For example, van Loon et al. (1986) have shown that 32 amino acids derived from yeast cytochrome Cl, an inter-membrane space protein, target mouse DHFR to the matrix. The addition of the next 19 amino acids which are uncharged cause DHFR to appear in the inter-membrane space. The interpretation is that there is a hierarchy of topogenic signals. First there is a translocation sequence which is analogous to the signal of matrix proteins, then the adjacent sequence effectively acts as a stop-transfer signal.

Targeting to the outer mitochondrial membrane involves a specific sequence located at the extreme N-terminus. Unlike the case with proteins that are fully translocated and imported into the mitochondria this outer membrane targeting sequence is not removed by cleavage. The signal, in fact, acts as the 'anchor' which retains the protein in the membrane (Hase *et al.*, 1984).

Transit peptides direct proteins into chloroplasts

Proteins that are destined for the chloroplast stroma have an N-terminal extension called the *transit peptide* which is removed by a two stage proteolysis. The first demonstration of the importance of the transit peptide came from studies of the small subunit of ribulose 1, 5 bisphosphate carboxylase (rubisco). A hybrid gene was constructed to produce a fusion protein containing the 57 amino acid transit peptide fused to the bacterial neomycin phosphotransferase. The gene was introduced into tobacco cells by *Agrobacterium*-mediated cell transformation and the distribution of the fusion protein in the cell was analysed in transformed callus cultures (see Chapter 6). The neomycin transferase was found in the stroma of the chloroplasts only when the transit peptide fusion protein was used and the transit peptide was proteolytically removed (Van den Broeck *et al.*, 1985).

The chloroplast transit peptides share some similarities with mitochondrial pre-sequences; they are rich in basic and hydroxylated amino acids and lack acidic amino acids and extended hydrophobic stretches. Recently, Hurt *et al.* (1986) have shown that the transit peptide and the mitochondrial pre-sequence are functionally interchangeable. They constructed a chimaeric gene where the 31 N-terminal residues of the rubisco small subunit transit peptide were fused to mouse dihydrofolate reductase or COXIV lacking its own pre-sequence. Both fusion proteins were targeted to the mitochondria indicating conservation in the targeting mechanisms although the chloroplast transit peptide was less efficient and was not cleaved. Presumably in plant cells additional mechanisms serve to distinguish between chloroplasts and mitochondria.

Transport to the thylakoid lumen requires two signals

The chloroplast has essentially six compartments, these are the outer membrane, the inner membrane, the inter-membrane space, the stroma, the thylakoid membrane and the thylakoid lumen (Figure 10.6). In order to identify the targeting signals, for each compartment two proteins have been compared. These are ferrodoxin which is found in the stroma and plastocyanin which is found in the thylakoid lumen, both derived from *Silene pratensis* (White Campion) (Smeekens *et al.*, 1986). Fusion proteins were constructed in which the transit peptides of the two proteins were exchanged. The fusion proteins were synthesised *in vitro* and their import and distribution in isolated pea chloroplasts was analysed. Both proteins were imported and were partially proteolytically processed but authentic mature proteins were not produced. The ferredoxin−plastocyanin fusion protein was found only in the stroma. In contrast the plastocyanin−ferredoxin fusion protein was

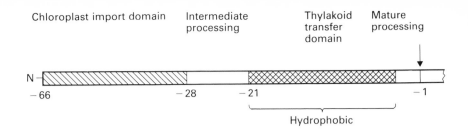

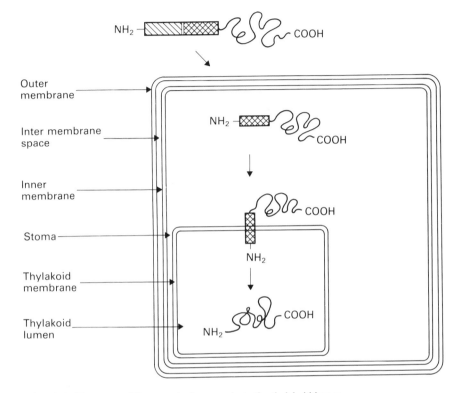

Fig. 10.6. Transport of *S. partensis* plastocyanin to the thylakoid lumen.

found in the stroma, in the thylakoid membrane and in the thylakoid lumen. These preliminary data suggest the presence of two separate signals in the plastocyanin transit peptide. Signal I directs proteins to the stroma and signal II, which is somewhat hydrophobic, directs proteins to the thylakoid membrane (Figure 10.6). Sequence comparisons between a number of transit peptides showed blocks of homology which might also indicate functional domains (Karlin-Neumann and Tobin, 1986). It is likely, therefore, that chloroplast transit peptides may have functional differences when compared to both mitochondrial pre-sequences and ER signal sequences.

Protein transport into the nucleus

Protein import into the nucleus does not require translocation across membrane because the nuclear envelope has permanent protein channels called

PROTEIN STRUCTURE AND FUNCTION

nuclear pores through which proteins enter and leave the nucleus. Certain proteins are much more concentrated in the nucleus than in the cytoplasm, indicating that proteins are not freely diffusable between cytoplasmic and nuclear compartments and there must, therefore, be a selective transport process. An early model for specific accumulation of nuclear proteins proposed that all proteins could freely enter the nucleus but that certain proteins would be specifically retained in the nucleus by virtue of binding to non-diffusable nuclear components such as laminin. More recent experiments in particular using recombinant-DNA approaches have shown that there is selective entry into the nucleus which is mediated by specific amino acid sequences that function in the mature polypeptide (reviewed by Dingwall and Laskey, 1986). These sequences are not removed by proteolytic cleavage; this is presumably to ensure that the proteins can be re-imported into the nucleus following the nuclear membrane breakdown and reformation during the cell cycle. The best characterised example is nuclear targeting of the SV40 large T antigen.

SV40 large T antigen is found predominantly in the nucleus of infected cells. It regulates viral gene expression and DNA replication and it is also responsible for neoplastic transformation. The protein is too large to freely diffuse through the nuclear pores, which exclude proteins of more than 60 kD. T antigen is a tetramer whose monomer molecular weight is 94 000 kD. Some initial studies of T antigen identified a region that was a potential DNA-binding domain and Kalderon et al. (1984a) prepared a series of single amino acid substitutions in this region by oligonucleotide-directed mutagenesis (see Chapter 8). Their technique is an example of a now classical approach to site-directed mutagenesis and is outlined in Figure 10.7. Two plasmids were constructed, one containing the wild-type early region and the other containing the early region that had a small deletion in the putative DNA-binding region. The plasmids were linearised and then denatured and re-annealed. In some cases a deleted single strand annealed to a wild-type strand to produce a gapped heteroduplex. A collection of synthetic 20-base oligonucleotides were prepared that would span the gap but which contained single base differences from the wild-type sequence. The oligonucleotides were synthesised as a mixture by including an equimolar mixture of all four nucleotides at five selected positions in the sequence. This generated 4^5 possible oligonucleotides of which theoretically one would be wild type and 15 would have single base changes. The gap-repaired molecules were transfected into. E. coli and colonies were screened for mutant plasmids. Some of the mutations that were isolated are shown in Figure 10.7.

The mutant plasmids were used to transfect Rat-1 cells and the localisation of T antigen was studied by immunofluorescent staining of permeabilised cells. The results from analysing the range of mutants and further mutants that had deletions and insertions in this region showed that lysine 128 was critical for nuclear localisation but that the surrounding amino acids also influenced nuclear location. The lys 128 mutant could still bind DNA and transform Rat-1 cells. However, viruses and mini-replicons carrying this mutation are replication-defective (Paucha et al., 1985).

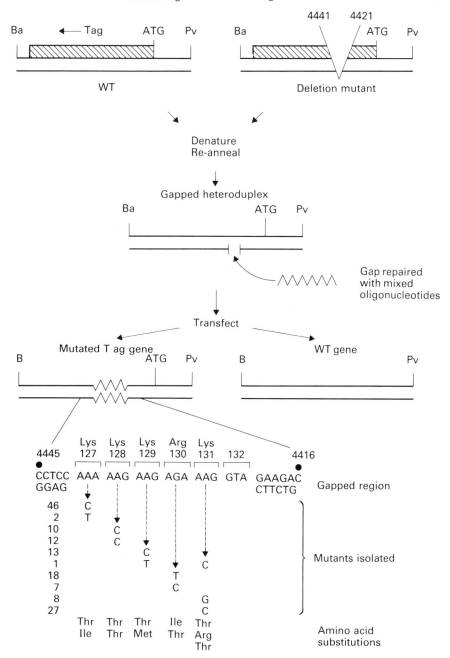

Fig. 10.7. Identification of a nuclear targeting signal in SV40 T antigen.

Similar approaches have been used to identify a number of nuclear targeting signals, some of which are shown in Table 10.1. Although sequences related to the SV40 sequence have been found in other predominantly viral proteins no single consensus sequence has emerged. It is possible that

PROTEIN STRUCTURE AND FUNCTION

Table 1. Nuclear targeting signals.

SV40 large T antigen	Pro Lys Lys[128] Lys Arg Lys Val	(1)
Influenza virus nucleoprotein	Ala[336] Ala Phe Glu Asp Leu Arg Val Leu Ser	(2)
Yeast ribosomal protein L3	Pro[18] Arg Lys Arg	(3)
Polyomavirus large T antigen	Val Ser Arg Lys[192] Arg Pro Arg Pro–	(4)
	–Pro Pro Lys Lys[282] Ala Arg Glu Asp	

(1) Kalderon *et al.* (1984b) (2) Davey *et al.* (1985) (3) Moreland *et al.* (1985) (4) Richardson *et al.* (1986).

multiple pathways for nuclear transport might exist but it is clear that specific amino acid signals are involved.

Receptor–ligand interactions

Many of the aspects of cell behaviour are controlled by interactions between external stimuli and cell-surface receptors. These receptors are integral membrane proteins and they fall into roughly two classes, those that span the membrane at multiple points and respond to a stimulus by creating an open channel in the membrane and those that span the membrane once and that appear to transmit the stimulus through the membrane. In both cases the receptors possess extracellular domains that interact with specific ligands. In this section we will discuss some of the experiments that use cloned receptor genes to characterise the ligand–receptor interaction and the mechanisms of signal transduction. Our examples are the acetyl choline receptor, which forms a channel, and the growth factor receptors, which transduce signals across membranes.

The acetyl choline receptor

The acetyl choline receptor (AChR) is an integral membrane protein that is involved in the propagation of electrical signals in the nervous system. Acetyl choline is released from nerve endings and interacts with AChR inducing the protein to form a channel that is selectively permeable to cations. This is called a *ligand-gated channel*. The receptor contains four different membrane-spanning subunits encoded by separate genes, these are alpha, beta, gamma and delta subunits. The alpha subunit binds acetyl choline and a single AChR has two alpha subunits and one each of the other subunits. The AChR has three main functions. These are, to bind acetyl choline, to open the channel (gating) and to allow the passage of cations such as sodium through the channel while excluding anions (permeation). Various gene transfer systems are being used to identify the molecular basis of these activities.

The cDNA for each of the receptor subunits has been cloned from the electric organ of the ray, *Torpedo californica*. These cDNAs were first expressed in heterologous cells using a combined COS cell and *Xenopus* oocyte expression system (Mishina *et al.*, 1984). The cDNAs were inserted into an SV40 mini-replicon so that they were expressed from the SV40 promoter. Individual cDNAs were then expressed transiently in COS cells to yield high levels of subunit specific mRNA. The mRNAs specific for the $\alpha-$, $\beta-$, $\gamma-$ and $\delta-$subunits derived from transfected COS cells were then combined in the same ratios as in the *T. californica* electric organ and microinjected into *Xenopus* oocytes. After three days the response of the oocytes to acetyl choline was examined by using standard electrophysiological techniques to assay intracellular potentials and membrane currents. A response typical of *T. californica* AChR was obtained. The amino acid sequences show a high degree of homology between the subunits but all four subunits are needed for full activity. This was shown by omitting each subunit-specific mRNA from the mixture that was injected into the oocytes. This study showed that a functional receptor could be assembled in a heterologous system and that all four subunits were required. The ability of the different combinations of mRNA to generate a receptor that bound the acetyl choline antagonist, α-bungarotoxin was also monitored. Omission of the β, γ and δ subunits had some effect but deletion of the α-subunit abolished α-bungarotoxin binding. This showed that only the α-subunit was essential for ligand binding.

In order to identify the functional domains of the α-subunit a series of small deletions and amino acid substitutions have been made in the α-subunit cDNA (Mishina *et al.*, 1985). In this study the bacteriophage SP6 promoter (Green *et al.*, 1983) was used to drive the synthesis of subunit-specific mRNA *in vitro*. The wild-type and mutant mRNA were capped and then microinjected with β-, γ- and δ-subunit-specific mRNAs into *Xenopus* oocytes. The oocytes were monitored for their response to acetyl choline. Figure 10.8 shows a diagram of the α-subunit and the sites of some of the key mutations (as summarised by Stevens, 1985). The amino terminal half of the polypeptide is extracellular and the carboxy terminus spans the membrane five times. There is a single disulphide bond in the extracellular portion and an N-glycosylation site at residue 141. Three classes of mutant were obtained: those that had no effect (N), those that could bind α-bungarotoxin but did not gate or permeate (P), and those that did not bind, gate or permeate (B). Some mutations did not yield useful information because the levels of protein that were expressed at the cell surface were very low. This may have been due to a disturbed conformation that blocked transport, as we have discussed earlier for other integral membrane proteins. The general conclusion is that mutations in the extracellular domain affect binding, and without binding there can be no gating or permeation. Mutations in the membrane-spanning segments affect permeation and mutations in the cytoplasmic region have no effect. Preventing N-glycosylation by converting residue 141 from an asparagine to aspartate abolished activity despite the presence of significant amounts of protein. Replacing the extracellular cysteines with serine also abolished binding in two cases and permeation in two cases. This places the binding site near cysteine 192.

The model for the function of the AChR that is emerging from these studies is that upon the binding of acetyl choline to the α-subunit there is a conformational change that results in the association of the membrane-spanning units of all the subunits. One of the membrane-spanning regions (MA) is an amphipathic α-helix, that is, once face of the helix displays

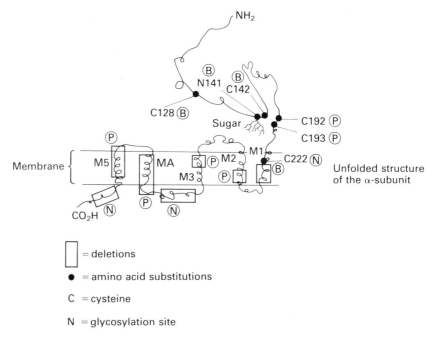

Fig. 10.8. Site directed mutagenesis of the acetylcholine receptor α-subunit. (Adapted from Stevens, 1985.)

charged amino acid side chains and the other displays hydrophobic; the others (M1−M5) are hydrophobic. The five subunits assemble into a pore which has the amphipathic helices at the centre stabilised by the hydrophobic helices packed around the outside.

Growth factor receptors

The growth and division of eukaryotic cells is controlled largely by a family of peptides referred to as growth factors that interact with specific receptors on the cell surface. There is currently tremendous interest in the genes for these growth factor receptors because many of them are *proto-oncogenes*. A working model for the altered growth characteristics of tumour cells is that there is an altered expression, which may be either qualitative or quantitative, of growth factors, their receptors or the intracellular mediators of the response to growth factor−receptor interactions that leads to a permanent stimulation of DNA synthesis and cell division (reviewed by Heldin and Westermark, 1984 and Weinberg, 1985; see Chapter 11 for a further discussion).

There appear to be three general classes of peptide growth factors, the platelet derived growth factor (PDGF) family, the epidermal growth factor (EGF) family and the insulin family. The receptors for these growth factors share a number of properties (Carpenter, 1987). They are all integral membrane proteins which have an external ligand-binding domain, a trans-membrane domain and a cytoplasmic domain. When the growth factor (the ligand) binds to the receptor the signal is transmitted across the membrane to induce changes in metabolism or to initiate DNA synthesis and cell division. The structures of the EGF receptor (EGFR) and the insulin receptor (IR) that have been deduced from their nucleotide sequences is shown in Figure 10.9 (Ullrich et al., 1985; Ebina et al., 1985; Ullrich et al., 1985).

The EGFR comprises a signal polypeptide chain whereas the IR has four polypeptide chains, linked by disulphide bonding, two α-subunits and two β subunits which are derived from a single precursor. All these receptors possess an unusual enzyme activity that phosphorylates tyrosine residues in the receptor itself (autophosphorylation) and in exogenous substrates. Tyrosine phosphorylation is relatively rare. The predominant substrates for protein phosphorylation in eukaryotic cells are threonine and serine residues and less than 0.1% of tyrosine residues are phosphorylated. The tyrosine kinase activity of EGFR or IR is located in the cytoplasmic domain and there is considerable homology between different receptors in this region. The EGFR and IR are also similar in that there are precisely 49 amino acids between the end of the transmembrane domain and tyrosine kinase domain. This may be a target for regulation of receptor activity because in the EGFR it contains a threonine residue (654) which is phosphorylated by protein kinase C resulting in receptor inactivation, there is a specific proteolytic cleavage site in this region of the insulin β chain. The receptors differ at their extreme C-termini; IR has a 90 amino acid tail and EGFR has a 240 amino acid tail, which is a major site of autophosphorylation and therefore, may regulate receptor activity. The membrane-spanning domain is a hydrophobic stretch of 23 amino acids. The extracellular domains show little homology except that they both have cysteine-rich regions. There are two of these in EGFR, which appear to result from a sequence duplication, and one in the IR-α subunit. There are 26 cysteines in this domain in IR-α and 27 in EGFR. Eighteen occupy homologous positions in the two receptors. Many could potentially form disulphide bridges and therefore the domain is called the *cross-linking domain*. It may be involved in receptor aggregation. Structure−function relationships in this domain are being studied by similar procedures to that described for the acetyl choline receptor.

One of the fundamental questions about these groups of receptors is how is the signal that results from specific ligand binding transduced across the membrane to effect the response. The similarity in some of the structural features of the receptors, i.e. the conserved cysteine-rich domains and the tyrosine kinase domain, suggests that some common mechanism might be involved even though the extracellular ligand-binding domains are different. To test this idea Riedel et al. (1986) have constructed a chimaeric receptor that has the insulin binding site and the epidermal growth factor intracellular domain (Figure 10.10). The cloned receptor cDNAs were manipulated to produce a chimaeric gene that had the N-terminal signal sequences and all of the α-subunit of insulin followed by part of the β subunit up to the transmembrane region which was then fused to the transmembrane domain and cytoplasmic domain of the EGFR. The receptor genes were inserted

OUTSIDE

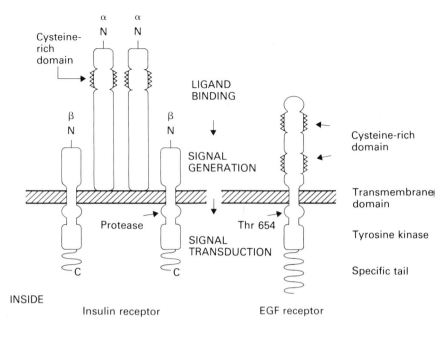

Fig. 10.9. The insulin and EGF receptors.

into an SV40 mini-replicon so that expression was directed by the SV40 early promoter and transfected into COS cells. At 44 hr after transfection expression of the wild-type and hybrid receptors could be demonstrated by binding experiments using radioactively labelled ligands. Stimulation of both EGF and insulin receptors can be assessed by monitoring autophosphorylation in the presence of ^{32}P-labelled ATP. The two receptors differ however, in the characteristics of their respective kinases. The insulin receptor kinase requires a substrate concentration of 100 micro-molar (μM) ATP and no activity is detected below these concentrations but only sub-picomolar (0.1 pM) concentrations of ATP are needed to detect EGF receptor kinase activity. The chimaeric receptor showed insulin-specific stimulation of autophosphorylation but the kinase was active at substrate concentrations of 0.1 pM. The chimaeric receptor, therefore, had the ligand-binding properties of the insulin receptor but the stimulus response of the EGF receptor. This experiment demonstrates that the two receptors must transduce the signal across the membrane by a common signal transfer mechanism. The only extracellular domains that are related in these two receptors are the cysteine-rich regions. This discovery that a common mechanism is involved that must be independent of the cytoplasmic domains and of the non-conserved extracellular domains narrows the search for the key region of the molecule. One idea for the mechanism of signal transduction is that the ligand induces a conformational change in the extracellular

　　　CHAPTER 10

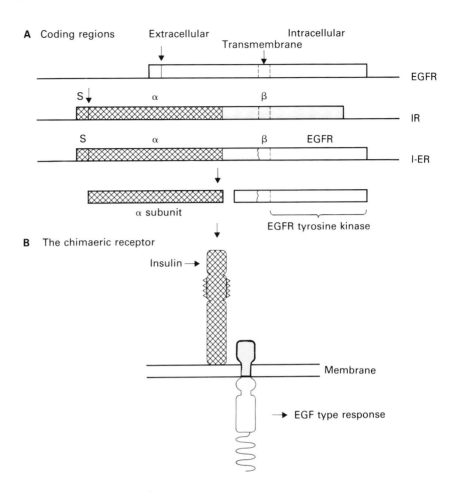

A Coding regions

Extracellular   Intracellular
Transmembrane

EGFR

S   α   β   IR

S   α   β   EGFR   I-ER

α subunit

EGFR tyrosine kinase

B   The chimaeric receptor

Insulin →

Membrane

→ EGF type response

Fig. 10.10. A chimaeric insulin–EGF receptor.

domain which triggers receptor dimerisation which then activates the tyrosine kinase. *In vitro* mutagenesis of the cysteine-rich blocks should provide some insight into this mechanism.

Analysing the immune response by gene transfer

The induction of an immune response and the effectiveness of the response depends upon the interaction of a number of different cell types. These interactions are mediated by cell-surface proteins and one important group of proteins is encoded by genes of the major histocompatibility complex (MHC). As we have already seen (Chapter 7) this is a large locus containing many different genes and there are numerous alleles of some of these genes.

A very simplified outline of the cellular interactions in the induction and

PROTEIN STRUCTURE AND FUNCTION

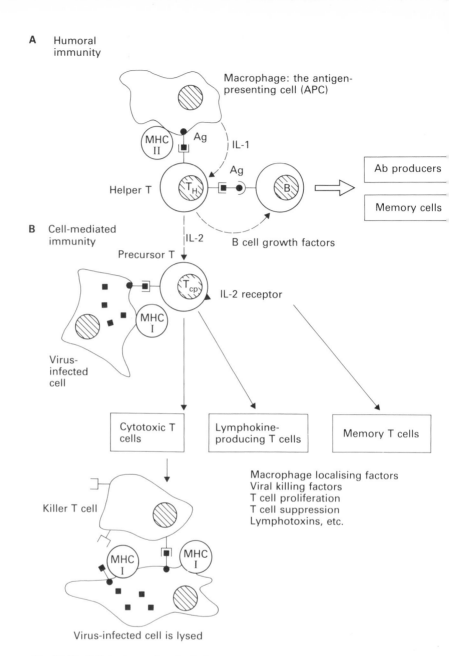

A Humoral immunity

Macrophage: the antigen-presenting cell (APC)

MHC II

Ag

IL-1

Ag

Helper T

T_H

B

Ab producers

Memory cells

B Cell-mediated immunity

IL-2

B cell growth factors

Precursor T

T_cp

IL-2 receptor

MHC I

Virus-infected cell

Cytotoxic T cells

Lymphokine-producing T cells

Memory T cells

Macrophage localising factors
Viral killing factors
T cell proliferation
T cell suppression
Lymphotoxins, etc.

Killer T cell

MHC I

MHC I

Virus-infected cell is lysed

Fig. 10.11. Cellular interactions in the immune response.

effector pathways of the immune response is shown in Figure 10.11 but more details can be found in Roitt (1988).

The ability to mount an antibody response to many antigens is dependent upon the cooperation between two classes of lymphocytes called B cells and T cells (see Chapters 1 and 5). The antigen is ingested by macrophages and displayed on the cell surface in combination with a class II histocompatibility antigen called Ia (letter I). A particular subset of T cells, called *helper T cells*, recognises the antigen/Ia complex via a T-cell surface receptor. Immature

B cells that display membrane-bound antibody that corresponds to another epitope of the same antigen also capture the antigen. The helper T cell forms a bridge between the macrophage and the B cell and this interaction induces the production of various soluble factors that stimulate B-cell proliferation to produce antibody-producing cells; this is *humoral immunity* (Figure 10.11A).

Antigen may also be presented at the surface of cells other than the specialised antigen-presenting macrophages. This occurs on the surface of cells infected with intracellular microorganisms such as viruses or leprosy bacilli or on tumour cells or cells from a genetically different member of the same species (*allogeneic cells*) as, for example, in organ transplantation. Immunity against such 'foreign' cells is mediated by the T cells, this is *cell-mediated immunity*. A simplified scheme of the development of cell-mediated immunity against a virus-infected cell is shown in Figure 10.11B. A subset of T lymphocytes called *cytotoxic precursor T cells (Tcp)* recognises antigens that are presented on the surface of a virus-infected cell but only when these are combined with a class I MHC antigen. The Tcp is then stimulated to proliferate by interleukin-2 which is produced by helper T cells. Several different T cell populations are produced. These include memory cells, lymphokine-producing cells and *killer T cells*. The killer T cells interact with and kill virally infected cells but only when the viral antigen is complexed with the same MHC antigen that was involved in inducing the initial response. These are, therefore, referred to as *MHC-restricted cytotoxic T cells* and the MHC antigen is referred to as the *restricting antigen*. Cytotoxic T cells are restricted by class I antigens and helper T cells are restricted by class II antigens.

The structure of the class I and II genes and the cell-surface receptors that they encode is outlined in Figure 10.12. The class I genes fall into three major groups in the mouse called K, D and L. These genes are highly polymorphic and, therefore, different inbred lines of mice have an individual set of these genes. This is called a *haplotype* and is denoted by a superscript, for example, C57BL/10 mice have the haplotype $H-2^b$ and BALB/C mice have the $H-2^d$ haplotype. The different groups of genes are also referred to with reference to haplotype, in BALB/C mice the genes are K^d L^d and D^d. The class II genes are found in the I (letter I) group of genes and this is divided into two regions, IA and IE. These genes are also polymorphic and are also discussed in the context of haplotype, e.g. BALB/C has IA^d genes.

The cell-surface receptors encoded by class I and II genes share similarities to antibodies in that they are comprised of functional domains. The membrane proximal domains share homology with immunoglobulin constant region domains (see the discussion on gene families in Chapter 1). The class I antigens consist of a single polypeptide chain that has three external domains called $\alpha1$, $\alpha2$ and $\alpha3$ (the analogous domains in the human class I antigens are N, C1 and C2), a transmembrane domain (TM) and a cytoplasmic domain (cyt). The molecule is closely associated with a second polypeptide called $\beta2$ microglobulin. The class II antigens are composed of two different polypeptide chains encoded by separate genes called α and β. Each chain has two external domains ($\alpha1$, $\alpha2$ or $\beta1$, $\beta2$), a transmembrane domain and

PROTEIN STRUCTURE AND FUNCTION

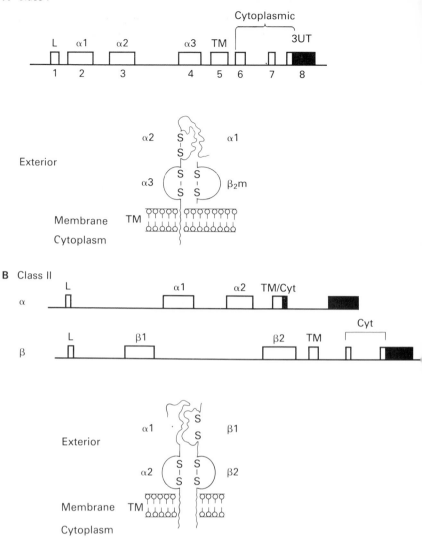

A Class I

B Class II

Fig. 10.12. Structure of the mouse class I and class II MHC genes and proteins.

a cytoplasmic domain. In both class I and class II proteins, the domains correspond to exons in the gene (Hood *et al.*, 1983; Germain and Malissen, 1986; Flavell *et al.*, 1986).

A number of questions about the function of these MHC antigens both in the induction and the execution of an immune response have been tackled by using gene transfer experiments. For example, is it possible to convert a cell into a target for cytotoxic T cells or into helper T cells simply by transferring class I or class II genes? Is it possible to mix α and β class II genes from different haplotypes and produce a functional antigen? What are the important domains in the MHC protein for recognition by T cells. Is it possible to mix domains from different haplotypes? What happens if hybrids between class I and class II genes are transferred? What are the key amino acid residues in the functional domains?

In one of the first gene transfer experiments with MHC genes Mellor et al. (1982) introduced an $H-2K^b$ gene into mouse Ltk$^-$ cells which have the $H-2K^k$ haplotype and showed that $H-2K^b$ antigen was expressed on the cell surface. Transfected and untransfected L cells were then infected with influenza virus and then mixed with influenza-specific T cells that were restricted to the $H-2K^b$ haplotype. Only the transfected L cells which expressed the new $H-2K^b$ antigens were killed. This type of experiment indicated that the class I MHC antigen was essential for target cell recognition by cytotoxic T cells. Similar studies of class II antigens showed that both the IAα and IAβ genes had to be introduced into recipient cells (Nacross et al., 1984) and that functional antigen was only generated when the α and β chains were derived from the same haplotype. In addition, helper T cells only respond to transfected cells that display the exact IAα:IA β pair with which they were initially induced (Leckler et al., 1986).

Because the domain structure of the MHC antigens parallels the exon structure of the genes a simple approach to constructing mutant proteins is by *exon-shuffling*. This simply means making recombinant genes that contain exons from different MHC genes. It is also easy to delete exons from individual genes. These types of experiments have shown that for class I genes the α1 and β2 domains are essential for their ability to function as restricting elements for cytotoxic T cells (Arnold et al., 1984; Reiss et al., 1983). When α1 and α2 domains are derived from different loci, e.g. K/D hybrids, entirely new antigens that are no longer recognised as 'self' and behave as foreign antigens are generated, indicating that the correct structure of the antigen is produced only by the interaction between the two domains (Allen et al., 1984; Bluestone et al., 1985). The cytoplasmic domains of the class I antigens are not important for recognition by cytotoxic T cells or for inducing cell lysis because truncated genes lacking exons 6, 7 and 8 (Figure 10.12) were still functional as T-cell targets (Murre et al., 1984). Similar experiments with class II antigens showed that there are multiple spatially distinct sites on the IA molecule that are recognised by helper T cells (e.g. Cohn et al., 1986). Recently the antigen presenting site on a class I antigen has been identified from the crystal structure and by site-directed mutagenesis (Bjorkman et al., 1987).

Hybrid class II/class I genes have been expressed on the cell surface to produce functional class II antigens (McCluskey et al., 1985). This type of experiment will be extended to produce a panel of cells expressing all the individual domains and various combinations with the aim of understanding the basis of molecular recognition between T cells and their targets.

Similar approaches are also being used to analyse other cell-surface proteins that are involved in the immune response. For example, the T-cell receptor (Ohashi et al., 1985; Saito et al., 1987) and various T-cell accessory antigens (Germain and Malissen, 1986).

Protein engineering

As we have discussed in the preceding sections it is possible to identify functional regions in a protein by observing the properties of a protein produced from a gene that has been specifically mutated *in vitro*. Once this has been done it is possible to 'tinker' with the functional regions with the aim of altering the function. For example, the specificity of an enzyme or antibody can be changed, the interaction with specific inhibitors can be abolished by mutating the binding site and the stability of proteins can be manipulated by making changes in other regions of the protein while preserving the integrity of the functional domain. This molecular tinkering is called *protein engineering* and it is forming the foundations of a new era of research into the properties of proteins and into the therapeutic and industrial applications of proteins. A number of eukaryotic proteins have now been engineered and to illustrate the approach we will describe the manipulation of the enzyme, trypsin. Additional examples such as the production of novel antibodies and the manipulation of some blood clotting factors and their inhibitors are discussed when we consider biotechnology in Chapter 12.

Redesigning trypsin

Trypsin is a member of a family of enzymes which hydrolyse proteins and which have a serine residue as part of their active site; they are therefore called the *serine proteases*. Some other serine proteases are elastase, chymotrypsin and enzymes involved in blood clotting such as kallikrein and tissue plasminogen activator. These enzymes are produced as inactive precursors called *zymogens* and they are activated by the cleavage of a prosegment of amino acids at the N-terminus. This family of enzymes has a common three-dimensional structure and essential amino acid residues for zymogen activation and catalysis have been identified. Although the catalytic mechanisms appear to be conserved amongst the serine proteases they use very different substrates and therefore have a wide range of activities. This variation appears to be due to the presence of different groups of amino acids at the substrate binding site in each of the enzymes. These amino acids are, therefore, targets for site-specific mutagenesis to discover the relationship between catalytic activity and substrate specificity. One goal of such studies is to produce enzymes with precisely tailored substrate specificity. For example by producing therapeutic proteases that are only active under specific conditions and on specific substrates this maximises the therapeutic potential and minimises non-specific degradation of tissues.

Craik *et al.* (1985) have made specific mutations in rat pancreatic trypsinogen II that influence substrate specificity. Trypsin, which is the active form of trypsinogen is specific for hydrolysis of a polypeptide at the peptide bond adjacent to an arginine or lysine residue. An entire rat trypsinogen cDNA was expressed in an SV40 vector in COS cells and the enzyme was secreted as a zymogen into the culture medium. Specific mutations were then made by the procedure of oligonucleotide-directed site-specific mutagenesis (see Chapter 8; Smith, 1985). The aim was to convert two glycine residues at positions 216 and 226 to alanine residues. The small glycines at these positions allow the entry of large amino acid side chains into the hydrophobic pocket which forms the active site of the enzyme. The mutant trypsinogen genes are inserted into an expression vectors which were introduced into COS cells and the mutant proteins were isolated.

The specific modifications each produced an enzyme that had a new set of kinetic parameters. The mutant 216 Gly to Ala had an enhanced specificity for arginine substrates; the affinity was the same as wild type but the catalytic efficiency was increased. Computer-graphic analyses and molecular modelling techniques indicate that the extra methyl group in alanine interferes with the binding of lysine substrates. The mutant 226 Gly to Ala has a 20-fold higher specificity for lysine than the wild type. Furthermore the conformation of this enzyme was subtly altered such that it was actually inactive unless specifically bound with the substrate. This preliminary study indicates the feasibility of redesigning enzymes and as the sophistication of computer-assisted molecular modelling is increased the target sites and types of mutations needed to achieve specific goals can be more precisely defined. At present there is still a certain amount of guesswork involved.

The topic of protein structure function prediction and protein engineering is large and is too advanced for this book. Further details can be found in Leatherbarrow and Fersht (1986), Doolittle (1986), Blundell *et al.* (1987), Robson and Garnier (1986) and Knowles (1987).

Summary

The use of specific mutagenesis, hybrid genes and gene transfer has allowed the identification of topogenic sequences that direct proteins to different sub-cellular compartments and to be secreted outside the cell. The future direction of research is to generate a larger number of specific amino acid changes in the signals to identify the key structure of a topogenic sequence that ensures correct targeting. It is also important to identify the cellular components of the targeting, sorting and translocation pathways. A key approach is to transfer genes which encode proteins with defective topogenic sequences and screen for mutant cells that overcome the defect. This is already being done to characterise the sorting pathway that directs proteins to the vacuole in *S. cerevisiae*.

The ability of cells to communicate with each other and with external factors is central to all aspects of the life of a eukaryote. Understanding the mechanisms of the communication is central to explaining, development and differentiation, to unravelling the complexities of immune response and to controlling and predicting disease. Genes for all types of surface proteins are being introduced into cells and mutagenesis and gene transfer are probably the key to a mechanistic explanation of many of these processes.

Characterising how proteins function is the prelude to controlling and manipulating function. This is the science of protein engineering which promises to revolutionise man's exploitation of nature. It will produce a range of proteins and organisms with completely new properties that will benefit us in areas that range from pollution control to health care.

11 Human Disease

Introduction

One of the exciting applications of gene transfer technology is to provide a route to understanding the genetic and biochemical causes of human diseases. As this greater understanding is achieved, the technology can then be extended to devise procedures for diagnosing and treating the diseases. Cloned oncogenes are being manipulated and returned to cells to throw light on the basis of tumourigenicity. Transgenic mouse lines have been created that mimic human diseases. Methods of replacing defective genes by gene therapy are being tested and gene transfer approaches offer promise for tackling the AIDS pandemic.

Understanding human disease

In this section we will describe gene transfer experiments that are being used to understand the causes of a number of human diseases. One of the most striking areas of progress is in our understanding of tumourigenesis. This stems from the discovery that certain normal cellular genes have the potential to induce tumours when they are activated. These genes have been extensively studied by re-introducing them into cultured cells and by creating transgenic mice that develop specific tumours. Transgenic mouse lines that display other diseases such as growth disorders, diabetes-like syndromes and neuromuscular disorders have also been created.

Oncogenes

There is a group of normal cellular genes that, when mutated or inappropriately expressed, will transform a normal cell into a tumour cell. When these genes are associated with tumours they are called *oncogenes* but in the normal cell they are usually referred to as *proto-oncogenes* or, more frequently, as *cellular oncogenes (c-onc)*. Cellular oncogenes have been identified by two routes. Firstly the genomes of acutely transforming retroviruses were found to contain genes that were essential for their neoplastic transforming activity. These are called *viral oncogenes (v-onc)*. Southern hybridisation experiments showed that viral oncogenes had closely homologous counterparts in normal cells and sequence analysis shows that they are indeed derived from cellular genes. The second route has been described in Chapter 7 and involves using DNA-mediated transfection to neoplastically transform a non tumour cell line using DNA derived from a tumour cell. This led to the discovery of oncogenes in tumour cell DNA. These are invariably mutated derivatives of normal cellular oncogenes and only a limited number of oncogenes can be detected by this technique (Varmus, 1984; Bishop, 1985a; Van Beveren and Verma, 1986).

Over 40 different cellular oncogenes have now been identified. Most of them are found in the genomes of all vertebrates and in some cases they are highly conserved between species. There are, for example, 40 known loci for oncogenes in the human genome (Verma, 1986b). A few cellular oncogenes are even found in invertebrates such as *Drosophila* and in microbial eukaryotes such as *Saccharomyces cerevisiae* (see Table 11.1). The same oncogene has often been found in different retroviruses which leads to some confusion in classification until the genes are sequenced. There are about 30 oncogenes that to date have not been found in retroviruses. These have been identified by transfection, e.g. N-*ras* (Shimizu *et al.*, 1983) or because they are expressed in tumour cells that arise following insertional mutagenesis by a retrovirus, e.g. *int*-1 (Brown *et al.*, 1986), or they are presumed to be

Table 11.1. Oncogenes*.
Group I: Protein kinases.

(A) Tyrosine kinases

Oncogene	Origin	Species	Comments
src**	RSV	Chicken	
abl**	Ab-MLV	Mouse	
fps/fes	FSV	Cat	
yes	Y73	Chicken	
fgr	GR-FSV	Cat	
ros	UR2	Chicken	
erb B**	AEV	Chicken	EGF receptor
neu	neuroblastoma	Rat	related to erb-B
fms	M-FSV	Cat	CSF-1 receptor
ets	AMVE26	Chicken	

(B) Serine kinase

mos	MoMSV	Mouse	

(C) Serine and theronine kinase

raf/mil/mht	3611MSV	Mouse	
	MH2	Chicken	

**Homologues found in *Drosophila*.

Group II: GTP binding proteins.

Oncogene	Origin	Species	Comments
Ha-ras†	Ha MSV	Rat	Frequently
Ki-ras	Ki MSV	Rat	isolated from
N-ras	neuroblastoma	Rat	spontaneous and chemically induced tumours.

†Homologues found in yeast, *Drosophila* and *Dictyostelium*.

HUMAN DISEASE

Group III: Nuclear proteins.

Oncogene	Origin	Species	Comments
myc	MC29	Chicken	
N-*myc*	neuroblastoma	Human	
L-*myc*	small cell lung carcinoma	Human	Often amplified
myb	AMV		
fos	FBJ-MSV		
ski	SKV		
SV-T	SV40		
Py-T	Polyoma		
E1A	Adenovirus		
E1B	Adenovirus		
v-*erb*A	AEV	Chicken	related to steroid hormone receptors

Group IV: Growth factors.

Oncogene	Origin	Species
sis	SSV	Monkey
	FeSV	Cat

Miscellaneous		
B-*lym*	B-cell lymphoma	Human
rel	REV-T	Turkey
kit	FeSV	Cat
int-1	Mammary tumour	Mouse
tx-1	Mammary tumour	Human
tx-2	pre-B-cell tumour	Human
T-*lym* 1	T-cell tumour	Human

*A more extensive list can be found in Bishop (1985b) and Weiss *et al.* (1985), see also Green and Chambon (1986); Carpenter (1987); Gilman (1987).

oncogenes because they are consistently amplified in tumour cells, e.g. N-*myc* or they are involved in tumour-specific chromosomal translocations, e.g. *bcr* (reviewed in Bishop, 1985a). There is another group of oncogenes that do not appear to have any cellular counterparts. These are viral oncogenes that are found in DNA tumour viruses, e.g. SV40, adenovirus and polyomavirus (Bishop, 1984).

The biochemical functions of most of the oncogenes is at present unknown but some of the cellular oncogenes are clearly involved in the normal regulation of cell growth and division. Many of the oncogenes can be broadly classified on the basis of biochemical characteristics and/or sub-cellular localisation (Weinberg, 1985; Sefton, 1986; Van Beveren and Verma, 1986). A list of some of the known oncogenes is given in Table 11.1.

How do oncogenes transform cells?

Studies of naturally occurring cancers in man and animals indicate that the development of a tumour is a multistep process (Klein and Klein, 1985). A simplified view is that there is an *initiation* step which predisposes a cell to become neoplastic. Secondary events are then required to convert this preneoplastic cell into a tumour cell. These are often called *progression* or *promotion* events. The two general stages have been identified in studies on animals with chemical carcinogens. Some chemicals which are applied to the skin produce no tumours unless a second chemical is subsequently applied. The second chemical alone does not produce a tumour (Boutwell, 1974). The chemicals are referred to as *tumour initiators* and *tumour promoters*, respectively. For example, mice treated with the initiator, DMBA (dimethylbenzanthracine) followed by the promoter TPA (12−0-tetradecanoylphorbol-13-acetate) develop papillomas at the site of application. Multiple stages leading to tumourigenesis can also be observed in cell cultures. We have already discussed the properties of cultured cells (Chapter 4) and seen that continuous cell lines can no longer be considered normal because they are capable of indefinite growth, they are *immortalised*. Their growth is, however, under control in that they are still contact-inhibited and they do not produce tumours when injected into animals. In some cases, however, when cell lines are passaged at high cell density, there is selection pressure for more rapidly growing cells and occasionally tumourigenic cells appear in these cultures. These display the full phenotype of a tumour cell, they grow in low serum, they are anchorage independent, i.e. they grow in soft agar, they reach very high cell densities and 'pile up' on the culture surface showing loss of contact inhibition and they produce tumours in nude mice. The difference between the two cell types is analogous to initiation and progression in chemical carcinogenesis.

These two changes in the behaviour of cells in culture can be induced in a single step by DNA tumour viruses (reviewed by Bishop, 1984). For example, infection of primary rodent fibroblasts by polyomaviruses both immortalises the cells so that they can grow indefinitely and confers tumourigenic properties. This is a direct consequence of the expression of viral genes. These are the early genes called middle T and large T antigens. Expression of large T antigen is required for immortalisation and middle T antigen produces morphological alterations and anchorage independence (Cuzin, 1984). These are generally referred to as *immortalising* and *transforming* genes, respectively. A similar picture emerges with adenovirus. The EIA gene product is required to immortalise primary cells and the EIB gene product produces the changes which are characteristic of malignant transformation. In the case of SV40, another widely studied DNA tumour virus, the immortalising and transforming functions are combined within a single protein, the large T antigen (Kriegler *et al.*, 1984). These studies with DNA tumour viruses further establish the principle that tumourigenesis proceeds in stages and these stages are a direct consequence of the expression of specific genes, in this case viral genes that determine immortalisation and malignant transformation.

A simple idea is that apparently spontaneous or carcinogen induced tumours arise through the inappropriate action of cellular oncogenes. Clearly when these genes are fulfilling their normal functions they do not cause tumours. They must, therefore, be activated in some way. In Figure 11.1 we have drawn a simplified scheme for tumourigenesis in three stages. Stage

I is *immortalisation*, the cell acquires the ability to grow indefinitely. Stage II is *malignant transformation*, the immortalised cell displays new phenotypic properties in culture and produces tumours in animals. Stage III represents the acquisition of properties that allow the tumour cell to invade surrounding tissues, to spread to new locations (*metastasis*) and to escape immune surveillance. This has been called *modulation* (Klein and Klein, 1985). These characteristics do not develop to the the same degree in all tumours. It is possible that cellular oncogenes could be activated at each stage, there could be classes of immortalising genes (stage I), transforming genes (stage II) and modulating genes (stage III).

There is also evidence for a class of '*tumour suppressors*' or '*anti-oncogenes*'. These are recessive genes which must be mutated at both alleles for a tumour to develop. The evidence for this comes from studies of conditions where there is a rare inherited predisposition to certain cancers. For example retinoblastoma (a tumour of retinal cells in the eye) cells always lack both alleles at the *rb-1* locus (Klein and Klein, 1985; Green and Wyke, 1985; Lee *et al.*, 1987).

This overall scheme provides a useful framework within which to think about cancer but it must be recognised to be a gross over-simplification at present. Many questions are prompted by considering this scheme, for example: (1) How do oncogenes become activated? (2) Is the activation of a single oncogene sufficient to produce the full tumour phenotype? (3) If multiple oncogenes are required do they act in sequence or in particular combinations and is their effect additive or synergistic? (4) Can the same type of tumour be produced by different oncogenes or is there tissue specificity? (5) What is the biochemical consequence of oncogene activation that results in the tumour phenotype? There is also an alternative hypothesis about oncogenes which should be borne in mind. This is that the activation of oncogenes is merely a consequence of tumourigenesis and not a cause. A balanced account of the possible role of oncogenes in cancer can be obtained by reading the following reviews: Duesberg, 1985; Bishop, 1985a; Land *et al.*, 1983a and Verma, 1986b.

In the next sections we will examine gene manipulation and gene transfer experiments that are being used to address some of the questions about activation and biochemical properties of specific oncogenes and discuss the evidence for multistep carcinogenesis via oncogene activation. In general, three mechanisms of activation have been discovered, these are *mutation*, *over-expression* and *deregulation* (Weinberg, 1985; Nusse, 1986). It is still too early in these investigations to establish 'rules' for oncogenesis and this may not be possible. For example, as we shall see some oncogenes can act alone, others always require additional factors and the same oncogenes can be activated by different mechanisms.

Protein kinases

The group I oncogenes (Table 11.1) are kinases, they do not show any overall homology but they are highly conserved in the functional kinase domains. The first oncogene to be biochemically analysed was *src* which was

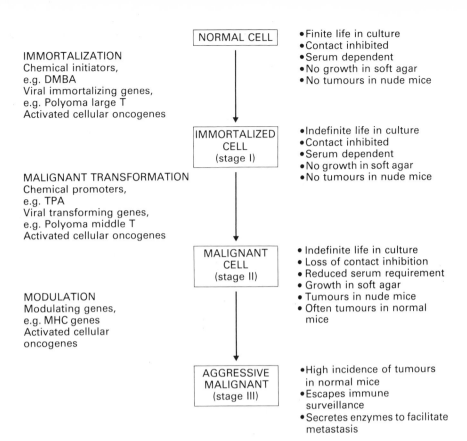

IMMORTALIZATION
Chemical initiators,
e.g. DMBA
Viral immortalizing genes,
e.g. Polyoma large T
Activated cellular oncogenes

MALIGNANT TRANSFORMATION
Chemical promoters,
e.g. TPA
Viral transforming genes,
e.g. Polyoma middle T
Activated cellular oncogenes

MODULATION
Modulating genes,
e.g. MHC genes
Activated cellular
oncogenes

NORMAL CELL
• Finite life in culture
• Contact inhibited
• Serum dependent
• No growth in soft agar
• No tumours in nude mice

IMMORTALIZED
CELL
(stage I)
• Indefinite life in culture
• Contact inhibited
• Serum dependent
• No growth in soft agar
• No tumours in nude mice

MALIGNANT
CELL
(stage II)
• Indefinite life in culture
• Loss of contact inhibition
• Reduced serum requirement
• Growth in soft agar
• Tumours in nude mice
• Often tumours in normal mice

AGGRESSIVE
MALIGNANT
(stage III)
• High incidence of tumours in normal mice
• Escapes immune surveillance
• Secretes enzymes to facilitate metastasis

Fig. 11.1. A simplified scheme for tumourigenesis.

found in Rous sarcoma virus (RSV). It encodes a 60 kD protein that is autophosphorylated and phosphorylates exogenous substrates. The protein is referred to as $pp60^{v-src}$. The kinase has the unusual property of phosphorylating tyrosine residues whereas most protein phosphorylation in the cell is on serine and threonine residues (Sefton, 1986). At least 12 loci have now been identified in birds and mammals that have tyrosine kinase activity and function as oncogenes. In normal cells tyrosine phosphorylation is involved in the response to the growth factors, EGF (epidermal growth factor), PDGF (platelet derived growth factor), CSF-l (colony stimulating factor) and somatomedin C (insulin-like growth factor). The receptors for these mitogenic peptides have intrinsic tyrosine kinase activity and two of them, EGF-receptor and CSF-l receptor are encoded by known cellular oncogenes, c-*erb B* and c-*fms* respectively (Carpenter, 1987). It is anticipated that other receptor genes will eventually be detected as activated oncogenes. The simple idea is that when the oncogenes are activated they display aberrant tyrosine kinase activity resulting in the inappropriate phosphorylation of regulatory proteins involved in signal transduction. This produces constant stimulation of the cells leading to proliferation.

It is possible that proteins are not the only targets for these kinases because signal transduction across membranes is a complex biochemical process. It often involves the phosphorylation of phosphatidylinositol lipids leading to the production of diacylglycerol which stimulates protein kinase C, a protein serine/threonine kinase which is involved in regulating signal transduction. Chemical promoters such as TPA are known to stimulate protein kinase C and therefore this enzyme may have a key role in tumourigenesis (Sefton, 1986; Bell, 1986; Edelman *et al.*, 1987; Gilman, 1987; Berridge, 1987).

Many of the group I oncogene products are membrane-associated which supports their involvement in the initial stages of signal transduction. Some of them are, however, found in the cytoplasm, e.g. *mos*, *mil/raf* and *fps/fes* and these may be involved in the response to the signal. Some of the kinases have serine and/or threonine specificity as does protein kinase C which may ultimately be discovered to be the product of an activated oncogene.

When cellular oncogenes are transduced by retroviruses they are always multiply mutated. They lack introns because retroviral replication always removes introns. Occasionally a partial intron sequence is retained at the 5′ end. Quite often not all of the exons from the cellular gene are incorporated, it is sometimes the 5′ and 3′ exons that are missing. The c-*onc* coding region is often 'embedded' in the viral genome as a result of 'recombination' between the viral genome and the cellular gene. The c-*onc* coding sequence is therefore frequently expressed as a fusion protein with a viral gene, usually a *gag* or *env*. There are also multiple single-base substitutions in the cellular gene, probably reflecting error-prone reverse transcription. In addition to these mutations, the v-*onc* is usually expressed at much higher levels when it is part of a retrovirus than the endogenous chromosomal gene. A major question therefore, is what is the major activating factor? Is it one or more of the mutations that produce a qualitatively different protein or is it the increased quantity of gene product, or perhaps both? (reviewed by Varmus, 1984).

The *src* oncogene has been intensively studied (reviewed by Hanafusa and Jove, 1987). A number of approaches have been used to show how the normal cellular homologue, c-*src*, of the RSV transforming gene v-*src* becomes activated and how the activated oncogene produces transformation.

The v-*src* gene differs from c-*src* at the C-terminus. Nineteen amino acids are deleted from c-*src* and replaced with 12 amino acids derived from a sequence that is found 900 nucleotides downstream from the natural 3′ end of c-*src*. This rearrangement may arise from recombination between two 8 bp repeats called α-sequences (Figure 11.2). Outside of the C-terminal region there are only eight amino acid differences out of 514 positions. To test whether these qualitative changes were important for transforming activity a number of hybrid genes have been constructed. The entire c-*src* has been placed under the control of the SV40 promoter and introduced into established cell lines, e.g. NIH3T3, (Shalloway *et al.*, 1984) or rat 2 cells (Parker *et al.*, 1984). Although the levels of pp60^{c-src} were considerably higher than the levels of pp60^{v-src} in RSV-transformed cells, the cellular homologue failed to induce any transformed foci. When exon 12 of c-*src* was replaced with the corresponding region of v-*src* and this hybrid c-*src*−v−*src* gene was expressed in NIH3T3 using the SV40 promoter, transformed foci

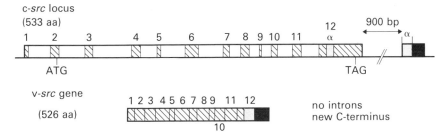

Fig. 11.2. Activation of c-*src* requires alteration at the 3' end.

were obtained at high frequency (Shalloway *et al.*, 1984). This showed that the alteration at the C-terminus was sufficient to activate the cellular oncogene. While this specific mutation may be the most significant activating factor additional studies have shown that increasing the quantity of normal pp60^{c-src} can produce some of the features of transformation. Johnson *et al.* (1985) fused *c*−*src* to the MoMSV-LTR promoter, which increases levels 10 fold. This construction induced transformed foci in NIH3T3 cells but at only 1% of the efficiency of v-*src* and although the cells showed low serum requirement and anchorage independent growth they were not tumourigenic in animals. Increased expression also augments the activity of v-*src*. Jacobovits *et al.* (1984) linked v-*src* to the mouse mammary tumour virus (MMTV) promoter to produce a hormone-responsive v-*src* gene. This was transfected into rat-2 cells selecting for a co-transformed *tk* gene and stable cell lines were produced. The cells were then treated with different concentrations of dexamethasone to induce expression from the MMTV promoter. The resulting cellular phenotype, i.e. whether it remained normal or became morphologically transformed was directly correlated with the level of expression of v-*src* induced by dexamethasone. This study therefore, indicates that maximum transformation potential requires a combination of qualitative and quantitative changes.

Similar studies have been carried out with a number of the protein kinase oncogenes and in general activation appears primarily to require mutation and in some cases is augmented by increasing the quantity of the gene product (Sefton, 1986). For example, *erb*-B is a truncated EGF receptor (Hayman, 1986) and activated *neu* has a val to glu change in the transmembrane domain (Bargmann *et al.*, 1986). There is, however, one exception in this class of oncogenes, mouse c-*mos*. This oncogene can be readily activated simply by increasing the levels of expression. The LTR from Moloney murine sarcoma virus MoMSV was fused to c-*mos* and the hybrid gene was introduced into NIH3T3 cells. Transformed foci were obtained whereas cloned cellular DNA containing c-*mos* does not produce foci in this assay (Blair *et al.*, 1981). Similar experiments have been undertaken with the human c-*mos* oncogene and this also transforms NIH3T3 cells but at a lower efficiency. There seems to have been selection for changes in the coding region of human c-*mos* that reduce its transforming potential. It is also interesting that a negative regulatory element has been found in the 5' flanking region of the human c-*mos* gene which protects it from activation by exogenous upstream elements, e.g. retroviral insertion (Blair *et al.*, 1986). The activation of protein kinase oncogenes by simply increasing gene expression appears to be less common than activation by mutation and this may reflect selection for mechanisms that restrict the catastrophic consequences of one-step gene activation mutations.

A major question about the protein kinase oncogenes is, what is the relationship between the kinase activity and the transforming potential? This question has been studied with v-*src*. Eight cellular polypeptides have been identified as newly phosphorylated on tyrosine in RSV-infected chicken cells. These are vinculin (phosphorylation of this could affect the cell cytoskeleton), enolase, phosphoglycerate mutase and lactate dehydrogenase (phosphorylation of these enzymes could perturb glucose metabolism which is a characteristic of tumour cells), and proteins of unknown function identified by their molecular weights as p81, p50, p42 and p36. Recent experiments have, however, suggested that none of these is the key substrate for the transforming properties of v-*src*. Kamps *et al.* (1986) constructed a mutation by oligonucleotide site-directed mutagenesis of the v-*src* gene to remove the second glycine residue. This residue is normally post-translationally modified by the addition of myristic acid, a 14-carbon fatty acid, that facilitates the interaction of v-*src* with the lipid bilayer. Viruses that carried this mutation were no longer able to transform chicken fibroblasts. The infected cells displayed none of the biochemical properties of transformed cells, i.e. there was no morphological change, no secretion of proteases, only minor perturbations in glucose metabolism and no loss of contact inhibition. When the phosphorylation of the eight substrates was compared between cells infected with mutant and wild type virus there was no difference. This study shows that interaction with the membrane is crucial for v-*src* to transform cells, suggesting that the protein may be involved in signal transduction. It also suggests that either tyrosine phosphorylation is not important or, and this is considered more likely, that the crucial substrate has not yet been identified.

The ras family of oncogenes

Transforming genes can be isolated from several spontaneous human tumour cell lines and from certain chemically induced mouse tumours when their DNA is used to transfect mouse NIH3T3 fibroblasts (see Chapter 7). The NIH3T3 cell line is a good first assay for transforming genes because it is 'hovering' between stages I and II (Figure 11.1) and is, therefore, easily pushed to stage II by exogenous factors. Of course when oncogenes are identified by transfection of NIH3T3 cells this does not necessarily mean that these genes are sufficiently potent to produce tumours in animals, it is merely an indication of their oncogenic potential. Many of the oncogenes that are isolated by transfection belong to the *ras* family (reviewed by Barbacid, 1987). This is a large and complex family of genes. There are, for example, at least seven distinct *ras* loci in the human genome. They fall into three categories, C-Ha-*ras* (*ras*^{H}), C-Ki-*ras* (*ras*^{K}) and N-*ras* (*ras*^{N}) which reflects their homology with the oncogenes found in the retroviruses, *H*arvey sarcoma virus, *K*irsten sarcoma virus and with the cellular oncogene, N-*ras*, identified in a *neuroblastoma* cell line. For example, the Ha-*ras*-oncogene has been isolated from two separate human bladder carcinoma cell lines, called T24 and EJ and Ki-*ras* oncogene has been isolated from human colon and lung carcinomas. As this is a modified gene it is often referred to as EJ*ras* or T24*ras* (see later). The *ras* genes are highly conserved both within and between species and homologues have been found in non-vertebrates, e.g. *Drosophila*, *Dictyostelium* and *Saccharomyces cerevisiae*.

The *ras* genes all encode a 21 kD protein called p21, they share maximum homology at the N-terminus, the first 86 amino acids are identical, but they vary markedly in the C-terminus. When the *ras* oncogenes that were isolated by transfection were sequenced they were always mutated when compared to normal cellular oncogenes. The mutations were however, not random but they were found at amino acid residues 12, 13 or 61. In addition, some of the activated viral *ras* proteins have a threonine at position 59 and this is autophosphorylated whereas normal cellular p21 is not phosphorylated. The normal functions of *ras* proteins are not known but several biochemical properties have been identified. The proteins bind guanine nucleotides with high affinity and they have a GTPase activity. They are also modified by the addition of lipid and become localised on the inner surface of the plasma membrane. Some of these properties are reminiscent of a class of proteins called *G-proteins*. These are membrane-associated proteins that are involved in the mechanism of signal transduction across membranes. They act to control the activity of adenylate cyclase to modulate the production of cyclic AMP (Gilman, 1987).

In order to assess the significance of the mutation in *ras* oncogenes for their oncogenic potential a number of mutagenesis studies have been undertaken (Figure 11.3). Fasano *et al.* (1984) performed random bisulphite mutagenesis on a cloned wild-type human H-*ras* gene and then assayed the mutant genes for their transforming potential in NIH3T3 cells. This technique will, of course, only produce mutations in codons that contain cytidine residues (see Table 8.1) and in this study mutations at the key position 12 were made but not at position 61. They analysed 76 different mutations scattered throughout the *ras* coding region. Only mutations at position 12, 13, 59 and 63 had significant transforming potential. These are all in the same region of the protein that are changed in oncogenes that have been activated *in vivo* although changes at positions 12, 13, 59 and 61 had only previously been seen. The nature of the amino acid change at position 12 does not appear to be important as Seeburg *et al.* (1984) showed that 18 different amino acid substitutions retained transforming potential. Der *et al.* (1986) undertook a similar study of position 61 and introduced 17 amino acid substitutions and only substitutions of proline and glutamate failed to activate the gene. In this case, however, there was a 1000 fold variation in the transforming potency of the mutant proteins. The activation of the *ras* oncogene and the potency of its transforming abilities does not seem to correlate with the known biochemical properties. Both wild-type and position 12 mutants display identical GTP binding and both localise at the plasma membrane (Der *et al.*, 1986). Mutants at position 61 show quantitative differences in the GTPase activities but there is no direct correlation

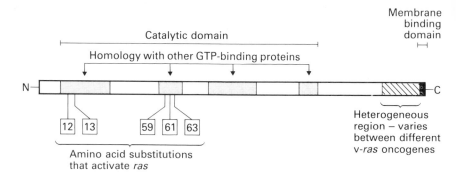

Fig. 11.3. Normal and activated *ras* proteins.

between the levels of enzyme activity and the transforming potential (Der *et al.*, 1986; Lacal *et al.*, 1986). The efficient activation of the *ras* oncogene is clearly correlated with highly specific mutations but the transforming effect of these mutations is not clearly mediated through their GTP binding or GTPase activities. It is possible that the mutations disturb the conformation of the protein (Srivastava *et al.*, 1985). As with a number of oncogenes massive over expression of the normal Ha-*ras* gene will result in transformation of normal cells (Chang *et al.*, 1982). Over expression also augments the weak transforming ability of some of the activated mutant *ras* oncogenes (Der *et al.*, 1986).

The discovery of *ras* homologues in the simple eukaryote *S. cerevisiae* appeared to offer a route to understanding the normal functions of these proteins. *S. cerevisiae* has two *ras* homologues, *RAS1* and *RAS2*, either of which is dispensable for viability but disruption of both genes is lethal (e.g. Tatchell *et al.*, 1984; Katoaka *et al.*, 1984). The yeast proteins bind and hydrolyse GTP and various genetic analyses have shown that the *RAS* proteins function as GTP dependent positive regulators of adenylate cyclase (Kataoka *et al.*, 1985; Toda *et al.*, 1985). Also, the expression of human p21 in yeast stimulates adenylate cyclase (Clark *et al.*, 1985). Unfortunately, in mammalian cells there is no evidence for adenylate cyclase activation by *ras* proteins (Becker *et al.*, 1985). It is possible that the mammalian genes have diverged significantly from their yeast counterparts, possibly to the point of acquiring new functions. There is some evidence that *ras* proteins affect the regulation of the cell cycle but this may be an indirect result of inducing the synthesis of growth factors (Levinson, 1986). It is also possible that *ras* proteins regulate phosphatydylinositol metabolism (Fleischman *et al.*, 1986).

The myc family of oncogenes

The *myc* oncogene was first discovered as a component of the avian retrovirus MC29 where it is expressed as a *gag*–*myc* fusion protein. The corresponding cellular oncogene c-*myc* is highly conserved between vertebrate species but no related gene has been found in non-vertebrates. A family of *myc* related genes has now been identified. These related genes have not appeared in retroviruses but were identified either by their homology with v-*myc* or because the loci were selectively amplified in specific tumours. For example, human N-*myc* is amplified in *n*euroblastoma cells, human L-*myc* is amplified in small-cell *l*ung carcinomas and human U-*myc* was identified by hybridisation. An additional gene, B-*myc* has been identified in the mouse genome.

The expression of N-*myc*, B-*myc* and L-*myc* is developmentally regulated in the mouse whereas c-*myc* is expressed throughout development and in a wide range of proliferating cells. This lack of tissue and stage specificity for expression of c-*myc* may explain why v-*myc* is the only virally coded oncogene that can induce tumours in all three major tissues, i.e. epithelial, mesenchymal and haemopoietic. The levels of c-*myc* expression are about 10-fold higher in proliferating cells but in some terminally differentiated cells which have lost the potential to divide e.g. B cells, c-*myc* expression is virtually absent. Expression of c-*myc* is further increased in response to a variety of mitogens, although this is only a transient increase. Many of the properties of c-*myc* suggest that it is a regulatory protein. The protein is found in the nucleus, it binds single and double-stranded DNA, and the c-*myc* mRNA has a very short half-life of about 10−30 min. The expression in response to mitogens and in proliferating versus quiescent cells is regulated both transcriptionally and post transcriptionally. The c-*myc* gene is unusual in having an extremely long 5′ untranslated leader exon which appears to contain an internal transcriptional activator and may affect translation and mRNA stability (Yang *et al.*, 1986; Cole, 1986). It is thought that the function of c-*myc* may be in determining the transition from $G0$ to $G1$ and the higher levels of expression may increase the probability of the cell entering S phase. The c-*myc* gene can be activated by any event that leads to its deregulation and this may or may not be associated with increased levels of expression. For example, c-*myc* is often involved in specific chromosome translocations in tumour cells such as plasmacytomas and Burkett lymphomas. In both types of tumour the translocation involves the immunoglobulin heavy-chain locus. In most cases the long untranslated exon of c-*myc* is disrupted; the net result of the translocation is to deregulate c-*myc* expression. There are often additional mutations in the c-*myc* coding regions but as this is not always true they may not be significant for activation. The c-*myc* gene can also be activated by proviral insertion which provides an enhancer which overrides the normal transcriptional controls (Nusse, 1986; Varmus, 1984; Cole, 1986; Kelly and Siebenlist, 1986).

Viruses which carry v-*myc* are often only weakly transforming and when either v-*myc* or c-*myc* is introduced into established cell lines no morphologically transformed foci are produced even though expression is directed by heterologous constitutive promoters. These observations suggest that activation of c-*myc* is not sufficient to produce a fully malignantly transformed cell. In this case, because there was no assay system a test of the idea that c-*myc* is activated by deregulation was not possible using the simple DNA-transfection approaches that we have described for studying *ras*, *mos* and *src* activation. This problem was solved by experiments by Land *et al.* (1983b) who showed that primary rat embryo fibroblasts were not transformed by expression vectors containing EJ*ras* or c-*myc* alone but when both oncogenes were used together transformed foci were obtained. Lee *et al.* (1985) used this co-transformation assay to prove that deregulation was required for c-*myc* activation and that additional changes in the gene had no effect. They introduced a range of c-*myc* constructions into rat embryo cells together with EJ*ras* which does not morphologically transform these cells without an additional oncogene (Figure 11.4). They observed no activity with the normal chicken and human c-*myc* alleles or with a human c-*myc* allele that had been amplified in a neuroectodermal tumour. They also failed to detect activity with c-*myc* alleles from bursal lymphomas that arose from proviral insertion. These alleles had lost the first exon and had point mutations in the coding region but this clearly did not activate the gene. They only found transforming activity when either normal c-*myc* alleles or mutated alleles were placed under the transcriptional control of avian or murine leukaemia virus LTRs. These studies therefore, support the idea that simply deregulating c-*myc* expression is sufficient to induce transformation.

Several nuclear oncogenes have now been discovered and these share some properties with c-*myc*. For example c-*fos*, which is present in FBJ murine osteosarcomavirus, is rapidly induced by mitogens but expression is only transient. The mRNA half-life is short and the stability and expression of the mRNA are controlled by sequences within the gene, this time at the 3′ end. Activation of the oncogene requires deletion or alteration of the carboxy terminus as well as over expression (Verma, 1986a). Unlike c-*myc* the activated *fos* gene is only found in bone tumours.

Some of the mechanisms of oncogene activation have been established but there are no hard and fast rules. Some oncogenes can be activated by mutation or over-expression and deregulation and in some cases the transforming potential may only be fully realised if there is both mutation and altered expression. Some of the information about oncogene activation is summarised in Table 11.2.

Multistep carcinogenesis

As we have seen in Figure 11.1 the development of a fully malignant tumour can be viewed as a multistep process. The contribution of different activated oncogenes to malignant transformation is being tested by using DNA-mediated transfection to introduce them into primary cells and established cell lines. Transfer into primary cells tests whether the oncogene is capable of immortalising a cell and transfer into established cells tests the transforming ability.

Cooperation between oncogenes was first demonstrated in studies with *ras*. A number of studies showed that when EJ*ras* was transfected into primary cells such as diploid hamster dermal fibroblasts and rat embryo cells (Land *et al.*, 1983b) no transformed foci developed. As we have seen, if c-*myc* was also introduced then fully transformed foci developed (Land *et al.*, 1983b). Other oncogenes have been shown to cooperate with *ras* in this way, e.g. adenovirus Ela (Ruley, 1983), polyomavirus large T antigen (Land *et al.*, 1983b) and N-*myc* (Yancopoulos *et al.*, 1985). It has been proposed that all these *ras*-cooperating oncogenes act as immortalising genes.

Normal chicken (ch) or human (hu) c-*myc*

c-*myc* from a bursal lymphoma activated by ALV proviral insertion

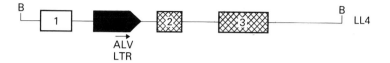

c-*myc* from a human neuroectodermal tumour

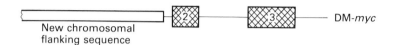

Allele or recombinant gene	Focus formation in the presence of EJ *ras*	Tumourigenicity
c-*myc* (ch)	−	−
c-*myc* (hu)	−	−
ALV-LTR-LL4 *myc*	25–100	+
DM *myc*	−	−
MLV-LTR-c-*myc* (hu)	50–500	+
MLV-LTR-DM *myc*	50–500	+

Fig. 11.4. c-*myc* is activated by deregulation.

The nuclear oncogenes such as c-*myc* and c-*myb* also cooperate with other oncogenes, e.g. *src*, *erb*B, *fes*, *yes*, *ros* and *mil/raf* (Weinberg, 1985). Those oncogenes such as *ras* and *src* which require cooperation with immortalising genes might be considered as *transforming oncogenes*.

Unfortunately this neat distinction between *immortalising* and *transforming* oncogenes does not stand up to close inspection. The behaviour of some oncogenes can be drastically influenced by the level of expression and the interpretation of the properties of an oncogene depends upon the assay that is used. This was first demonstrated for *ras* by Spandidos and Wilkie (1984). They linked transcriptional enhancers from SV40 and MoMSV to T24*ras* and the normal Ha-*ras*-1 gene. The hybrid genes were then transferred to early-passage rat embryo, muscle and skin cells and low-passage chinese hamster lung cells. In this study, when T24*ras* was over-expressed, primary cells were rescued from senescence. The normal Ha-*ras* oncogene also rescued cells from senescence but only in the presence of enhancers. This data shows that the *ras* oncogene can immortalise cells if it is either mutated or over-expressed but that mutation and over-expression are required for full malignant transformation. The immortalising properties of *ras* are also indicated from its interaction with chemical carcinogens. Activated Ha-*ras* will act as an initiator for the development of TPA induced tumours *in vivo*

HUMAN DISEASE

Table 11.2. Activation of oncogenes.

Oncogene	Ref.	Mechanism of activation
myc	(1)	Deregulation of transcription and stabilisation of the transcript. Augmented by over-expression.
ras	(2)	Point mutations at single amino acid residues 12, 13, 59, 61 or 63. (C-*Ha-ras* has immortalising activity if massively over-expressed).
src	(3, 4)	Deletion or replacement of the C-terminus. Augmented by interaction with polyoma middle T antigen and by over-expresson (c-*src* has some transforming activity when massively over-expressed).
fos	(5, 6)	Alteration of the 3′ end of the mRNA and deletion of the protein C-terminus. Augmented by over-expression and by polyomavirus large T.
erb-B	(7)	Deletion of the N-terminal ligand binding domain.
neu	(8)	Single-point mutation at residue 664 in the transmembrane domain.

(1) Cole, 1986; (2) Levinson, 1986; (3) Sefton, 1986; (4) Johnson *et al.*, 1985; (5) Verma 1986a; (6) Jenuwein *et al.*, 1985; (7) Hayman, 1986; (8) Bargmann *et al.*, 1986.

(Brown *et al.*, 1986) and *in vitro* (Dotto *et al.*, 1985). The *myc* which has been tentatively classed as an immortalising oncogene does not, however, function as an initiator (Dotto *et al.*, 1985). Over-expression of a number of other oncogenes extends their activity. For example, when the expression of ElA, p53 and c-*myc* genes is driven by strong promoters they will induce a number of changes in established cell lines that are characteristic of malignant transformation, e.g. low serum dependence and anchorage independent growth, but without overtly producing morphological changes (e.g. Kilekar and Cole, 1986). In some cases, however, injection of these transformed cells produces tumours in nude mice (Eliyahu *et al.*, 1985). These 'immortalising genes' clearly behave as transforming genes under these conditions. Clearly single oncogenes can induce the full transforming phenotype when transfected into primary cells. It is also likely that some tumours in animals are the result of the activation of single oncogenes. This is best demonstrated by studies of environmentally induced tumours. In many cases tumours produced by a known carcinogen always contain the same activated oncogene (Table 11.3; Bishop, 1985b). The high incidence of retroviruses carrying two different oncogenes indicates, however, that cooperation is often required.

The third stage of tumourigenesis that we have described in Figure 11.1 is the least well understood but some clues are emerging. Firstly we have seen that over-expression augments the activity of oncogenes. It is a common observation that highly malignant tumours have amplified oncogenes. In some cases where tumour progression has been monitored oncogene amplification occurs late and is highly correlated with high malignancy. This presence of amplified oncogenes can be used as a prognostic indication of tumour severity (Slamon *et al.*, 1987). Over-expression of an oncogene may, therefore, represent the last step in a multistep process. It is also clear that malignant tumours escape immune surveillance. This correlates with a reduction in the expression of class I MHC antigens on the tumour cell

Table 11.3. Oncogenes consistently activated by known carcinogens.

Oncogene	Carcinogen	Tumour
C-Ha-*ras*	Dimethylbenzanthracene	Mouse papilloma
C-Ha-*ras*	Nitrosomethylurea	Mouse breast carcinoma
C-*ras*	S-methycholanthrene Benzopyrene N-methyl-N-nitrosoguanidine Diethyl nitrosamine	Transformed foetal guinea pig cells
C-Ki-*ras*	γ-irradiation	Mouse thymic lymphoma
N-*ras*	Nitrosomethylurea	Mouse thymic lymphoma
C-*neu*	Ethylnitrosourea	Rat neuroblastoma and glioblastoma

Source of information, Bishop (1985b).

surface (Goodenow *et al.*, 1985). The importance of the class I antigens in tumourigenesis has been elegantly demonstrated by re-introducing class I genes into tumour cells by transfection and showing loss of tumourigenicity (Wallick *et al.*, 1985). Understanding the factors that determine gene amplification and MHC gene expression may help to devise ways of controlling tumour progression. The ability to vary the expression of oncogenes by inserting them into different expression vectors and then introduce them in different combinations into different cell types is clearly a powerful tool for the dissection of the biochemical basis of tumourigenesis. The interactions are, however, complex and simple rules and hierarchies of interactions are not yet apparent. Some of the possible sites of action of known cellular oncogenes are summarised in Figure 11.5.

Animal models for human diseases

Carcinogenesis

The discovery and characterisation of viral and cellular oncogenes coupled with the ability to produce transgenic mice has led to an exciting new approach to studying cancer. When oncogenes are injected into mouse embryos the resulting transgenic mice develop tumours. These tumours occur reliably and they are usually restricted to one or a few specific tissues. This restriction reflects the properties of enhancer elements in the vector that is used to direct the expression of the oncogene. This means that different oncogenes can induce tumours in a range of tissues depending upon the associated enhancer sequence. There is usually a delay in the production of tumours until the mice are several months old; this means that they can be bred to produce lines of mice where tumourigenesis is inherited as a dominant trait.

A series of different tumours have been induced by expressing large T antigen in transgenic mice. In a typical study Hanahan (1985) fused 520 bp of the upstream region of the rat insulin II gene to the early region of SV40.

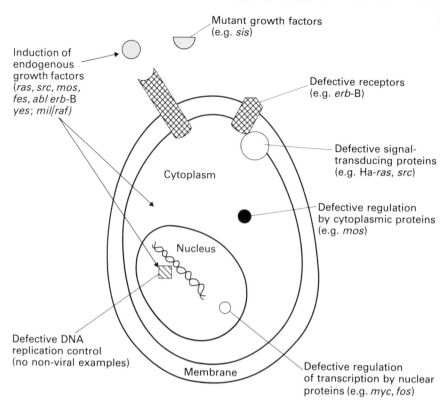

Induction of endogenous growth factors (*ras*, *src*, *mos*, *fes*, *abl* erb-B *yes*; *mil/raf*)

Mutant growth factors (e.g. *sis*)

Defective receptors (e.g. *erb*-B)

Cytoplasm

Defective signal-transducing proteins (e.g. Ha-*ras*, *src*)

Defective regulation by cytoplasmic proteins (e.g. *mos*)

Nucleus

Defective DNA replication control (no non-viral examples)

Membrane

Defective regulation of transcription by nuclear proteins (e.g. *myc*, *fos*)

Fig. 11.5. Sites of action of activated cellular oncogenes.

The insulin upstream region contained an mRNA cap site and a tissue-specific enhancer and the SV40 early region contained the entire large T coding sequence but lacked all the SV40 promoter and enhancer elements (Figure 11.6). Insulin is only synthesised in the pancreas which is composed of isolated islands of endocrine cells (the islets of Langerhans) embedded in the exocrine mass of the pancreas. The islets consist of four cells types arranged in a characteristic pattern. These are the B cells which secrete insulin and the α and pp cells which respectively secrete glucagon, soma-tostatin and pancreatic polypeptide. The tissue specificity of insulin expression *in vivo* has been confirmed in transient transfection studies which show expression in insulinoma cells but not in fibroblasts (Walker *et al.*, 1983; see Chapter 8).

The hybrid gene was injected into fertilised mouse embryos to produce several, fully transgenic offspring. These were bred to produce heterozygous transgenic lines but many of the mice died prematurely after about 12 weeks. These mice had abnormal pancreatic histology, the cells in all the islets were more densely packed, many of the islets had increased in size (hyperplasia) and some of the islets contained solid highly vascularised β cell tumours. When sections of normal pancreas are examined by immuno-histochemical staining the islets stain predominantly with α-cell and β-cell specific antiserum and show a characteristic peripheral distribution of α-cells: also, none of these cells express T antigen. When sections of the

CHAPTER 11

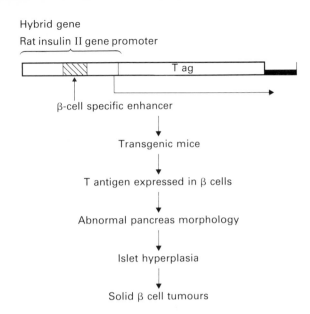

Fig. 11.6. Producing transgenic mice with pancreatic β-cell tumours.

pancreas from any of the transgenic mice were examined either after natural death or when they were killed at eight weeks, characteristic abnormalities were always observed. T antigen expression was detected in the nuclei of β-cells at all stages and at eight weeks, i.e. before any sign of disease, the islets were densely packed with β-cells and the other cell types were either excluded or the normal relationships between the cell types was disrupted. At 12 weeks most of the islets showed hyperplasia and solid tumours developed at between 10 to 20 weeks. The progression from hyperplasia to full-blown tumour only occurred in about 5% of the islets. The tumours only contained β-cells and large T antigen expression was detected only in β-cells and in no other major tissue in the mice. The tumourigenic phenotype was 100% heritable as a dominant trait.

These data indicate that in this case tumourigenesis is a multistep process where the expression of the T antigen is necessary but not sufficient. T antigen expression produces proliferation leading to hyperplasia but additional events are required because not all cells expressing T antigen develop into tumours. This event could occur in a rare cell in which case the tumour will be clonal, i.e. arise from a single altered cell or there could be an islet-specific change that affects all β-cells in the islet.

The importance of the enhancer sequence in determining the type of tumour that is produced is illustrated in Figure 11.7 (for supporting references, see Palmiter and Brinster, 1986). This shows the results of experiments where the SV40 T antigen has been fused to a range of different enhancers including its own. When the enhancer from the elastase I gene is used, pancreatic tumours are also produced but in this case the acinar cells are transformed. When the SV40 T antigen is introduced with its own enhancer, specific tumours of the choroid plexus (epithelial lining of the ventricles of the brain) develop in nearly all the mice within three months of

HUMAN DISEASE

birth. This tissue specificity is clearly a property of the SV40 enhancer, because when an enhancerless T antigen gene was fused with a mouse metallothionein−human growth hormone (MT-hGH) gene a different tissue distribution of neoplasia was detected. In particular, hepatocytes and Schwann cells (the insulating sheath around axons) were affected. This is characteristic of metallothionein and growth hormone gene expression and is further evidence that enhancers may also be contained within the coding region of the hGH gene (see Chapters 4 and 8). The levels of T antigen expression in choroid plexus tumours was extremely high as compared to unaffected brain tissue and the T antigen gene had amplified and rearranged in the tumours.

The cellular oncogene c-*myc* has also been used to produce tumours in transgenic mice. As we have already discussed this gene is thought to be one of the nuclear proteins that mediates the cellular response to growth factors. Its expression is normally strictly regulated (see above, Kelly *et al.*, 1983). Stewart *et al.* (1984b) fused the coding region of c-*myc* to the tissue specific promoter/enhancer in the LTR of mouse mammary tumour virus (MMTV). The MMTV-LTR promoter is induced by prolactogen and glucocorticoids and MMTV produces breast tumours. Transgenic mice produced with this hybrid gene developed mammary adenocarcinomas at the second or third pregnancy. This characteristic was inherited by all of their offspring. Transcripts corresponding to the fusion c-*myc* gene were detected in all breast tissue and regularly in salivary gland and intestine and in a few lines transcripts were detected in a wide range of tissues. Tumours developed in only a few areas of the breast, but never in the salivary gland or lung. In lines where transcripts were detected in a range of tissues, tumours were found in testes, mast cells, pre-B cells, B cells and T cells (Leder *et al.*, 1986). Other similar studies have shown that when the c-*myc* gene is linked to immunoglobulin gene enhancers transgenic mice develop B-cell leukaemia within several months (Adams *et al.*, 1985). This observation is particularly relevant to human tumours such as Burkitts lymphoma which are characterised by a translocation which brings c-*myc* under the influence of immunoglobulin enhancers (see Chapter 1 for a discussion of gene rearrangements in cancer).

In all the examples we have discussed it is clear that although the expression of the oncogene is necessary for a tumour to develop it is not sufficient. This is indicated by the fact that not all cells that express the gene become neoplastic. This observation supports the idea that cancer is a multistage process. The presence of the expressed oncogene clearly predisposes to neoplasia but additional events are required to convert the precancerous cell to a cancer cell. In the case of SV40 T antigen the rearrangements and amplification in the choroid plexus may increase expression or alter the gene. In the case of MMTV−c-*myc* the deregulated high expression may contribute to the next stage and other hormonal influences during pregnancy may also be important. It is possible that rare mutations may occur in the oncogene, increasing its transforming potential. The importance of mutation is emphasised by studies with Ha-*ras*. The normal cellular Ha *ras* does not produce tumours in transgenic mice. When the mutant on cogenic form, with the glycine to valine substitution at amino acid 12, is

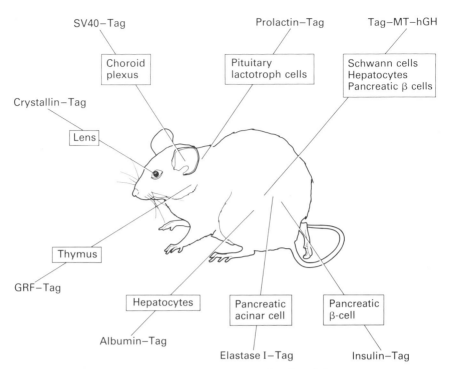

SV40–Tag

Prolactin–Tag

Tag–MT–hGH

Choroid plexus

Pituitary lactotroph cells

Schwann cells
Hepatocytes
Pancreatic β cells

Crystallin–Tag

Lens

Thymus

GRF–Tag

Hepatocytes

Pancreatic acinar cell

Pancreatic β-cell

Albumin–Tag

Elastase I–Tag

Insulin–Tag

Fig. 11.7. Tumours included by different promoter – T antigen fusion genes.

fused to the elastase I gene enhancer all the pancreatic acinar cells in transgenic mice became transformed within days of their appearance during foetal development. Mice are born with pancreatic tumours and die within a few days (cited in Palmiter and Brinster, 1986). This indicates that 'single hit' carcinogenesis is possible with a mutant oncogene but *in vivo* the mutation would be a rare event.

The ability to produce transgenic mouse lines that are predisposed to the development of specific tumours at a high frequency should vastly accelerate our understanding of cancer. The pre-cancerous state and all stages up to the appearance and progression of the tumour can be analysed in defined tissue. Previously researchers had to use chemical carcinogens or viruses that produce tumours at a much lower frequency and with a random time of appearance. It should be possible to use transgenic lines that are predisposed to neuroplasia to assay for the 'second-hit' factors. Different oncogenes can be introduced into embryos of these predisposed mice and their effects on the rate of tumour development and tumour distribution and progression can be studied. It will also be possible to establish more reliable animal models for anti-cancer therapies.

Transgenic mice produced with immortalising oncogenes also provide useful reagents in that the oncogene immortalises the transformed cells so that they can be cultured *in vitro*. This allows the production of specialised cell lines which can be used to produce large quantities of proteins, e.g. insulin from an insulinoma. Different oncogenes can also be introduced into the cell lines by DNA-mediated transfection as a first test of oncogene interactions before moving to transgenic mouse studies.

HUMAN DISEASE

Other diseases

Transgenic mouse models have been established for a number of human diseases (Table 11.4). In particular, growth control disorders have been extensively studied because the first transgenic mice were created with the rat growth hormone gene (rGH) (Palmiter and Brinster, 1986, Chapter 5). An inbred mouse line carrying the mutation, *little*, mimics human pituitary dwarfism. When transgenic derivatives of *little* mice were created that carried an *Mt*-rGH gene, they grew to three times the size of their non-transgenic litter mates (Hammer *et al.*, 1984). This indicates that a growth defect can be corrected but other more complex consequences of growth hormone gene expression were also seen. For example, all the *Mt*-rGH transgenic females had reduced fertility (Hammer *et al.*, 1985a). This appears to be due to the effect of growth hormone on the differentiation of the liver which determines steroid hormone synthesis (Norstedt and Palmiter, 1984). Growth can also be affected by introducing the gene for growth hormone releasing factor (GRF) (Hammer *et al.*, 1985b) which in turn stimulates the production of somatomedin C (insulin-like growth factor (IGF-l)). It is anticipated that by expressing these polypeptide growth factors, i.e. GH, GRF and IGF-l, at different levels and in different tissues in transgenic mice, for example, by using the albumin enhancer to target to the liver, the complex cascade of reactions that control growth and sexual differentiation can be analysed and specific therapies for growth defects can be devised.

The immune response is also being analysed in transgenic mice. Certain strains of mice fail to express the class II MHC antigen Eα because the gene has a large deletion. These mice cannot produce antibodies to Eα-restricted antigens (see Chapter 10 for a brief discussion of MHC class restriction). When a cloned Eα gene is used to produce a transgenic line from these deficient mice the immune response to Eα-restricted antigens is restored (Le Meur *et al.*, 1985; Pinkert *et al.*, 1985). This approach shows that MHC antigens can be expressed in transgenic mice and opens the way for analysing the complex regulatory interactions between components of the immune response.

Table 11.4. Some transgenic animal models for human diseases.

Transgenic configuration	Disease model	Ref
MT-rGH/*little* mice	Pituitary dwarfism	1
Human insulin	Diabetes	2
β-globin/β-globin defective mice	β-thalassemia	3
Human Hbˢ/β-globin defective mice	Sickle-cell anaemia	3
MHC antigen genes/(1−E)/ immune deficient mice	Immunodeficiency diseases	4
Papovavirus JC early region	Progressive multifocal leucoencephalopathy	6
MBP/Shiverer mice	Muscular dystrophy	7

(1) Hammer *et al.*, 1984; (2) Selden *et al.*, 1986; (3) Constantini *et al.*, 1986; (4) Le Meur *et al.*, 1985; (5) Pinkert *et al.*, 1985; (6) Small *et al.*, 1986; (7) Redhead *et al.*, 1987.

Recently an exciting model has been developed that holds promise for understanding neuromuscular disorders such as muscular dystrophy (Redhead *et al.*, 1987; Popko *et al.*, 1987). The conduction of nerve impulses along axons is facilitated by the properties of the myelin membranes which form a multilayered insulating sheath. The myelin membranes contain a family of proteins called *myelin basic proteins (MBP)* which account for 30−40% of the protein content of the myelin membranes. There is an autosomal recessive mutation in the mouse called *shiverer (shi)*. Shiverer mice have a severe muscular disorder. They literally shiver or tremor soon after birth and develop severe muscular seizures and have a shortened life-span. These mice have very low levels of MBP and the genetic lesion is, in fact, a deletion at the *shi* locus. Redhead *et al.* (1987) have inserted the normal mouse MBP gene into the germ line of shiverer mice by microinjecting embryos with a 37 kb genomic clone. The homozygous transgenic shiverer mice no longer shivered or died prematurely. A series of different genetic crosses between different MBP transgenic mice and with other mutant strains of mice has led to the production of mouse lines which have different degrees of myelin deficiency and shivering phenotype (Popko *et al.*, 1987). This appears to be correlated with the levels of MBP produced from multiple copies of the transferred gene and/or differences in expression in different chromosomal locations. These different strains can now be used to assess the role of MBP and myelin in nerve function and to assess potential therapies for treating or ameliorating demyelinating diseases.

Another model for demyelinating diseases is to use the papovavirus, JC, genome to produce transgenic mice. The expression of the early region of JC virus produces a severe progressive demyelination in the central nervous system but not in the peripheral nervous system (Small *et al.*, 1986).

Some other disease models include the analysis of glucose regulation in transgenic mice expressing human insulin genes as a model for diabetes (Selden *et al.*, 1986). Hypogonadism in *hpg* mice has been reversed by the introduction of gonadotrophin-releasing hormone gene into the germ line (Mason *et al.*, 1986). There is a thalassemia-like syndrome in mice due to a lack of the β^{maj}-globin gene. This has been ameliorated or completely corrected by introducing the cloned β^{maj}-globin gene into the germ line (Constantini *et al.*, 1986). This system will clearly provide insight into the regulation of β-globin synthesis and provide approaches to treating thalassemia. The system is also being extended to produce a model for sickle cell anaemia by introducing the mutant human sickle-cell β-globin gene. This will provide a system for testing various anti-sickling drugs.

Clearly the potential of the transgenic mouse to model human diseases is enormous and future research will increase our understanding of the bio-chemical and physiological factors in a variety of human diseases and allow a more rational basis for designing and testing new therapies.

Gene therapy

Over 2000 human diseases have already been shown to result from a defect in a single gene and some of these genes have now been cloned (Belmont and Caskey, 1986; Table 11.5). Many loci on the human genome have now been mapped using a variety of techniques (de la Chapelle, 1985; Orkin, 1986; White *et al.*, 1986; Gusella, 1986) and genes involved in such severe diseases as cystic fibrosis and muscular dystrophy are cloned (Little, 1986; Estivill *et al.*, 1987; Koenig *et al.*, 1987; Porteous and van Heyningen, 1986). It may be possible to treat patients suffering from single gene disorders by gene therapy (Anderson, 1984). The aim is to re-introduce the wild-type gene into somatic cells so that the normal gene product can be synthesised and therefore compensate for the defective gene. The most suitable recipient cells are the hematopoietic stem cells as these are a self-renewing population. They can easily be removed from the bone marrow, manipulated *in vitro* and then returned to the patient. As we have discussed in Chapter 5 these techniques have already been established in mice using retroviral vectors to ensure a high enough transfection frequency to guarantee that the rare pluripotent stem cells receive the gene. It is also possible to apply electroporation techniques which allow simple plasmid vectors to be used (Toneguzzo and Keating, 1986). The possibility of using other host cells such as liver and skin is also being studied.

The clinical success of gene therapy is anticipated by the finding that several genetic disorders can be treated by tissue transplantation (Pyeritz, 1984). This is particularly successful with bone marrow transplants for diseases where the defective gene is normally expressed in haematopoietic cells (Parkman, 1986). It is likely that gene therapy would, however, have significant advantages over conventional bone marrow transplantation because it would involve the patients own cells rather than cells from a donor. There are always problems associated with finding a suitable donor and preventing either rejection or a reaction called graft versus host disease (GVHD) where the donor T lymphocytes attack the recipients cells.

The first target disease for gene therapy is likely to be the extremely rare severe combined immunodeficiency syndrome (SCID) caused by the lack of the enzyme adenosine deaminase (ADA). The effects of this deficiency are primarily in the hematopoietic cells where the accumulation of deoxyadenosine and dATP causes the destruction of the immature T cells. ADA is normally expressed constitutively and only small amounts are required to prevent this accumulation. This means that there will be no requirement to develop vectors that ensure sophisticated regulation of the transferred gene.

Gene therapy will be far more difficult for genes whose products are required in non-lymphoid cells and/or in precisely regulated quantities. For example, another rare disease that has been suggested as an early target is Lesch–Nyhan syndrome. This results from a defect in hypoxanthine phosphoribosyl transferase (HPRT) which is manifested in the hematopoietic cells and in the central nervous system. It is unlikely that engineered bone marrow cells expressing HPRT will affect the CNS deficiencies of the disease. Other more common single gene defects, for example, the haemoglobinopathies such as sickle cell anaemia and the thalassemias (Weatherall, 1982) are not yet suitable targets for gene therapy. This is because the regulation of globin gene expression is complex, to ensure that only adult globins are synthesised in adult cells and that a 1:1 ratio between α and β-

Table 11.5. Some human diseases caused by single-gene defects

Disease	Affected cell	Defective protein
Immune deficiency	Hematopoietic	Purine nucleoside phosphorylase
Combined immune deficiency	Hematopoietic	Adenosine deaminase
		Hypoxanthine phosphoribosyl
Lesch–Nyhan	Hematopoietic and CNS	
		Carbonic anhydrase II
Osteopetrosis	Hematopoietic	Glucocerebrosidase
Gaucher's	Hematopoietic	α-galactosidase
Fabry's	Hematopoietic	α-globin; β-globin
α, β-thalassemia	Hematopoietic	Factor VIII
Hemophilia	Liver	LDL receptor
Familial hypercholesterolemia	Liver	
		Pyruvate carboxylase
Lactic acidosis	Liver	α-L-Iduronidase
Hurler's	Hematopoietic	Arginosuccinate lyase
Hyperammonemia	Liver	

globin is maintained. Any imbalance in the quality and quantity of globin synthesised from transferred genes could in itself produce disease.

All the emphasis in gene therapy is being placed on the manipulation of somatic cells but as we have discussed above, it is possible to correct genetic defects in animals by transferring genes into the germ line. It is highly unlikely that there will be any desire to attempt these experiments in man. The technical limitations are considerable because any introduced gene must not perturb the normal structure and function of the genome. All the procedures that we have discussed so far (Chapters 4 and 5) have involved a large random element. The transferred DNA integrates randomly, the number of copies may vary, the appropriate expression is not always guaranteed, transferred genes may become inactivated and transferred DNA has a high mutation frequency. The only way that germ-line gene therapy could be contemplated is if it could be guaranteed to precisely replace the defective gene with the normal gene, i.e. gene targeting in mammalian cells would have to be as reliable as gene targeting in yeast (Chapter 3). Alternatively, it may be possible to develop a non-integrating stable plasmid vector that would be maintained and expressed without disturbing the genome. This may be possible using, for example, BPV vectors. The ethical considerations of gene therapy are discussed by Walters (1986) and Chargaff (1987) and the practical considerations are discussed below.

Retrovirus vectors

Experiments with mice have shown that it is possible to introduce genes into pluripotent stem cells by using retrovirus vectors (Chapter 5). There are, however, several technical problems to be overcome before these techniques can be used in humans. The first problem is that the expression of the transferred gene can be highly variable and in some cases there is no expression. For example, the simple vector pLPL-*hprt* (Willis *et al.*, 1984) contains the human *hprt* gene inserted into MLV provirus such that it is expressed via the full-length genomic transcript and the 5' LTR promoter. Recombinant LPL-*hprt* viruses were used to infect a B-lymphoblastoid cell line that had been cultured from a patient with Lesch—Nyhan disease, i.e. the cells were *hprt*⁻. A number of *hprt*⁺ clones were isolated showing that this metabolic defect could indeed be corrected by gene transfer, but the levels of expression were low and variable, ranging between 4 to 23% of normal values. In addition, after prolonged culture, about nine months under non-selective conditions, many of the clones reverted due to suppression of the 5' LTR promoter or rearrangements and deletions of the provirus (Willis *et al.*, 1984).

Recently a range of retrovirus vectors with different configurations of transcriptional units have been analysed to try to identify the optimum vector to ensure expression in stem cells (Magli *et al.*, 1987; Stewart *et al.*, 1987). Some general conclusions are emerging from these studies: both the 5' LTR promoter and SV40 promoters are highly susceptible to suppression and should not therefore, be used for gene therapy. The highest levels of expression were obtained by using the HSV-*tk* promoter. The basis for promoter suppression in retrovirus vectors is not known and although in some cases inactivated vectors are hypermethylated (Jaenisch and Jahner, 1984) this is not consistently observed (Stewart *et al.*, 1987).

If retrovirus vectors are to be used for human stem cell transformation it is essential that there is no risk of any retrovirus-induced disease. The first requirement, that no infectious retroviruses must contaminate the defective recombinant retroviruses, can be achieved by using helper-free packaging cell lines such as PA317 (Chapter 5; Hock and Miller, 1986). It is also important that infectious retroviruses are not generated by recombination between the vector and endogenous retrovirus related elements. This seems unlikely because no such elements have been detected in human DNA. To minimise any theoretical risk retrovirus vectors are being constructed with the minimal viral sequences that are required to form an integrated provirus. This minimizes the target for any homologous recombination.

A third potential hazard is that one of the properties of retroviruses when they integrate into host genomes is that they can activate adjacent cellular genes. If they integrate near a cellular proto-oncogene this could initiate malignant transformation (Nusse, 1986). This risk must be minimised and 'handicapped' vectors are being constructed with defective LTR enhancers to reduce the potential for gene activation (Hawley *et al.*, 1987; Chang *et al.*, 1987). These have deletions in the 3'LTR which remove the viral promoter and enhancer sequence. When the virus replicates it is the 3'LTR which is copied to produce the 5' control signals of the provirus (see Chapter 4). The proviral form of these handicapped vectors, therefore, has no potential for expression. This means that it may be less likely to activate cellular genes.

To evaluate the potential for gene therapy a mouse model for human Lesch—Nyhan disease has been established (Hooper *et al.*, 1987; Kuehn *et al.*, 1987). HPRT deficient embryonal stem (ES) cells (Chapter 5) have been selected in culture by growing in the presence of the toxic purine analogues such as 8-azaguanidine (see Figure 4.5) which kill *hprt*⁺ cells. The *hprt*⁻ ES cells were then injected into blastocysts to produce germ-line ES chimaeras that produced female offspring that are heterozygous for HPRT deficiency. Male progeny of these mice are viable and totally lack HPRT activity. The long term effect of manipulating stem cells with retrovirus vectors carrying *hprt* genes can now be analysed. Some of the more recent work on developing models for gene therapy using retrovirus vectors includes the correction of the defect in glucocerebrosidase in fibroblasts from Gauscher's disease patients (Sorge *et al.*, 1987) and the expression of the human adenosine diaminase gene in diploid skin fibroblasts from an ADA deficient human (Palmer *et al.*, 1987).

Gene targeting in mammalian cells

Even if retrovirus vectors can be developed that are engineered to minimise their pathogenic potential they still integrate randomly and therefore the cosequences of gene transfer cannot be guaranteed. There is, therefore, considerable effort being directed towards developing high frequency gene targeting vectors that will allow direct gene replacement. The first suggestion that it might be possible to modify a specific chromosomal gene *in vivo* came from an experiment by Smithies *et al.* (1985). Their experiment is outlined in Figure 11.8. They constructed a test plasmid that contained part of the human β-globin locus, the G418 resistance cassette from pSV2neo (Chapter 4) and the *E. coli* amber suppressor *supF*. A unique BstXI site within the β-globin DNA is used to linearise the plasmid before transformation to enhance homologous recombination. This is similar to the approach used to target genes in *S. cerevisiae*. Correct targeting by homologous recombination with the chromosomal locus generates the modified locus shown in Figure 11.9. This contains a 7.7 kb XbaI fragment that is not found in either the test plasmid or the normal locus and is, therefore, diagnostic for the correct recombination event.

The linear test plasmid was introduced into a human cell line by electroporation. The human cell line HuTl was derived from human diploid fibroblasts fused to HAT^R mouse erythroleukaemia cells to produce a convenient easily grown and clonable cell line which expressed human β-globin. Transformed cells were selected in medium containing G418 and were grown as pools of transformed cells derived from about 1000 independent colonies. DNA was isolated from the pools and digested with XbaI. The digest was size-selected by gel electrophoresis to enrich for 5.5–8.5 kb fragments which were ligated with the double-amber-mutant, phage cloning vector, charon 3A-Xba, packaged *in vitro* and plated on to a non-suppressor *E. coli* host. Only phage carrying *supF* forms plaques. The plaques are screened by two separate filter hybridisations using a probe A that detects the human β-globin sequences in the test plasmid and a probe B that detects the adjacent sequences in the chromosomal locus. Phages that hybridise to both are picked for detailed restriction mapping to confirm that they contain the specifically modified human β-globin gene. Phages of the predicted stucture were detected at a frequency of 1 per 300 to 1100 independent clones. When the experiment was done with a human EJ bladder carcinoma

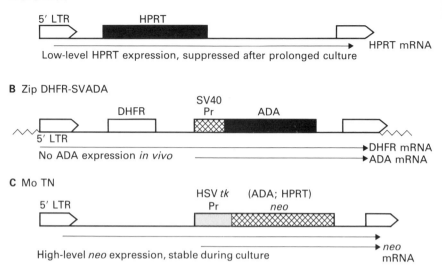

A LPL-HPRT

5′ LTR   HPRT

HPRT mRNA

Low-level HPRT expression, suppressed after prolonged culture

B Zip DHFR-SVADA

SV40
DHFR   Pr   ADA

5′ LTR

DHFR mRNA

No ADA expression *in vivo*   ADA mRNA

C Mo TN

HSV *tk*   (ADA; HPRT)
5′ LTR   Pr   *neo*

neo

High-level *neo* expression, stable during culture   mRNA

Fig. 11.8. Retrovirus vectors for human gene therapy. Pr = promoter.

cell line that does not express β-globin, targeting was only detected when G418 selection was not imposed. One explanation is that when the test plasmid is effectively targeted to an inactive β-globin locus the *neo* gene is not expressed and therefore these transformants are killed during G418 selection. The frequency of specific targeting was, however, low of the order of 10^{-2} to 10^{-3}.

Thomas *et al.* (1986) have used direct injection into the nucleus to study gene targeting. They first created a cell line containing a mutated *neo* gene that had an amber mutation in the N-terminus that did not confer G418 resistance. They then injected a plasmid carrying a second *neo* gene that was mutated in a different place resulting in deletion of 52 amino acids at the C-terminus. The transformants were selected for resistance to G418 and theoretically only these cells which had recombined the two mutant *neo* genes to produce a wild type gene would survive. About one in 1000 injected cells gave rise to G418[R] transformants. When they analysed the structure of the repaired *neo* gene they discovered that in half the cases the mutant *neo* gene had been replaced by a wild type gene and no vector sequences had been introduced. This must, therefore, involve either a double recombination event or a gene conversion. A single cross-over would generate a duplication of the *neo* locus flanking the vector sequences (see Figure 3.3). The remaining G418[R] transformants in these experiments still contained the original amber mutation but each gene now harboured a compensating mutation that allowed a functional protein to be produced. The frequency of second mutations was clearly higher than the spontaneous mutation frequency and mutants were not produced if unrelated DNA or the original amber mutant *neo* gene was used in the transfection. The interpretation is that the introduction of a homologous piece of DNA with at least one nucleotide mismatch causes high frequency mutation of the

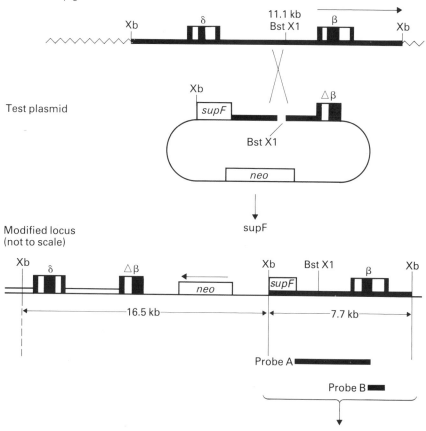

The human β globin locus

Test plasmid

Modified locus
(not to scale)

16.5 kb

7.7 kb

Probe A

Probe B

Fig. 11.9. Gene targeting in human cells. Fragments carrying *SupF* are rescued in the double amber vector Charon 3AXba. Only those derived from the correctly modified gene will contain homology to both probes A and B. If the test plasmid integrates at other loci it will only hybridize with A. The identification of a 7.7 kb Xba fragment in an amber vector that has A and B homology is proof of homologous recombination.

chromosomal target gene by heteroduplex formation and aberrant mismatch correction (Thomas and Capecchi, 1986). While this technique, if it works with normal cellular genes, will be of great value for introducing mutations into mammalian chromosomes, it acts as a cautionary note for human gene therapy. Clearly, when genes are transferred into human cells to restore a genetic defect it is undesirable to increase the probability of producing new mutations. Another cautionary note about attempting to replace defective human genes is that even though the frequency of 10^{-2} to 10^{-3} might be improved the high level of illegitimate recombination that occurs in mammalian cells will still allow additional random integration. It is also likely that the transferred gene could be mutated during transfection (see Chapter 7), further reducing the chances of successful therapy. Finally, transferred genes, e.g. β-globin, are often expressed at much lower levels than endogenous genes and there may be insufficient gene product to correct the deficiency although sequences that seem to allow normal expression of transfected globin genes have been identified (Grosveld *et al.*, 1987).

Despite the considerable technical limitations of effective human somatic cell gene therapy (Lewin, 1986; Robertson, 1986; Liskay, 1986) it is likely that this will be attempted soon for severe life threatening diseases such as ADA deficiency (Belmont and Caskey, 1986).

Recombinant DNA approaches to the AIDS problem

Acquired immune deficiency syndrome (AIDS) is currently a major world health problem. The disease is caused by a retrovirus called HIV-l (human immunodeficiency virus) which was only discovered in 1983 (Barre-Sinoussi *et al.*, 1983; Popovic *et al.*, 1984). The virus used to be called human T cell lymphotropic virus (HTLV-III) or lymphadenopathy virus (LAV). AIDS appears to be a new disease of man probably appearing in Africa in the mid-1950s. The most likely original source of infection was from monkeys (Quinn *et al.*, 1986). At the time of writing there is no vaccine against HIV infection and no drugs which inhibit the virus and allow complete recovery. The disease is, therefore, usually fatal and is spreading through the population by sexual transmission. A number of approaches that involve gene manipulation and gene transfer are being used to understand the molecular biology and pathogenesis of the virus and to produce potential vaccines and therapies.

HIV genome organisation

The genomic organisation of HIV-l is shown in Figure 11.9. The virus is clearly a retrovirus but a number of aspects of its pathogenesis (see later) and genome structure indicate that it is probably a member of the lentivirus sub-family (Sonigo *et al.*, 1985; Gonda *et al.*, 1986). The other retroviruses that we have discussed in this book belong to the oncovirus sub-family, i.e. they have oncogenic potential. HIV-l has three large coding regions in common with oncoviruses; these are *gag* (the core antigen), *pol* (replication and integration enzymes) and *env* (the external envelope glycoprotein) (see Chapter 4). In addition, there are several smaller open reading frames which encode proteins that are involved in the control of HIV gene expression (Ratner *et al.*, 1985; Wain-Hobson *et al.*, 1985; Robson and Martin, 1985; Chen, 1986). The *tat* gene has two coding exons and encodes a small, 86 amino acid protein that enhances viral gene expression. It appears to function as a *trans*-activating enhancer of transcription (hence the acronym *tat*) (Sodroski *et al.*, 1985, 1986a) but it may also affect mRNA stability, transcription termination and/or translation (Cullen, 1986; Okamoto and Wong-Staal, 1986; Kao *et al.*, 1987). Sequences within the R region of the 5' LTR of HIV called TAR are required for the viral gene expression to respond to the *tat* protein (Wright *et al.*, 1986).

The *art* (*trs*) gene is expressed via the same mRNA as the *tat* gene but using a different reading frame (Rosen *et al.*, 1986; Feinberg *et al.*, 1986). Therefore, it encodes an entirely different protein. This is 116 amino acids in length and is also involved in increasing viral gene expression and replication. The mechanism of action of this protein is not clear. It appears to antagonise repression of the translation of viral mRNA encoding *gag–pol* and *env*, hence the acronym *art* (anti-repressor of translation) (Rosen *et al.*, 1986). There is, however, also evidence that it enhances the production of *gag*, *pol* and *env* by regulating splicing to ensure the appropriate balance of mRNAs, hence the acronym *trs* (*trans*-acting regulation of splicing) (Feinberg *et al.*, 1986). The *sor* gene (short open reading frame) encodes a protein of about 22.5 kD which can be detected in HIV infected cells (Kan *et al.*, 1986). The 3' *orf* encodes a 27 kD protein. The functions of these proteins is unknown but neither of them appear to be essential for viral replication (Sodroski *et al.*, 1986b).

HIV life cycle

HIV is the most complex retrovirus analysed to date although the basic gene expression strategy is similar to that of the oncogenic retroviruses such as MLV (see Figures 4.21 and 4.22). There is a genomic length mRNA that acts as both viral RNA and mRNA for the production of the *gag* precursor protein and a *gag—pol* fusion protein. This *gag—pol* fusion is probably produced via a translational frame-shifting as described for other retroviruses (Jacks and Varmus, 1985; Jacks *et al.*, 1988). The viral *env*, *sor*, *tat*, *art* and 3' *orf* proteins are produced from sub-genomic spliced mRNAs (which have not yet been fully characterised) (Aldovini *et al.*, 1986; Feinberg *et al.*, 1986). The life cycle of HIV is outlined in Figure 11.10 and potential target stages for therapeutic drugs have been identified (Table 11.6) (Mitsuya and Broder, 1987; Klatzmann and Gluckman, 1986).

HIV replicates predominantly in a specific subset of helper T lymphocytes that carry a surface receptor that is recognised by the T4 (CD-4) monoclonal antibody. There is some evidence that B cells may also be infected and that the virus also replicates in the brain. The clinical symptoms of HIV infection reflect the tissue tropism of the virus. These are primarily a profound immunosuppression and lymphadenopathy due to destruction of helper T cells. This led to a description of the disease as Acquired Immuno-Deficiency Syndrome (AIDS). There are also B-cell lymphomas and brain lesions in many patients. Once inside the T cell, the virus replicates via the classical retrovirus strategy. The RNA genome is copied into a double-stranded DNA intermediate by the viral reverse transcriptase and the DNA either remains as a circular intermediate or becomes integrated into the T-cell genome as a provirus. The provirus is usually not expressed immediately but seems to require an activation step which is affected by a wide variety of lymphocyte-stimulating factors (Klatzmann and Gluckman, 1986; Gluckmann *et al.*, 1986). Following activation, viral gene expression is extremely efficient because it is enhanced by the viral *trans*-acting factors which amplify the production of new virions. The viral structural proteins, e.g. *gag* and *env* are post-translationally modified. The *gag* polyprotein is processed from a pre-cursor (p55) to three proteins, p24, p17 and p15 by the viral protease which is encoded by the *pol* gene. The gag proteins are myristilated and the *env* proteins are glycosylated prior to assembly into infectious virions, which occurs at the cell membrane. The viruses are released from the membrane by budding.

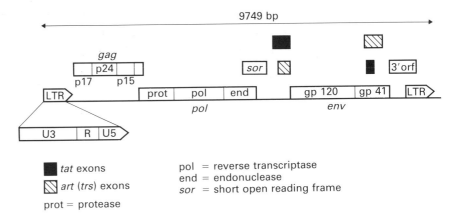

9749 bp

gag
p24
p17   p15
sor   3'orf
LTR   prot   pol   end   gp 120   gp 41   LTR
pol   *env*

U3   R   U5

■ *tat* exons

▨ *art* (*trs*) exons

prot = protease

pol = reverse transcriptase
end = endonuclease
sor = short open reading frame

Fig. 11.10. The genome organization of HIV-1.

Table 11.6. Targets for anti-HIV therapies.

Stage in life cycle	Potential means of intervention
Virus attachment to target cell	Antibodies against *env* Antibodies against T4 antigen
Virus entry	Drugs to block fusion or uncoating
Virus genome replication	Inhibit reverse transcriptase Inhibit RNase H
Provirus formation	Inhibit integrase
Virus gene expression	Antisense constructs to block *tat* and *trs* function
Virus assembly	Inhibit viral protease Inhibit cellular myristylation and glycosylation
Virus release	Inhibit with interferon

As a result of viral replication the T cells are destroyed. This is mediated by the viral *env* protein which is inherently T-cell toxic. An interaction between a T-cell receptor protein on uninfected cells and the *env* protein displayed on the surface of an infected T cell results in cell-to-cell fusion. This produces large syncitia that eventually lyse (Sodroski *et al.*, 1986c; Lifson *et al.*, 1986).

Prevention and therapy

Prevention of HIV infection will require a vaccine and a number of approaches are being taken. The first step is to identify potential immunising antigens that induce protective antibodies and cell-mediated immunity. HIV proteins have been expressed in a number of systems, for example, *env*, *tat* and *sor* have been expressed in *E. coli* (Chang *et al.*, 1985; Aldovini *et al.*, 1986; Kan *et al.*, 1986) *gag* and *pol* have been expressed in yeast (Kramer *et al.*, 1986) and *env* proteins have been expressed in vaccinia virus

410   CHAPTER 11

vectors (e.g. Chakrabarti *et al.*, 1986) and yeast (Adams *et al.*, 1987c). All of these recombinant-DNA-derived antigens are being tested for their ability to induce neutralising antibodies. The general recombinant DNA approaches to producing a vaccine are the same for most viruses and are discussed in detail in Chapter 12. One of the major problems, however, with developing a vaccine based on the *env* protein is that this shows significant variation between isolates and antibodies raised against one strain of virus do not always neutralise other strains (e.g. Wiley *et al.*, 1986).

The rationale behind developing therapeutic agents to control or cure AIDS is to identify a stage in the virus life cycle that can be inhibited without serious side effects on normal cell metabolism. Certain aspects of the virus life cycle are unique to the virus and are, therefore, good candidates for therapeutic intervention (Table 11.6). One approach is to construct model systems. For example, Kramer *et al.* (1986) have expressed part of the HIV genome in *S. cerevisiae*. They showed that the viral *gag* and *pol* genes were expressed and that the protease was functional and cleaved *gag* into the mature proteins. The viral protease is essential for the production of infectious virus and the yeast system can be used to test various anti-protease drugs. The reverse transcriptase also functions in yeast and rapid inhibition assays can be developed with this system (Barr *et al.*, 1987).

Another use of gene transfer technology is to study the factors that control gene expression. It is likely that if the activities of *tat* and *art* could be inhibited then virus replication would be severely or completely inhibited. For example, to analyse the *tat* protein Wright *et al.* (1986) constructed a mouse cell line that expressed the *tat* protein by using a BPV-based expression vector (Figure 11.11). The *tat* protein was expressed using the mouse metallothionein promoter and stable cell lines expressing *tat* were selected on the basis of resistance to cadmium conferred by a human metallothionein gene. The cell line was then transfected with second plasmids that had the HIV LTR or portions of the LTR fused to the bacterial CAT reporter gene. The expression of CAT was 40 to 300 fold higher in the *tat*-expressing cell line than a control cell line that contained only the BPV vector, showing that there was *trans*-activation. Transcription directed by the LTR requires sequences upstream of -91 and *trans*-activation is abolished if sequences between $+38$ and $+83$ are deleted. These are in the 5' untranslated region of the mRNA and the formation of a hairpin loop may influence expression (Muesing *et al.*, 1987). It is hoped that an increased understanding of the protein — nucleic acid interactions that mediate transcription and *trans*-activation will lead to the design of new anti-viral drugs.

Another idea is to use '*antisense*' RNA to block viral replication. A number of studies have shown that gene expression can be blocked by introducing an RNA into a cell which is of the opposite polarity to the gene to be blocked. It is thought that the formation of specific double-stranded duplexes inhibits translation (e.g. Izant and Weintraub, 1984). These anti-sense nucleic acids could be introduced into the cells by constructing a non-pathogenic T-cell tropic virus vector that would effectively interfere with HIV replication (Haase, 1986).

Although there is as yet no vaccine or cure one fact is certain: without

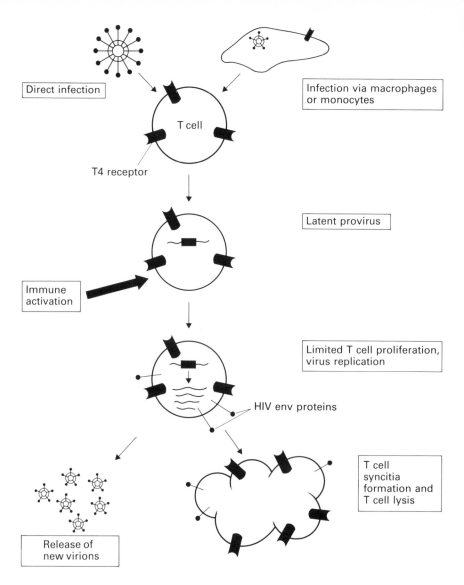

Fig. 11.11. The life cycle of HIV-1.

the application of recombinant DNA and gene transfer techniques our knowledge of the life cycle and pathogenesis of HIV would be severely limited. The prospects for vaccination and therapy will now increase with the unravelling of the molecular biology of HIV infection.

Summary

Recombinant DNA and gene transfer techniques have provided the means to test new theories about the causes and cures of human diseases without using patients as the experimental system. The pace of discovery will therefore accelerate and a wider range of new drugs can be tested in rational animal models. One of the most dramatic impacts of these new technologies

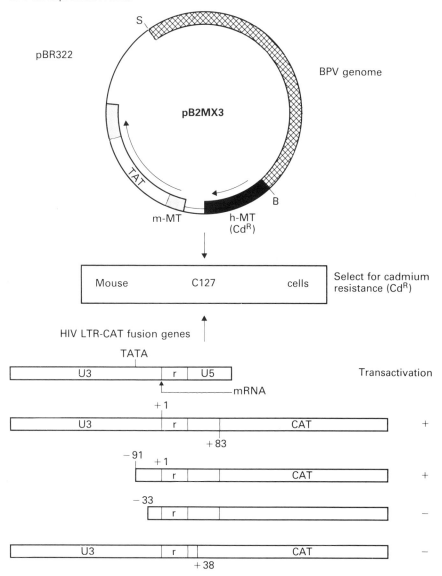

Fig. 11.12. Understanding the regulation of HIV gene expression.

is in unlocking the black box marked 'cancer' to reveal a potentially logical scheme of genetic events that turn a normal cell into a tumour cell via the activation of oncogenes. It is anticipated that understanding the biochemical functions of normal and activated oncogenes will result in the production of prognostic, diagnostic and therapeutic agents. The idea that diseases caused by single-gene defects might be curable can now be entertained because of the development of specialised retrovirus vectors for gene therapy. Finally, when a new disease such as AIDS appears, recombinant DNA technologies allow rapid progress towards understanding the disease and provide hope for prevention and cure.

12 Biotechnology

Introduction

The traditional definition of biotechnology is that it is the exploitation of the biochemical potential of microorganisms for medical, agricultural and industrial purposes. Brewing, bread making, cheese production, penicillin synthesis and sewage treatment are all examples but, the ability to transfer genes back into living organisms has dramatically extended the scope of biotechnology. This new gene transfer technology means that genes can be rapidly altered and they can be transferred between species, opening up completely new possibilities for increasing biochemical versatility. Biotechnology now involves plants, animal cells and animals in addition to microorganisms.

One major impact of the new technology has been the ability to convert cells into 'factories' to synthesise compounds that were previously available only in limited quantities. Examples of such compounds are peptide hormones, antiviral and anti-tumour proteins and growth factors. The technology also provides new routes to the achievement of traditional goals, for example in the production of antigens as vaccines. The ability rapidly to manipulate the genotype also opens up possibilities for extending the traditional capabilities of an organism. For example, yeasts can be modified to produce low-carbohydrate beer and the yield of crops can be increased by producing herbicide and pest-resistant strains. Biotechnology can also be used to solve environmental problems, such as the disposal of waste products, and generation of alternative energy sources by the production of fuel alcohol.

In this chapter we will not cover all the goals and achievements of biotechnology but we will pick selected examples to illustrate the impact of gene manipulation and gene transfer.

Table 12.1 lists some of the applications of biotechnology.

The pharmaceutical industry

A number of human proteins that are naturally involved in determining resistance to disease or in maintaining correct physiological balances can now be used as specific therapeutic agents. This is because the genes can be cloned and expressed to produce sufficient quantities of protein for clinical

Table 12.1. Some applications of biotechnology

Pharmaceutical industry
Vaccines
Diagnostic reagents
Anti-tumour agents
New drugs

Agricultural industry
Disease-resistant plants and animals
Improved feed conversion in animals
Increased nutritional value of animals and plants
Pest-resistant plants

Food industry
New food ingredients, e.g. sweeteners
New processes, e.g. in cheese production
Single-cell protein

Waste treatment
Degradation of milk whey
Enhanced recovery of minerals from waste

Chemical industry
Alternative feedstocks, e.g. ethanol from cellulose
New enzymes, e.g. for organic solvent environments

testing and use. Before gene manipulation techniques were available it was not possible to obtain many of these proteins in significant quantities and in fact in some cases, such as peptide hormones, the proteins are produced in such minute quantities that they could only be studied when the genes were cloned. The proteins that are receiving the most attention are lymphokines, growth factors, peptide hormones and 'blood proteins'. 'Second generation' drugs are also being developed using protein engineering to modify active sites, to build compound drugs or to increase pharmacological effectiveness, for example by increasing stability. Another important application of gene manipulation technology in the phamaceutical industry is in the production of new vaccines. Expensive, sometimes dangerous, processes can be replaced by methods which are safer and which often produce more effective vaccines. In addition there is now the possibility of producing vaccines for diseases such as AIDS for which no effective vaccine exists at present.

The production of proteins for use in humans poses the most stringent test for gene manipulation technologies. The process must be economically feasible and the product must pass strict controls. For example, before a product can be marketed for human use in the USA it must be manufactured and tested under stringent Food and Drug Administration (FDA) regulations that have been established to cover human drugs. The FDA must then be convinced that the new product is both effective and safe. This means that there must be no dangerous side-effects. If for example a natural human protein is to be used as a drug it must not be immunogenic as this might result in the destruction of natural processes as well as making the drug ineffective after repeated administration. The recombinant product must not be contaminated with anything dangerous, such as pathogenic viruses or tumourigenic nucleic acid. In addition the production process must be absolutely predictable.

Vaccines

There are still many life-threatening infectious diseases of man, such as hepatitis B, malaria, Burkitts lymphoma, rabies and AIDS. There are also many diseases which although not fatal, have dramatic social or economic importance, for instance genital herpes virus and influenza virus infections. Furthermore, there are many important infectious diseases of farm and domestic animals such as foot and mouth disease, and vesicular stomatitis. There is no effective cure for these diseases and the best route to their control is vaccination to prevent the initial infection. One of the major impacts of gene manipulation technology is in the production of vaccines against some of these diseases. The first commercial production of a recombinant DNA product from eukaryotic cells was a vaccine against human hepatitis B virus produced in yeast.

The design of vaccines

The proteins on the surface of an infectious agent, for example the envelope proteins of a virus, are recognised as antigens by the immune system and specific antibodies are produced during an infection. These antibodies neutralise the virus and prevent further infection of susceptible cells. The appearance of these *neutralising antibodies* often correlates with recovery from the disease but in some infections the primary response occurs too late to prevent death or serious disease. Vaccination involves eliciting the production of neutralising antibodies so that on infection a rapid secondary response will limit the extent of the disease at a very early stage and so prevent overt symptoms.

There are three classical types of vaccines. These are: (1) *live vaccines* — live infectious agents which have been *attenuated* so that upon infection they elicit an antibody response but they do not cause disease; (2) killed vaccines — infectious agents, usually the virulent disease-causing organism, that have been inactivated by extreme treatments such as exposure to heat, low pH or formalin; and (3) *subunit vaccines* — pure preparations of antigen(s) derived from the virulent organism that elicits a neutralising antibody response. There are advantages and disadvantages with each type of vaccine. Live vaccines are usually effective because the organism replicates, providing high, local concentrations of antigen over a prolonged period, which maximally stimulates the immune response. Live vaccines also stimulate cell-mediated immunity. The disadvantage is that there could be a reverse mutation to virulence making the vaccine strain itself dangerous. In many situations there may be no option for a live vaccine as the organisms cannot be grown or because they cannot be attenuated.

Killed and subunit vaccines cannot produce disease but each batch must be extensively tested (innocuity testing) to ensure no contamination with live organisms. Very large amounts of killed and subunit vaccines must be administered to ensure a high enough antigenic mass to elicit an antibody response and several injections are, therefore, usually required. This means the preparations must be highly purified because such large doses of any contaminating products might be toxic. Subunit and killed vaccines are usually very expensive and in addition may be less immunogenic than live vaccines because the major antigens have been altered during preparation. They also involve handling large quantities of live infectious agent in the manufacturing process (reviewed by Brown, 1984).

The application of recombinant DNA techniques can solve many of the technical limitations of conventional vaccine production as well as opening up possibilities for vaccines for diseases where no alternative currently exists and for producing novel types of vaccines. In this section we will discuss recombinant subunit vaccines, live recombinant virus vaccines and novel 'epitope-carrier' vaccines.

Recombinant subunit vaccines

The general approach to preparing recombinant subunit vaccines is to clone the gene which specifies the neutralising antigen and then express the gene in an appropriate host-vector system. The requirements for the host-vector system are that the protein is still antigenic and that the yields are high enough to generate a commercially viable process. The antigen must still be purified but no innocuity testing will be required as there is no possibility of contamination with live agent. No infectious agents need to be handled in the manufacturing process and the degree of purification may be low if the host cell itself is considered non-toxic. The following sections describe the successful production of a subunit vaccine against hepatitis B virus (HBV) and the continuing efforts to produce a subunit vaccine against influenza.

A RECOMBINANT SUBUNIT VACCINE AGAINST HBV

At least 2 million people worldwide are infected with HBV. In Far East Asia and tropical Africa at least 10% of the population are affected and in North America there are 200 000 new cases each year. The consequences of infection range from an asymptomatic carrier state through to acute and chronic hepatitis and liver cirrhosis which are often fatal. In addition, infection with HBV is highly correlated with the subsequent development of a specific liver cancer (hepatic cellular carcinoma, HCC) which accounts for 300 000 deaths each year. The virus is highly host-specific, it only infects man and chimpanzee and it cannot be cultured in tissue culture cells *in vitro*. This means that the only significant source of virus as a vaccine is from infected people.

During HBV infection virus particles are released into the blood. When the virus is actively replicating in the liver whole virions called *Dane particles* are found (Figure 12.1). These are about 42 nm in diameter and consist of a lipoprotein envelope surrounding a virus capsid containing the viral DNA genome. The envelope contains the major viral antigen called the *hepatitis B surface antigen, HBsAg*, in addition to host-cell-derived lipid and carbohydrate. The capsid is a single viral protein HBcAg and may also play some role in determining immunity. In asymptomatic patients the virus is present but there is no disease. These individuals are called carriers.

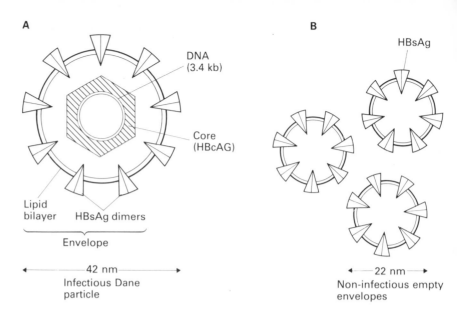

A — DNA (3.4 kb)

Core (HBcAG)

Lipid bilayer   HBsAg dimers

Envelope

← 42 nm →
Infectious Dane particle

B — HBsAg

← 22 nm →
Non-infectious empty envelopes

Fig. 12.1. Hepatitis B virus particles in human serum.

Very few virions are found in their serum but there are large numbers of smaller, 22 nm, particles which are empty envelopes carrying HBsAg. These particles are non-infectious but immunogenic, they can be purified from the plasma of healthy chronic carriers and form an effective and safe vaccine. Why then is there a need for another approach to vaccination against HBV? The reasons are: (1) human serum is not always readily available and it is costly to collect and (2) HBsAg is a *conformational antigen*, i.e. the protein is only fully immunogenic when it is assembled as dimers into the particles. This means that the preparations cannot be freed from contaminating infectious particles by any extreme denaturing treatments because this would destroy the immunising antigen. Therefore, purification procedures based on size fractionation are required and all preparations must be extensively tested in chimpanzees. The collecting, purification and testing of the vaccine is therefore extremely costly and at about 100 US$ per treatment this is too expensive for mass vaccination in Third World countries where it is most needed. In addition, despite the purity of the plasma-derived vaccine there is increasing concern about using products derived from human blood because of fear of contamination with the AIDS virus. There was therefore a strong impetus for developing an alternative vaccine. The breakthrough began in 1978–79 when a number of research groups cloned and sequenced the viral genome allowing the coding regions for the viral antigens to be characterised (reviewed by Tiollais *et al.*, 1985; Garem and Varmus, 1987).

The organisation of the S region coding for HBsAg is shown in Figure 12.2. The surface antigens exist in glycosylated (gp) and non-glycosylated (p) forms. The virion contains 300–400 molecules of the major protein which carries the group (a) and subtype (d, r, y, w) antigens and is encoded by the S gene. Two other proteins, the large and middle proteins, are found

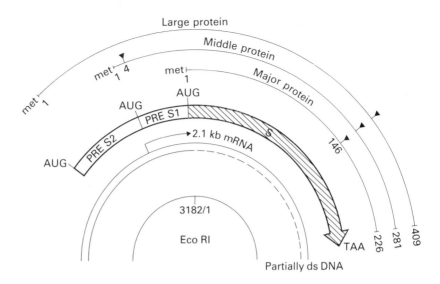

Major protein: 226 aa; GP27 and P24
Middle protein: 281 aa; GP33 and GP36
Large protein: 409 aa; GP 42 and P32.

Fig. 12.2. The S region of HBV.

in smaller amounts. These have additional sequences at the N terminus referred to as preS2 and preS1 respectively. The preS1 region carries a highly immunogenic antigenic determinant (an *epitope*) and a receptor for polymerised human serum albumin (pHSA). This mediates the attachment of HBV to hepatocytes and antibodies to this epitope appear to be important in determining recovery from infection.

The S region was initially expressed in an *E. coli* host but although some protein was made, the yields were extremely low and the protein did not assemble into immunogenic particles (e.g. Pasek *et al.*, 1979). Attention was therefore turned to producing the HBsAg in yeast and animal cells. To produce HBsAg in yeast, Valenzuela *et al.* (1982) initially used an 835 bp fragment of the HBV genome that spanned the S gene but lacked the preS2 and most of the preS1 region. This was inserted into a 2 μm based *ADH* promoter expression vector (Chapter 3) and HBsAg was expressed in yeast at about 25 μg per litre of culture. When yeast extracts were examined by electron microscopy, particles could be seen that resembled the 22 nm particles in human serum (Figure 12.1). The particles were slightly smaller and more heterogeneous and contained lipid but no carbohydrate. The HBsAg is synthesised in yeast as a 23 kD unglycosylated monomer which is insoluble, but after mechanical extraction the monomers and membrane fragments aggregate to produce soluble particles (Hitzeman *et al.*, 1983). Using new high-efficiency expression vectors, for example driving expression with the *GAPDH* promoter, yields of at least 2.5 mg l^{-1} have been obtained (Valuenzuela *et al.*, 1985a) and the particles are purified by an immuno-affinity procedure (McAleer *et al.*, 1984). Yeast-derived particle preparations were used to vaccinate chimpanzees and provided complete protection against subsequent infection with live hepatitis B virus (Murray *et al.*, 1984; McAleer *et al.*, 1984). A new vaccine had been produced! The yeast expression system has now been scaled up to produce the first generation of recombinant hepatitis vaccines which have been approved for human use. The second generation is already in sight. There is increasing evidence that the preS1 epitope is required for full immunity (Milich *et al.*, 1985) and yeast strains expressing a preS1−S gene have now been produced (Valenzuela *et al.*, 1985a).

HBsAg particles have also been made in animal cells. One of the most promising systems uses a *dhfr* amplicon vector in *dhfr⁻* CHO cells (Chapter 4). The expression of the S gene is driven by the SV40 early promoter and methotrexate-resistant cells expressing 1.2 μg 10^{-6} cells can be selected (Patzer *et al.*, 1986). Animal cell production may have some advantages over the yeast system. The major one is that the particles are secreted into the culture medium; this means that purification is simplified and that the production process is continuous because the stable cell line secretes HBsAg for days. It also appears that the animal cell particles are more immunogenic than the yeast particles (Patzer *et al.*, 1986). This may reflect the fact that they are glycosylated or that purification procedures that do not involve cell disruption preserve the antigenicity. The preS1−S particles have also been produced in animal cells (Michel *et al.*, 1984) and animal cell derived particle vaccines are currently in clinical trials.

RECOMBINANT SUBUNIT VACCINES AGAINST OTHER VIRUSES

The approaches described for producing HBsAg can be adapted for the production of any viral antigen. A significant effort is currently being put into developing a subunit vaccine against influenza virus type A. This causes the familiar respiratory disease 'flu' which can range in severity from an upleasant upper respiratory tract infection to a severe, sometimes fatal, infection involving the whole respiratory tract and lungs and occasionally the central nervous system. The virus has an extremely high attack rate such that the cost to society in terms of health care and working days lost can be enormous. In addition variant viruses arise periodically that can cause extremely severe disease with a very high mortality, particularly amongst the very young and very old. Preparing an effective vaccine against flu presents a real challenge and, although initial attempts at immunising mice took place in 1935, there is still no really effective vaccine. Full immunity requires an antibody and cell-mediated response. Therefore a live vaccine is preferable but there is no reliable marker for attenuation so such a vaccine cannot be developed. Killed whole vaccines elicit severe reactions particularly in children and subunit vaccines have reduced immunogenicity due to alterations to the proteins. Superimposed on these practical problems is a major problem arising from the extreme genetic variability of the virus. There are two surface proteins, the haemagglutinin (HA) and neuraminidase (N). Antibodies to the HA seem to be most important in immunity but protection is significantly improved if antibodies against the neuraminidase are also produced. So far 13 different HA genes have been identified, each encoding antigenically distinct proteins called subtypes. Antibodies against one HA subtype will not protect against another. In addition there are nine subtypes of neuraminidase and viruses can occur with any combination of HA and N subtypes. A particular subtype of virus may circulate for several years and suddenly be replaced by a completely new subtype. This is called *antigenic shift*. There will be no immunity in the population against this new subtype and vaccines prepared the year before will be useless. In addition the viral antigens can change gradually by point mutations within the HA gene to produce an antigen which is less effectively neutralised. This is called *antigenic drift*. Antigenic variation is outlined in Figure 12.3 and a full discussion can be found in Palese and Kingsbury (1983), Stuart-Harris *et al.* (1985) and Wiley and Skehel (1987).

A number of different host–vector systems have been used to express influenza virus surface antigens. The structure of the HA protein has been discussed in Chapter 10. It is an integral membrane protein with an N-terminal signal sequence and a hydrophobic C-terminal membrane anchor domain. Gething and Sambrook (1982) inserted a modified HA gene lacking the C-terminal coding region into an SV40 late region replacement vector (Chapter 4). This transient expression system produced 10^9 molecules of HA per cell during a 60 hr lytic infection. The protein was secreted into the culture medium and assembled into trimers. To achieve these yields in a permanent cell line a high-copy number BPV_{69T} vector (Chapter 4) had to be used (Sambrook *et al.*, 1985) and the cell line and promoter had to be optimised. The mouse metallothionein promoter was 50-fold more efficient than the SV40 early promoter and there was a several orders of magnitude difference in yields between different cell lines. The highest yields were 10^7 molecules per cell per 24 hr l^{-1}. Influenza virus neuraminidase has also been expressed in monkey cells using SV40 viral vectors (Davis *et al.*, 1983). Complete HA and HA minus secretory signal polypeptides have also been expressed in yeast (Jabbar *et al.*, 1985) and insect cells (Kuroda *et al.*, 1986). Insect cells lack a 'flu HA receptor and because HA molecules will not therefore be re-absorbed from the culture medium after secretion they may have advantages for large-scale purification.

Live recombinant virus vaccines

The most impressive example of vaccination with a live non-recombinant virus has been the use of vaccinia virus to protect against smallpox. Following a world-wide campaign, the World Health Organisation (WHO) declared on 26 October 1979 that smallpox had been completely eradicated from the world and that further vaccination was unnecessary (reviewed by Behbehani, 1983). The potential of vaccinia virus is now being extended by using recombinant DNA techniques to produce hybrid vaccinia viruses that express the genes of unrelated pathogens. The general approach is to insert the coding region for the neutralising antigen of a pathogen into the vaccinia virus genome to produce a recombinant virus that still replicates in human cells but in addition to expressing vaccinia antigens also expresses the additional antigen. This system has all the advantages of a live vaccine

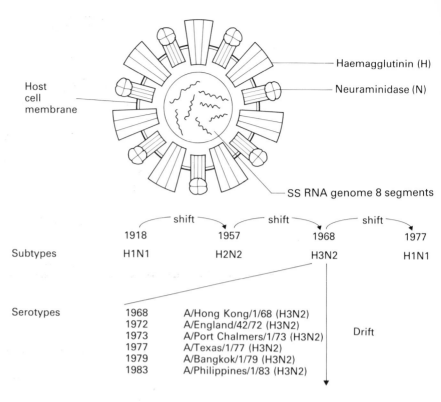

Fig. 12.3. Antigenic variation in influenza virus type A.

without any of the dangers of using the pathogen itself (Mackett *et al.*, 1982; Panicalli and Paoletti, 1982).

The vaccinia virus genome is large, 180 kb, and non-infectious because unlike other DNA viruses it replicates entirely in the cytoplasm and therefore encodes its own replication and transcription enzymes, which must be carried into the cell in the virus particle. The genome is therefore not suitable as a direct cloning vector and foreign genes are inserted in a two-stage process. The foreign genes are inserted into a plasmid vector called a *recombination vector* and then inserted into the virus genome via homologous recombination *in vivo* (this is similar to procedures described for manipulating Ti plasmids and baculoviruses, Chapters 6 and 4). A general procedure has been described by Mackett *et al.* (1984) and is outlined in Figure 12.4. Vaccinia virus has a thymidine kinase (*tk*) gene which is not essential for infectivity. The *tk* gene was sub-cloned on a Hind III fragment into pBR328 (a relative of pBR322) and then disrupted by the insertion of a 275 bp fragment carrying the promoter and transcription start site from a second vaccinia gene that codes for an early 7.5 kD protein. This promoter (p7.5) is efficient and constitutive and is used to drive the expression of any foreign coding sequence inserted at the unique BamHI or SmaI sites. This plasmid, pGS20, is introduced into *tk*⁻ host cells such as human 143 cells by standard CaPO₄ transfection. At the same time the cells are also infected with wild-type *tk*⁺ vaccinia virus. Homologous recombination occurs in the

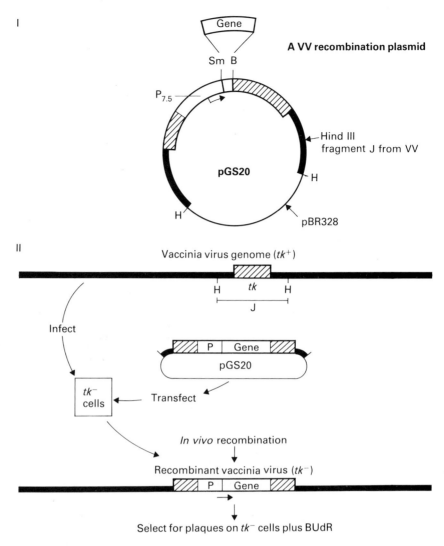

I

Gene

Sm B

A VV recombination plasmid

P$_{7.5}$

pGS20

Hind III
fragment J from VV

H

H

pBR328

II

Vaccinia virus genome (tk^+)

H tk H

J

Infect

tk^-
cells

Transfect

P Gene

pGS20

In vivo recombination

Recombinant vaccinia virus (tk^-)

P Gene

Select for plaques on tk^- cells plus BUdR

Fig. 12.4. A general method for constructing recombinant vaccinia viruses (VV). The vaccinia genome is the thick dark line; the *tk* gene is hatched.

cell between pGS20 and the wild type viral genome across the *tk* region resulting in a recombinant virus that now has a defective *tk* gene and contains the foreign gene. At least 25 kb of foreign DNA can be inserted into the viral genome (Smith and Moss, 1983). The virus plaques are harvested and the lysate is used to infect fresh *tk*$^-$ cells but this time in the presence of BUdR (see Chapter 4). Only *tk*$^-$ virus will replicate and so recombinant viruses are selected and can be propagated to produce high titre (8×10^8 p f.u. ml^{-1}) stocks. The foreign gene is expressed transiently in these cells with peak production at about 20 hr p.i. before lysis occurs. The system can therefore also be used for protein production *in vitro*. The virus stock is used to vaccinate animals with the foreign antigen. Potential vaccines against a number of important viral diseases are being developed using the vaccinia virus system. These include rabies, which is a severe and usually fatal diseases of the nervous system (Kireny *et al.*, 1984; Blancou *et al.* 1986); malaria, this is one of the most severe global health problems and is caused by protozoa of the *Plasmodium* spp (Ozaki *et al.*, 1983; Smith *et al.* 1984); herpes type I (cold sore) and type II (genital herpes) infections (Cremer *et al.*, 1985) influenza infections (Panicalli *et al.*, 1983); hepatitis E (Moss *et al.*, 1984), Epstein–Barr virus, the causative agent of Burkitt's lymphoma (Mackett, 1985); vesicular stomatitis virus (VSV) which causes an economically damaging contagious disease in cattle, horses and pigs (Mackett *et al.*, 1985); and the envelope protein of HTLVIII/LAV, the AIDS virus (Chakrabarti *et al.*, 1986; Hu *et al.*, 1986). It is also possible to produce multivalent recombinant vaccinia viruses, for example, the HSV-1gD, HBsAg and influenza HA genes have been inserted into a single vaccinia genome. Rabbits innoculated with this recombinant produced antibodies that were reactive to all three antigens (Perkus *et al.*, 1985). To date there is a 100% success rate with the expression of foreign antigens in vaccinia virus vectors although significant immunity in laboratory animals has not been demonstrated in all cases (reviewed by Moss and Flexner, 1987; Tartaglia and Paoletti, 1988).

General considerations about live recombinant virus vaccines

Although a significant proportion of the human population has been vaccinated with vaccinia in the smallpox eradication programme it is not thought that this will significantly affect the potency of the recombinant vaccines. It seems that even limited replication at the vaccination site will deliver sufficient antigen to provide immunity. The next generation of vaccinia virus vectors will drive the foreign antigen gene expression with more powerful promoters to ensure a high level of antigenic stimulation even after repeated vaccinations, which would be required for example to protect against influenza epidemics each year.

There does not seem to be any technical barrier to the production of live recombinant vaccinia viruses expressing the surface proteins of any pathogen. This is, therefore, at present a most attractive route to effective vaccination in the Third World because once the recombinant virus is produced no sophisticated techniques are needed to prepare the vaccine. It would be cheap and easy to distribute and administer and a single vaccination would provide life-long immunity. There are however, some serious concerns about the advisability of using these viruses. There is the general concern that affects all recombinant organisms and that is, is it ethical to release them into the environment? World Health Organisation is also concerned that after the massive effort to eradicate smallpox it would be unwise to distribute a closely related organism throughout the population in case a mutation to increased virulence could produce a new smallpox-like disease. The recombinant vaccines appear to be less virulent than vaccinia virus in laboratory animals but as the factors that determine the virulence of pathogens are largely unknown this objection cannot be ignored. There is also a concern that the introduced DNA could broaden the host range of vaccinia virus even further leading to new and possibly serious diseases if the virus enters cells or hosts from which it was previously barred and/or shows increased toxicity to these cells. The advantages and disadvantages of live vaccinia virus vectors are discussed in more detail by Brown *et al.* (1986) and Keus (1986).

The approaches discussed above can be applied to any virus that can be genetically manipulated and that can be attenuated to act as an innocuous carrier. For example human adenoviruses have been manipulated to express HBsAg (Davis *et al.*, 1985; Ballay *et al.*, 1985) and this may prove more acceptable than vaccinia virus. The broad host range of adenoviruses makes them particularly useful for the production of veterinary vaccines.

Recombinant DNA techniques might also be used to produce conventional attenuated live vaccines. This would be done by replacing a virulence gene(s) with an inactive gene. The limitation of this approach is that much more research is needed to identify the virulence genes in most viruses (Chanock and Lerner, 1984).

Self-assembling peptide-carrier vaccines

Effective vaccination using a subunit or antigenic peptide requires the correct presentation of the antigen to the immune system. Ideally the antigens should be exposed in multiple copies on the surface of a high-molecular-weight complex so that they are available to maximally stimulate an immune response. They should also be in a conformation that approximates to the native conformation. One solution to this problem has been provided by *antigen engineering*. The antigen is embedded into a carrier protein that is itself capable of assembling into a stable multimeric complex. This is done by constructing a hybrid gene between the coding region for the carrier protein and the coding region for an antigenic determinant and then expressing this gene in a suitable host cell. The HBsAg has been tested as a carrier protein because it readily assembles into the 22 nm particle. The idea is to modify the HBsAg gene so that other antigens are exposed on the surface of the particle. Valenzuela *et al.* (1985b) inserted a 900 nucleotide fragment of the gene encoding herpes simplex virus IgD protein into the preS1 gene of HBV to preserve the correct reading frame and the hybrid gene was inserted into a high-efficiency yeast expression vector. Particles were produced by yeast and they reacted with both anti-HBsAg and anti-HSV1gD antibodies indicating that a hybrid antigen had been produced but these particles were not tested for their abilities to induce neutralising antibody to HSV1. A similar approach has been used to present the poliovirus neutralising epitope in HBsAg particles produced in mammalian cells (Delpeyroux *et al.*, 1986). In this study however, the hybrid particles were much less effective as immunogens than intact poliovirions. The antibody response to HBsAg can interfere with the response to a second antigen. The HBsAg is said to exert *immunodominance* which may explain the low immunogenicity of the polio-HBsAg particles.

A new epitope carrier system has been developed based on the virus-like particles that are produced by the yeast retrotransposon Ty (Kingsman *et al.*, 1987b; Adams *et al.*, 1987a, b; Chapter 8). The coding region for the

major Ty particle protein pl was fused to a region of a human immunodeficiency virus (HIV) gene for the envelope glycoprotein (gp120) and the
hybrid construction was expressed in the efficient vector pMA91 (Figure
12.5). The hybrid p1-HIV gp120 protein assembled into particles which
were morphologically distinct from normal Ty virus-like particles. These
hybrid particles reacted with anti-HIV antiserum and induced a significant
antibody response to HIV in rabbits. Ty-VLPs are produced in very high
yields filling the yeast cell completely (Mellor *et al.*, 1985c). They therefore
provide a very simple and economical carrier for potentially any epitope.

Clearly, the technology is now available to produce a new generation of
subunit and live vaccines against many of the major life-threatening or
distressing diseases of man and animals. The key factors in the use of these
vaccines will now be economical and ethical. Is it economical for companies
to market cheap vaccines for the Third World? Is it ethical to distribute
recombinant organisms into the environment? Can world-wide eradication
policies be coordinated as with the smallpox campaign?

Lymphokines

When lymphocytes are stimulated with antigen a number of proteins are
secreted which augment the host defensive response against infectious agents
or endogenous tumours. These compounds are called lymphokines and
some of them are listed in Table 12.2. One of the first targets of biotechnology
was to clone and produce large quantities of these natural defence proteins
in the expectation that they could be used to treat viral infections and
cancers. Human interferon-alpha was the first lymphokine to be expressed
in heterologous cells (reviewed by Kingsman and Kingsman, 1983; see
articles in Pestka, 1986). Interferon was first described by Isaacs and
Lindenmann (1957) as a substance that interfered with viral replication and
it was subsequently shown to inhibit the growth of tumour cells *in vitro*.
Interferon could almost be considered as the protein that launched the new
biotechnology industry with promise of a 'cure-all' that convinced the stock
markets of the world to invest in the new technology. In fact, although IFN
alone may not be a broad spectrum anti-cancer agent, lymphokine research

Table 12.2. Lymphokines.

Interferons	(IFN)
Alpha	(leucocyte)
Beta	(fibroblast)
Gamma	(immune)
Interleukin 1 (IL-1, T-cell activating factor)	
Interleukin 2 (IL-2, T-cell growth factor, TCGF)	
Interleukin 3 (IL-3, haematopoietic growth factor)	
Interleukin 4 (IL-4, a B-cell growth factor)	
Granulocyte colony stimulating factors (G-CSF)	
Granulocyte−macrophage colony stimulating factor (GM-CSF)	
Macrophage activating factor (MAF, identical to IFN-gamma)	
T-cell replacing factor (TRF)	
Migration inhibition factor (MIF)	
Tumour necrosis factor (TNF)	

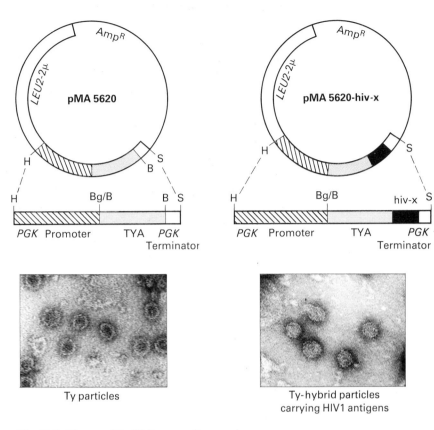

A Ty-VLP vector

B Hybrid Ty-VLP vector

Ty particles

Ty-hybrid particles
carrying HIV1 antigens

Fig. 12.5. The use of Ty-VLPs as an epitope carrier.

has proved to be even more exciting than predicted because a whole range of different compounds have now been discovered.

Interferon is, in fact, a complex mixture of three main classes of protein, interferon-alpha (IFN$-\alpha$), interferon-beta (IFN$-\beta$) and interferon-gamma (IFN$-\gamma$). The interferons have both anti-viral and anti-proliferative activity. Interferon-alpha has been approved for the treatment of some forms of leukaemia (review by Pestka *et al.*, 1987). Recently interleukin-2 (IL-2) has been shown to have significant anti-tumour activity in clinical trials because it stimulates the proliferation of T cells that specifically destroy certain tumour cells (reviewed by Rosenberg and Lotze, 1986). Another promising lymphokine is tumour necrosis factor (TNF) (reviewed by Old, 1985). This is secreted by antigen-stimulated macrophages and causes the necrosis of tumour cells in culture but not normal cells. Its activity is augmented by the interferons, and recombinant TNF is now in clinical trials. The interactions between these lymphokines and their main target cells, the B and T cells of the immune system, is shown in Figure 12.6 and reviewed by Old (1987).

Many of the lymphokines can be successfully produced in *E. coli* for

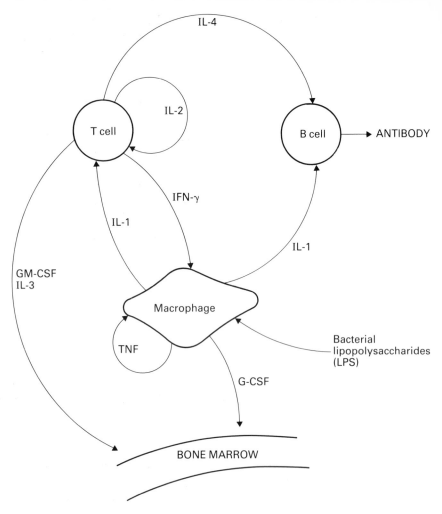

Fig. 12.6. Interactions between some lymphokines and cells of the immune system (adapted from Old, 1987).

Table 12.3. The production of some lymphokines in heterologous eukaryotic cells.

Lymphokine	Host	Vector systems	
IFN-α	Yeast	*ADH*; ARS	(1)
	Silkworm larvae	AcNPV	(2)
IFN-β	Insect cells	AcNPV	(3)
	Mouse C127 fibroblasts	BPV	(4)
IFN-γ	Yeast	*PGK*; 2 μm	(5)
	Hamster (*dhfr⁻*)	*dhfr* amplicon	(6)
	CHO		
IL-2	Yeast	αF secretion	(7)
	Insect cells	AcNPV	(8)

References: (1) Hitzeman *et al.*, 1981; (2) Maeda *et al.*, 1985; (3) Smith *et al.* (1983b); (4) Mitrani-Rosenbaum *et al.*, 1983; (5) Derynck *et al.*, 1983; (6) Haynes and Weissman, 1983; (7) Brake *et al.*, 1984; (8) Smith *et al.*, 1985.

example interferon-alpha (e.g. Remaut *et al.*, 1986) and interleukin−2 (Rosenberg *et al.*, 1984b) and therefore 'gene manipulation in eukaryotes' has not been necessary. Several of the lymphokines are however produced more readily in eukaryotic cells. In addition synthesis in eukaryotic cells may be particularly important for those which are glycosylated, e.g. IFN−γ, IFN−β and IL−2. In addition, eukaryotic secretion systems can be used to produce authentically processed proteins. A summary of some of the eukaryotic cell production systems for lymphokines is given in Table 12.3.

Although some promising results have been obtained with lymphokines in human clinical trials there are problems to be overcome before an effective anti-cancer therapy is produced. The most significant limitation to date has been the occurrence of unpleasant and often dangerous side effects. In the case of TNF, extreme toxicity, often lethality, has been observed in clinical trials with the recombinant product (Old, 1985). Manipulating the lymphokines to achieve beneficial effects without the deleterious effects awaits a deeper understanding of their individual mechanisms of action and of the importance of interactions between different agents.

Peptides

There are many important natural human peptides that mediate physiological functions, for example, glycoprotein hormones (Pierce and Parsons, 1981) neuroendocrine peptides (Douglass *et al.*, 1984) and growth factors (James and Bradshaw, 1984). There is tremendous potential in manipulating these peptides to produce new drugs which either augment or antagonise natural processes. Some of the applications of these natural peptides are shown in Table 12.4. Already, significant progress has been made in producing some of these peptides by gene manipulation. We cannot cover all the studies here but some examples are listed in Table 12.5 to illustrate the diversity of systems that have been used. There are some specific problems associated with some of the peptides. For example, the glycoprotein hormones, luteinising hormone (LH), follicle-stimulating hormone (FSH), chorionic gonadotrophin (GH) and thyrotropin (TH) consist of two chains, the α−chain is common to all four and the β-chain is unique and confers receptor specificity and biological activity. Dual expression vectors have to be used to produce both chains in the same cell, for example, the SV40 late expression vector of Reddy *et al.* (1985), described in Chapter 4. In addition, glycosylation is essential for the interaction of the peptide with its cognate cell receptor (e.g. Sairam and Bhargavi, 1985). It is not yet known whether recombinant glycoprotein hormones produced in other cells have the full spectrum of biological activity and several host cells may have to be tested.

Many peptides are produced as precursors which are secreted and usually require several proteolytic steps in the maturation pathway to produce fully active peptides. This proteolytic maturation is often highly tissue-specific (see Chapter 10), which imposes some constraints on the host-vector system. For example the neuroendocrine peptide, enkephalin is produced from a complex precursor polyprotein, preproenkephalin. The expression of preproenkephalin has been studied using a vaccinia virus transient expression

Table 12.4. Some applications of natural peptides.

Peptide	Application
Epidermal growth factor (EGF)	Wound healing
Insulin	Control of diabetes
Growth hormone (GH)	Treating growth disorders
Glycoprotein hormones (LH, FSH, CG)	Management of fertility
Atrial natriuretic factor (ANF)	Management of hypertension
Calcitonin	Treating osteoporosis
Erythropoietin	Treating chronic anaemia

Table 12.5. The production of some natural peptides in heterologous eurkaryotic cells.

Peptide	Host cell	Vector	
Human growth hormone	Mouse fibroblasts (C127)	BPV	(1)
Human insulin	Mouse pituitary cells (AtT20)	Simple plasmid integration	(2)
Epidermal growth factor	Yeast	αF secretion	(3)
Enkephalin	Mouse pituitary cells (AtT20)	Vaccinia	(4)
Human chorionic gonadotrophin (α and β chain)	Monkey cells (CV-1)	SV40 dual late replacement	(5)
Erythropoietin	COS cells	SV40 replicon	(6)
Bovine luteinising hormone (α and β chain)	Hamster cells (*dhfr⁻* CHO1)	Two *dhfr* amplicons	(7)
β-endorphin	Yeast	αF secretion	(8)
Somatostatin	COS cells	SV40 replicon	(9)
Human parathyroid hormone	Rat pituitary cells (GH4)	Retrovirus	(10)
Atrial natriuretic factor	Yeast	αF secretion	(11)

References: (1) Pavlakis and Hamer, 1983; (2) Moore *et al.*, 1983; (3) Brake *et al.*, 1984; (4) Thomas *et al.*, 1986; (5) Reddy *et al.*, 1985; (6) Jacobs *et al.*, 1985; (7) Kaetzel *et al.*, 1985; (8) Bitter *et al.*, 1984; (9) Warren and Shields, 1984; (10) Hellerman *et al.*, 1984; (11) Vlasuk *et al.*, 1986.

system (Thomas *et al.*, 1986) in a number of cell types. These were AtT20 (mouse anterior pituitary), BSC40 (African green monkey kidney) and L*tk*⁻ (mouse fibroblasts). All the cells secreted significant amounts of metenkaphalin but only the AtT20 line produced the correct processed enkaphalin peptides. Similarly proinsulin is not processed to insulin in AGMK cells (Gruss and Khoury, 1981) nor in mouse fibroblasts (Moore *et al.*, 1983). Clearly either the correct cell line must be used or the coding sequences for the peptides must be manipulated such that only the mature peptide is produced.

There are a number of peptides that have been shown to affect cell growth. For example, epidermal growth factor (EGF) platelet-derived growth factor (PDGF), insulin-like growth factors (IGF−1 and IGF−2) and nerve growth factor (NGF). These appear to be relatively easy to produce in large quantities in simple systems such as the alpha factor secretion system in yeast (Brake *et al.*, 1984; Chapter 4).

Blood proteins

A large number of potentially therapeutic proteins occur in human plasma and some of those being actively investigated are listed in Table 12.6. The most economically important blood proteins are factor VIII, which is the clothing factor required by males suffering from haemophilia B, serum albumin which is used extensively as a blood volume expander in cases of severe blood loss and thrombolytic agents such as t-PA.

Clotting factors

Blood clotting is a complex process because it must be highly controlled. Inappropriate clotting would block blood vessels resulting in death but there must be rapid clotting at sites of tissue injury to prevent haemorrhage. A simplified outline of the clotting process is shown in Figure 12.7. There are two pathways for blood clotting; the *intrinsic pathway* is triggered when the blood contacts an abnormal surface and the *entrinsic pathway* is triggered by components released into the blood from damaged tissues. Both pathways involve the activation of a cascade of proteins called *factors* from inactive (e.g. XII) to active (e.g. XIIa) forms which in both pathways finally results in the activation of factor X. This then initiates the formation of the fibrin clot. Tissue damage initiates a cascade of reactions each of which involves the conversion of an inactive protein into an active protease. Factors VIII and V are co-factors that are required for the activation of factors X and prothrombin respectively and tissue factor activates factor VII. Clotting is accelerated by thrombin which activates factors VIII and V. A deficiency in a single protein in either pathway results in a clotting disorder. Haemophiliacs suffer from uncontrolled bleeding. In most cases this is due to a deficiency in the production of factor VIII (haemophilia A). A rarer form of the disease (haemophilia B) is due to a lack of factor IX (Christmas factor).

The only hope for survival for haemophilia patients is the ready availability of the clotting factors which have to be purified from human blood. It costs in the region of 10 000 US$ per annum to supply a single patient and so this treatment is clearly only available in affluent societies.

The genes for factor VIII (Gitschier *et al.*, 1984; Wood *et al.*, 1985; Vehar *et al.*, 1984; and Toole *et al.*, 1984) and factor IX (Choo *et al.*, 1982;

Table 12.6. Therapeutic uses of blood proteins.

Serum albumin (maintaining blood-tissue balance; transporting hormones, etc.)
Fibronectin (Removing particles from blood)
α1-antitrypsin (Elastase and collagenase inhibitor.
               Possible use in emphysema and arthritis).

Clotting factors
Factor XIII
Factor IX (Christmas factor)
Fibrinogen (Factor I)
Prothrombin

Anticoagulants
Antithrombin III
Tissue plasminogen activator (t-PA)
Kidney plasminogen activator/urokinase (k-PA/u-PA)

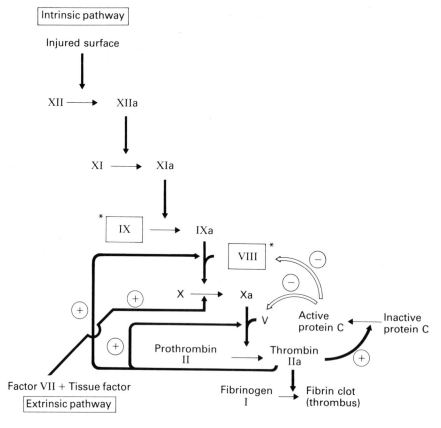

Fig. 12.7. Blood clotting. Asterisk indicates defective genes in X-linked haemophilia.

Kurachi and Davies, 1982) have been cloned and expressed in mammalian cells. In the case of factor VIII the cloning was extremely demanding because the gene is in fact one of the largest human genes characterised. The coding sequence is dispersed amongst 26 exons that span 186 kb of DNA and so chromosome walking had to be used. The mRNA is also large, about 9 kb and therefore a full-length cDNA could not be obtained in a single clone. Factor VIII is synthesised as a 2351 amino acid protein that is secreted and extensively glycosylated at 25 sites. The folding of the protein is critical to produce an active molecule as it involves 17 disulphide bridges (random pairing would produce 2.2×10^{10} different molecules!). The protein is activated by cleavage with thrombin (factor IIa) at three sites to generate three fragments that are linked together. The linkage involves a calcium ion bridge. The protein contains a domain that is homologous to the copper binding serum protein ceruloplasmin (domain A) which is repeated three times, a duplicated region (domain C) and an internal domain (B) that is lost after cleavage (Figure 12.8). To produce such a large protein that may have stringent requirements to achieve the precise tertiary conformation, a mammalian cell expression system is used. This should be economically feasible as factor VIII is a very high value product and is not required in vast quantities. Some success has been achieved with a *dhfr* amplicon in CHO cells (Lawn and Vehar, 1986).

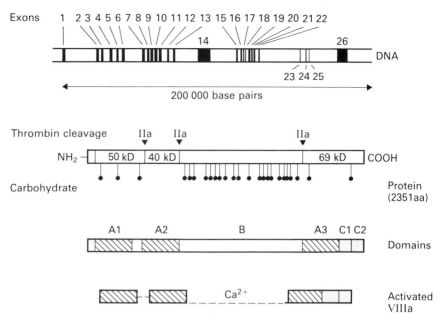

Fig. 12.8. Factor VIII: gene and protein.

Factor IX also has extensive post-translational modifications that are essential for biological activity. It is produced as a preproenzyme and its activation involves two proteolytic cleavages mediated by factor XI. The first 12 glutamate residues in the mature protein are carboxylated via a vitamin-K dependent enzyme located in the microsomes to produce γ-carboxyglutamate residues which bind calcium ions and mediate the interaction with phospholipid that is required for activity. There are also several essential sialic carbohydrates linked to the protein chain and aspartate residue 64 is hydroxylated. Because of these extensive post-translational modifications mammalian cell systems have been used to express the gene. There are however differences between different host cells. For example, using a vaccinia virus transient system fully biologically active factor IX was produced in a human hepatoma cell line, but factor IX produced in similar yields from mouse fibroblasts was three times less active (de la Salle *et al.*, 1985). Factor IX has also been expressed using simple plasmid vectors in hamster BHK cells (Busby *et al.*, 1985) and rabbit hepatoma cells (Anson *et al.*, 1985). To date the yields in mammalian cell culture supernatents have been lower than detected in serum and therefore significant improvements are needed before a commercial process is developed.

Anti-thrombolytic agents

There are a number of diseases caused by the blockage of blood vessels by intravascular clots, i.e. *thrombosis*. This can produce coronary heart disease, strokes and pulmonary embolism. Thrombosis is the commonest cause of death in the Western World. Normally clots are broken down by the action

433

of the enzyme *plasmin*. This is present in the inactive form, *plasminogen*
which is activated by a group of serine proteases called *plasminogen activators*
(Figure 12.9). One of these, called *urokinase*, can be isolated from human
urine and is already used to disperse thromboses. However, a single course
of treatment could require urokinase produced from 5000 litres of urine and
may cost 8000 US$. In addition urokinase is not very specific and cleaves free
plasminogen molecules in the circulation, releasing large amounts of plasmin
which increases the risk of haemorrhage.

A second class of activator, called *tissue plasminogen activator (t-PA)*, is
produced by a number of cell types. This activator binds to fibrin and is
selective for plasminogen when it is bound to fibrin thrombi. In other words
it only works where it is needed. t-PA is produced in far too small amounts
in normal tissue to be purified for clinical use although a number of tumour
cell lines secrete increased amounts (Gronow and Bliem, 1983). The t-PA
gene has been cloned (Pennica *et al.*, 1983). The mature protein is 522
amino acids long and has a complex tertiary structure. There are 17 disulphide
bridges and the protein is folded into four domains, a presumed fibrin-
binding or finger domain, an epidermal growth factor domain that is common
to many serine proteases, two disulphide triple-looped structures called
kringles and a serine protease domain. The molecule is also often cleaved to
produce a two-chain protein. There are four potential glycosylation sites but
the presence of carbohydrate has been confirmed on only three in native t-
PA (Figure 12.10). When t-PA is expressed in bacterial and yeast cells,
yields of active protein are low and because of the complex tertiary structure,
mammalian cells seem the most promising host system for production. The
precise expression systems being used for preparing recombinant t-PA for
clinical trials have not been fully published but good yields have been
obtained using *dhfr* amplicon vectors in CHO cells (e.g. Kaufman *et al.*,
1985). Second-generation recombinant t-PA is also being produced by
expressing individual domains of the protein and various mutant derivatives

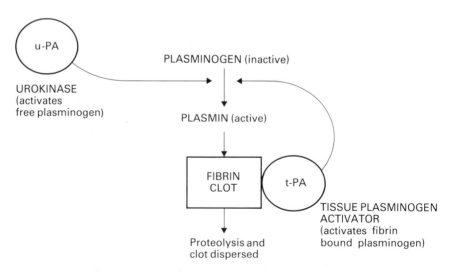

Fig.12.9. The dispersal of blood clots.

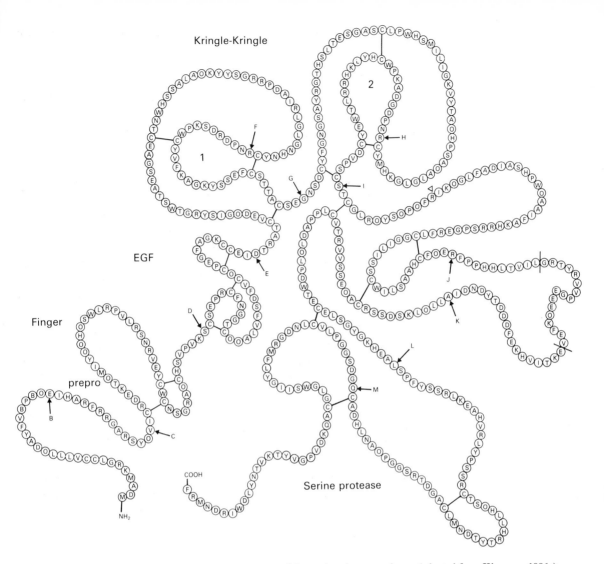

Fig. 12.10. The structure of tissue plasminogen activator (adapted from Klausner, 1986a).

(van Zonneveld *et al.*, 1986; Klausner, 1986a). For example, the finger domain is being modified in an attempt to increase its binding to fibrin in order to increase fibrin selectivity. Removal of the kringle domains appears to reduce the interaction of natural inhibitors with the protein and it is possible that the EGF domain may be dispensable so that a smaller molecule can be used. If a minimal active t-PA is defined then the requirements for expression may be less stringent and simple systems such as yeast may ultimately be used.

Protease inhibitors

Blood contains a number of protease inhibitors. The major one is alpha-1-antitrypsin. This inhibits trypsin, thrombin and plasmin but its main target is elastase in the lung. In the absence of α-l-antitrypsin there is cumulative degradation of lung tissue known as *emphysema*. This often fatal condition can be partially relieved by the injection of partially purified human α-l-antitrypsin. There are, however, general problems in addition to the expense and difficulty of extracting protein from human tissues. Firstly, extremely large quantities, of the order of 4 g, are required per treatment and secondly the protein is readily inactivated by oxidation. (This probably explains the increase in emphysema in smokers because free radicals generated by cigarette smoke inactivate the α-l-antitrypsin in the lung). To overcome these problems the coding sequence for α-l-antitrypsin has been expressed in yeast and the protein has been engineered to increase oxidation resistance (Rosenberg *et al.*, 1984b). This has been done by using site-directed mutagenesis to substitute a valine residue for the oxidation-sensitive methionine 358 residue at the active centre of the protein. Several hundred milligrams of fully biologically active α-l-antitrypsin can be produced per litre of yeast culture and this is all oxidation resistance. As yeast can be grown very cheaply in vast fermentations this is likely to provide an economic process to generate sufficient α-l-antitrypsin for clinical use. In addition anti proteases with new specificities are being constructed with this system. For example replacing the active centre methionine with arginine destroys the anti-elastase activity but creates an anti-thrombin III type specificity (reviewed by Carrell, 1984).

Other inhibitors, such as antithrombin III, are likely to be future targets for biotechnology.

Enzymes as therapeutic agents

There are many enzymes, other than those we have discussed as blood products, that have clinical application, particularly if they can be readily immobilised for presentation to body fluids (Klein and Langer, 1986). For example, bacterial urease is already being used to accelerate the removal of urea from the blood during dialysis in cases of kidney failure. A number of potentially useful enzymes are derived from eukaryotic cells and gene manipulation techniques can be used to increase yields and to produce more effective enzymes by site-directed mutagenesis. Some of these enzymes and their potential applications are listed in Table 12.7. Some progress has already been made in this area. For example, very high levels of the potential anti-arthritic agent, superoxide dismutase, have been produced in yeast (Hallewell *et al.*, 1987).

Antibody engineering

Monoclonal antibody technology is having a massive impact on all aspects of biology and biotechnology. A detailed account is outside the scope of this book but the technique, originally devised by Köhler and Milstein (1975) is outlined in Figure 12.11 and technical details can be found in Langone and Vanakis (1986). When mice are immunised with a complex antigen, a mixture of antibodies is produced that react with the different surface epitopes. Antibodies are produced by B cells which synthesise only one type of

Table 12.7. Potential clinical applications of enzymes derived from eukaryotic cells.

	Enzyme Source	Application
Superoxide dismutase	Bovine liver	Anti-inflammatory: prevents the accumulation of oxygen-free radicals.
Catalase	Bovine liver	Anti-inflammatory: prevents the formation of hydroxy radicals.
Tyrosinase	Mushrooms	Prevents the accumulation of phenols in liver failure.
α-galactosidase	Figs	Prevents accumulation of ceramides in Fabry's disease.
UDP-glucoronyl transferase	Rabbit liver	Conjugates excess toxic bilirubin in jaundice and liver disease.
Uricase	Hog liver	Prevents hyperuricaemia in gout.
β-glucuronidase	Bovine liver	Metabolises mucopolysaccharides in mucopolysaccharidosis VII disease.

antibody molecule, that is, molecules having only one combining specificity (see Chapters 1 and 4). The spleen contains a mixture of all these B cells but they cannot be separated and they will not grow readily *in vitro*. However, B cells can be fused with mouse myeloma cells using inactivated Sendai virus. Myeloma cells are immortal and they secrete large quantities of antibody. A myelema cell line that is deficient in purine metabolism is used so that it can be selected against in HAT medium. The B cells are resistant to HAT medium but do not proliferate, the myeloma cells proliferate but die in HAT. The fused cells, called *hybridoma cells*, are HAT-resistant and immortal and therefore they can be selected. The hybridomas are cloned in microtitre dishes by diluting the population of cells to less than one cell per well. Each clone secretes a single type of antibody so that now all the possible antibody producing cells have been separated from the original mixture. The mouse antibodies produced from the cloned hybridomas are therefore *monoclonal antibodies*. The small clone of cells is amplified by culturing the cells in roller bottles or in the peritoneal cavities of mice where they grow in suspension as tumour cells called *ascites* cells. The panel of monoclonal antibodies that is produced after injecting a complex antigen, for example a virus, can be used to identify the key antigenic site. This has been done, for example, to locate the influenza virus haemagglutinin epitopes (Gerhard *et al.*, 1981).

'Monoclonals' can be used to separate cells with specific surface markers from mixed populations. Some tumour cells have unique surface antigens

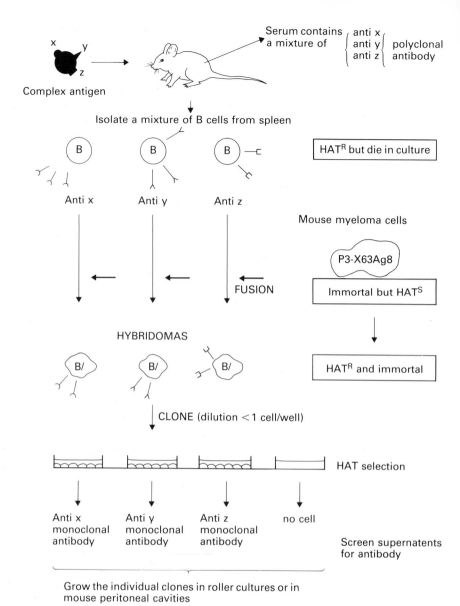

Fig. 12.11. The production of monoclonal antibodies.

and monoclonals that bind to these can be used to identify tumours in biopsy tissue using immunoreagents or *in situ* using radio-labelled antibodies to image the tumour so that they can be detected by X-rays. They can be used for tissue typing, blood grouping, serotyping of microorganisms and probing the function of cell surface molecules (reviewed by Scott, 1985). The potential uses of monoclonal antibodies is enormous and now gene manipulation is being used to extend their usefulness even further. There are two basic applications of gene manipulation technology to this field. The first is to produce novel antibodies with new effector functions and the second is the production of human monoclonal antibodies (Williams, 1988).

The production of antibodies with novel effector functions

A generalised plan of the functional domains in an antibody molecule is shown in Figure 12.12. (The structure of antibody genes is introduced in Chapter 1.) The protein is broadly split into an *antigen combining* site and an *effector region* (roughly corresponding to the Fab and Fc domain originally identified by Porter, 1958). The antigen combining site consists of framework regions which are fairly homologous between antibodies and *complementarity determining regions* (CDR) which give the antibody molecule the unique specificity of binding. The CDRs are generated by somatic mutations which occurs at a high frequency in two hypervariable regions of the V gene and by variable recombination in the D/J region. The effector region is composed of heavy chain only and the different classes of antibody have different effector functions. The antigen binding site functions independently of the effector region.

Heavy- and light-chain genes introduced into B-cell tumour lines by standard procedures are expressed to produced functional antibodies (Neuberger, 1983; Ochi *et al.*, 1983; Oi *et al.*, 1983; Chapter 8). This opens up the possibility of manipulating the genes *in vitro* to alter the properties of the antibody that is expressed. This process of *antibody engineering* can be used to replace the effector region with a variety of protein domains that are completely unrelated to antibodies. For example Neuberger *et al.* (1984) replaced most of the constant exons of a mouse immunoglobulin heavy-chain gene with the coding region of a DNase from *Staphylococcus aureus* (Figure 12.13.). The gene was originally cloned from a hybridoma raised against the simple chemical (referred to as a *hapten*) 4−hydroxy−3 nitro-phenacetyl (NP) and therefore the variable region specifies anti NP binding. The hybrid antibody−nuclease gene was inserted into pSV2*gpt* and transfected into a plasmacytoma line that expressed lambda light chains but no heavy chains. Antibody molecules were secreted that reacted specifically with NP but when these antibodies were mixed with single-stranded DNA they also digested the DNA!

Recombinant antibodies have a variety of uses and some of these are illustrated in Figure 12.14. An exciting possibility is the construction of specific anti-tumour drugs. At present cytotoxic drugs do not discriminate against tumour cells and normal cells and therefore the most powerful drugs cannot be used. The new idea is to raise a panel of monoclonals against a tumour cell, and identify a monoclonal that only reacts with tumour cells (this may not be possible with all tumours). The antibody gene is then cloned and the effector region is replaced with a cytotoxic drug (Klausner, 1986b). The hybrid antibodies will specifically 'target' to tumour cells and deliver the drug where it is most needed. For example the castor oil bean contains a protein called *ricin* which is highly toxic to eukaryotic cells. The ricin coding sequence can now be genetically fused to the V region of a tumour-specific monoclonal antibody gene to produce a tumour-specific drug. This type of drug is often called a *magic bullet* or an *immunotoxin*. Any cytotoxic agent could be used, e.g. tumour necrosis factor, interleukin 2, abrin (reviewed by Vitetta and Uhr, 1985). The value of recombinant

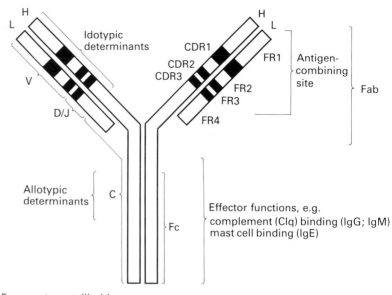

Fc = Fragment crystallizable
Fab = Fragment antibody binding
CDR = Complementarity determining region
FR = Framework region
L   = Light chain
H   = Heavy chain
V   = Variable domain
D/J = Joining region
C   = Constant region

Fig. 12.12. Functional domains in an antibody molecule.

antibodies in anti-cancer therapy is now being assessed in research groups all over the world.

Recombinant antibodies can also be used as immunoreagents. For example, the standard diagnostic *ELISA test (enzyme-linked immunosorbent assay)* can be simplified by creating an antibody—enzyme fusion that can produce a colour reaction with the cognate antigen by simply mixing the sample, the recombinant antibody and substrates for the enzyme.

The production of recombinant antibodies with novel effector functions will probably work best for independently folding protein domains and proteins that have significant activity as monomers. For more complex proteins the yields and activity may be reduced. For example *c−myc* protein could not be produced by this approach (Neuberger *et al.*, 1985).

Producing human monoclonal antibodies

It is extremely difficult to produce human monoclonals. The procedure outlined in Figure 12.11 is hardly generally applicable to man! Human tumour cell lines that produce antibodies secrete much lower levels than analogous mouse cells and so it is difficult to produce large quantities of human antibodies by this route. If monoclonal antibodies are to achieve their full potential as therapeutic agents it will be important to use essentially

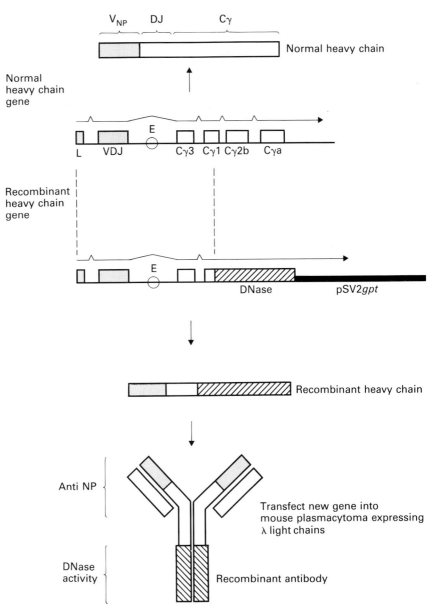

Fig. 12.13. Producing recombinant antibodies with novel effecter functions. E = heavy chain enhancer (see Chapter 8); NP = 4-hydroxy-3-nitrophenacetyl; DNAse = *Staphylococcus aureus* nuclease.

human antibodies. This is because antibodies are themselves antigens. Humans injected with mouse antibodies rapidly develop anti-mouse antibody antibodies directed against antigens in both the C and V regions. This will limit the effectiveness of any therapy because repeated injections may induce immunity to the drug. To overcome this problem chimaeric human–mouse antibodies have been constructed. In the first experiments hybrid genes that expressed mouse antigen-binding domains fused to human constant-region domains were produced (Morrison *et al.*, 1984; Boulianne *et al.*, 1984). A

A Magic bullets

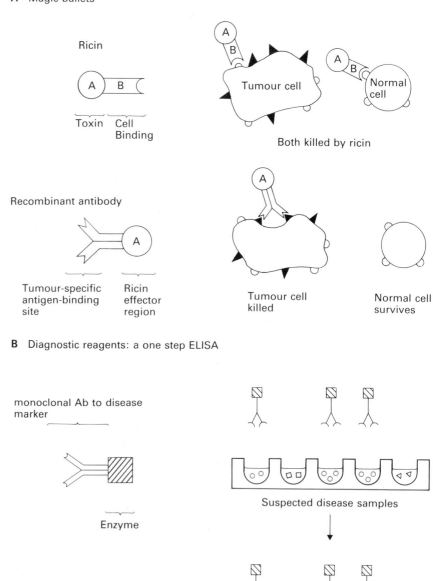

Ricin

Toxin | Cell Binding

Tumour cell

Normal cell

Both killed by ricin

Recombinant antibody

Tumour-specific antigen-binding site | Ricin effector region

Tumour cell killed

Normal cell survives

B Diagnostic reagents: a one step ELISA

monoclonal Ab to disease marker

Enzyme

Suspected disease samples

Add enzyme reagents

Colour test

+   −   +   +   −

442

Fig. 12.14. Applications of recombinant antibodies.

sophisticated approach to engineering the V region has been described by Jones *et al.* (1986). The CDRs from the heavy-chain variable region of a mouse antibody gene were introduced into the V_H domain of a human antibody gene. This allowed the production of a monoclonal antibody that had a human framework and a mouse antigen-binding site. The engineered antibodies are less antigenic than mouse monoclonal antibodies indicating that these 'humanised' monoclonals might have increased therapeutic potential.

If monoclonals are to be used as therapeutic and *in vivo* diagnostic agents very large quantities will be required. Advances in mammalian cell vector technology may provide these yields. Other approaches have been tried. For example, functional antibodies have been secreted from yeast transformed with two co-selected expression plasmids for the heavy- and light-chain genes (Wood *et al.*, 1985). Yields of antibody would however need to be significantly improved to make this a commercial process. It may, however, be possible to produce engineered antibodies that contain only the combining site and these might be produced in higher yields in yeast (van Brunt, 1986).

Purifying recombinant proteins

Producing a cell that synthesises large amounts of a desired protein is only the first stage in achieving a useful process. It is important to be able to recover the protein by a simple, economical method that results in high yields of biologically active product. A discussion of the traditional tools of the protein chemist is outside the scope of this book but there are a number of recombinant DNA approaches to aiding protein purification. A general approach is to fuse the gene to a coding region for which an affinity matrix is readily available. For example, a protein found on the surface of the bacterium *Staphylococcus aureus*, called protein A, binds to immunoglobulins via their Fc receptor domain. The coding region for the protein A–immunoglobulin binding domain can be fused via a linker to any other coding region. For example, if the insulin-like growth factor (IGF) is fused to protein A via a gene fusion, then the resulting fusion protein can be harvested by passing the cell extract or culture supernatent through a column consisting of IgG coupled to agarose, i.e. an *affinity matrix*. The fusion protein binds to the IgG via protein A and all contaminating proteins flow through. The bound material is eluted and then treated to release the IGF from the protein A by treating with a protease that cleaves at the linker. The linker sequence is designed so that a protease can be used that does not cleave within the IGF coding region. The contaminating protein A portion is then removed by passing the extracts over a second IgG affinity column to bind the protein A while the IGF passes through.

Recombinant antibodies also provide a new approach to purifying valuable proteins. B-cell tumour cells secrete very large quantities of antibody, up to 20 mg l^{-1}, and any antibody fusion protein may be produced in similar yields. The protein can be purified using an affinity column carrying the cognate antigen. The protein can be liberated from the column by proteolytic cleavage using a natural target site or one that has been specifically engineered into the fusion (e.g. a factor Xa cleavage site, Nagai and Thoreson, 1984) (Figure 12.15).

In another purification method the recombinant protein, in this example urogastrone, is given a positively charged 'tail' by adding a run of synthetic arginine codons to the 3′ end of the coding sequence (Brewer and Sassenfeld 1985). Arginine is the most basic naturally occurring amino acid and it is always positively charged at pH values below 12. The poly-arginine fusion protein therefore, binds strongly to a negatively charged ion-exchange matrix (cation exchanger). The bulk of the contaminating proteins will not bind but there will be some positively charged contaminants. The bound material is eluted, using a salt gradient to displace the proteins from the matrix, and the eluate is treated with carboxypeptidase B. This enzyme is specific for arginine or lysine residues and sequentially digests them from the C-terminus. This removes the positive charge from the recombinant protein and in a second round of cation exchange the contaminating positively charged amino acids are retained but the recombinant protein no longer binds and is therefore obtained in a highly pure form. Purifying recombinant proteins is discussed in more detail by Sofer (1986) and Bonnerjea et al. (1986).

Choosing a host−vector system

Many different host−vector systems have been discussed in this book and their disadvantages and advantages have been noted. Choosing a system is dictated by convenience, cost, yields and safety of the product. In some cases there is a range of systems that can be used. For example, high yield of fully biologically active interferon can be obtained in E. coli (Remaut et al., 1986), S. cerevisiae (Mellor et al., 1983), insect cells (Smith et al., 1983b) and mammalian cells (Derynck et al., 1983) (reviewed in Pestka, 1986; Pestka et al., 1987). Despite the effectiveness of all these systems, however, a Japanese group have described the production of interferon by injecting a recombinant insect virus into silk worms and harvesting the haemolymph after allowing virus replication and expression for four days. Interferon expression was driven by the polyhedrin promoter (Chapter 4) and it was produced at high yields of 50 μg per larva (Maeda et al., 1985). This is clearly a convenient production system because it is linked to an established industry, i.e. silk production, but it is unlikely to be widely used. Another example of choosing a host−vector system based on an established industry is the modification of brewers yeast to produce non-toxic but valuable proteins at the end of the beer fermentation (Dixon, 1986).

For many proteins the factors of convenience, cost, and yields take second place to the need to produce a fully authentic product. For example, as we have seen, many of the blood proteins require complex post-translational modifications which may only occur in mammalian cells. The best guarantee that an authentic protein is produced is to use the natural cell of origin. This is usually not feasible because the cells cannot be conveniently cultured. An alternative approach is to use genetically engineered

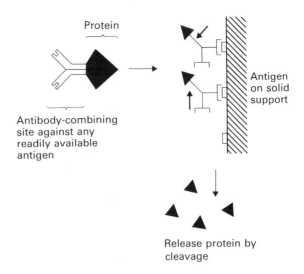

Protein

Antibody-combining
site against any
readily available
antigen

Antigen
on solid
support

Release protein by
cleavage

Fig. 12.15. A novel affinity purification procedure using hybrid antibodies.

tumours (Kaplan *et al.*, 1985). As we have seen in Chapter 11, the expression of an oncogene can be targeted to different cells by adding a tissue-specific enhancer to the recombinant gene. This can result in a tumour which will over-produce its normal cell product, e.g. excessive insulin production by an insulinoma. Whether this approach is feasible on a large scale is not clear.

One significant problem with recombinant proteins is the extent and quality of glycosylation. Many pharmaceutically important proteins are glycosylated and in some cases, e.g. the glycoprotein hormones, the carbohydrate moiety is critical. In the case of viral antigens, the carbohydrate moiety may contribute to the immunogenicity (Berman and Laskey, 1985). In general, the complexity of the carbohydrate side chains is correlated with the complexity of the organism. Simple high mannose side chains are found in *S. cerevisiae* as compared to the more complex branched structures and different glycosyl residues of mammalian cells. If complex glycosylation is required for the full biological activity of a protein then the simple production of that protein in yeast would not be useful. In some cases aberrant O-linked glycosylation has been seen on proteins secreted from *S. cerevisiae*, e.g. when GM-CSF is secreted using α-factor vectors (Ernst *et al.*, 1987). Similarly, glycosylation in insect cells is often the high-mannose type and there are no sialic acid residues on insect cell glycoproteins. The answer, however, is not just to use a mammalian cell system because glycosylation is species-specific and to some extent cell-specific. A human pharmaceutical should theoretically be produced in a human cell if glycosylation is critical. The only way this can be achieved is to use a tumour cell line; however a characteristic feature of tumour cells is altered glycosylation! One solution to this problem is to remove the carbohydrate residues from recombinant proteins and then chemically add the correct residues (Van Brunt, 1986). This makes *S. cerevisiae* a more attractive host system because the proteins are glycosylated at the correct sites and because the carbohydrate is simple and hence may be easy to remove. The initial core glycosylation is identical to that in mammalian cells and may therefore provide a good 'peg' on which to attach the correct complex residues. In some cases the complex glycosylation is in itself a problem because the proteins are more likely to be cleared from the blood. This can be overcome by adding sialic acid residues to the carbohydrate to block filtration in the kidney and non-specific cell binding.

Another dilemma is whether to use a secretion system or to produce the protein intracellularly. During intracellular production yields can increase to about 10% of total cell protein but this intracellular concentration, which may be about 200μMolar, is far in excess of the optimium concentration (about 1μMolar) for protein folding. The protein therefore tends to fold incorrectly and forms a denatured insoluble precipitate. The intracellular yields are often high but the protein must be renatured *in vitro* which may not be possible or it may be time consuming and expensive. The advantage, however, is that the yields are often much higher than yields of secreted proteins and the precipitates can be purified easily often to 95% purity, by simple centrifugation procedure.

Proteins produced intracellularly will not be glycosylated and this may or may not be a problem. In some cases, e.g. in the production of a fungal endoglucanase (Van Arsdell *et al.*, 1987) and the Epstein−Barr envelope glycoprotein (Schultz *et al.*, 1987) the secretion of proteins by *S. cerevisiae* results in hyperglycosylation. This, in fact, far from being a problem, improved the thermal stability of the endoglucanase but for some pharmaceuticals the excess carbohydrate may be undesirable.

Secreted recombinant proteins must be recovered from large volumes of culture medium which are likely to contain contaminating proteins. Purification is therefore usually multistep and usually involves several chromatography stages. A major advantage of secretion systems is that the N-terminus is likely to be authentic as a result of cleavage at the signal peptide whereas intracellular proteins may retain the N-terminal methionine (Chapter 2).

Whichever host−vector system is chosen the yields of recombinant proteins can be maximised by optimising each of the stages of gene expression outlined in Chapter 1 and discussed in more detail in Chapter 8. This has been reviewed for yeast by Kingsman *et al.* (1985) and for the production of mammalian proteins in mammalian cells by Bebbington and Hentschel (1985). In general, increasing the copy number of the gene results in increased yields. Therefore, high-copy number vectors such as the 2μm vector in yeast and viral mini-replicons and amplicons in mammalian cells are widely used. High-efficiency promoters are also important. For example the *PGK* (yeast), adenovirus *MLP* (mammalian cells) and polyhedrin (insect cells) promoters all drive high-level expression. Proteins that are toxic to host cells can be produced by using regulated promoters. Translational efficiency can also be improved particularly in mammalian cells, by the provision of *trans*-acting factors such as VA1 RNA. It is not yet clear how much, if any, improvement to translation can be achieved by altering codon usage but translation initiation environment can be optimised.

There are a few general considerations about host–vector systems. For example, proteins that are needed in tonne quantities such as human serum albumin, α-1 antitrypsin and bovine growth hormone are likely to be produced in microbial systems. Highly active proteins such as the interleukins or factor VIII that are needed in kilogram quantities could be made economically in cell culture. The final consideration for the commercial production of any recombinant protein has to be economics but this is often not simple. It depends upon competing technologies, public acceptance of different products and political factors. For example, the concern that human growth hormone derived from the human pituitaries was contaminated with Jacob Creutzfield virus resulted in the appearance of a market for the more expensive recombinant product. For countries in the Third World the access to recombinant DNA-derived proteins such as vaccines and diagnostic reagents is going to be critically dependent upon the development of the cheapest and simplest systems. These are likely to be microbial or robust continuous cell lines.

Agriculture and food industries

Agriculture and food industries are major exploiters of traditional biotechnology. Much of the initial impact of gene manipulation and gene transfer is, therefore, in improving established procedures rather than in creating entirely novel processes. In this section we discuss the production of improved plant varieties that are resistant to herbicides, viruses and insect pests, the farming of transgenic animals, the manipulation of yeast to improve the brewing process and novel approaches to the disposal of waste products.

The production of herbicide-resistant plants

The broad spectrum herbicide glyphosphate (phosphonomethyl glycine) inhibits the shikimic acid pathway in plants. The same pathway is found in bacteria such as *Salmonella typhimurium*, but it is not found in mammalian cells. Glyphosphate is an inhibitor of the enzyme 5-enolopyruvyl-shikimate-3-phosphate (EPSP) synthase. It therefore blocks the production of chorismic acid which is an essential precursor for the synthesis of the aromatic amino acids and essential aromatic metabolites such as p-amino benzoate (Figure 12.16A). The only drawback to glyphosphate is that it is equally effective on valuable crop plants and on weeds! It would be a great advantage to produce crop plants that were resistant to glyphosphate so that contaminating, and therefore competing, weeds could be destroyed effectively without ruining the crop. Two mechanisms of resistance to glyphosphate have been identified from studies with cultured plant cells and with bacteria. Plant cells that have been selected for growth in glyphosphate have 40-fold elevated levels of ESPS synthase indicating that over-production of the enzyme overcomes inhibition. Glyphosphate-resistant mutants of *S. typhimurium* can be isolated by standard mutagenesis. Many of these have alterations in the *aro*A gene that codes for ESPS synthase. The mutant

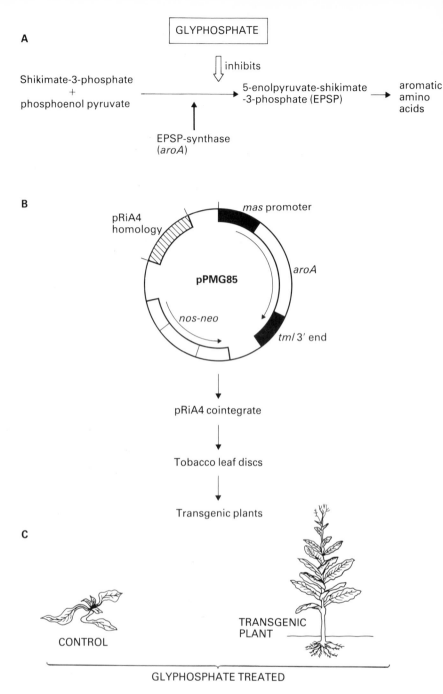

Fig. 12.16. Herbicide-resistant transgenic plants.

EPSP-S has a single amino acid substitution of a proline to a serine which causes a decreased affinity for glyphosphate but which does not affect the specific activity of the enzyme (Comai *et al.*, 1983, 1985).

The mutant *aro*A gene was introduced into tobacco plants using an Ri plasmid cointegrate vector and the leaf disc transformation technique

(Chapter 6). The expression of the bacterial gene in plants was directed by the promoter from the mannopine synthase gene (*mas*) (Figure 12.16B). Transgenic plants were regenerated from roots and were selected on the basis of resistance to kanamycin. The plants were sprayed with glyphosphate at concentrations normally applied in the field. The transgenic plants containing the mutant *aro*A gene continued to flourish but control plants transformed with the vector alone all died (Figure 12.16C). Similar results were obtained by introducing a chimaeric EPSP synthase gene into *Petunia* cells; the coding sequence was derived from *Petunia* but expression was driven to high levels by using the efficient CaMV-35S promoter which produced resistance (Shah *et al.*, 1986). In this study the EPSP-synthesis was targeted to the chloroplast because the natural transit peptide was preserved (see Chapter 10). As the chloroplast is the major site of synthesis of aromatic amino acids the presence of a transit peptide is considered desirable although, as the data with the bacterial *aro*A protein show, it is not essential. These two studies provide a basis for designing strategies to produce selective herbicide tolerance in important crop plants.

A new method for producing virus-resistant plants

Viral infections cause massive losses in yields of many crops. For example, tomato mosaic virus (TMV) infection destroys 50 million US dollars worth of tomatoes each year. One approach to preventing this is to laboriously inject each plant with milder strains of the virus which offers cross-protection against severe infection by more virulent strains. This is called induced resistance and has also been observed in bacterial, fungal and viroid infections (Dean and Kuć, 1985). A more desirable approach is to identify genes for virus resistance and produce plants that carry these genes as a heritable trait.

Very little is known about the mechanism of disease resistance and the basis of cross-protection. One model however, suggests that the coat protein of the pre-infecting virus is important in determining cross-protection (Sherwood and Fulton, 1982). This idea has formed the basis of a procedure to produce transgenic virus resistant plants (Abel *et al.*, 1986). The coat protein (CP) cDNA from tobacco mosaic virus (TMV) was inserted into tobacco plants using the split end Ti vector system and leaf disc transformation (Chapter 6). The CP cDNA was expressed using the strong CaMV-35S promoter, transformed cells were selected on the basis of kanamycin resistance and were regenerated into mature transgenic plants. These contained 1–5 copies of the CP cDNA and TMV-CP was detected as 0.1% of the total extracted leaf protein. Seedlings produced from these transgenic plants were infected with TMV and in all cases the appearance of disease symptoms was significantly delayed as compared to control plants. A proportion of the plants, from 4 to 47% depending on the transgenic line, did not develop any symptoms of infection. At present the molecular basis for the resistance is still unclear. Some suggestions are: (1) the capsid protein repackages the superinfecting RNA rendering it non-infectious; (2) the CP-RNA competes for the replication system of the infecting virus or (3) it may interfere with viral maturation. Similar results have been obtained by inserting the alfalfa mosaic virus coat protein gene into tobacco and tomato plants (Tumer *et al.*, 1987). Other likely targets are the grains, which suffer 1.5 to 2 billion dollars loss per year due to diseases such as barley mosaic virus, wheat viruses and corn mosaic virus (Bialy and Klausner, 1986).

Plant resistance to insect pests via the transfer of bacterial toxin genes

Microorganisms can be used as insecticides because they infect and kill the pest (Klausner, 1984). The most widely applied example of this approach is the use of the bacterium *Bacillus thuringiensis* to destroy caterpillars. When these bacteria sporulate on the surface of leaves they produce crystals which are converted to toxic peptides by proteases in the midgut of insect larvae

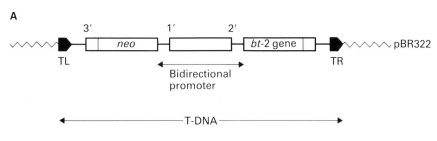

Fig. 12.17. A, T-DNA vector for the expression of the *B. thuringiensis* toxin gene *bt-2*. B, Transgenic plants containing the *bt-2* genes are resistant to insects. Left = control. Right = *bt-2* transgenic plants. (Photograph courtesy of Dr M. Vaeck.)

that feed on leaves. The caterpillars which injest the toxins are paralysed. *B. thuringiensis* spores containing the toxin can be sprayed directly on to plants and will protect against the larval stages of many insect pests on over 200 important crops. These microbial insecticides, although effective and broad spectrum are 25% more expensive than chemical insecticides and require more care in administration. They are, however, considered less harmful than most chemicals for products for human consumption because they are highly specific for insect larvae. Recently the bacterial gene coding for the *B. thuringiensis* toxin has been introduced into tobacco cells to generate transgenic plants that express the toxin (Figure 12.17; Vaeck *et al.*, 1987). The toxin gene (*bt*) was inserted into a mini-T-DNA vector that contained a bidirectional promoter to drive both the *bt* gene and the selectable *neo* marker. This was then transferred to *A. tumafaciens* to form a cointegrate plasmid by recombination with a derivative of an octopine Ti plasmid (see Chapter 6). Transgenic tobacco plants were obtained by leaf disc transformation. The plants were infested with larvae of the lepidopteran, *Manduca secta*; these were all killed within three days of eating the leaves of the *bt*-transgenic plants and caused minimal damage to the plants. The control plants were however completely destroyed. Other *Bacillus* spp produce toxins that are effective against different larvae and therefore there is the prospect of producing plants that are genetically resistant to a number of different insect pests (Kristiansen and Lewis, 1986).

Farming transgenic animals

The techniques of embryo microinjection that were developed to generate transgenic mice (Chapter 5) have now been extended to domestic animals. Some technical modifications were necessary to achieve success because the pronuclei are often difficult to see because of the opacity of the eggs of these species. Hammer *et al.* (1985c) solved the problem by centrifuging the eggs at 15 000 *g* for 3 min which separates the cytoplasm into phases and makes the nuclear structures more visible. A mouse metallothionein−human growth hormone gene fusion was injected into sheep and pig embryos. The frequency of integration was good with pig embryos (about 11.0%) but very poor with sheep (about 1.3%). Only one transgenic sheep was produced and in this animal the transferred DNA had rearranged. Gene copy numbers in pigs ranged from 1 to 490 and in many cases the gene fusion was preserved intact and human growth hormone was expressed and detected in plasma samples. In these initial experiments there appeared to be no physiological effect of human growth hormone expression on the development of the pigs.

There are several reasons for pursuing this technology. It may be possible that improvements in expression of growth hormone genes might produce increased growth in animals as was seen with mice. Perhaps a more important achievement would be altering the quality of the meat. For example, injection of bovine growth hormone into cows and lambs improves the efficiency of food conversion and increases the ratio of protein to fat to produce leaner meat. Growth hormones also increase the milk yields. It may also be possible to produce animals with increased ovulation rates by increasing gonadotrophin levels or by introducing a specific gene from sheep, called the *Booroolla* gene, which is involved in determining ovulation rates. This will increase productivity. It may also be possible to alter the properties of wool or hide by introducing new keratin genes (Lovell-Badge, 1985). Major improvements of the growth and characteristics of an animal are however likely to be several years in development because much greater understanding of the normal physiology is required. The hormonal status of an animal is normally finely balanced and it may be simplistic to expect the alteration of the levels of one hormone to override these balances. It is also likely that alterations in endogenous levels of some hormones may affect the levels of others. For example, high levels of growth hormone reduced the fertility of some transgenic mice (Hammer *et al.*, 1984).

A more immediate use for transgenic animals may be to use them as 'bio-factories' to produce some of the new pharmaceutical products that were discussed above. As we have seen in Chapter 11 it is possible to target the expression of a transferred gene to a specific tissue by using a tissue-specific enhancer. It is not difficult to imagine an approach to produce transgenic cattle or sheep that secret novel peptides and proteins into their milk via targeted expression in mammary glands. This would give a supply of easily purified proteins carrying post-translational modifications. Animal serum is already harvested to produce growth factors and antibodies and milk collection is clearly a well established technology (reviewed by Clark *et al.*, 1987).

Gene manipulation in the brewery

Brewing beer is one of the ancient of biotechnology industries. The key organism is, of course, *S. cerevisiae* which is now amenable to the most sophisticated level of gene manipulation (Chapter 3). It is now possible to produce new strains of *S. cerevisiase* by gene manipulation to improve or diversify the traditional brewing process (Stewart *et al.*, 1984a) (Figure 12.18). The malt wort consists of a mixture of mono-, di-, tri- and polysaccharides which are derived from the starch present in barley. *S. cerevisiae* can ferment most of these except the polysaccharides called dextrins which constitute about 22% of the carbohydrate. Another related strain of yeast called *S. diastaticus* secretes enzymes which break down dextrins into glucose and therefore the starch is completely fermented to alcohol. Unfortunately, this strain produces unpleasant tasting beer. To overcome this the *DEX* gene from *S. diastaticus* which encodes amylo α1,4 glucosidase has been introduced into *S. cerevisiae* (Tubb, 1986). This produces a strain which can now produce low-carbohydrate beer by fermenting most of the dextrins to produce ethanol and, most importantly, the beer tastes good. Further improvements can be made by introducing enzymes that can degrade the very high molecular weight branched dextrins. For example, the *GAI* gene from the filamentous fungus *Aspergillus awamori* encodes a glucoamylase that will degrade raw starch by attacking both α1,4- and α1-6 glycosidic bonds. When the *GAI* cDNA was introduced into *S. cerevisiae* in a high-

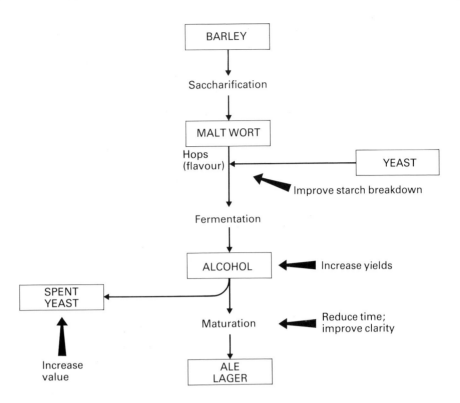

Fig. 12.18. Gene manipulation in the brewery.

efficiency expression vector the enzyme was secreted using its own signal sequence and the new *S. cerevisiae* strain could ferment starch (Innes *et al.*, 1985). The expression vector contained the *ENO1* (enolase) gene promoter which is useful because it is efficient and its activity is not inhibited by glucose whereas the glucoamylase gene promoters are usually repressed by glucose.

It is also likely that the properties of these enzymes will be modified by *in vitro* mutagenesis. For example the *A. awamori* glucoamylase is not destroyed by the normal beer pasteurisation process. This means that on prolonged storage any residual dextrins are converted to glucose, producing sweet beer. It should be possible to increase the lability of the enzyme by appropriate amino acid changes. Other activities which might be engineered into brewing yeasts are the ability to produce proteases such as papain which maintain the clarity of beer, and an ability to tolerate higher ethanol concentrations.

The brewing process is wasteful in that the quantity of yeast increases by about 5-fold during the process and about three quarters of this is discarded. In this state it constitutes a low value product. Some is used to produce yeast extract but it is generally regarded as a waste disposal problem rather than as a useful by-product. Recently gene manipulation techniques are being used to increase the value of the spent yeast. The aim is to introduce expression vectors into brewing strains. These vectors are engineered to contain highly regulated promoters which are switched off during brewing but which can be switched on when the spent yeast is harvested and re-suspended in an inducing medium. In this way, large quantities of useful proteins which are needed in bulk, such as enzymes used in food production, e.g. chymosin or inert blood proteins such as serum albumin, can be produced (Dixon, 1986).

Gene manipulation and cheese production

Cheese is produced by the action of milk clotting enzymes which are traditionally extracted from the fourth stomach of suckling calves as a preparation called rennet. The major enzyme, chymosin, cleaves the major milk protein casein and causes coagulation which is the first stage in cheese making, i.e. the production of curds. Chymosin is required in large quantities and production from suckling calfs may eventually be uneconomical and/or unacceptable. The gene for calf chymosin has been cloned and expressed in *S. cerevisiae* to produce high yields of enzyme that displays all the characteristics of the natural enzyme including the ability to coagulate milk (Mellor *et al.*, 1983; R.A. Smith *et al.*, 1985). Genetically manipulated yeast may therefore, provide an economical alternative source of chymosin.

The cheese industry generates a problematic waste product whey, which is the liquid remaining after curd production. It has a high carbohydrate content and is readily metabolised by microorganisms which would create a huge oxygen demand if it were discharge directly into rivers. Whey contains 4–5% lactose, a small amount of protein and a high mineral and vitamin content. It is therefore worth using rather than discarding. One approach is to degrade the lactose into glucose and galactose which can then be fermented by *S. cerevisiae* to produce ethanol and single cell protein. The degradation requires β-galactosidase which until recently had to be added to whey to produce a hydrolysate because *S. cerevisiae* cannot grow on lactose. To overcome this problem the genes for β-galactosidase and lactose permease derived from *Klyveromyces lactis* have been transferred into *S. cerevisiae* (Sreekrishna and Dixon, 1986). Although improvements in expression of the enzymes have still to be made such a genetically reprogrammed yeast will ultimately be able to use whey as a substrate to produce fuel alcohol, potable alcohol and biomass that could be used to supplement animal feeds during the process of disposal of a major pollutant (Russell, 1986).

The exploitation of cellulose

Cellulose is the most abundant organic molecule on the planet. It is a renewable resource because it is the major component of plants. About 4×10^{10} tonnes of cellulose are synthesised annually by photosynthesis and the total world resource is 7×10^{11} tonnes. Cellulose is a linear polymer of anhydrous glucose molecules joined by β1–4 linkages. The recurring unit is called cellobiose. The complete degradation of cellulose to glucose would provide a major source of food, fuel and chemical feedstocks (Coughlan, 1985). In addition, the ability to degrade cellulose economically would reduce pollution from the waste products of forestry and agriculture and from the paper, lumber and textile industries.

Cellulose is difficult to degrade because it usually exists as insoluble fibres and crystalline arrays. The fibres associate with other polysaccharides such as hemicellulose and pectin and are often surrounded by lignin which prevents degradative enzymes reaching the cellulose. The natural degradation of plant matter is carried out by microorganisms such as filamentous fungi, e.g. *Trichoderma reesei* which secrete massive amounts (20 mg protein ml-in culture) of cellulolytic enzymes (Montencourt, 1983). A mixture of enzymes particularly endocellulases, exocellulases and β-glucosidases is secreted. The productivity of individual enzymes is low, they have a low specific activity particularly for cystalline cellulose and their synthesis is often repressed by glucose. In addition, several different organisms must coexist to produce the full range of enzymes. This means that natural organisms cannot easily be used as the basis of an economical cellulose degradation process.

The genes for the various cellulolytic enzymes from a range of bacteria and filamentous fungi have been cloned. A simple idea is to introduce these genes into the yeast *S. cerevisiae* (e.g. Skipper *et al.* 1985) *S. cerevisiae* ferments glucose to produce ethanol but does not degrade cellulose. If it could be engineered to secrete cellulases it would both degrade the substrate to produce glucose and then ferment the glucose to give a single-step ethanol production process (reviewed by Knowles *et al.*, 1987).

Efficient cellulolytic microbes would be of great value, but also potentially of great danger. Additional manipulations would have to be made to ensure that these organisms could not damage the cellulose based industries by degrading the valued product rather than the waste products.

Summary

Recombinant DNA and gene transfer technology have provided the basic tools to develop entirely new processes and products for benefiting society. These range from the production of vaccines against global and life threatening deseases such as hepatitis to increasing productivity on the farm. In the future we will see biotechnologists develop new therapies for diseases such as cancer and heart disease and new diagnostic tools for predicting and preventing disease. Biotechnology can help solve some world food problems by creating plants that thrive in 'hostile' environments so putting the food where it is needed, i.e. in the Third World. In the more affluent areas of the world where there is self-sufficiency and at times excess, the increased efficiency and productivity generated by biotechnology will release land back to nature giving improved possibilities for conservation and leisure. Biotechnology can also help to control pollution by creating organisms with novel metabolic capacity to destroy waste products. Biotechnology is therefore a civilising technology.

The design and development of all these new products and processes relies upon exploiting the basic information about the biology of the different organisms that we have discussed in previous chapters. Exploiting knowledge about gene expression and genome organisation and replication allows new expression vectors to be developed. The information gained about how proteins are targeted and modified post-translationally can be used to select host cells and to design secretion systems for producing proteins and to modify organelles, as in chloroplast engineering to produce herbicide-resistant plants. A knowledge of the structure—function relationship in proteins is helping biotechnologists design and produce new therapeutic proteins and industrial enzymes. Each new advance in our understanding of fundamental principles can be used to extend the potential of biotechnology.

It is to be hoped that this pursuit of pure knowledge and its wise application will continue indefinitely.

References

Abel, P.P., Nelson, R.S., De, B., Hoffmann, N., Rogers, S.G., Fraley, R.T. and Beachy, R.N. (1986) Delay of disease development in transgenic plants that express the tobacco mosaic virus coat protein gene. *Science* **232**, 738–743.

Adams, J.M. (1985) Oncogenic intelligence: oncogene activation by fusion of chromosomes in leukaemia. *Nature* **315**, 542.

Adams, J.M., Harris, A.W., Pinkert, C.A., Corcoran, L.M., Alexander, W.S., Cory, S. Palmiter, R.D. and Brinster, R.L. (1985) The *c-myc* oncogene driven by immunoglobulin enhancers induces lymphoid malignancy in transgenic mice. *Nature* **318**, 533–538.

Adams, S.E., Kingsman, S.M. and Kingsman, A.J. (1987a) The yeast Ty element: recent advances in the study of a model retro-element. *Bioessays* **7**, 3–9.

Adams, S.E., Dawson, K.M., Gull, K., Kingsman, S.M. and Kingsman, A.J. (1987c) The expression of hybrid HIV: Ty virus-like particles in yeast. *Nature* **329**, 68–70.

Adams, S.E., Mellor, J., Gull, K., Sim, R.B., Tuite, M.F., Kingsman, S.M. and Kingsman, A.J. (1987b) The functions and relationships of Ty-VLP protein in yeast reflect those of mammalian retroviral proteins. *Cell* **49**, 111–119.

Ahlquist, P., French, R. and Bujarski, J.J. (1987) Molecular studies of brome mosaic virus using infectious transcripts from cloned cDNA. *Adv. Virus Res.* **32**, 215–242.

Ahlquist, P., French, R., Junda, M. and Loesch-Fries, L.S. (1984) Multicomponent RNA plant virus infection derived from cloned viral cDNA. *Proc. Nat. Acad. Sci. USA* **81**, 7066–7070.

Air, G.M. and Laver, W.G. (1986) The molecular basis of antigenic variation in influenza virus. *Adv. Virus Res.* **31**, 53–102.

Alberts, B., Bray, D. Lewis, J., Raff, M., Roberts, K. and Watson, J.D. (1983) *The Molecular Biology of the Cell*. Garland Publishing, New York.

Aldovini, A., Debouck, C., Feinberg, M.B., Rosenberg, M., Arya, S.K. and Wong-Staal, F. (1986) Synthesis of the complete *trans*-activation gene product of human T-lymphotropic virus type III in *Escherichia coli*: Demonstration of immunogenicity *in vivo* and expression *in vitro*. *Proc. Natl Acad. Sci. USA* **83**, 6672–6676.

Allen, H., Wraith, D., Pala, P., Askonas, B. and Flavell, R.A. (1984) Domain interactions of H-2 class 1 antigens alter cytotoxic T-cell recognition sites. *Nature* **309**, 279–281.

Alt, F.W. (1986) Antibody diversity: new mechanism revealed. *Nature* **322**, 772–773.

Amara, S.G., Jonas, R., Rosenfeld, M., Ong, E.S. and Evans, R.M. (1982) Alternative RNA processing in calcitinin gene expression generates mRNAS encoding different polypeptide products. *Nature* **298**, 240–244.

Anderson, M.A., Cornish, E.C., Man, S-L., Williams, E.G., Hoggart, R., Atkinson, A., Bonig, I., Grego, B., Simpson, R., Roche, P.J., Haley, J.D., Penschow, J.D., Niall, H.D., Tregear, G.W., Coghlan, J.P., Crawford, R.J. and Clarke, A.T. (1986) Cloning of cDNA for a stylar glycoprotein associated with expression of self compatability in *Nicotiana alata*. *Nature* **321**, 33–38.

Anderson, W.F. (1984) Prospects for human gene therapy. *Science* **226**, 401–409.

Anderson, W.F., Killos, L., Sanders-Haigh, L., Kretschmer, P.J. and Diacumakos, E.G. (1980) Replication and expression of thymidine kinase and human globin genes micro-injected into mouse fibroblasts. *Proc. Natl Acad. Sci. USA* **77**, 5399–5403.

Andrews, B.J., Proteau, G.A., Beatty, L.G. and Sadowski, P.D. (1985) The FLP recombinase of the 2μ circle of DNA of yeast: interaction with its target sequences. *Cell* **40**, 795–803.

Androphy, E.J., Schiller, J.T. and Lowy, D.R. (1985) Identification of the protein encoded by the E6 transforming gene of bovine papillomavirus. *Science* **230**, 442–445.

Anson, D.S., Austen, D.E.G. and Brownlee, G.G. (1985) Expression of active human clotting factor IX from recombinant DNA clones in mammalian cells. *Nature* **315**, 683–685.

REFERENCES

Ares, M., Mangin, M. and Weiner, A.M. (1985) Orientation dependent transcriptional activation upstream of a human U2 snRNA gene. *Molec. Cell. Biol.* **5**, 1560−1570.

Ariga, H., Hane, T. and Iguchi-Ariga, S.M.M. (1987) Autonomous replicating sequences from mouse cells which can replicate in mouse cells *in vivo* and *in vitro*. *Molec. Cell. Biol.* **7**, 1−6.

Arnold, B., Burgert, H-G., Hamann, U., Hammerling, G., Kees, U. and Kvist, S. (1984) Cytolytic T cells recognise the two amino terminal domains of H-2 K antigens in tandem in influenza A infected cells. *Cell* **38**, 79−87.

Avvedimento, V.E., Musti, A.M., Obici, S., Cocozza, S. and Di Lauro, R. (1984) Sequence Organisation of the 3′ half of the rat thyroglobulin gene. *Nucl. Acids Res.* **12**, 3461−3472.

Babiss, L.E., Herbst, R.S., Bennett, A.L. and Darnell, J.E. (1987) Factors that interact with the rat albumin promoter are present both in hepatocytes and other cell types. *Genes Develop.* **1**, 256−267.

Bachmair, A., Finley, O. and Varshavsky, A. (1986) *In vivo* half-life of a protein is a function of its amino terminal residue. *Science* **234**, 179−186.

Baer, R., Bankier, A.T., Biggin, M.D. Deininger, P.L., Farrell, P.J., Gibson, T.J., Hatfull, G., Hudson, G.S., Satchwell, S.C., Séguin, C., Tuffnell, P.S. and Barrell, B.G. (1984) DNA sequence and expression of the B95-8 Epstein−Barr virus genome. *Nature* **310**, 207−211.

Baim, S.B., Pietras, D.F., Eustice, D.C. and Sherman, F. (1985) A mutation allowing an mRNA secondary structure diminishes translation of *Saccharomyces cerevisiae* iso-1-cytochrome C. *Molec. Cell. Biol.* **5**, 1839−1846.

Baker, S.M. and Platt, T. (1986) Pol 1 transcription: which comes first, the end or the beginning. *Cell* **47**, 839−840.

Ballance, D.J., Buxton, F.P. and Turner, G. (1983) Transformation of *Aspergillus nidulans* by the orotidine-5′-phosphate decarboxylase gene of *Neurospora crassa*. *Biochem. Biophys. Res. Comm.* **112**, 284−289.

Ballance, D.J. and Turner, C. (1985) Development of a high-frequency transforming vector for *Aspergillus nidulans*. *Gene* **36**, 321−331.

Ballay, A., Levrero, M., Buendia, M-A. Tiollais, P. and Perriccaudet, M. (1985) *In vitro* and *in vivo* synthesis of the hepatitis B virus surface antigen and of the receptor. For polymerised human serum albumin from recombinant adenoviruses. *EMBO J.* **4**, 3861−3865.

Balazs, E., Bouzoubaa, S., Guilley, H., Jonasd, G., Paszkowski, J. and Richards, K. (1985) Chimeric vector construction for higher plant transformation. *Gene* **40**, 343−348.

Ballou, C.E. (1982) Yeast cell wall and cell surface. In *Molecular Biology of the Yeast Saccharomyces, Vol.2: Metabolism and Gene Expressions*, pp. 335−336 (eds J. Strathern, E.W., Jones & J.R. Broach), Cold Spring Harbor Laboratory, New York.

Banerji, J., Rusconi, S. and Schaffner, W. (1981) Expression of a β-globin gene is enhanced by remote SV40 DNA sequences. *Cell* **27**, 299−308.

Bankaitis, V.A., Johnson, L.A. and Emv, S.D. (1986) Isolation of yeast mutants defective in protein targeting to the vacuole. *Proc. Natl Acad. Sci. USA* **83**, 9075−9079.

Bannerjee, A.K. (1980) 5′-terminal cap structure in eukaryotic messenger ribonucleic acid. *Microbiol. Rev.* **44**, 175−205.

Baron, N., Lapidot, A. and Manor, H. (1987) Unusual sequence element found at the end of an amplicon. *Molec. Cell. Biol.* **7**, 2636−2640.

Barbacid, M. (1987) *ras* genes. *Ann. Rev. Biochem.* **56**, 779−829.

Barberis, A., Superti-Furga, G. and Busslinger, M. (1987) Mutually exclusive interaction of the CCAAT-binding factor and of a displacement protein with overlapping sequences of a histone gene promoter. *Cell* **50**, 347−359.

Bargiello, T.A. and Young, M.W. (1984) Molecular genetics of a biological clock in *Drosophila*. *Proc. Natl Acad. Sci. USA* **81**, 2142−2146.

Bargmann, C.I., Hung, M-C. and Weinberg, R.A. (1986) Multiple independent activations of the *neu* oncogene by a point mutation altering the transmembrane domain of p185. *Cell* **45**, 649−657.

Barnes, D.A. and Thorner, J. (1986) Genetic manipulation of *Saccharomyces cerevisiae* by use of the *LYS2* gene. *Molec. Cell. Biol.* **6**, 2828−2838.

Baron, M.H. and Maniatis, T. (1986) Rapid reprogramming of globin gene expression in transient heterokaryons. *Cell* **46**, 591−602.

Barr, P.J., Power, M.D., Lee-Ng, C.T., Gibson, H.L. and Luciw, P.A. (1987) Expression of active human immuno-deficiency virus reverse transcriptase in *Saccharomyces cerevisiae*. *Bio/Technol.* **5**, 486−489.

Barré-Sinoussi, F., Chermann, J.C., Rey, F., Nugeyre, M.T., Chamaret, S., Gruest J., Dauguet, C., Axler-Blin-C., Vézinet-Brun, A., Rouzioux, F., Rozenbaum, W. and Montegnier, L. (1983) Isolation of a T-lymphotropic retrovirus from a patient at risk of acquired immune deficiency syndrome (AIDS). *Science* **220**, 868−870.

Barton, K.A. and Brill, W.J. (1983) Prospects in plant genetic engineering. *Science* **219**, 671–676.

Barker, R.F., Thompson, D.V., Talbot, D.R., Swanson, J. and Bennetzen, J.L. (1984) Nucleotide sequence of the maize transposable element Mu1. *Nucl. Acids Res.* **12**, 5955–5967.

Bateman, A.J. (1975) Simplification of palindromic telomere theory. *Nature* **253**, 379.

Baumann, G., Raschke, E., Bevan, M. and Schoffl, F. (1987) Functional analysis of sequences required for transcriptional activation of a soybean heat shock gene in transgenic tobacco plants. *EMBO J.* **6**, 1161–1166.

Baylies, M.K., Bargiello, T.A., Jackson, F.R. and Young, M.W. (1987) Changes in the abundance or structure of the *per* gene product can alter periodicity of the *Drosophila* clock. *Nature*, **326**, 390–392.

Bazett-Jones, D.P., Yeckel, M. and Gottesfeld, J.M. (1985) Nuclear extracts from globin synthesising cells enhance globin transcription *in vitro*. *Nature* **317**, 824.

Beach, D. and Nurse, P. (1981) High frequency transformation of the fission yeast *Schizosaccharomyces pombe*. *Nature* **290**, 140–142.

Beachy, P.A., Helfand, S.L. and Hognen, D.S. (1985) Segmental distribution of bithorax complex proteins during *Drosophila* development. *Nature* **33**, 545–551.

Bebbington, C. and Hentschel, C. (1985) The expression of recombinant DNA products in mammalian cells. *Trends Biotech.* **3**, 314–317.

Beck, E., Ludwig, G., Auerswald, E.A., Reiss, B. and Schaller, H. (1982) Nucleotide sequence and exact localisation of the neomycin phosphotransferase gene from transposon Tn5. *Gene* **19**, 327–336.

Becker, P.B., Gloss, B., Schmid, W., Strahle, U. and Schutz, G. (1986) *In vivo* protein –DNA interactions in a glucocorticoid response element require the presence of the hormone. *Nature* **324**, 686–688.

Beckner, S.K., Hattori, S. and Shih, T. Y. (1985) The *ras* oncogene product is not a regulatory component of adenylate cyclase. *Nature* **317**, 71–72.

Beggs, J.D. (1978) Transformation of yeast by a replicating hybrid plasmid. *Nature* **275**, 104–109.

Beggs, J.D., Van den Berg, J., Van Ooyen, A. and Weissman, C. (1980) Abnormal expression of chromosomal rabbit β-globin gene in *Saccharomyces cereviciae*. *Nature* **283**, 835–840.

Behbehani, A.M. (1983). The smallpox story: Life and death of an old disease. *Microbiol. Rev.* **47**, 455–509.

Bell, R.M. (1986) Protein kinase C activation by diacyl glycerol second messengers. *Cell* **45**, 631.

Belmont, J.W. and Caskey, C.T. (1986). Developments leading to human gene therapy. In *Gene Transfer* (ed. R. Kucherlapatic) pp. 411–441. Plenum Press, New York.

Belsham, G.J., Barker, D.G. and Smith, A.E. (1986) Expression of polyoma virus middle – T antigen in *Saccharomyces cerevisiae*. *Eur. J. Biochem.* **156**, 413–421.

Bender, W., Akurn, M., Karch, F., Beachy, P.A., Perfer, M. Spierer, P., Lewis, E.B. and Hogness D.S. (1983) Molecular genetics of the bothorax complex in *Drosophila melanogaster*. *Science* **221**, 23–29.

Bendig, M.M. and Williams, J.G. (1983) Replication and expression of *Xenopus laevis* globin genes injected into fertilized *Xenopus* eggs. *Proc. Natl Acad. Sci. USA* **80**, 6197–6201.

Bennett, J.W. and Lazure, L.L. (1985) *Gene Manipulatious in Fungi*. Academic Press, London.

Bennetzen, J.L. and Hall, B.D. (1982a) Codon selection in yeast. *J. Biol. Chem.* **257**, 3029–3031.

Bennetzen, J.L. and Hall, B.D. (1982b) The primary structure of the *Saccharomyces cerevisiae* gene for alcohol dehydrogenase 1. *J. Biol. Chem.* **257**, 3018–3025.

Benoist, C. and Chambon, P. (1980) Deletions covering the putative promoter region of early mRNAs of simian virus 40 do not abolish T-antigen expression. *Proc. Natl Acad. Sci. USA* **77**, 3865–3869.

Benoist, C. and Chambon, P. (1981) *In vivo* sequence requirements of the SV40 early promoter region. *Nature* **290**, 304–311.

Berg, J.M. (1986) Potential metal-binding domains in nucleic acid binding proteins. *Science* **232**, 485–487.

Berg, L., Lusky, M., Stenlund, A. and Botchan, M.R. (1986a) Repression of bovine papillomavirus replication is mediated by a virally encoded traumacting factor. *Cell* **46**, 753–762.

Berg, L.J., Singh, K. and Botchan, M. (1986b) Complementation of a BPV low copy number mutant: evidence for a temporal requirement of the complementing gene. *Molec. Cell. Biol.* **6**, 859–869.

Berget, S.M. and Robberson, B.L. (1986) U1, U2 and U4/U6 small nuclear ribonucleoproteins are required for *in vitro* splicing but not polyadenylation. *Cell* **46**, 691–696.

Berk, A.J. (1986) Adenovirus promoters and E1A transactivation. *Ann. Rev. Genet.* **20**, 45–81.

459   REFERENCES

Berk, A.J. and Sharp, P.A. (1977) Sizing and mapping of early adenovirus mRNAS by gel electrophoresis of S1 endonuclease digested hybrids. *Cell* **12**, 721−732.

Berman, J., Tachibana C.Y. and Tye, B.K. (1986) Identification of a telomere-binding activity from yeast. *Proc. Natl Acad. Sci. USA* **83**, 37B.

Berman, P.W. and Laskey, L.A. (1985) Engineering glycoproteins for use as pharmaceuticals. *Trends Biotech.* **3**, 51−53.

Bernstein, L.B., Mount, S.M. and Weiner, A.M. (1983) Pseudogenes for human small nuclear RNA U3 appear to arise by integration of self primed reverse transcripts of the RNA into new chromosomal sites. *Cell* **32**, 461−472.

Berridge, M.J. (1987) Inositol triphosphate and diacyl glycerol: two interacting second messengers. *Ann. Rev. Biochem.* **56**, 159−195.

Bevan, M. (1984) Binary *Agrobacterium* vectors for plant transformation. *Nucl. Acids Res.* **12**, 8711−8721.

Bialy, H. and Klausner, A. (1986) A new route to virus resistance in plants. *Bio/Technol.* **4**, 96.

Bieker, J.J., Martin, P.L. and Roeder, R.G. (1985) Formation of a rate-limiting intermediate in 5S RNA gene transcription. *Cell* **40**, 119−127.

Bienz, M. and Pelham, H.B. (1986) Heat shock regulating elements function as an inducible enhancer in the *Xenopus hsp70* gene and when linked to a heterologous promoter. *Cell* **45**, 753−760.

Bingham, P.M., Levis, R. and Rubin, G.M. (1981) Cloning of DNA sequences from the *white* locus of *D. melanogaster* by a novel and general method. *Cell* **25**, 693−704.

Birchmeier, C., Folk, W. and Birnstiel, M. (1983) The terminal RNA stem−loop structure and 80 bp of spacer DNA are required for the formation of 3′ termini of sea urchin H2A mRNA. *Cell* **35**, 433−440.

Birchmeier, C., Grosschedl, R. and Birnstiel, M.L. (1982) Generation of authentic 3′ termini of an H2A mRNA *in vivo* is dependent on a short inverted DNA repeat and on spacer sequences. *Cell* **28**, 739−745.

Birchmeier, C., Schuielperli, D., Sionzo, G. and Birnstice, M.L. (1984) 3′ editing of mRNAs: sequence requirements and involvement of a 60-nucleotide RNA in maturation of histone mRNA precursors. *Proc. Natl Acad. Sci. USA* **81**, 1057−1061.

Birnstiel, M.L., Busslinger, M. and Strub, K. (1985) Transcription termination and 3′ processing: the end is in site! *Cell* **41**, 349−359.

Bishop, J.M. (1984). Viral oncogenes. *Cell* **42**, 23−38.

Bishop, J.M. (1985a) Trends in oncogenes. *Trends Genet.* **1**, 245−250.

Bishop, J.M. (1985b) Viral oncogenes. *Cell* **42**, 23−38.

Bitter, G.A., Chen, K.K., Banks, A.R. and Lai, P-H. (1984) Secretion of foreign proteins from *Saccharomyces cerevisiae* directed by α-factor gene fusions. *Proc. Natl Acad. Sci. USA* **81**, 5330−5304.

Bitter, G.A. and Egan, K.M. (1984) Expression of heterologous genes in *Saccharomyces cerevisiae* from vectors utilising the glyceraldehyde-3-phosphate dehydrogenase gene promoter. *Gene* **32**, 263−274.

Bjorkman, P.J., Saper, M.A., Samraoui, B., Bennett, W.S., Strominger, J.L. and Wiley, D.C. (1987) Structure of the human class I histocompatibility antigen HLA-A2. *Nature* **329**, 506−512.

Black, D.L. and Steutz, J.A. (1986) Pre-mRNA splicing *in vitro* requires intact U4/U6 small nuclear ribonucleoprotein. *Cell* **46**, 697−704.

Blackburn, E.H. and Gall, J.G. (1978) A tandemly repeated sequence at the termini of the extrachromosomal ribosomal RNA genes in *Tetrahymena*. *J. Molec. Biol.* **120**, 33−53.

Blackburn, E.H. and Szostak, J.W. (1984) The molecular structure of centromeres and telomeres. *Ann. Rev. Biochem.* **53**, 163−194.

Blair, D.G., Oskarsson, M.K., Seth, A., Dunn, K.J., Dean, M., Zweig, M., Tainsky, M.A. and Van de Woude, G.F. (1986) Analysis of the transforming potential of the human homolog of *mos*. *Cell* **46**, 785−794.

Blair, D.G., Oskarsson, M., Wood, T.G., McClements, W.L., Fischinger, P.J. and Van de Woude, G.G. (1981) Activation of the transforming potential of a normal cell sequence: a molecular model for oncogenesis. *Science* **212**, 941−943.

Blake, C.C.F. (1985) Exons and the evolution of proteins. *Int. Rev. Cytol.* **93**, 149−185.

Blancou, J., Kieny, M.P., Lathe, R. Lecocq, J.P., Pastoret, P.P., Soulebot, J.P. and Desmettre, P. (1986) Oral vaccination of the fox against rabies using a live recombinant vaccinia virus. *Nature* **322**, 373−375.

Blobel, G. (1980) Intracellular protein topogenesis. *Proc. Natl Acad. Sci. USA* **77**, 1496−1500.

Blobel, G. and Dobberstein, B. (1975) Transfer of protein across membranes II Reconstitution of functional rough microsomes from heterologous components. *J. Cell. Biol.* **67**, 852−862.

Bloom, K.S. and Carbon, J. (1982) Yeast centromere DNA is in a unique and highly ordered structure in chromosomes and small circular micrchromosomes. *Cell* **29**, 305−317.

Bloom, K.S., Fitzgerald-Hayes, M. and Carbon, J. (1983) Structural analysis and sequence organisation of yeast centromeres. *Cold Spring Harb or Symp. Quant. Biol.*, Vol. 617, pp. 1175–1185. Cold Spring Harbor, New York.

Bloom, K., Hill, A. and Yeh, E. (1986) Structural analysis of a yeast centromere. *Bio Essays* **4**, 100–104.

Blumenthal, A.M., Kriegstein, H.J. and Hogness, D.S. (1974) The units of DNA replication in *Drosophila melanogaster* chromosomes. *Cold Spring Harbor. Symp. Quant. Biol.* Vol. 38, pp. 205–251. Cold Spring Harbor, New York.

Bluestone, J.A., Foo, M., Allen, H., Segal, D. and Flavell, R.A. (1985) Allospecific cytolytic T lymphocytes recognise conformational determinants on hybrid mouse transplantation antigens *J. Expl Med.* **162**, 268–281.

Blundell, T.L., Sibanda, B.L., Sternberg, M.J.E. and Thornton, J.M. (1987) Knowledge based prediction of protein structures and the design of novel molecules. *Nature* **326**, 347–352.

Boast, S., La Mantia, G., Lania, L. and Blasi, F. (1983) High efficiency of replication and expression of foreign genes in SV40-transformed human fibroblasts. *EMBO J.* **2**, 2327–2331.

Boeke, J.D., Garfinkel, D.J., Styles, C.A. and Fink, G.R. (1985) Ty elements transpose through an RNA intermediate. *Cell* **40**, 491–500.

Boeke, J.D., LaCroute, F. and Fink, G.R. (1984) A positive selection for mutants lacking orotidine-5′-phosphate decarboxylate activity in yeast: 5-fluoro-orotic acid resistance. *Molec. Gen. Genet.* **197**, 345–347.

Bogenhagen, D.F. (1985) The intragenic control region of the *Xenopus* 5S gene contains two factor A binding domains that must be aligned properly for efficient transcription initiation. *J. Biol. Chem.* **260**, 6466–6471.

Bohmann, D. Keller, W., Dale, T., Scholer, H.R., Tebb, G. and Mattaj, I.W. (1987) A transcription factor which binds to the enhancer of SV40, immunoglobulin heavy and U2 snRNA genes. *Nature* **325**, 268–272.

Botchan, M., Stringer, J., Mitchison, T. and Sambrook. J. (1980) Integration and excision of SV40 DNA from the chromosome of a transformed cell. *Cell* **20**, 143–152.

Bond, V.C and Wold, B. (1987) Poly-L-Ornithine-mediated transformation of mammalian cells. *Molec. Cell. Biol.* **7**, 2286–2293.

Boulianne, G.L., Hozumi, N. and Shulman, M.J. (1984) Production of functional chimaeric mouse/human antibody. *Nature* **312**, 643–646.

Bonnerjea, J., Oh, S., Hoare, M. and Dunnill, P. (1986) Protein purification: the right step at the right time. *Bio/Technol.* **4**, 954–958.

Borlaug, N.E. (1983) Contributions of conventional plant breeding to food production. *Science* **219**, 689–694.

Borrelli, E., Hen, R. and Chambon, P. (1984) Adenovirus-2 E1A products repress enhancer induced stimulation of transcription. *Nature* **312**, 608–612.

Boshart, M., Weber, F., Jahn, G., Dorsch-Hasler, K., Fleckenstein, B. and Schaffner, W. (1985) A very strong enhancer is located upstream of an immediate early gene of human cytomegalovirus. *Cell* **41**, 521–530.

Botchan, M., Stringer, J., Mitchison, T. and Sambrook, J. (1980) Integration and excision of SV40 DNA from the chromosome of a transformed cell. *Cell* **20**, 143–152.

Botstein, D. and Shortle, D. (1985) Strategies and applications of *in vitro* mutagenesis. *Science* **229**, 1193–1201.

Boutry, M., Nagy, F., Poulsen, C., Aoyagi, K. and Chua, N-H. (1987) Targeting of bacterial chloramphenicol acetyl transferase to mitochondria in transgenic plants. *Nature* **328**, 340–342.

Boutwell, R.K. (1974) The function and mechanism of promoters of carcinogenesis. *CRC Crit. Rev. Toxicol.* **2**, 419–431.

Brady, G., Funk, A., Mattern, J., Schutz, G. and Brown, R. (1985) Use of gene transfer and a novel cosmid rescue strategy to isolate transforming sequences. *EMBO J.* **4**, 2583–2588.

Brake, A.J., Merryweather, J.P., Loit D.G., Heberleia, U.A., Masriarz, F.R., Mullenbach, G.T., Urdea, M.S., Valenzuela, P. and Barr, P.J. (1984) α-Factor directed synthesis and secretion of mature foreign proteins in *Saccharomyces cerevisiae*. *Proc. Natl Acad. Sci. USA* **81**, 4642–4646.

Bram, R.J. and Kornberg, R.D. (1985) Specific protein binding to far upstream activating sequences in polymerase II promoters. *Proc. Natl Acad. Sci. USA* **82**, 43–47.

Bram, R.J. and Kornberg, R.D. (1987) Isolation of a *Saccharomyces cerevisiae* centromere DNA-binding protein, its human homolog, and its possible role as a transcription factor. *Molec. Cell. Biol.* **7**, 403–409.

Brand, A.H., Breeden, L., Abraham, J., Sternglaz, R. and Nasmyth, K. (1985) Character-ization of a 'silencer' sequence in yeast: a DNA sequence with properties opposite to those of a transcriptional enhancer. *Cell* **41**, 41–48.

Brash, D.E., Reddel, R.R., Quanrud, M., Yang, K., Farrell, M.R. and Harris, C.C. (1987) Strontium phosphate transfection of human cells in primary culture: stable expression of the

Simian viris 40 large T antigen gene in primary human bronchial epethelial cells. *Molec. Cell. Biol.* **7**, 2031–2034.

Breitbart, R.E., Andreades, A. and Nadal-Ginard, B. (1987) Alternative splicing: a ubiquitous mechanism for the generation of multiple protein isoforms from single genes. *Ann. Rev. Biochem.* **56**, 467–497.

Breitbart, R.E. and Nadal-Genard, B. (1987) Developmentally induced, muscle-specific *trans* factors control the differential splicing of alternative and constitutive troponin T exons. *Cell* **49**, 793–803.

Brent, R. (1985) Repression of transcription in yeast. *Cell* **42**, 3–4.

Brent, R. and Ptashne, M. (1984) A bacterial repressor protein or a yeast transcription terminator can block upstream activation of a yeast gene. *Nature* **312**, 612–615.

Brent, R. and Ptashne, M. (1985) A eukaryotic transcriptional activator bearing the DNA specificity of a prokaryotic repressor. *Cell* **43**, 729–736.

Brewer, B. and Fangman, W. (1987) The localisation of relication origins on ARS plasmids in *Saccharomyces cerevisiae*. *Cell* **51**, 463–471.

Brewer, S.J. and Sassenfeld, H.M. (1985) The purification of recombinant proteins using C-terminal polyarginine fusions. *Trends Biotech.* **3**, 119–125.

Briggs, M.K., Kadonaga, J.T., Bell S.P. and Tjian, R. (1986) Purfication and biochemical characterisation of the promoter-specific transcription factor Sp1. *Science* **234**, 47–52.

Brinster, R.L., Ritchie, K.A., Hammer, R.E., O'Brien, R.L., Arp, B. and Storb, U. (1983) Expression of a microinjected immunoglobulin gene in the spleen of transgenic mice. *Nature* **306**, 332–336.

Brisson, N., Paszkowski, J., Penswick, J.R., Gronenborn, B., Potrykus, I. and Hohn, T. (1984) Expression of a bacterial gene in plants by using a viral vector. *Nature* **310**, 511–514.

Broach, J.R. (1982) The yeast plasmid 2μ circle. *Cell* **28**, 203–204.

Broach, J.R., Li Y-Y, Chen L-C. and Jayaram, M. (1983a) Vectors for high level inducible expression of cloned genes in yeast. In *Experimental Manipulation of Gene Expression* (ed. M. Inouye), pp. 82–117. Academic Press, New York.

Broach, J.R., Li, Y-Y., Feldman, J., Jayaram, M., Abraham, J., Nasmyth, K. and Hicks, J. (1983b) Localisation and sequence analysis of yeast origins of DNA replication. *Cold Spring Harbor Symp. Quant. Biol.* **47**, 1165–1173. Cold Spring Harbor, New York.

Brock, M.L. and Shapiro, D.J. (1983) Estrogen stabilises vitellogenin mRNA against cytoplasmic degradation. *Cell* **34**, 207.

Brody, E. and Abelson, J. (1985) The "splicesome": yeast pre-messenger RNA associates with a 40S complex in a splicing-dependent reaction. *Science* **228**, 963–967.

Brown, A.M.C., Wildin, R.S., Prendergast, T.J. and Varmus, H.E. (1986) A retrovirus expressing the putative mammary oncogene *int*-1 causes partial transformation of a mammary epithelial cell line. *Cell* **46**, 1001–1009.

Brown, D.D. (1982) How a simple animal gene works. *Harvey Lectures* (series), **76**, 27–44.

Brown, D.D. (1984) The role of stable complexes that repress and activate eukaryotic genes. *Cell* **37**, 359–365.

Brown, F. (1984) Synthetic viral vaccines. *Ann. Rev. Microbiol.* **3**, 221–235.

Brown, F., Schild, G.C. and Ada, G.L. (1986) Recombinant vaccinia viruses as vaccines. *Nature* **319**, 549–550.

Brown, K., Quintanilla, M., Ramsden, M., Kerr, I.B., Young, S. and Balmain, A. (1986) V-*ras* genes from Harvey and BALB marine Sarcoma eiruses can act as initiators of two stage mouse skin carcinogenesis. *Cell* **46**, 447–456.

Brown, M.S., Anderson, R.G.W. and Goldstein, J. (1983) Recycling receptors: the round trip itinerary of migrant membrane proteins. *Cell* **32**, 663–667.

Bucchini, D., Reynaud, C-A., Ripoche, M-A., Grimal, H., Jami, J. and Weill, J-C. (1987) Rearrangement of a chicken immunoglobulin gene occurs in the lymphoid lineage of transgenic mice. *Nature* **326**, 409–411.

Buchanan-Wollaston, V., Passiatore J.E. and Cannon, F. (1987) The *mob* and *ori* T mobilization functions of a bacterial plasmid promote its transfer to plants. *Nature* **328**, 172–175.

Burgess, T.L., Crack, C.S. and Kelly, R.B. (1985) The exocrine protein trypsinogen is targeted into the secretary granules of an endocrine cell line: studies by gene transfer. *J. Cell. Biol.* **101**, 639–645.

Burgess, T.L. and Kelly, R.B. (1987) Constitutive and regulated secretion. *Ann. Rev. Cell. Biol.* **3**, 243–295.

Burhans, W.C., Selegue, J.E. and Heintz, N.G. (1986) Isolation of the origin of replication associated with the amplified Chinese hamster dihydrofolate reductase domain. *Proc. Natl Acad. Sci. USA* **83**, 7790–7794.

Burke, D.T., Carle, G.F. and Olsen, M.V. (1987) Cloning of large segments of exogenous DNA into yeast by means of artificial chromosome vectors. *Science* **236**, 806–813.

Busby, S., Kumar, A., Joseph, M., Halfpap, L., Insley, M., Berkner, K., Kurachi, K. and Woodbury, R. (1985) Expression of active human factor IX in transfected cells. *Nature* **316**, 271–273.

Butler, J.P.G. (1983) The folding of chromatin. *CRC Crit. Rev. Biochem.* **15**, 57−91.

Buxton, F.P. and Radford, A. (1984) The transformation of mycelial spheroplasts of *Neurospora crassa* and the attempted isolation of an autonomous replicator. *Molec. Gen. Genet.* **196**, 339−344.

Butt, T.R., Sternberg, E., Herd, J. and Crooke, S.T. (1984) Cloning and expression of a yeast copper metallothionein gene. *Gene* **27**, 23−33.

Byers, B. (1981) Cytology of the yeast life cycle. In *Molecular Biology of the Yeast* Saccharomyces, *Vol. 1: Life cycle and inheritence* (eds J. Strathern, E.W. Jones and J.R. Broach) pp. 59−97, Cold Spring Harbor Laboratory, New York.

Cabezon, T., De Wilde, M., Herion, P., Loriau, R. and Bollen, A. (1984) Expression a human α1-antitirypsin cDNA in the yeast *Saccharomyces cerevisiae*. *Proc. Natl Acad. Sci. USA* **81**, 6594−6598.

Calos, M.P., Lebkowski, J.S. and Botchan, M.R. (1983) High mutation frequency in DNA transfected into mammalian cells. *Proc. Natl Acad. Sci. USA* **72**, 1392−1396.

Camier, S., Gabrielsen, O., Baker, R. and Sentenac, A. (1985) A split binding site for transcription factor τ on the tRNA$_3^{GLU}$ gene, *EMBO J.* **4**, 491−500.

Campbell, J.L. (1986) Eukaryotic DNA replication. *Ann. Rev. Biochem.* **55**, 733−771.

Campo, M.S. (1985) Bovine papillomavirus DNA: a eukaryotic cloning vector. In *DNA Cloning: a Practical Approach*, Vol. II, pp. 213−239, IRL Press, Oxford.

Cannon, R.E., Wu, G.J. and Railey, J.F. (1986) Functions of and interactions between the A and B blocks in adenovirus type 2-specific VA RNA1 gene. *Proc. Natl Acad. Sci. USA* **83**, 1285.

Cantoval, J.M, Diez, B., Barredo, J.L., Alvarez, E. and Martin, J.F. (1987) High frequency transformation of *Penecillium chrysogenum*. *Bio/Technol.* **5**, 494−499.

Capecchi, M.R. (1980) High efficiency transformation by direct microinjection of DNA into cultured mammalian cells. *Cell* **22**, 479−488.

Caplan, A., Herrera-Estrella, L., Inze, D., Van Haute, E., Van Montagu, M., Schell, J. and Zambryski, P. (1983) Introduction of genetic material into plant cells. *Science* **222**, 815−821.

Cappello, J., Handelsman, K. and Lodish, H.F. (1985) Sequence of Dictyostelium DIRS-1: an apparent retrotransposon with inverted terminal repeats and an internal circle junction sequence. *Cell* **43**, 105−115.

Carbon, J. (1984) Yeast centromeres: Structure and function. *Cell* **37**, 351−353.

Carle, G.F., Frank, M. and Olsen, M.V. (1986) Electrophoretic separations of large DNA molecules by periodic inversion of the electric field. *Science* **232**, 65−68.

Carlson, M. and Botstein, D. (1982) Two differentially regulated mRNAs with different 5′ ends encode secreted and intracellular forms of yeast invertase. *Cell* **28**, 145−154.

Carpenter, G. (1987) Receptors for growth factors and other polypeptide mitogens. *Ann. Rev. Biochem.* **56**, 881−915.

Carrell, R. (1984) Therapy by instant evolution. *Nature* **312**, 14.

Carroll, S.M., Gaudray, P., De Rose, M.L., Emery, J.F., Meinhoth, J.L., Nakkem, E., Subler, M., Von Hoff, D.D. and Wahl, G.M. (1987) Characterisation of an episome produced in hamster cells that amplify a transfected CAD gene at high frequency: functional evidence for a mammalian replication origin. *Molec. Cell. Biol.* **7**, 1740−1750.

Carson, M.J. and Hartwell, L. (1985) CDC17: An essential gene that prevents telomere elongation in yeast. *Cell* **42**, 249−257.

Case, M.E., Schweizer, M., Kushner, S.R. and Giles, N.H. (1979) Efficient transformation of *Neurospora crassa* by utilising hybrid plasmid DNA. *Proc. Natl Acad. Sci. USA* **76**, 5259−5263.

Cavalier-Smith, T. (1974) Palindromic base sequences and replication of eukaryotic chromosome ends. *Nature* **250**, 467−470.

Cavalier-Smith, T. (1982) Skeletal DNA and the evolution of genome size. *Ann. Rev. Biophys. Bioeng.* **11**, 273−302.

Celniker, S.E. and Campbell, J.L. (1982) Yeast DNA replication *in vitro*: initiation and elongation events mimic *in vivo* processes. *Cell* **31**, 201.

Celniker, S.E., Sweder, K., Srienc, F., Bailey, J.E. and Campbell, J.L. (1984) Deletion mutations affecting autonomously replicating sequence ARS1 of *Saccharomyces cerevisiae*. *Molec. Cell. Biol.* **4**, 2455−2466.

Cepko, C.L., Roberts, B.E. and Mulligan, R.C. (1984) Construction and applications of a highly transmissible retrovirus shuttle vector. *Cell* **37**, 1053−1062.

Chada, K., Magram, J., Raphael, K., Radice, G., Lacy, E. and Costantini, F. (1985) Specific expression of a foreign β-globin gene in erythroid cells of transgenic mice. *Nature* **314**, 377−380.

Chakrabarti, S., Robert-Guroff, M., Wong-Staal, F., Gallo, R.C. and Moss, B. (1986) Expression of the HTLV-III envelope gene by a recombinant vaccinia virus. *Nature* **320**, 535−537.

Chaleff, R.S. (1983) Isolation of agronomically useful mutants from plant cell cultures. *Science* **219**, 676−681.

Chan, C.S.M. and Tye, B-K. (1980) Autonomously replicating sequences in *Saccharomyces cerevisiae. Proc. Natl Acad. Sci. USA* **77**, 6329−6333.

Chang, A.C.Y., Nunberg, J.H., Kaufman, R.J., Erlich, H.A., Schimke, R.T. and Cohen, S. (1978) Phenotypic expression in *E. coli* of a DNA sequence coding for mouse dihydrofolate reductase. *Nature* **275**, 617−624.

Chang, E.H., Furth, M.E., Scolnick, E.M. and Lowy, D.R. (1982) Tumorigenic transformation of mammalian cells induced by a normal human gene homologous to the oncogene of Harvey murine sarcoma virus. *Nature* **297**, 479−483.

Chang, N.C., Matteucci, M., Perry, J., Wulf, J.J., Chen, C.Y. and Hitzeman, R.A. (1986) *Saccharomyces cerevisiae* secretes and correctly processes human interferon hybrid proteins containing yeast invertase signal peptides. *Molec. Cell. Biol.* **6**, 1812−1819.

Chang, S.M.W., Wager-Smith, K., Tsao, T.Y., Henkel-Tigges, J., Vaishnav, S. and Caskey, C.T. (1987) Construction of a defective retrovirus containing the human hypoxanthine phosphoribosyl transferase cDNA and its expression in cultured cells and mouse bone marrow. *Molec. Cell. Biol.* **7**, 854−863.

Chang, T.W., Kato I., McKinney, S., Chanda, P., Barone, A.D., Wong-Staal, F., Gallo, R.C. and Chang, N.T. (1985) Detection of antibodies to human T-cell lymphotropic virus III (HTLV-III) with an immunoassay employing a recombinant *Escherichia coli*-derived viral antigenic peptide. *Biotechnol.* **3**, 905−909.

Chanock, R.M. and Lerner, R.A. (eds) (1984) *Modern Approaches to Vaccines: Molecular and Chemical Bases of Virus Virulence and Immunogenicity.* Cold Spring Harbor, New York.

Chen, C. and Okayama, H. (1987) High efficiency transformation of mammalian cells by plasmid DNA. *Molec. Cell. Biol.* **7**, 2745−2752.

Chen, E., Howley, P., Levenson, A. and Seeburg, P. (1982) The primary structure and genetic organisation of the bovine papilloma virus type 1 genome. *Nature* **299**, 529−534.

Chen, W. and Struhl, K. (1985) Yeast mRNA initiation sites are determined primarily by specific sequences and not by the distance from the TATA element. *EMBO J.* **4**, 3273−3280.

Childs, G., Maxson, R., Cohn, R.H. and Kedes, L. (1981) Orphons: dispersed genetic elements derived from tandem repetitive genes of eukaryotes. *Cell* **23**, 651−663.

Choi, Y.D., Grabowski, P.J., Sharp, P.A. and Dreyfuss, G. (1986) Heterogeneous nuclear ribonucleoproteins: role in RNA splicing. *Science* **231**, 1534−1539.

Choo, K.H., Gould, K.G., Rees, D.J.G. and Brownlee, G.G. (1982) Molecular cloning of the gene for human antihaemophilic factor IX. *Nature* **299**, 178−180.

Chargaff, E. (1987) Engineering molecular nightmare. *Nature* **327**, 199−200.

Chen, I.S.Y. (1986) Regulation of AIDS virus expression. *Cell* **47**, 1−2.

Chu, G., Hayakawa, H. and Berg, P. (1987) Electroporation for the efficient transfection of mammalian cells with DNA. *Nucl. Acids Res.* **15**, 1312.

Chu, G. and Sharp, P.A. (1981) A gene chimaera of SV40 and mouse β-globin is transcribed and properly spliced. *Nature* **289**, 372−382.

Church, G.M. and Gilbert, W. (1984) Genomic sequencing. *Proc. Natl Acad. Sci. USA* **80**, 3963−3965.

Ciejek, E.M., Tsai, M-J and O'Malley, B.W. (1983) Actively transcribed genes are associated with the nuclear matrix *Nature* **306**, 607−609.

Ciliberto, G., Rangei, G., Costanzo, F., Dente, L. and Cortese, R. (1983) Common and interchangeable elements in the promoters of genes transcribed by RNA polymerase III. *Cell* **32**, 725−733.

Citri, Y., Colot, H.V., Jacquier, A.C., Yu, Q., Hall, J.C., Baltimore, D. and Rosbash, M. (1987) A family of unusually spliced biologically active transcripts encoded by a *Drosophila* clock gene. *Nature* **326**, 42.

Christman, J.K., Gerber, M., Price, P.M., Flordellis, C., Edelman, J. and Acs, G. (1982) Amplification of expression of hepatatis B surface antigen in 3T3 cells cotransfected with a dominant-acting gene and cloned viral DNA. *Proc. Natl Acad. Sci. USA* **79**, 1815−1819.

Clare, J. and Farabarugh, P. (1985) Nucleotide sequence of yeast Ty element: evidence for an unusual mechanism of gene expression. *Proc. Natl Acad. Sci. USA* **82**, 2829.

Clark, A.J., Simons, P., Wilmut, I. and Lathe, R. (1987) Pharmaceuticals from transgenic live stock. *Trends Biotech.* **5**, 20−25.

Clark, S.G., McGrath, J.P. and Levinson, A.D. (1985) Expression of normal and activated human *Ha-ras* cDNAs in *Saccharomyces cerevisiae. Molec. Cell. Biol.* **5**, 2746−2752.

Clarke, L., Amstutz, H., Fishel, B. and Carbon, J. (1986) Analysis of centromeric DNA in fission yeast *Schizosaccharomyces pombe. Proc. Natl Acad. Sci. USA* **83**, 8253−8258.

Clarke, L. and Carbon, J. (1980) Isolation of a yeast centromere and construction of functional small circular chromosomes. *Nature* **287**, 504−509.

Clarke, L. and Carbon, J. (1983) Genomic substitutions of centromeres in *Saccharomyces cerevisiae. Nature* **305**, 23−28.

Clarke, L. and Carbon, J. (1985) The structure and function of yeast centromeres. *Ann. Rev. Genet.* **19**, 29−56.

REFERENCES

Clemens, M.J. (1987) A potential role for RNA transcribed from B2 repeats in the regulation of mRNA stability. *Cell* **49**, 157−158.

Cohn, L.E., Glimcher, L.H., Waldman, R.A., Smith, J., Ben-Nun, A., Seidman, J.G. and Choi, E. (1986) Location of functional regions of the I-A^6 molecule by site directed mutagenesis. *Proc. Natl Acad. Sci. USA* **83**, 747−751.

Cochran, M.D. and Weissmann, C. (1984) Modular structure of the β-globin and the TK promoters. *EMBO J.* **3**, 2453−2459.

Cockerill, P.N. and Garrard, W.T. (1986) Chromosomal loop anchorage of the Kappa immunoglobulin gene occurs next to the enhancer in a region containing topoisomerase II sites. *Cell* **44**, 273−282.

Cocking, E.C. (1960) A method for the isolation of plant protoplasts and vacuoles. *Nature* **187**, 917−929.

Cocking, E.C. (1985) Protoplasts from root hairs of crop plants. *Bio/Technol.* **3**, 1104−1106.

Cocking, E.C. and Davey, M.R. (1987) Gene transfer in cereals. *Science* **236**, 1259−1262.

Cohen, J.D., Eccleshall, T.R., Needleman, R.B., Federoff, H., Buchferer, B.A. and Marmur, J. (1980) Functional expression of the *Escherichea coli* plasmid gene coding for chloramphenicol acetyl transferase. *Proc. Natl Acad. Sci. USA* **77**, 1078−1082.

Colbère-Garapin, F., Horodniceanu, F., Kourilsky, P. and Garapin, A-C. (1981) A new dominant hybrid selective marker for higher eukaryotic cells. *J. Molec. Biol.* **50**, 1−14.

Colberg-Poley, A.M., Voss, S.D., Chaudhury, K., Stewart, C.L., Wagner, E.F. and Gruss, P. (1985) Clustered homeo boxes are differentially expressed during murine development. *Cell* **43**, 39−45.

Cole, M.D. (1986) The *myc* oncogene: its role in transformation and differentiation. *Ann. Rev. Genet.* **20**, 361−384.

Colman, A. and Robinson, C. (1986) Protein import into organelles: hierarchical targeting signals. *Cell* **46**, 321−322.

Comai, L., Facciotti, D., Hiatt, W.R., Thompson, G., Rose, R.T. and Stalker, D.M. (1985) Expression in plants of a mutant *aro A* gene from *Salmonella typhimurium* confers tolerance to glyphosphate. *Nature* **317**, 741−744.

Comai, L., Sen, L.C. and Stalker, D.M. (1983) An altered *aro A* gene product confers resistance to the herbicide glyphosphate. *Science* **221**, 370−371.

Cone, R.D. and Mulligan, R.C. (1984) High efficiency gene transfer into mammalian cells: Generation of helper-free recombinant retrovirus with broad mammalian host range. *Proc. Natl Acad. Sci. USA* **81**, 6349−6353.

Cone, R.D., Weber-Benarons, A., Baorto, D. and Mulligan, R.C. (1987) Regulated expression of a complete human β-globin gene encoded by a transmissible retrovirus vector. *Molec. Cell. Biol.* **7**, 887−897.

Costantini, F., Chada, K. and Magram, J. (1986) Correction of murine β-thalassemia by gene transfer into the germ line. *Science* **233**, 1192−1194.

Costantini, F. and Lacy, E. (1982) Introduction of a rabbit β-globin gene into the mouse germ line. *Nature* **294**, 92−94.

Coughlan, M.P. (1985) The properties of fungae and bacterilal cellulases with comment on their production and application. In *Bio/Technology and Genetic Engineering Reviews, Vol. 3* (ed. G.E. Russell), pp. 39−109. Intercept Ltd., Newcastle upon Tyne.

Covarrubias, L., Nishida, Y. and Mintz, B. (1986) Early post implantation embryo lethality due to DNA rearrangements in a transgenic mouse strain. *Proc. Natl Acad. Sci. USA* **83**, 6020−6024.

Covarrubias, L., Nishida, Y., Terao, M., D'Eustachio, P. and Mintz, B. (1987) Cellular DNA rearrangements and early developmental arrest caused by DNA insertion in transgenic mouse embryos. *Molec. Cell. Biol.* **7**, 2243−2247.

Craik, C.S., Largman, C., Fletcher, T., Roczmak, S., Barr, P.J., Fletterick, R. and Rutter, W.J. (1985) Redesigning trypsin: Alteration of substrate specificity. *Science* **226**, 291−297.

Crawford, N.M., Campbell, W.H. and Davis, R.W. (1986) Nitrate reductase from squash: cDNA cloning and nitrate regulation. *Proc. Natl Acad. Sci. USA* **83**, 8073−8076.

Cregg, J.M., Barringer, K.J., Hessler, A.Y. and Madden, K.R. (1985) *Pichia pastoris* as a host system for transformation. *Molec. Cell. Biol.* **5**, 3376−3385.

Cregg, J.M., Tschopp, J.F., Stillman, C., Siegel, R., Akong, M., Craig, W.S., Buckholz, R.G., Madden, K.R., Killaris, S.A., Davis, G.R., Smiley, B.L., Cruze, J., Torregrossa, R., Velicelebi, G. and Thill, G.P. (1987) High level expression and efficient assembly of hepatitis B suface antigen in the methylotrophic yeast *Pichia pastoris*. *Bio/Technology* **5**, 479−486.

Cremer, K.J., Mackett, M., Wohlenberg, C., Notkins, A.L. and Moss, B. (1985) Vaccinia virus recombinant expressing herpes simplex virus typ 1 glycoprotein D prevents latent herpes in mice. *Science* **228**, 737−740.

Crothers, D.M. (1987) Gel electrophoresis of protein−DNA complexes. *Nature* **325**, 464−465.

Crouse, G.F., Simonsen, C.C., McEwan, R.N. and Schimke, R.T. (1982) Structure of

465   REFERENCES

amplified normal and variant dihydrofolate reductase genes in mouse sarcoma S180 cells. *J. Biol. Chem.* **257**, 7887–7897.

Cullen, B.R. (1986) *Trans*-activation of human immunodeficiency virus occurs via a bimodal mechanism. *Cell* **46**, 973–982.

Cullen, D., Gray, G.L., Wilson, L.J., Hayenga, K.J., Lamsa, M.H., Rey, M.W., Norton, S. and Berka, R.M. (1987) Controlled expression and secretion of bovine chymosin in *Aspergillus nidulans*. *Bio/Technol.* **5**, 369–376.

Cuzin, F. (1984) The polyoma virus oncogenes. *Biochem. Biophys. Acta* **781**, 193–204.

Dani, G.M. and Zakian, V.A. (1983) Mitotic and meiotic stability of linear plasmids in yeast. *Proc. Natl Acad. Sci. USA* **80**, 3406–3410.

Danner, D. and Leder, P. (1985) Role of an RNA cleavage/poly (A) addition site in the production of membrane bound and secreted IgM mRNA. *Proc. Natl Acad. Sci. USA* **82**, 8658–8662.

Davey, J., Dimmock, N.J. and Colman, A. (1985) Identification of the sequence responsible for the nuclear accumulation of the influenza virus nucleoprotein in *Xenopus* oocytes. *Cell* **40**, 667–675.

Davies, J. and Smith, D.I. (1978) Plasmid-determined resistance to antimicrobial agents. *Ann. Rev. Microbiol.* **32**, 469–518.

Davis, A.R., Bos, T.J. and Nayak, D.P. (1983) Active influenza virus neuraminidase is expressed in monkey cells from cDNA cloned in simian virus 40 vectors. *Proc. Natl Acad. Sci. USA* **80**, 3976–3980.

Davis, A.R., Kostek, B., Mason, B.B., Hsiao, C.L., Morin, J., Dheer, S.K. and Hung, P.P. (1985) Expression of hepatitis B surface antigen with a recombinant adenovirus. *Proc. Natl Acad. Sci. USA* **82**, 7560–7564.

Dean, R.A. and Kuć, J. (1985) Induced systemic protection in plants. *Trends Biotechnol.* **3**, 125–129.

Deans, R.J., Denis, K.A., Taylor, A. and Wall, R. (1984) Expression of an immunoglobulin heavy chain gene transfected into lymphocytes. *Proc. Natl Acad. Sci. USA* **81**, 1292–1296.

Deb, S., DeLucia, A.L., Koff, A., Tsui, S. and Tegtmeyer, P. (1986) The adenine–thymine domain of the simian virus 40 core origin directs DNA bending and coordinately regulates DNA replication. *Molec. Cell. Biol.* **6**, 4578–4584.

De Banzie, J.S., Sinclair, L. and Lis, J.T. (1986) Expression of the major heat shock gene of *Drosophila melanogaster* in *S. cerevisiae*. *Nucl. Acids Res.* **14**, 3587–3601.

DeBenedetti, A. and Baglioni, C. (1984) Inhibition of mRNA binding to ribosomes by localised activation of ds RNA-dependent protein kinase. *Nature* **311**, 79–81.

De Blas, A.L. and Cherwinski, H.M. (1983) Detection of antibodies on nitrocellulose paper: immunoblots with monoclonal antibodies. *Anal. Biochem.* **113**, 214–219.

De Block, M., Herrera-Estrella, L., Van Montagu, M., Schell, J. and Zambryski, P. (1984) Expression of foreign genes in regenerated plants and in their progeny. *EMBO J.* **3**, 1681–1689.

De Block, M., Schell, J. and Van Montagu, M. (1985) Chloroplast transformation by *Agrobacterium tumefaciens*. *EMBO J.* **4**, 1367–1372.

De Ciccio, D.V. and Spradling, A. (1984) Localisation of a *cis*-acting element responsible for the developmentally regulated amplification of *Drosophila* chorion genes. *Cell* **38**, 45–54.

De Franco, D. and Yamamoto K. (1986) Two different factors act separately or together to specify functionally distinct activities at a single transcriptional enhancer. *Molec. Cell. Biol.* **20**, 305–321.

De Greve, H., Dhaese, P., Seurinck, J., Lemmess, H., Van Montagu, M. and Schell, J. (1982a) Nucleotide sequence and transcript map of the *Agrobacterium tumefaciens* Ti plasmid encoded octopine synthase gene. *J. Molec. Appl. Genet.* **1**, 499–512.

De Greve, H., Leemans, J., Hernalsteens, J.P., Thia-Toong, L., De Beuckeleer, M., Willmitzer, L., Otten, L., Van Montagu, M. and Schell, J. (1982b) Regeneration of normal and fertile plants that express octopine synthase from tobacco crown gall after deletion of tumour-controlling functions. *Nature* **300**, 752–755.

De la Chapelle, A. (1985) Mapping hereditary disorders. *Nature* **317**, 473.

De la Pena, A., Lorz, A. and Schell, J. (1987) Transgenic rye plants obtained by injecting DNA into young floral tillers. *Nature* **325**, 274–276.

De la Salle, H., Altenburger, W., Elkaim, R., Dott, K., Dieterlé, A., Tolstoshev, P. and Lecocq, J-P. (1985) Active γ-carboxylated human factor IX expressed using recombinant DNA techniques. *Nature* **316**, 268–270.

Delpeyroux, F., Chenuner, N., Lim, A., Malpièce, Y., Blondel, B., Crainic, R., Van der Werf, S. and Strieck, R.E. (1986) A poliovirus neutralisation epitope expressed on hybrid hepatitis B surface antigen particles. *Science* **233**, 472–475.

Denniston, K.J., Yoneyama, T., Hoyer, B.H. and Gerin, J.L. (1984) Expression of hepatitis B virus surface and e antigen genes cloned in bovine papillomavirus vectors. *Gene* **32**, 357–368.

REFERENCES

Depicker, A., Stachel, S., Dhaese, P., Zambryski, P. and Goodman, A.M. (1982) Nopaline synthase: transcript mapping and DNA sequence. *J. Molec. Appl. Genet.* **1**, 561−574.

Der, C.J., Finkel, T. and Cooper, G.M. (1986) Biological and biochemical properties of human ras^H genes mutated at codon 61. *Cell* **44**, 167−176.

Derynck, R., Singh, A. and Goeddel, D.V. (1983) Expression of human interferon-α cDNA in yeast. *Nucl. Acids Res.* **11**, 1819−1837.

De Saint Vincent, B.R., Delbruck, S., Eckhart, W., Meinkoth, J., Vitto, L. and Wahl, G. (1981) The cloning and reintroduction into animal cells of a functional CAD gene, a dominant amplifiable genetic marker. *Cell* **27**, 267−77.

Deshayes, A., Herrera-Estrella, L. and Caboche, M. (1985) Liposome-mediated transformation of tobacco mesophyll protoplasts by an *Escherichia coli* plasmid. *EMBO J.* **4**, 2731−2737.

De Wet, J.R., Wood, K.V., DeLuca, M., Helinski, D.R. and Subramani, S. (1987) Firefly luciferase gene: structure and expression in mammalian cells. *Molec. Cell. Biol.* **7**, 725−737.

Dhawale, S.S., Paietta, J.V. and Marzluff, G.A. (1984) A new, rapid and efficient transformation procedure for *Neurospora. Curr. Genet.* **8**, 77−79.

Dick, J.E., Magli, M.C., Huszar, D., Phillips, R.A. and Bernstein, A. (1985) Introduction of a selectable gene into primitive stem cells capable of long-term reconstitution of the hemopoietic system of W/W^V mice. *Cell* **42**, 71−79.

Dickerson, R.E., Drew, H.R., Conner, B.N., Wing, R.M., Fratini, A.V. and Kopka, M.L. (1982) The anatomy of A-, B- and Z-DNA. *Science* **216**, 475−485.

Dierks, P., Van Ooyen, A., Cochran, M.D., Dobkin, C., Reiser, J. and Weissmann, C. (1983) Three regions upstream from the cap site are required for efficient and accurate transcription of the rabbit β-globin gene in mouse 3T6 cells. *Cell* **32**, 695−706.

Di Maio, D., Corbin, V., Sibley, E. and Maniatis, T. (1984) High-level expression of a cloned HLA heavy chain gene introduced into mouse cells on a bovine papillomavirus vector. *Molec. Cell. Biol.* **4**, 340−350.

DiMaio, D., Treisman, R. and Maniatis, T. (1982) Bovine papillomavirus vector that propagates as a plasmid in both mouse and bacterial cells. *Proc. Natl Acad. Sci. USA* **79**, 4030−4034.

DiNardo, S., Kuner, J.M., Theis, J. and O'Farrell, P.H. (1985) Development of embryonic pattern in *D. melanogaster* as revealed by the accumulation of the nuclear engrailed protein. *Cell* **43**, 59−69.

Dignam, J.D., Martin, P.L., Shastry, B.S. and Roeder, R.G. (1983) Eukaryotic gene transcription with purified components. *Meth. Enzymol.* (eds) Wu, R., Grossman, L. and Moldave, K **101**, 582.

Dingwall, C. and Laskey, R.A. (1986) Protein import into the cell nucleus. *Ann. Rev. Cell. Biol.* **2**, 367−390.

Dixon, B. (1986) Recycling brewer's yeast to make drugs. *Bio/Technol.* **4**, 936.

Dobson, M.J., Futcher, A.B. and Cox, B.S. (1980) Loss of 2μm DNA from *Saccharomyces cerevisiae* transformed with the chimaeric plasmid pJDB219. *Curr. Genet.* **2**, 201−205.

Dobson, M.J., Mellor, J., Fulton, A.M., Roberts, N.A., Bowen, B.A., Kingsman, S.M. and Kingsman, A.J. (1984) The identification and high level expression of a protein encoded by the yeast Ty element. *EMBO J.* **3**, 1115.

Dobson, M.J., Tuite, M.F., Roberts, N.A., Kingsman, A.J., Kingsman, S.M., Perkins, R.E., Conroy, S.C., Dunbar, B. and Fothergill, L.A. (1982) Conservation of high efficiency promoter sequences in *Saccharomyces cerevisiae. Nucl. Acids Res.* **10**, 2625−2637.

Doefler, W. (1983) DNA methylation and gene activity. *Ann. Rev. Biochem.* **52**, 93.

Donis-Keller, H. *et al.* (1987) A genetic linkage map of the human genome. *Cell* **51**, 319−337.

Doolittle, R.F. (1986) *Of Urfs and Orfs: a primer on how to analyse derived amino acid sequences.* University Science Books, Mill Valley, California.

Doolittle, W.F. and Sapienza, C. (1980) Selfish genes, the phenotype paradigm and genome evolution. *Nature* **284**, 601−603.

Doring, H.P. and Starlinger, P. (1984) Barbara McClintock's controlling elements now at the DNA level. *Cell* **39**, 253−259.

Dotto, G.P., Parada, L.F. and Weinberg, R.A. (1985) Specific growth response of *ras*-transformed embryo fibroblasts to tumour promoters. *Nature* **318**, 472−475.

Douglass, J., Civelli, O. and Herbert, E. (1984) Polyprotein gene expression: generation of diversity of neuroendocrine peptides. *Ann. Rev. Biochem.* **53**, 665−715.

Downie, J.A. and Johnston, A.W.B. (1986) Nodulation of legumes by Rhizobium: the recognised root? *Cell* **47**, 153−154.

Doyle, C., Roth, M.G., Sambrook, J. and Gething, M-J. (1985) Mutations in the cytoplasmic domain of the influenza haemagglutinin affect different stages of intracellular transport. *J. Cell. Biol.* **100**, 704−714.

Draetta, G., Brizuela, L., Potashkin, J. and Beach, D. (1987) Identification of p34 and p13, human homologues of the cell cycle regulators of fission yeast encoded by *cdc2*$^+$ and *suc1*$^+$. *Cell* **50**, 319−325.

Dranginis, A.M. (1986) Regulation of cell type in yeast by the mating-type locus. *Trends Biochem.* **11**, August, 328–331.

Dreyfuss, G. (1986) Structure and function of nuclear and cytoplasmic ribonucleoprotein particles. *Ann. Rev. Cell. Biol.* **2**, 459–498.

Duesberg, P. (1985) Activated proto-onc genes: sufficient or necessary for cancer. *Science* **228**, 669–677.

Dunn, B., Szauter, P., Pardue, M.L. and Szostak, J.W. (1984) Transfer of yeast telomeres to linear plasmids by recombination. *Cell* **39**, 191–201.

Durnam, D.M. and Palmiter, R.D. (1981) Transcriptional regulation of the mouse metallothionein-I gene by heavy metals. *J. Biol. Chem.* **256**, 5712–5716.

Durnham, D.M., Perrin, F., Gannon, F. and Palmiter, D. (1980) Isolation and characterisation of the mouse metallothionein-I gene. *Proc. Natl Acad. Sci. USA* **77**, 6511–6515.

Dynan, W.S. and Tjian, R. (1985) Control of eukaryotic mRNA synthesis by sequence specific DNA binding proteins. *Nature* **316**, 774–777.

Early, R., Rogers, J., Davis, M., Calame, K., Bond, M., Wall, R. and Hood, L. (1980) Two mRNAs can be produced from a single immunoglobulin μ gene by alternative RNA processing pathways. *Cell* **20**, 313–319.

Earnshaw, W.C., Halligan, B., Cooke, C.A., Heck, M.M.S. and Liu, L.F. (1985) Topoisomerase II is a structural component of mitotic chromosome scaffolds. *J. Cell. Biol.* **100**, 1716–1725.

Ebina, Y., Ellis, L., Jarnagin, K., Edery, M., Graf, L., Clauser, E., Ou, J-L., Masiarz, F., Kan, Y.W., Goldfine, I.D., Roth, R.A. and Rutter, W.J. (1985) The human insulin receptor cDNA: The structural basis for hoormone-activated transmembrane signalling. *Cell* **40**, 747–758.

Echols, H. (1986) Multiple DNA–protein interaction governing high precision DNA transactions. *Science* **233**, 1050–1056.

Ecker, J.R. and Davis, R.W. (1986) Inhibition of gene expression in plant cells by expression of antisense RNA. *Proc. Natl Acad. Sci. USA* **83**, 5372–5376.

Edelman, A.M., Blumenthal, D.K. and Krebs, E.G. (1987) Protein serine/threonine kinases. *Ann. Rev. Biochem.* **56**, 615–651.

Eglitis, M.A., Kantoff, P., Gilboa, E. and Anderson, W. F. (1985) Gene expression in mice after high efficiency retroviral-mediated gene transfer. *Science* **230**, 1395–1398.

Eilers, M and Schatz, G. (1986) Binding of a specific ligand inhibits import of a purified precursor protein into mitochondria. *Nature* **322**, 228–232.

Eilers, M. and Schatz, G. (1988) Protein unfolding and the energetics of protein translocation across biological membranes. *Cell* **52**, 481–483.

Eissenberg, J.C, Cartwright, L.L., Thomas, G.H. and Elgin, S.C.R. (1985) Selected topics in chromatin structure. *Ann. Rev. Genet.* **19**, 485–536.

Elgin, S.C.R. (1984) Anatomy of hypersensitive sites. *Nature* **309**, 213–214.

Eliyahu, D., Michalovitz, D. and Oren, M. (1985) Over production of p53 antigen makes established cells highly tumourigenic. *Nature* **316**, 158–160.

Emerman M. and Temin, H.M. (1984a) High frequency deletion in recovered retrovirus vectors containing exogenous DNA with promoters. *J. Virol.* **50**, 42–49.

Emerman, M. and Temin, H.M. (1984b) Quantitative analysis of gene suppression in integrated retrovirus vectors. *Molec. Cell. Biol.* **6**, 792–800.

Emr, S.D., Schauer, I., Hansen, W., Esmon, P. and Schekman, R. (1984) Invertase–β-galactosidase hybrid proteins fail to be transported from the endoplasmic reticulum in *Saccharomyces cerevisiae*. *Molec. Cell. Biol.* **4**, 2347–2355.

Engels, W.R. (1983) The P family of transposable elements in *Drosophila*. *Ann. Rev. Genet.* **17**, 315–344.

Engleman, D.M. and Steitz, T.A. (1981) The spontaneous insertion of proteins into and across membranes: the helical hairpin hypothesis. *Cell* **23**, 411–422.

Enver, T. (1985) A pulling out of fingers. *Nature* **317**, 385.

Ephrussi, A., Church, G.M., Tonegawa, S. and Gilbert, W. (1985) B lineage-specific interaction of an immunoglobulin enhancer with cellular factor *in vivo*. *Science* **227**, 134–140.

Ernst, J.F., Mermod, J-J., Delamarter, J.F., Mattaliano, R.J. and Moonen, P. (1987) O-glycosylation and novel processing events during secretion of α-factor/GM-CSF fusions by *Saccharomyces cerevisiae*. *Bio/Technol.* **5**, 831–834.

Estivill, X. *et al.* (1987) A candidate for the cystic fibrosis locus isolated by selection for methylation-free islands. *Nature* **326**, 840–845.

Etcheverry, T., Forrester, W. and Hitzeman, R. (1986) Regulation of the chelatin promoter during the expression of human serum albumin or yeast phosphoglycerate kinase in yeast. *Bio/Technol.* **4**, 726–730.

Etkin, L.D. and Roberts, M. (1983) Transmission of integrated sea urchin histone genes by nuclear transplantation in *Xenopus laevis*. *Science* **221**, 67–69.

Evans, M.J. and Kaufman, M.H. (1981) Establishment in culture of pluripotential cells from mouse embryos. *Nature* **292**, 154–156.

Falkner, F.G., Mocikat, R. and Zachau, H.G. (1986) Sequences closely related to an immuno-globulin gene promoter/enhancer element occur also upstream of other eukaryotic and of prokaryotic genes. *Nucl. Acids Res.* **14**, 8819–8826.

Fangman, W.L., Hice, R.H. and Chlebowicz-Sledziewska, E. (1983) ARS replication during the yeast S phase. *Cell* **32**, 831–838.

Farnham, P.J. and Schimke, R.T. (1986) *In vivo* transcription and delimitation of promoter elements of the murine dihydrofolate reductase gene. *Molec. Cell. Biol.* **6**, 2392.

Fasano, O., Aldrich, T., Tamanoi, F., Taparowsky, E., Furton, M. and Wigler, M. (1984) Analysis of the transforming potential of the human H-*ras* gene by random mutagenesis. *Proc. Natl. Acad. Sci. USA* **71**, 4008–4012.

Fawcett, D.H., Lister, C.K., Kellett, E. and Finnegan, D.I. (1986) Transposable elements controlling I–R hybrid dysgenesis in *D. melanogaster* are similar to mammalian LINEs. *Cell* **47**, 1007–1015.

Federoff, N.V. (1983) Controlling elements in maize In *Mobile Genetic Elements*. (ed. J.A. Shapiro), pp. 1–65. Academic Press, New York.

Federspiel, N.A., Beverley, S.M., Schilling J.W and Schimke, R.T. (1984) Novel DNA rearrangements are associated with dihydrofolate reductase gene amplification. *J. Biol. Chem.* **259**, 9127–9140.

Feinberg, M.B., Jarrett, R.F., Aldovini, A., Gallo, R.C. and Wong-Staal, F. (1986) HTLV-III expression and production involve complex regulation at the levels of splicing and translation of viral RNA. *Cell* **46**, 807–817.

Feinstein, S.C., Ross, S.R. and Yamamoto, K.R. (1982) Chromosomal position effects determine transcriptional potential of integrated mammary tumour virus DNA. *J. Molec. Biol.* **156**, 239–248.

Felsenfeld, G. and McGhee, J.D. (1986) Structure of the 30 nm chromatin fiber. *Cell* **44**, 375–377.

Ferro-Novick, S., Novick, P., Field, C. and Schekman, R. (1984) Yeast secretory mutants that block the formation of active cell surface enzymes. *J. Cell. Biol.* **98**, 35–43.

Fleischman, L.F., Chahwala, S.B. and Cantley, L. (1986) *Ras*-transformed cells: altered levels of phosphatidyl-inositorl-4, 5-bisphosphate and catabolites. *Science* **231**, 407–410.

Fields, B.N. and Krupe, D.M. (1986) *Fundamental Virology*. Raven Press, New York.

Fincham, J.R.S., Day, P.R. and Radford, A. (1979) *Fungal Genetics, Botanical Monographs, Vol. 4*. Blackwell Scientific Publications, Oxford.

Fink, G.R. (1986) Translational control of transcription in eukaryotes. *Cell* **45**, 155–156.

Finkelstein, D.B. and Strausberg, S. (1983) Heat shock regulated production of *Escherichia coli* β-galactosidase in *Saccharomyces cerevisiae*. *Molec. Cell. Biol.* **3**, 1625–1633.

Finnegan, D.J. (1985) Transposable elements in eukaryotes. *Int. Rev. Cytol.* **93**, 281–326.

Fischer, J.A. and Maniatis, T. (1986) Regulatory elements in *Drosophila Aclh* gene expression are conserved in divergent species and separate elements mediate expression in different tissues. *EMBO J.* **5**, 1275–1289.

Fitzgerald, M. and Shenk, T. (1981) The sequence 5'-AAUAAA-3' forms part of the recognition site for polyadenylation of late SV40 mRNAs. *Cell* **24**, 251–260.

Fitzgerald-Hayes, M., Clarke, L. and Carbon, J. (1982) Nucleotide sequence comparisons and functional analysis of yeast centromere DNAs. *Cell* **29**, 235–244.

Fire, A. (1986) Integrative transformation of *Caenorhabditis elegans*. *EMBO J.* **5**, 2673–2680.

Fujita, N., Nelson, N., Fox, T.D., Claudio, T., Lindstrom, J., Rierman, H. and Hess, G.P. (1986) Biosynthesis of the *Torpedo california* acetyl choline receptor α subunit in yeast. *Science* **231**, 1284–1287.

Flavell, R.A., Allen, H., Burkly, L.C., Sherman, D.H., Waneck, G.L. and Widera, G. (1986) Molecular biology of the H-2 histocompatibility complex. *Science* **233**, 437–443.

Fluhr, R., Kuhlemeier, C., Nagy, F. and Chua, N-H. (1986) Organ-specific and light-induced expression of plant genes. *Science* **232**, 1106–1112.

Flores, H.E., Hoy, M.W. and Pickard, J.J. (1987) Secondary metabolites from root cultures. *Trends Biotechnol.* **5**, 64–69.

Folger, K.R., Thomas, K. and Capecchi, M. (1985a) Nonreciprocal exchanges of information between DNA duplexes coinjected into mammalian cell nuclei. *Molec. Cell. Biol.* **5**, 59–69.

Folger, K.R., Thomas, K. and Capecchi, M.R. (1985b) Efficient correction of mismatched bases in plasmid heteroduplexes injected into cultured mammalian cell nuclei. *Molec. Cell. Biol.* **6**, 70–74.

Folger, K.R., Wong, E.A., Wahl, G. and Capecchi, M. (1982) Patterns of integration of DNA microinjected into cultured mammalian cells: Evidence for homologous recombination between injected plasmid DNA molecules. *Molec. Cell. Biol.* **2**, 1372–1387.

Foulkes, D.M. and Shenk, T. (1980) Transcriptional control region of the adenovirus VA1 RNA gene. *Cell* **22**, 405–413.

Fraley, R.T., Rogers, S.G., Horsch, R.B., Sanders, P.R., Flick, J.S., Adams, S.P., Bittner, M.L., Brand, L.A., Fink, C.L., Fry, J.S., Gallupi, G.R., Goldberg, S.R., Hoffmann, N.L.

and Woo, S.C. (1983) Expression of bacterial genes in plant cells. *Proc. Natl Acad. Sci. USA* **80**, 4803–4807.

Freeling, M. (1984) Plant transposable elements and insertion sequences. *Ann. Rev. Plant Physiol.* **35**, 277–298.

Friedlander, M. and Blobel, G. (1985) Bovine opsin has more than one signal sequence. *Nature* **318**, 338–342.

French, R., Janda, M. and Ahlquist, P. (1986) Bacterial gene inserted in an engineered RNA virus: Efficient expression in monocotyledonous plant cells. *Science* **231**, 1294–1297.

Freshney, R.I. (1983) *Culture of Animal Cells. A Manual of Basic Technique.* A.R. Liss, New York.

Fried, M. and Crothers, P.M. (1981) Equilibria and kinetics of *Lac* repressor–operator interactions by polyacrylamide gel electrophoresis. *Nucl. Acids Res.* **9**, 6505–6525.

Fromm, M.E., Taylor, L.P. and Walbot, V. (1985) Expression of genes transferred into monocot and dicot plant cells by electroporation. *Proc. Natl Acad. Sci. USA* **82**, 5824–5828.

Fromm, M.E., Taylor, L.P. and Walbot, V. (1986) Stable transformation of maize after gene transfer by electroporation. *Nature* **319**, 791–793.

Fukunaga, R., Sokawa, Y. and Nagata, S. (1984) Constitutive production of human interferons by mouse cells with bovine papillomavirus as a vector. *Proc. Natl Acad. Sci. USA.* **81**, 5086–5090.

Fujita, T., Ohno, S., Yasumitsu, H. and Taniguchi, T. (1985) Delimitation and properties of DNA sequences required for the regulated expression of human interferon-β gene. *Cell* **41**, 489–496.

Futcher, A.B. (1986) Copy number amplification of the 2 μm circle plasmid of *Saccharomyces cerevisiae. J. Theoret. Biol.* **119**, 197–204.

Gall, J.G. (1981) Chromosome structure and the C-value paradox. *J. Cell. Biol.* **91**, 3s–14s.

Ganem, D. and Varmus, H.G. (1987) The molecular biology of the hepatitis B viruses. *Ann. Rev. Biochem.* **56**, 651–695.

Garabedian, M.J., Shepherd, B.M. and Wensink, P.C. (1986) A tissue-specific transcription enhancer from the *Drosophila* yolk protein 1 gene. *Cell* **45**, 859–867.

Garber, R.W., Kuroiwa, A. and Gehring, W.J. (1983) Genomic and cDNA clones of the homeotic locus *Antennapedia* in *Drosophila. EMBO J.* **2**, 2027–2036.

Garcia, J.V., Bich-Thuy, L.T., Stafford, J. and Queen, C. (1986) Synergism between immunoglobulin enhancers and promoters. *Nature* **322**, 383–385.

Garfinkel, D.J., Boeke, J.D. and Fink, G.R. (1985) Ty element transposition: reverse transcriptase and virus-like particles. *Cell* **42**, 507–517.

Garoff, H. (1985) Using recombinant DNA techniques to study protein targeting in the eukaryotic cell. *Ann. Rev. Cell. Biol.* **1**, 403–445.

Gasser, S.M. and Laemmli, U.K. (1986) Cohabitation of scaffold binding regions with upstream/enhancer elements of three developmentally regulated genes of *D. melanogaster. Cell* **46**, 521–530.

Gaudet, A. and Fitzgerald-Hayes, M. (1987) Alterations in the adenine-plus-thymine rich region of CEN3 affect centromere function in *Saccharomyces cerevisiae. Molec. Cell. Biol.* **7**, 68–75.

Gebara, M.M., Drevon, C., Hascourt, S.A., Steingrimsdoddir, H., James, M.R., Burke, J.F., Arlelt, C.F. and Hehmann, A.R. (1987) Inactivation of a transfected gene in human fibroblasts can occur by deletion, amplification, phenotypic switching or methylation. *Molec. Cell. Biol.* **7**, 1459–1464.

Gehring, W.J. (1985) The homeo box: a key to the understanding of development. *Cell* **40**, 3–5.

Gehring, W.J. (1987) Homeo boxes in the study of development. *Science* **236**, 1245–1252.

Gerard, R.D. and Gluzman, Y. (1985) A new host cell system for regulated simian virus 40 DNA replication *Molec. Cell. Biol.* **5**, 3231–3240.

Gerard, R. and Gluzman, Y. (1986) Functional analysis of the role of the A+T-rich region and upstream flanking sequences in siman virus 40 DNA replication. *Molec. Cell. Biol.* **6**, 4570–4577.

Gerhard, W., Yewdell, J., Frankel, N.E. and Webster, R. (1981) Antigenic structure of influenza virus haemagglutinin defined by hybridoma antibodies. *Nature* **290**, 713–717.

Germain, R.N. and Malissen, B. (1986) Analysis of the expression and function of class II major histocompatibility complex-encoded molecules by DNA-mediated gene transfer. *Ann. Rev. Immunol.* **4**, 281–315.

Gerster, T., Matthias, P. Thali, M., Jwicny, J. and Schaffner, W. (1987) Cell type specificity element of the immunoglobulin heavy chain gene enhancers. *EMBO J.* **6**, 1323–1330.

Gething, M.J., MaCammon, K. and Sambrook, J. (1986) Expression of wild type and mutant forms of influenza hemagglutinin: the role of folding in intracellular transport. *Cell* **46**, 939–950.

Gething, M.J. and Sambrook, J. (1981) Cell surface expression of influenza haemagglutinin from a cloned DNA copy of the RNA gene. *Nature* **293**, 620–625.

Gething, M.J. and Sambrook, J. (1982) Construction of influenza haemagglutinin genes that code for intracellular and secreted forms of the protein. *Nature* **300**, 598−603.

Gidoni, D., Dynan, W.S. and Tjian, R. (1984) Multiple specific contacts between a mammalian transcription factor and its cognate promoters. *Nature* **302**, 409−413.

Gidoni, D., Kadonaga, J.T., Barrera-Saldana, H., Takahashi, K., Chambon, P. and Tjian, R. (1985) Bidirectional transcription mediated by tandem Spl binding interactions. *Science* **230**, 511−517.

Giguere, V., Hollenberg, S.M., Rosenfeld, M.G. and Evans, R.M. (1986) Functional domains of the human glucoccorticoid receptor. *Cell* **46**, 645−652.

Gil, A. and Proudfoot, N.J. (1987) Position-dependent sequence elements downstream of AAUAAA are required for efficient rabbit β-globin mRNA 3′ end formation. *Cell* **49**, 399−406.

Gilman, A.G. (1987) G proteins: transducers of receptor-generated signals. *Ann. Rev. Biochem.* **56**, 615−651.

Giniger, E., Rarnum, S.M. and Patshne, M. (1985) Specific DNA binding of GAL4, a positive regulatory protein of yeast. *Cell* **40**, 767−774.

Ginsberg, A.M., King, B.O. and Roeder, R.G. (1984) Xenopus 5S gene transcription factor, TFIIIA: characterisation of a cDNA clone and measurement of RNA levels through development. *Cell* **39**, 479−489.

Giri, I. and Danos, O. (1986) Papillomavirus genomes: from sequence data to biological properties. *Trends Genet.* **2**, 227−232.

Gitschier, J., Wood, W.I., Goralka, T.M., Wion, K.L., Chen, E.Y., Eaton, D.H., Vehar, E.G.A., Capon, D.J. and Lawn, R.M. (1984) Characterization of the human factor VIII gene. *Nature* **312**, 326−330.

Glikin, G.C., Gargiulo, G., Rena-Descalzi, L. and Worcel, A. (1983) *E. coli* single strand binding protein stabilises specific denatured sites on superhelical DNA. *Nature* **303**, 771.

Gluckmann, J.C., Klatzmann, D. and Montagnier, L. (1986) Lymphadenopathy associated virus infection and acquired immunodeficiency syndrome. *Ann. Rev. Immunol.* **4**, 97−119.

Gluzman, Y. (1981) SV40-transformed simian cells support the replication of early SV40 mutants. *Cell* **23**, 175−182.

Gluzman, Y., Kuff E.L. and Winocour, E. (1977) Recombination between endogenous and exogenous simian virus 40 genes 1. Rescue of a simian virus 40 temperature-sensitive mutant by passage in permissive transformed monkey lines. *J. Virol.* **24**, 534−540.

Godowski, P.J., Rusconi, S., Miesfeld, R. and Yamamoto, K.R. (1987) Glucocorticoid receptor mutants that are constitutive activators of transcriptional enhancement. *Nature* **325**, 365−367.

Goebi, M.G. and Petes, T.D. (1986) Most of the yeast genomic sequences are not essential for cell growth and cell division. *Cell* **46**, 983−992.

Goldberg, D.A., Posakony, J.W. and Maniatis, T. (1983) Correct developmental expression of a cloned alcohol dehydrogenase gene transduced into the *Drosophila* germ line. *Cell* **34**, 59−73.

Goldfarb, M., Shimizu, K., Perucho, M. and Wigler, M. (1982) Isolation and preliminary characterization of a human transforming gene from T24 bladder carcinoma cells. *Nature* **296**, 404−409.

Goff, C.G., Moir, D.T., Kohno, T., Gravius, T.C., Smith, R.A., Yamasaki, E. and Taunton-Rigby, A. (1984) The expression of calf prochymosin in *Saccharomyces cereviciae*. *Gene* **27**, 35−46.

Gonda, M.A., Braun, M.J., Clements, J.E., Pyper, J.M., Wong-Staal, F., Gallo, R.C. and Gilden, R.V. (1986) Human T-cell lymphotropic virus type III shares sequence homology with a family of pathogenic lentiviruses. *Proc. Natl Acad. Sci. USA* **83**, 4007−4011.

Goodbourn, S., Burstein, H. and Maniatis, T. (1986) The human β-interferon gene enhancer is under negative control. *Cell* **45**, 601−610.

Goodenow, R.S., McMillan, M., Orn, A., Nicolson, M., Davidson, N., Frelinger, J.A. and Hood, L. (1982) Identification of a BALB/C H-2L^d gene by DNA-mediated gene transfer. *Science* **215**, 677−679.

Goodenow, R.S., Vogel, J.M. and Linsk, R.L. (1985) Histocompatibility antigens on murine tumours. *Science* **230**, 777−783.

Gordon, J.W. and Ruddle, F.H. (1985) DNA mediated genetic transformation of mouse embryos and bone marrow − a review. *Gene* **33**, 121−136.

Gordon, J.W., Scangos, G.A., Plotkin, D.J., Barbosa, J.A. and Ruddle, R.H. (1980) Genetic transformation of mouse embryos by microinjection of purified DNA. *Proc. Natl Acad. Sci. USA* **77**, 7380−7384.

Gorman, C., Padmanabhan, R. and Howard, B. (1983) High efficiency DNA-mediated transformation of primate cells. *Science* **221**, 551−553.

Gorman, C.M. and Howard, B.H. (1983) Expression of recombinant plasmids in mammalian cells is enhanced by sodium butygrate. *Nucl. Acids Res.* **11**, 7631−7648.

Gorman, C.M., Merlino, G.T., Willingham, M.C., Pastan, I. and Howard, B. (1982a) The

Rous sarcoma virus long terminal repeat is a strong promoter when introduced into a variety of eukaryotic cells by DNA-mediated transfection. *Proc. Natl Acad. Sci. USA* **79**, 6777–6781.

Gorman, C.M., Moffat, L.F. and Howard, B.H. (1982b) Recombinant genomes which express chloramphenicol acetyl transferase in mammalian cells. *Molec. Cell. Biol.* **2**, 1044–1051.

Gorman, C.M., Rigby, P.W.J. and Lane, D.P. (1985) Negative regulation of viral enhancers in undifferentiated embryonic stem cells. *Cell* **42**, 519–526.

Graf, L.H., Urlaub, G. and Chasin, L.A. (1979) Transformation of the gene for hypoxanthine phosphoribosyl transferase. *Somat. Cell. Genet.* 1031–1044.

Graham, C.F. and Wareing, P.F. (1984) *Developmental Control in Animals and Plants*. Blackwell Scientific Publications, Oxford.

Graham, F.L. and Van der Eb, A.J. (1973) A new technique for the assay of infectivity of human adenovirus 5 DNA. *Virology* **52**, 456–467.

Grant, D.M., Lambowitz, A.M., Rambosek, J.A. and Kinsey, J.A. (1984) Transformation of *Neurospora crassa* with recombinant plasmids containing the cloned glutamate dehydrogenase (*am*) gene: evidence for autonomous replication of the transforming plasmid. *Molec. Cell. Biol.* **4**, 2041–2051.

Green, A.R. and Wyke, J.A. (1985) Anti-oncogenes. A subset of regulatory genes involved in carcinogenesis. *Lancet* **III**, 475–477.

Green, M.R., Maniatis, T. and Melton, D.A. (1983) Human β-globin pre-mRNA synthesised *in vitro* is accurately spliced in *Xenopus* oocyte nuclei *Cell* **32**, 681–694.

Green, S. and Chambon, P. (1986) A superfamily of potentially oncogenic hormone receptors. *Nature* **324**, 615–616.

Green, S. and Chambon, P. (1987) Oestradiol induction of a glucocorticoid responsive gene by a chimaeric receptor. *Nature* **325**, 75–78.

Green, M.R. and Roeder, R.G. (1980) Definition of a novel promoter for the major adenovirus-associated virus mRNA. *Cell* **22**, 231–242.

Greenwald, I. (1985) *lin-12*, A nematode homeotic gene is homologous to a set of mammalian proteins that includes epidermal growth factor. *Cell* **43**, 583–590.

Greenwald, I.S., Sternberg, P.W. and Horvitz, H.R. (1983) The lin-12 locus specifies cell fates in *Caenorhabditis elegans*. *Cell* **34**, 435–444.

Grimsley, N., Hohn, T., Davies, J.W. and Hohn, B. (1987) *Agrobacterium*-mediated delivery of infectious maize streak virus into maize plants. *Nature* **325**, 177–179.

Gritz, L. and Davies, J. (1983) Plasmid encoded hygromycin B resistance: the sequence of hygromycin B phosphotransferase and its expression in *Escherichia cori* and *Saccharomyces cerevisiae*. *Gene* **25**, 179–188.

Gronenborn, B., Gardner, R.C., Schaefer, S. and Shepherd, R.J. (1981) Propagation of foreign DNA in plants using cauliflower mosaic virus as vector. *Nature* **294**, 773–776.

Gronow, M. and Bliem, R. (1983) Production of human plasminogen activators by cell culture. *Trends Biotechnol.* **1**, 26–29.

Grosschedl, R. and Baltimore, D. (1985) Cell type specificity of immunoglobulin gene expression is regulated by at least three DNA sequence elements. *Cell* **41**, 885–897.

Grosschedl, R., Weaver, D., Baltimore, D. and Costantini, F. (1984) Introduction of a μ immunoglobulin gene into the mouse germline: specific expression in lymphoid cells and synthesis of functional antibody. *Cell* **38**, 647–658.

Grosveld, G.C., de Boer, E., Shewmaker, C.K. and Flavell, R.A. (1982) DNA sequences necessary for transcription of the rabbit β-globin gene *in vivo*. *Nature* **294**, 120–126.

Grosveld, F., van Assendelft, G.B. Greaves, D.R. and Kollias, G. (1987) Position independent, high level expression of the human β-globin gene in transgenic mice. *Cell* **51**, 975–985.

Gruss, P. and Khoury, G. (1981) Expression of simian virus 40–rat preproinsulin recombinants in monkey kidney cells: use of preproinsulin RNA processing signals. *Proc. Natl Acad. Sci. USA* **78**, 133–137.

Guan, J-L. and Rose, J.K. (1984) Conversion of a secretory protein into a transmembrane protein results in its transport to the Golgi complex but not to the cell surface. *Cell* **37**, 779–787.

Guarente, L. (1984) Yeast promoters: positive and negative elements. *Cell* **36**, 799–800.

Guarente, L. (1988) UASs and enhancers: common mechanisms of transcriptional activation in yeast and mammals. *Cell* **52**, 303–305.

Guarente, L. and Hoar, E. (1984) Upstream activation sites of the *CYC1* gene of *Saccharomyces cerevisiae* are active when inverted but not when placed downstream of the TATA box. *Proc. Natl Acad. Sci USA* **81**, 7865–7869.

Guarente, L., Yocum, R.R. and Gifford, P. (1982) A *GAL10-CYC1* hybrid yeast promoter identifies the *GAL4* regulatory region as an upstream site. *Proc. Natl Acad. Sci. USA* **79**, 7410–7414.

Gurdon, J.B. (1974) *The Control of Gene Expression in Animal Development*. Clarendon Press, Oxford.

Gurdon, J.B. and Melton, D.A. (1981) Gene transfer in amphibian eggs and oocytes. *Ann. Rev. Genet.* **15**, 189−218.

Gusella, J.F. (1986) DNA polymorphisms and human disease. *Ann. Rev. Biochem.* **55**, 831−855.

Guthrie, C. and Abelson, J. (1982) Organization and expression of tRNA genes in *Saccharomyces cerevisiae.* In *The Molecular Biology of the Yeast Saccharomyces* (eds J.N. Strathern, E.W. Jones and J.R. Broach), pp. 487−525. Cold Spring Harbor Laboratory, New York.

Gwynne, D.I., Buxton, F.P., Williams, S.A., Garven, S. and Davies, R.W. (1987) Genetically engineered secretion of active human interferon and a bacterial endoglucanase from *Aspergillus nidulans. Bio/Technol.* **5**, 713−719.

Haase, A. (1986) Pathogenesis of lentivirus infections. *Nature* **322**, 130−136.

Haber, D.A., Beverley, S.M., Kiely, M.L. and Schimke, L.T. (1981) Properties of an altered dihydrofolate reductase encoded by amplified genes in cultured mouse fibroblasts. *J. Biol. Chem.* **256**, 9501−9510.

Haenlin, M., Steller, H., Pirrotta, V. and Momer, E. (1985) A 43 kb Cosmid P transposon rescues the fs(1)K10 morphogenetic locus and three adjacent *Drosophila* developmental mutants. *Cell* **40**, 827−837.

Haguenaur-Tsapis, R. and Hinnen, A. (1984) A deletion that includes the signal petidase cleavage site impairs processing, glycosylation and secretion of cell surface yeast acid phosphatase. *Molec. Cell. Biol.* **4**, 2668−2675.

Hahn, S., Hoar, E.T. and Guarente, L. (1985) Each of three 'TATA elements' specifies a subset of the transcription initiation sites at the *CYC-1* promoter of *Saccharomyces cerevisiae. Proc. Natl Acad. Sci. USA* **82**, 8562−8566.

Hall, B.D., Clarkson, S.G. and Tocchini-Valentini, G. (1982) Transcription initiation of eukaryotic transfer RNA genes. *Cell* **29**, 3−5.

Hallewell, R.A., Mills, R. Tekamp-Olsen, P., Blacker, R., Rosenberg, S., Otting, F., Masiarz, F.R. and Scandella, C.J. (1987) Amino terminal acetylation of authentic human Cu, Zn superoxide dismutase produced in yeast. *Bio/Technol.* **5**, 363−369.

Hamada, H., Petrino, M.G. and Kakunaga, T. (1982) A novel repeat element with Z-DNA forming potential is widely found in evolutionarily diverse eukaryotic genomes. *Proc. Natl Acad. Sci. USA* **711**, 6465−6469.

Hamer, D.H. (1986) Metallothionein. *Ann. Rev. Biochem.* **55**, 913−953.

Hamilton, R., Watanabe, C.K. and de Boer, H.A. (1987) Compilation and comparison of the sequence context around the AUG start codons in *Saccharomyces cerevisiae* mRNAs. *Nucl. Acids Res.* **15**, 3581−3593.

Hammer, R.E., Brinster, R.L. and Palmiter, R.D. (1985a) Use of gene transfer to increase animal growth. *Cold Spring Harbor Symp. Quaut. Biol.*, Vol. **50**, pp. 379−387. Cold Spring Harbor, New York.

Hammer, R.E., Brinster, R.L., Rosenfeld, M.G., Evans, R.E. and Mayo, K.E. (1985b) Expression of human growth hormone releasing factor in transgenic mice results in increased somatic growth. *Nature* **315**, 413−416.

Hammer, R.E., Krumlauf, R., Camper, S.A., Brinster, R.L. and Tilghman, S.M. (1987) Diversity of alpha-fetoprotein gene expression in mice is generated by a combination of separate enhancer elements. *Science* **235**, 53−58.

Hammer, R.E., Palmiter, R.D. and Brinster, R.L. (1984) Partial correction of murine hereditary growth disorder by germ line incorporation of a new gene. *Nature* **311**, 65−67.

Hammer, R.E., Pursel, V.G., Rexroud, C.E., Wall, R.J., Bolt, D.J., Ebert, K.M., Palmiter, R.D. and Brinster, R.L. (1985c) Production of transgenic rabbits, sheep and pigs by microinjection. *Nature* **315**, 680−683.

Hammer, R.E., Swift, G.H., Ornitz, D.M., Quaife, C.J., Palmiter, R.D., Brinster, R.L. and MacDonald, R.J. (1987) Rat elastase I regulatory element is an enhancer that directs correct cell specificity and developmental onset of expression in transgenic mice. *Molec. Cell. Biol.* **7**, 2956−2967.

Hanafusa, H. and Jove, R. (1987) Cell transformation by the viral *src* oncogene. *Ann. Rev. Cell. Biol.* **3**, 31−56.

Hanahan, D. (1985) Heritable formation of pancreatic β-cell tumours in transgenic mice expressing recombinant insulin/simian virus 40 oncogenes. *Nature* **315**, 115−122.

Hand, R. (1978) Eukaryotic DNA: organisation of the genome for replication. *Cell* **15**, 317−325.

Hansen, W., Garcia, P.D. and Walter, P. (1986) *In vitro* protein translocation across the yeast endoplasmic reticulum: ATP-dependent post-translational translocation of the pre pro α factor: *Cell* **45**, 397−406.

Harashima, S. and Hinnebusch, A.G. (1986) Multiple GCD genes required for repression of *GCN4*, a transcriptional activator of amino acid biosynthetic genes in *Saccharomyces cerevisiae. Molec. Cell. Biol.* **6**, 3990−3998.

Harbers, K., Jähner, D. and Jaenisch, R. (1981) Microinjection of cloned retroviral genomes into mouse zygotes: integration and expression in the animal. *Nature* **293**, 540–542.

Harbers, K., Kuehn, M., Deluis, H. and Jaenisch, R. (1984) Insertion of retrovirus into the first intron of α1 (I) collagen gene leads to embryonic lethal mutation in mice. *Proc. Natl. Acad. Sci. USA* **81**, 1504–1508.

Harland, R.M. and Laskey, R.A. (1980) Regulated replication of DNA microinjected into eggs of *Xenopus laevis. Cell* **21**, 761–771.

Hart, C.P., Awgulewitsch, A., Fainsod, A., McGinnis, W. and Ruddle, F.H. (1985) Homeo box gene complex on mouse chromosome II: Molecular cloning, expression in embryogenesis and homology to a human homeo box locus. *Cell* **43**, 9–18.

Hauber, J. and Cullen, B. (1988) Mutational analysis of the *trans*-activation responsive region of the human immunodeficiency virus type I long terminal repeat. *Molec. Cell. Biol.* **62**, 673–679.

Hartley, J.L. and Donelson, J.E. (1980) Nucleotide sequence of the yeast plasmid. *Nature* **286**, 860–865.

Hartshorne, T.A., Blumberg, H. and Young, E.T. (1986) Sequence homology of the yeast regulatory protein *ADR1* with *Xenopus* transcription factor TFIII A. *Nature* **320**, 283–287.

Hartung, S., Jaenisch, R. and Breindl, M. (1986) Retrovirus insertion inactivates mouse α1 (I) collagen gene by blocking initiation of transcription. *Nature* **320**, 365–367.

Hase, T., Muller, V., Riezman, H. and Schatz, G. (1984) 70-kd protein of the yeast mitochondrial outer membrane is targeted and anchored via its extreme amino-terminus. *EMBO J.* 3, 3157–3184.

Hashimoto, C. and Stertz, J.A. (1986) A small nuclear ribonucleoprotein associates with the AAUAAA polyadenylation signal *in vitro. Cell* **45**, 581–591.

Hasnain, S.E., Manarathu, E.K. and Leung, W-C. (1985) DNA mediated transformation of *Chlamydomonas reinhardi* cells: use of ammunoglycoside 3′ phosphotransferase as a selectable marker. *Molec. Cell. Biol.* 5, 3647–3650.

Hauser, C.A., Joyner, A.L., Kleen, R.D., Learned, T.K., Martin, G.R. and Tjian, R. (1985) Expression of homologous homeo box containing genes in differentiated human teratocarcinoma cells and mouse embryos. *Cell* **43**, 19–28.

Hawley, R.G., Covrarrubias, L., Hawley, T. and Mintz, B. (1987) Handicapped retroviral vectors efficently transduce foreign genes into hematopoietic stem cells. *Proc. Natl Acad. Sci. USA* **84**, 2406–2410.

Hay, R., Böhni, P. and Gasser, S. (1984) Protein import into mitochondria. *Biochem. Biophys. Acta* **779**, 65–87.

Hayman, M.J. (1986) erb-B: growth factor receptor turned oncogene. *Trends Genet.* 2, 260–262.

Haynes, J. and Weissmann, C. (1983) Constitutive long term production of human interferons by hamster cells containing multiple copies of a cloned interferon gene. *Nucl. Acids Res.* 11, 687–707.

Hayward, W.S., Neel, B.A. and Astrin, S.W. (1981) Activation of a cellular *onc* gene by promoter insertion in ALV-induced lymphoid leukosis. *Nature* **290**, 475–480.

Hazelrigg, T., Levis, R. and Rubin, G.M. (1984) Transformation of *white* locus DNA in *Drosophila*: Dosage compensation, *Zeste* interaction and position effects. *Cell* **36**, 469–481.

Hegemann, J.H., Pridmore, R.D., Schneider, R. and Philippsen, P. (1986) Mutations in the right boundary of *Saccharomyces cerevisiae* centromere 6 lead to non functional or partially functional centromeres. *Molec. Gen. Genet.* **205**, 305–311.

Heidecker, G. and Messing, J. (1986) Structural analysis of plant genes. *Ann. Rev. Plant Physiol.* **37**, 439–466.

Heilman, C., Engel, L., Lowy D. and Howley, P. (1982) Virus specific transcription in bovine papillomavirus transformed cells. *Virology* **119**, 22–34.

Hernalsteens, J-P., Thia-Toong, L., Schell, J. and Van Montagu, M. (1984) An *Agrobacterium*-transformed cell culture from the monocot *Asparagus officinalis. EMBO J.* 3, 3039–3041.

Heldin, C-H. and Westermark, B. (1984) Growth factors: mechanism of action and relation to oncogenes. *Cell* **37**, 9–20.

Hellerman, J.G., Cone, R.C., Potts, J.T., Rich, A., Mulligan, R.S. and Kronenberg, H.M. (1984) Secretion of human parathyroid hormone from rat pituitary cells infected with a recombinant retrovirus encoding preproparathyroid hormone. *Proc. Natl Acad. Sci. USA* **81**, 5340–5344.

Hen, R., Borrelli, E. and Chambon, P. (1985) Repression of the immunoglobulin heavy chain enhancer by the adenovirus-2 E1A products. *Science* **230**, 1391–1395.

Henikoff, S. and Cohen, E.H. (1984) Sequences responsible for transcription termination on a gene segment in *Saccharomyces cerevisiae. Molec. Cell. Biol.* **4**, 1515–1520.

Henikoff, S. and Furlong, C. (1983) Sequences of a *Drosophila* DNA segment that functions in *Saccharomyces cerevisiae* and its regulation by a yeast promoter. *Nucl. Acids Res.* **11**, 789–800.

Hereford, L.M. and Rosbash, M. (1977) Number and distribution of polyadenylated RNA sequences in yeast. *Cell* **10**, 453–462.

Herr, W. and Clarke, J. (1986) The SV40 enhancer is composed of multiple functional elements that can compensate for one another. *Cell* **45**, 461–470.

Herrera-Estrella, L., Depicker, A., van Montagu, M. and Schell, J. (1983a) Expression of chimaeric genes transferred into plant cells using a Ti plasmid-derived vector. *Nature* **303**, 209–213.

Herrera-Estrella, L., De Block, M., Messens, E., Hernalsteens, J-P., Van Montagu, M. and Schell, J. (1983b) Chimeric genes as dominant selectable markers in plant cells. *EMBO J.* **2**, 987–995.

Herskowitz, I. and Oshima, Y. (1981) Control of cell type in *Saccharomyces cerevisiae*: Mating type and mating type interconversion. In *The Molecular Biology of the Yeast Saccharomyces: Vol. I Life Cycle and Inheritance* (eds J.W. Strathern, T.W. Jones and J.R. Broach), pp. 181–211, Cold Spring Harbor, New York.

Heyer, W-D., Sipiczki, M. and Kohl, J. (1986) Replicating plasmids in *Schizosaccharomyces pombe*: improvement of symmetric segregation by a new genetic element. *Molec. Cell. Biol.* **6**, 80–89.

Hicks, J. and Fox, C.F. (1986) *Yeast Cell Biology*, A.R. Liss, New York.

Hieter, P., Mann, C., Snyder, M. and Davis, R.W. (1985b) Mitotic stability of yeast chromosomes: a colony colour assay that measures non-disjunction and chromosome loss. *Cell* **40**, 393–403.

Hieter, P., Pridmore, D., Hegemann, J.H., Thomas, M., Davis, R.W. and Philippsen, P. (1985a) Functional selection and analysis of yeast centromeric DNA. *Cell* **42**, 913–921.

Hill, A. and Bloom, K. (1987) Genetic manipulation of centromere function. *Molec. Cell. Biol.* **7**, 2397–2405.

Hinnebusch, A.G. (1984) Evidence for translational regulation of the activator of general amino acid control in yeast. *Proc. Natl Acad. Sci. USA* **81**, 6442–6446.

Hinnen, A., Hicks, J.B. and Fink, G.R. (1978) Transformation of yeast. *Proc. Natl Acad. Sci. USA* **75**, 1929–1933.

Hirschberg, C.B. and Snider, M.D. (1987) Topography of glycosylation in the rough endoplasmic reticulum and Golgi apparatus. *Ann. Rev. Biochem.* **56**, 63–89.

Hirschberg, J. and McIntosh, L. (1983) Molecular basis of herbicide resistance in *Amaranthus hybridus. Science* **222**, 1340–1349.

Hitzeman, R.A., Chen, C.Y., Hagie, F.E., Patzer, E.J., Lui, C-C., Estell, D.A., Miller, J.V., Yaffe, A., Kleid, D.G., Levinson, A.D. and Oppermann, H. (1983a) Expression of hepatitis B virus surface antigen in yeast. *Nucl. Acids Res.* **11**, 2745–2763.

Hitzeman, R.A., Hagie, F.E., Levine, H.L., Goeddel, D.W., Ammerer, G. and Hall, B.D. (1981) Expression of human gene for interferon in yeast. *Nature* **293**, 717–722.

Hitzeman, R.A., Leung, D.W., Perry, L.J., Kohr, W.J., Levine, H.L. and Goeddel, D.R. (1983b) Secretion of human interferon by yeast. *Science* **219**, 620–625.

Hock, R.A. and Miller, A.D. (1986) Retrovirus-mediated transfer and expression of drug resistance genes in human haematopoietic progenitor cells. *Nature* **320**, 275–277.

Hodgson, C.-P., Elder, P.K., Ono, T., Foster, D.N., Getz, M.J. (1984) Structure and expression of mouse VL30 genes. *Molec. Cell. Biol.* **3**, 2221–2231.

Hofer, E., Hofer, R.H-W. and Darnell, Y. (1982) Globin RNA transcription: a possible termination site and demonstration of transcriptional control correlated with altered chromatin structure. *Cell* **29**, 887–893.

Hoffman, E.P., Brown, R.H. and Kunkel, L.M. (1987) Dystrophin: the protein product of the Duchenne muscular dystrophy locus. *Cell* **51**, 919–928.

Hoffman, J.W., Steffen, D., Gusella, J., Tabin, C., Bird, S., Cowing, D. and Weinberg, R.A. (1982) DNA methylation affecting the expression of murine leukaemia proviruses. *J. Virol.* **44**, 144–157.

Hoffman-Liebermann, B., Liebermann, D., Kedes, L.H. and Cohen, S. (1985) TU elements: a heterogeneous family of modularly structured eukaryotic transposons. *Molec. Cell. Biol.* **5**, 991–1001.

Hoekema, A., Hirsch, P.R., Hooykaas, P.J.J. and Schilperoort, R.A. (1983) A binary plant vector strategy based on separation of the *vir* and T-region of the *Agrobacterium tumefaciens* Ti-plasmid. *Nature* **303**, 179–180.

Hogan, B.L.M. (1985) How is the mouse segmented? *Trends Genet.* **1**, 67–75.

Hogan, B.L.M., Costantini, F.C. and Lacy, E. (1986) *Manipulation of the Mouse Embryo: a Laboratory Manual.* Cold Spring Harbor Laboratory, New York.

Hollenberg, S.M., Giguere, V., Segui, P. and Evans, R.M. (1987) Colocalization of DNA-binding and transcriptional activation functions in the human glucocorticoid receptor *Cell* **49**, 39–46.

Holsters, M., Villarroel, R., van Montagu, M. and Schell, J. (1982) The use of selectable markers for the isolation of plant-DNA/T-DNA junction fragments in a cosmid vector. *Molec. Gen. Genet.* **185**, 283–289.

Hood, L., Kronenberg, M. and Hunkapillar, T. (1985) T cell antigen receptors and the immunoglobulin supergene family. *Cell* **40**, 225–229.

475   REFERENCES

Hood, L., Steinmetz, M. and Goodenow, R. (1982) Genes of the major histocompatibility complex. *Cell* **28**, 685–687.

Hood, L., Steinmetz, M. and Malissen, B. (1983) Genes of the major histocompatibility complex of the mouse. *Ann. Rev. Immunol.* **1**, 529–568.

Hookyaas, P.J.J. and Schilperoort, R.A. (1984) The molecular genetics of crown gall tumourigenesis. *Adv. Genet.* **22**, 210–283.

Hooper, M., Hardy, K., Handyside, A., Hunter, S. and Monk, M. (1987) HPRT-deficient (Lesch-Nyhan) mouse embryos derived from germline colonisation by cultured cells. *Nature* **326**, 292–295.

Hooykaas-Van Slogteren, G.M.S., Hooykaas, P.J.J. and Schilperoort, R.A. (1984) Expression of Ti plasmid genes in monocotyledenous plants infected with *Agrobacterium tumeficiens*. *Nature* **311**, 763–764.

Hope, I. and Struhl, K. (1986) Functional dissection of a eukaryotic transcriptional activator protein. *Cell* **46**, 885–894.

Horsch, R.B., Fraley, R.T., Rogers, S.G., Sanders, P.R., Lloyd, A. and Hoffmann, N. (1984) Inheritance of functional foreign genes in plants. *Science* **237**, 496–498.

Horsch, R.B., Fry, J.E., Hoffmann, N.L., Eichholtz, D., Rogers S.G. and Fraley, R.T. (1985) A simple general method for transferring genes into plants. *Science* **227**, 1229–1231.

Horwich, A.L., Kalonsek, F., Fenton, W.A., Pollock, R.A. and Rosenberg, L.E. (1986) Targeting of pre-ornithine transcarbamylase to mitochondria: definition of critical regions and residues in the leader peptide. *Cell* **44**, 451–459.

Horwich, A.L., Kalonsek, F., Mellman, I. and Rosenberg, L.E. (1985) A leader peptide is sufficient to direct cytosolic dihydrofolate reductase into the mitochondrial matrix. *EMBO J.* **4**, 1129–1135.

Hsiao, C-L. and Carbon, J. (1981) Direct selection procedure for the isolation of functional centromeric DNA. *Proc. Natl Acad. Sci. USA* **78**, 3760–3764.

Hu, S-L., Kosowski, S.G. and Dalrymple, J.M. (1986) Expression of AIDS virus envelope gene in recombinant vaccinia viruses. *Nature* **320**, 537–540.

Hudson, L.D., Erbe, R.W. and Jacoby, L.B. (1980) Expression of human arginosuccinate synthetase gene in hamster transferents. *Proc. Natl Acad. Sci. USA* **77**, 4234–4238.

Huffman, G.A., White, F.F., Gordon, M.R. and Nester, E.W. (1984) Hairy-root-inducing plasmid: Physical map and homology to tumour-inducing plasmids. *J. Bacteriol.* **157**, 269–276.

Huttner, K.M., Barbosa, J.A., Scangos, G.A., Pratcheva, D.D. and Ruddle, F.H. (1981) DNA-mediated gene transfer without carrier DNA. *J. Cell. Biol.* **91**, 153–156.

Huttner, K., Scangos, G.A. and Ruddle, F.H. (1979) DNA-mediated gene transfer of a circular plasmid into murine cells. *Proc. Natl Acad. Sci. USA* **76**, 5820–5824.

Hurt, E.C., Pesold-Hurt, B. and Schatz, G. (1984) The cleavable prepiece of an imported mitochondrial protein is sufficient to direct cytosolic dihydrofolate reductase into the mitochondrial matrix. *FTBS Lett*, **178**, 306–310.

Hunkapiller, T. and Hood, L. (1986) The growing immunoglobulin gene superfamily. *Nature* **323**, 15–16.

Hurt, E.C., Pesold-Hurt, B., Suda, K., Oppliger, W. and Schatz, G. (1985) The first twelve amino acids (less than half of the pre-sequence) of an imported mitochondrial protein can direct mouse cytosolic dihydrofolate reductase into the yeast mitochondrial matrix. *EMBO J.* **4**, 2061–2068.

Hurt, E.C. and Schatz, G. (1987) A cytosolic protein contains a cryptic mitochondrial targeting signal. *Nature* **325**, 499–503.

Hurt, E.C., Soltanifar, N., Goldschmidt-Clermont, M., Rochaix, J-D. and Schatz, G. (1986) The cleavable pre-sequence of an imported chloroplast protein directs attached polypeptides into yeast mitochondria. *EMBO J.* **5**, 1343–1350.

Huttmark, D., Klemenz, R. and Gehring, W.J. (1986) Translational and transcriptional control elements in the untranslated leader of the heat shock gene *hsp 22*. *Cell* **44**, 429–438.

Ikemura, T. (1982) Correlation between the abundance of yeast transfer RNAs and the occurrence of the respective codons in protein genes. *J. Molec. Biol.* **158**, 573–597.

Innis, M.A., Holland, M.J., McCabe, P.C., Cole, G.E., Wittman, V.P., Tal, R., Watt, K.W.K., Gelfand, D.H., Holland, J.P. and Meade, J.H. (1985) Expression, glycosylation, and secretion of an *Aspergillus* glucoamylase by *Saccharomyces cerevisiae*. *Science* **228**, 21–26.

Irani M., Taylor, W.E. and Young, E.T. (1987) Transcription of the *ADH2* gene in *Saccharomyces cerevisiae* is limited by positive factors that bind competitively to its intact promoter region on multicopy plasmids. *Molec. Cell. Biol.* **7**, 1233–1241.

Isaacs A. and Lindenmann, J. (1957) Virus interference. I. The interferon *Proc. R. Soc. London, Ser. B*, **147**, 258–267.

Ito, H., Fukuda, Y., Murata, K. and Kimura, A. (1983) Transformation of intact yeast cells treated with alkali cations. *J. Bacteriol.* **153**, 163–168.

Izant, J.G. and Weintraub, H. (1984) Inhibition of thymidine kinase gene expression by antisense RNA: A molecular approach to genetic analysis. *Cell* **36**, 1007–1015.

Jabbar, M.A., Sivasubramanian, N. and Nayak, D.P. (1985) Influenza viral (A/WSN/33) hemagglutinin is expressed and glycosylated in the yeast *Saccharomyces cerevisiae*. *Proc. Natl Acad. Sci. USA* **82**, 2019−2023.

Jacks, T. and Varmus, H.E. (1985) Expression of the Rous Sarcoma virus *pol* gene by ribosomal frameshifting. *Science* **230**, 1237.

Jacks, T., Power, M.D., Masiarz, F.R., Luciw, P., Barr, P.J. and Varmus, H.E. (1988) Characterisation of ribosomal frameshifting in HIV-I *gag-pol* expression. *Nature* **331**, 280−283.

Jackson, I.J. (1986) Do mammals need P elements. *Nature* **321**, 658−657.

Jacquier, A., Rodriguez, J.R. and Rosbash, M. (1985) A quantitative analysis of the effects of 5′ junction and TACTAAC box mutants and mutant combinations on yeast mRNA splicing. *Cell* **43**, 423−430.

Jacobs, K., Shoemaker. C., Rudersdorf, R., Neill, S., Kaufman, R.J., Mufson, A., Seehra, J., Jones, S.S., Hewick, R., Fritsch, E.F., Kawakita, M., Shimizu, T. and Miyake, T. (1985) Isolation and characterisation of genomic and DNA clones of human erythropoietin. *Nature* **313**, 806−809.

Jaenisch, R. (1985) Mammalian neural crest cells participate in normal embryonic development on microinjection into post-implantation embryos. *Nature* **318**, 181−183.

Jaenisch, R., Harbers, K., Schnieke, A., Höhler, J., Chumakov, I., Jahner, D., Grotkopp, D. and Hoffmann, E. (1983) Germline integration of Moloney murine leukaemia virus at the *Mov13* locus leads to recessive lethal mutation and early embryonic death. *Cell* **32**, 209−216.

Jaenisch, R. and Jähner, D. (1984) Methylation, expression and chromosomal position of genes in mammals *Biochim. Biophys. Acta* **782**, 1−9.

Jaenisch, R., Jähner, D., Nobis, P., Simmon, I., Löhler, J., Harbers, K. and Grotkopp, D. (1981) Chromosomal position and activation of retroviral genomes inserted into the germ line of mice. *Cell* **24**, 519−529.

Jähner, D. and Jaenisch, R. (1985) Retrovirus induced *de novo* methylation of flanking host sequences correlates with gene inactivity. *Nature* **315**, 594−597.

Jähner, D., Stuhlmann, H., Stewart, C.L., Harbers, K., Löhler, J., Simmon, I. and Jaenisch, R. (1982) *De novo* methylation and expression of retroviral genomes during mouse embryogenesis. *Nature* **298**, 623−628.

Jakobovits, E.B., Majors, J.E. and Varmus, H.E. (1984) Hormonal regulation of the Rous sarcoma virus *src* gene via a heterologous promoter defines a threshold dose for cellular transformation. *Cell* **38**, 757−765.

James, R. and Bradshaw, R.A. (1984) Polypeptide growth factors. *Ann. Rev. Biochem.* **53**, 259−293.

Jayaram, M., Li, Y.-Y. and Broach, J.R. (1983) The yeast plasmid 2μ circle encodes components required for its high copy propagation. *Cell* **34**, 95−104.

Jayaram, M., Sutton, A. and Broach, J.R. (1985) Properties of REP3: a *cis*-acting locus required for stable propagation of the *Saccharomyces cerevisiae* plasmid 2 μm circle. *Molec. Cell. Biol.* **5**, 2466−2475.

Jaynes, J.M., Yang, M.S., Espinoza, N. and Dodds, J.H. (1986) Plant protein improvement by genetic engineering: use of synthetic genes. *Trends Biotechnol.* **4**, 314−320.

Jazwinski, S.M. and Edelman, G.M. (1984) Evidence for the participation of a multiprotein complex in yeast DNA replication *in vitro*. *J. Biol. Chem.* **259**, 6852.

Jazwinski, S.M., Niedzwiecka, A. and Edelman, G.M. (1983) *In vitro* association of a replication complex with a yeast chromosomal replicator, *J. Biol. Chem.* **258**, 2754.

Jeffreys, A.J., Wilson, V. and Thein, S.L. (1985) Hypervariable minisatellite regions in human DNA. *Nature* **314**, 67−73.

Jen, G.C. and Chilton, M-D. (1986) The right border region of pTiT37 T-DNA is intrinsically more active than the left border region in promoting T-DNA transformation. *Proc. Natl Acad. Sci. USA* **83**, 3895−3899.

Jenh, C-H., Deng, T., Li, D., De Wille, J. and Johnson, L.F. (1986) Mouse thymidylate synthase messenger RNA lacks a 3′ untranslated region. *Proc. Natl Acad. Sci. USA* **83**, 8482−8487.

Jenkins, N. and Copeland, N.G. (1985) High frequency germline aquisition of ecotropic MuLV proviruses in SWR/J-RF/J hybrid mice. *Cell* **43**, 811−819.

Jensen, E.O., Marcker, K.A. and Villadsen, S. (1986) Heme regulates the expression in *Saccharomyces cerevisiae* of chimaeric genes containing 5′-flanking soybean leghemoglobin sequences. *EMBO J.* **5**, 843−847.

Jensen, J.S., Marcker, K.A., Otten, L. and Schell, J. (1986) Nodule-specific expression of a chimaeric soybean leghaemoglobin gene in transgenic *Lotus corniculatus*. *Nature* **321**, 669−674.

Jenuwein, T., Muller, D., Curran, T. and Muller, R. (1985) Extended life span and tumorigenicity of non-established mouse connective tissue cells transformed by the *fos* oncogene of FBR-MuSV. *Cell* **41**, 629−637.

Jigami, Y., Muraki, M., Harada, N. and Tanaka, H. (1986) Expression of synthetic human-lysozyme gene in *Saccharomyces cerevisiae*: use of a synthetic chicken-lysozyme signal sequence for secretion and processing. *Gene* **43**, 273–279.

Jiminez, A. and Davies, J. (1980) Expression of a transposable antibiotic resistance element in *Saccharomyces*. *Nature* **287**, 869–871.

Johnson A.D. and Herskowitz, I. (1985) A repressor (MAT α2 product) and its operator control expression of a set of cell type specific genes in yeast. *Cell* **42**, 237–247.

Johnson, L.M., Bankaitis, V.A. and Emr, S.D. (1987) Distinct sequence determinants direct intracellular sorting and modification of a yeast vacuolar protease. *Cell* **48**, 875–885.

Johnson, P.J., Coussens, P.M., Danko, A.V. and Shalloway, D. (1985) Overexpressed $pp60^{c-src}$ can induce focus formation without complete transformation of NIH 3T3 cells. *Molec. Cell. Biol.* **5**, 1073–1083.

Johnson, P.F., Landschulz, W.H., Graves, B.J. and McKnight, S.L. (1987) Identification of a rat liver nuclear protein that binds to the enhancer core element of three animal viruses. *Genes Develop.* **1**, 133–146.

Johnston, S. and Hopper, J.E. (1982) Isolation of the yeast regulatory gene *GAL4* and analysis of its dosage effects on the galactose-melibiose region. *Proc. Natl Acad. Sci. USA*, **79**, 6971–6975.

Jolly, D.T., Okayama, H., Berg, P., Esty, A.C., Filpula, D., Bohlen, P., Johnson, G.G., Shively, J.E., Hunkapillar, T. and Friedmann, T. (1983) Isolation and characterisation of a full-length expressible cDNA for human hypoxanthine phosphoribosyl transferase. *Proc. Natl Acad. USA* **80**, 477–481.

Jolly, D.J., Willis, R.C. and Friedmann, T. (1986) Variable stability of a selectable provirus after retroviral gene transfer into human cells. *Molec. Cell. Biol.* **6**, 1141–1147.

Jones, J.D.G., Dunsmuir, P. and Bedbrook, J. (1985) High level expression of introduced chimaeric genes in regenerated transformed plants. *EMBO J.* **4**, 2411–2418.

Jones, K.A., Yamamoto, K.R. and Tjian, R. (1985) Two distinct transcription factors bind to the HSV thymidine kinase promoter *in vitro*. *Cell* **42**, 559–572.

Jones, P.T., Dear, P.H., Foou, J., Neuberger, M. and Winter, G. (1986) Replacing the complementarity determining regions in a human antibody with those from a mouse. *Nature* **321**, 522–525.

Jones, R.A., Kadonaga, J.T., Rosenfeld, P.J., Kelly, T.J. and Tjian, R. (1987) A cellular DNA-binding protein that activates eukaryotic transcription and DNA replication. *Cell* **48**, 79–89.

Joos, H., Timmerman, B., Van Montagu, M. and Schell, J. (1983) Genetic analysis of transfer and stabilisation of *Agrobacterium* DNA in plant cells. *EMBO J.* **2**, 2151–2160.

Joyner, A.L., Keller, G., Phillips, R.A. and Bernstein, A. (1983) Retrovirus transfer of a bacterial gene into mouse haematopoietic progenitor cells. *Nature* **305**, 556–558.

Joyner, A.L. and Bernstein, A. (1983) Retrovirus transduction: generation of infectious retroviruses expressing dominant and selectable genes is associated with *in vivo* recombination and deletion events. *Molec. Cell. Biol.* **3**, 2180–2190.

Joyner, A.L., Kornberg, T., Coleman, K.G., Cox, D.R. and Martin, G.R. (1985) Expression during embryogenesis of a mouse gene with sequence homology to the *Drosophila engrailed* gene. *Cell* **43**, 29–37.

Julius, D., Blair, L., Brake, A., Sprague, G. and Thorner, J. (1983) Yeast α factor is processed from a larger precursor polypeptide: the essential role of a membrane bound dipeptidyl amino peptidase. *Cell* **32**, 839–852.

Juluis, D., Schekman, R. and Thorner, J. (1984) Glycosylation and processing of prepro-α-factor through the yeast secretory pathway. *Cell* **36**, 309–318.

Kaddurah-Daouk, R., Greene, J.M., Baldwin, A.S. and Kingston, R.E. (1987) Activation and repression of mammalian gene expression by the c-*myc* protein. *Genes Develop.* **1**, 347–357.

Kadonaga, J.T., Jones, K.A. and Tjian, R. (1986) Promoter-specific activation of RNA polymerase II transcription by Sp1. *Trends Biochem.* **11**, 20–23.

Kaetzel, D.M., Browne, J.K., Wondisford, F., Nett, T.M., Thomason, A.R. and Nilson, J.H. (1985) Expression of biologically active bovine luteinizing hormone in Chinese hamster ovary cells. *Proc. Natl Acad. Sci. USA* **82**, 7280–7283.

Kaiser, C.A. and Botstein, D. (1986) Secretion-defective mutations in the signal sequence for *Saccharomyces cerevisiae* invertase. *Molec. Cell. Biol.* **6**, 2382–2391.

Kaiser, C.A., Preuss, D., Grisafi, P. and Botstein, D. (1987) Many random sequences functionally replace the secretion signal sequence of yeast invertase. *Science* **235**, 312–317.

Kalderon, D., Richardson, W.D., Markham, A.F. and Smith, A.E. (1984a) Sequence requirements for nuclear location of simian virus 40 large T antigen. *Nature* **311**, 33–39.

Kalderon, D., Roberts, B.L., Richardson, W.D. and Smith, A.E. (1984b) A short amino acid sequence able to specify nuclear location. *Cell* **39**, 499–509.

Kamps, M.P., Buss, J.E. and Sefton, B.M. (1986) Rous sarcoma virus transforming protein lacking myristic acid phosphorylates known polypeptide substrates without inducing transformation. *Cell* **45**, 105–112.

478   REFERENCES

Kan, N.C., Franchini, G., Wong-Staal, F., DuBois, G.C., Robey, W.G., Lauten berger, J.A. and Papas, T.S. (1986) Identification of HTLV III/LAV *sor* gene product and detection of antibodies in human sera. *Science* **231**, 1553–1555.

Kao, F-T. (1985) Human genome structure. *Int. Rev. Cytol.* **96**, 51–88.

Kao, S-Y., Calman, A.F., Luciw, P.A. and Peterlin, B.M. (1987) Anti-termination of transcription within the long terminal repeat of HIV-I by *tat* gene product. *Nature* **330**, 489–493.

Kaplan, D.R., Bockus, B., Roberts, T.M., Bolen, J., Israel, M. and Schaffhausen, B.S. (1985) Large-scale production of polyoma middle T antigen by using genetically engineered tumours. *Molec. Cell. Biol.* **5**, 1795–1799.

Karch, F., Weiffenbach, B., Peifer, M., Bender, W., Duncan, I., Celniker, S., Crosby, M. and Lewis, E.B. (1985) The abdominal region of the bithorax complex. *Cell* **43**, 81–96.

Karess, R.E. (1985) P element mediated germ line transformation of *Drosophila*. In *DNA Cloning: a Practical Approach, Vol. II* pp. 121–141 (ed. D.M. Glover), IRL Press, Oxford.

Karess, R.E. and Rubin, G.M. (1984) Analysis of P transposable element functions in *Drosophila*. *Cell* **38**, 135–146.

Karin, M., Haslinger, A., Holtgreve, H., Cathala, G., Slater, E. and Baxter, J.D. (1984) Activation of a heterologous promoter in response to dexamethasone and cadmium by metallothionein gene 5′ flanking DNA. *Cell* **36**, 371–379.

Karin, M., Najarian, R., Haslinger, A., Valenzuela, P., Welch, J. and Fogel, S. (1984) Primary structure and transcription of an amplified genetic locus: the *CUP1* locus of yeast. *Proc. Natl Acad. Sci. USA* **81**, 337–341.

Karin, M. and Richards, R.I. (1982) Human metallothionein genes — primary structure of the metallothionein-II gene and a related processed gene. *Nature* **299**, 797–802.

Karlin-Neumann, G.A. and Tobin, E.M. (1986) Transit peptides of nuclear-encoded chloroplast proteins share a common amino acid framework. *EMBO J.* **5**, 9–13.

Karlsson, S., Humphries, R.K., Gluzman, Y. and Nienhuis, A.W. (1985) Transfer of genes into hematopoietic cells using recombinant DNA viruses. *Proc. Natl Acad. Sci. USA* **82**, 158–162.

Karlsson, S. and Nienhuis, A.W. (1985) Developmental regulation of human globin genes. *Ann. Rev. Biochem.* **54**, 1071–1108.

Karlsson, S., Papayannopoulos, T., Schweiger, S.G., Stamatoyannopoulos, G. and Nienhuis, A.W. (1987) Retroviral-mediated transfer of genomic globin genes leads to regulated production of RNA and protein. *Proc. Natl Acad. Sci. USA* **84**, 2411–2415.

Karpov, V.L., Preobrazhenskaya, O.V. and Mirzabekov, A.D. (1984) Chromatin structure of hsp70 genes activated by heat shock: Selective removal of histones from the coding region and their absence from the 5′ region. *Cell* **36**, 423–43.

Kataoka, T., Powers, S., Cameron, S., Fasano, O., Goldfarb, M., Broach, J. and Wigler, M. (1985) Functional homology of mammalian and yeast *RAS* genes. *Cell* **40**, 19–26.

Kataoka, T., Powers, S., McGill, U.C., Fasano, O., Strathern J., Broach, J. and Wigler, M. (1984) Genetic analysis of yeast *RAS1* and *RAS2* genes. *Cell* **37**, 437–445.

Kaufer, N.F., Simanis, V. and Nurse, P. (1985) Fission yeast *Schizosaccharomyces pombe* correctly excises a mammalian RNA transcript intervening sequence. *Nature* **318**, 78–80.

Kaufman, R.J. (1985) Identification of the components necessary for adenovirus translational control and their utilisation in cDNA expression vectors. *Proc. Natl Acad. Sci. USA* **82**, 689–693.

Kaufman, R.J. and Sharp, P.A. (1982) Construction of a modular dihydrofolate reductase cDNA gene: analysis of signals utilised for efficient expression. *Molec. Cell. Biol.* **9**, 1304–1319.

Kaufman, R.J. and Sharp, P.A. (1983) Growth-dependent expression of dihydrofolate reductase mRNA from modular cDNA genes *Molec. Cell. Biol.* **3**, 1598–1608.

Kaufman, R.J., Wasley, L.C., Spiliotes, A.J., Gossels, S.D., Latt, S.A., Larsen, G.R. and Kay, R.M. (1985) Coamplification and coexpression of human tissue-type plasminogen activator and murine dihydrofolate reductase sequences in Chinese hamster ovary cells. *Molec. Cell. Biol.* **5**, 1750–1759.

Kavathas, P. and Herzenberg, L.A. (1983) Stable transformation of mouse L cells for human membrane T-cell differentiation antigens, HLA and β2-microglobulin: selection by fluorescence activated cell sorting. *Proc. Natl Acad. Sci. USA* **80**, 524–528.

Kavathas, P., Sukhatme, V.P., Herzenberg, L.A. and Parnes, J.R. (1984) Isolation of the gene encoding the human T lymphocyte differentiation antigen Leu-2 (T8) by gene transfer and cDNA subtraction. *Proc. Natl Acad. Sci. USA* **81**, 7688–7692.

Kawai, S. and Nishizawa, M. (1984) New procedure for DNA transfections with polycation and dimethyl sulfoxide. *Molec. Cell. Biol.* **4**, 1172–1174.

Kay, M.A. and Jacobs-Lorena, M. (1985) Selective translational regulation of ribosomal protein gene expression during early development of *Drosophila melanogaster*. *Molec. Cell. Biol.* **3**, 3563–3591.

Kay, R. (1983) Cyclic AMP and development in the slime mould. *Nature* **301**, 659.

REFERENCES

Kearsey, S.E. (1984) Structural requirements for the function of a yeast chromosomal replicator. *Cell* **37**, 299–307.

Kearsey, S.E. (1986) Replication origins in yeast chromosomes. *Bioessays* **4**, 157–161.

Kearsey, S.E. (1983) Analysis of sequences conferring autonomous replication in bakers yeast. *EMBO J.* **2**, 1571–1575.

Kedes, L. and Maxson, R. (1981) Histone gene organisation: paradigm lost. *Nature* **294**, 11–12.

Keegan, L., Gill, G. and Ptashne, M. (1986) Separation of DNA binding from the transcription-activating function of a eukaryotic regulatory protein. *Science* **231**, 699–705.

Keene, M.A. and Elgin, S.C.R. (1984) Patterns of DNA structural polymorphisms and their evolutionary relationships. *Cell* **36**, 121–129.

Keith, B. and Chua, N-H. (1986) Monocot and dicot pre-mRNAs are processed with different efficiencies in transgenic tobacco. *EMBO J.* **5**, 2419–2426.

Keller, G., Paige, C., Gilboa, E. and Wagner, E.F. (1985) Expression of a foreign gene in myeloid and lymphoid cells derived from multipotent haematopoietic precursors. *Nature* **318**, 149–154.

Kelley, D., Coleclough, C. and Perry, R.P. (1982) Functional significance and evolutionary development of the $5'$ terminal regions of immunoglobulin variable region genes. *Cell* **29**, 681–689.

Kelly, J.H. and Darlington, G.J. (1985) Hybrid genes: molecular approaches to tissue-specific gene regulation. *Ann. Rev. Genet.* **19**, 273–296.

Kelly, K., Cochran, B.H., Stiles, C.D. and Leder, P. (1983) Cell specific regulation of the *c-myc* gene by lymphocyte mitogens and platelet-derived growth factor. *Cell* **35**, 603–610.

Kelly, K. and Siebenlist, U. (1986) The regulation and expression of *c-myc* in normal and malignant cells. *Ann. Rev. Immunol.* **4**, 317–339.

Kessler, S.W. (1975) Rapid isolation of antigens from cells with a staphylococcal protein A-antibody adsorbent: parameters of the interaction of antibody-antigen complexes with protein A. *J. Immunol.* **115**, 1617–1624.

Keus, J.A.R. (1986) An accidental human trial of recombinant vaccinia virus: a step towards the acceptance of live recombinant vaccines. *Trends Biotechnol.* **4**, 105–106.

Kieny, M.P., Lathe, R., Drillien, R., Spehner, D., Skory, S., Schmitt, D., Wiktor, T., Koprowski, H. and Hecocq, J.P. (1984) Expression of rabies virus glycoprotein from a recombinant vaccinia virus. *Nature* **312**, 163–166.

Kikuchi, Y. (1983) Yeast plasmid requires a *cis*-acting locus and two plasmid proteins for its stable maintenance. *Cell* **35**, 487–493.

Kilekar, A. and Cole, M.D. (1986) Tumourigenicity of fibroblast lines expressing the adenovirus E1a, cellular p53 or normal c-*myc* genes. *Molec. Cell. Biol.* **6**, 7–14.

Kim, S., Mellor, J., Kingsman, A.J. and Kingsman, S.M. (1986) Multiple control elements in the *TRP1* promoter of *Saccharomyces cervisiae*. *Molec. Cell. Biol.* **6**, 4251–4258.

Kingsman S.M. and Kingsman, A.J. (1983) The production of interferon in bacteria and yeast. In *Interferons, Society for General Microbiology Symp. Vol. 35* (eds D.C. Burke and A. Morris), pp. 211–254. Cambridge University Press.

Kingsman, S.M., Kingsman, A.J., Dobson, M.J., Mellor, J. and Roberts, N.A. (1985) Heterologous gene expression in *Saccharomyces cerevisiae*. In *Biotechnology and Genetic Engineering Reviews* (ed. G.E. Russell) Vol. 3, pp. 377–416, Intercept, Newcastle upon Tyne.

Kingsman, S.M., Kingsman, A.J. and Mellor, J. (1987a) The production of mammalian proteins in *Saccharomyces cerevisiae*. *Trends Biotechnol.* **5**, 53–57.

Kingsman, A.J., Mellor, J., Adams, S., Rathjen, P.D., Malim, M.H., Fulton, A.M., Wilson, W. and Kingsman, S.M. (1987b) The genetic organisation of the yeast Ty element. In *Virus Replication and Genome Interactions*, the Seventh John Innes Symposium (ed. J.W. Davies, R. Hull, K.F. Chater, T.H.N. Ellis, G.P. Lomonossoff and H.W. Woolhouse). *J. Cell. Sci.* **155**, supplement 7. The Company of Biologists Ltd, Cambridge.

Kingston, R.E., Schuetz, T.J. and Larin, Z. (1987) Heat-inducible human factor that binds to a human *hsp*70 promoter. *Molec. Cell. Biol.* **7**, 1530–1534.

Kioussis, D., Wilson, F., Daniels, C., Leveton, C., Taverne, J. and Playfair, J.H.L. (1987) Expression and rescuing of a cloned human tumour necrosis factor gene using an EBV-based shuttle cosmid vector. *EMBO J.* **6**, 355–361.

Kioussis, D., Wilson, F., Khazaie, K. and Grosveld, F.G. (1985) Differential expression of human globin genes introduced in K562 cells. *EMBO J.* **4**, 927–931.

Kit, S., Dubbs, D.R., Piekarski, L.J. and Hsu, T.C. (1963) Deletion of thymidine kinase activity from L cells resistant to bromodeoxy uridine. *Exptl Cell Res.* **31**, 297–312.

Kitajewski, J., Schneider, R.J., Safer, B., Munemitsu, S.M., Samuel, C.E., Thimmappaya, B. and Shenk, T. (1986) Adenovirus VA1 RNA antagonises the antiviral action of interferon by preventing activation of the interferon-induced eIF-2α kinase. *Cell* **45**, 195–201.

Klar, A. (1987) Determination of yeast cell lineage. *Cell* **49**, 433–435.

Klatzmann, D. and Gluckman, J.C. (1986) HIV infection: facts and hypotheses. *Immunol. Today* 7, 291–296.

Klausner, A. (1984) Microbial insect controls using bugs to kill bugs. *Bio/Technol* 2, 408–419.

Klausner, A. (1986a) Researchers probe second generation t-PA. *Bio/Technol.* 4, 706–711.

Klausner, A. (1986b) Taking aim at cancer with monoclonal antibodies. *Bio/Technol.* 4, 185–194.

Klee, H., Horsch, R. and Rogers, S. (1987) *Agrobacterium*-mediated plant transformation and its further applications to plant biology. *Ann. Rev. Plant Physiol.* 38, 467–486.

Klein, G. and Klein, E. (1985) Evolution of tumours and the impact of molecular oncology. *Nature* 315, 190–195.

Klein, M.D. and Langer, R. (1986) Immobilised enzymes in clinical medicine: an emerging approach to new drug therapies. *Trends Biotechnol.* 4, 179–186.

Klein, T.M., Wolf, E.D., Wu, R. and Sanford, J.C. (1987) High-velocity microprojectiles for delivering nucleic acids into living cells. *Nature* 327, 70–93.

Klemenz, R., Hultmark, D. and Gehring, W.J. (1985) Selective translation of heat shock mRNA in *Drosophila melanogaster* depends on sequence information in the leader. *EMBO J.* 4, 2053–2060.

Klessig, D.F., Brough, D.E. and Cleghon, V. (1984) Introduction, stable integration and controlled expression of a chimeric adenovirus gene whose product is toxic to the recipient human cell. *Molec. Cell. Biol.* 4, 1354–1362.

Kmiec, E.B., Razvi, F. and Worcel, A. (1986) The role of DNA-mediated transfer of TFIII A in the concerted gyration and differential activation of the *Xenopus* 5S RNA genes. *Cell* 45, 209–218.

Kniskern, P.J., Hagopian, A., Montgomery, D.L., Burke, P., Dunn, N.R., Hoffman, K.J., Miller, W.J. and Ellis, R.W. (1986) Unusually high-level expression of a foreign gene (hepatitis B virus core antigen) in *Saccharomyces cerevisia Gene* 46, 135–141.

Knowles, J.R. (1987) Tinkering with enzymes: what are we learning? *Science* 236, 1252–1258.

Koenig, M., Hoffman, E.P., Bartleson, C.J., Monaco, A.P., Feener, C. and Kunkel, L.M. (1987) Complete cloning of the Duchenne muscular dystrophy (DMD) cDNA and preliminary genomic organisation of the DMD gene in normal and affected individuals *Cell* 50, 509–517.

Khöhler, G. and Milstein, C. (1975) Continuous culture of fused cells secreting antibody of predefined specificity. *Nature* 256, 495–497.

Kollias, G., Wrighton, N., Hurst, J. and Grosveld, F. (1986) Regulated expression of human $^A\gamma$-, β- and hybrid γβ-globin genes in transgenic mice: manipulation of the developmental expression patterns. *Cell* 46, 89–94.

Konarska, M.M., Padgett, R.A. and Sharp, P.A. (1985) *Trans* splicing of mRNA precursors *in vitro Cell* 42, 157–164.

Konarska, M.M. and Sharp, P.A. (1987) Interactions between small nuclear ribonucleoprotein particles in formation of splicesomes. *Cell* 49, 763–774.

Kondorosi, E. and Kondorosi, A. (1986) Nodule induction on plant roots by *Rhizobium*. *Trends Biol. Sci.* 11, 296–299.

Kornberg, A. 1980. *DNA Replication*. W.H. Freeman, San Francisio.

Kornberg, A. (1982) *1982 Supplement to DNA Replication*. W.H. Freeman, San Francisio.

Kornberg, R. (1981) The location of nucleosomes in chromatin: specific or statistical. *Nature* 292, 579–580.

Kornberg, T., Siden, I., O'Farrell, P. and Simon, M. (1985) The engrailed locus of *Drosophila*: *In situ* localisation of transcripts reveals compartment specific expression. *Cell* 40, 45–53.

Kornblihtt, A.R., Umezawa, K., Vibe-Pedersen, K. and Baralle, F.E. (1985) Primary structure of human fibronectin: differential splicing may generate at least 10 polypeptides from a single gene. *EMBO J.* 4, 1755–1759.

Kornfeld, R. and Kornfeld, S. (1985) Assembly of asparagine-linked oligosaccharides. *Ann. Rev. Biochem.* 54, 631–664.

Koshland, D., Kent, J.C. and Hartwell, L.H. (1985) Genetic analysis of the mitotic transmission of mini chromosomes. *Cell* 40, 393–403.

Kovesdi, T., Reichel, R. and Nevins, J.R. (1986) Identification of a cellular transcription factor involved in E1A transactivation. *Cell* 45, 219–228.

Kozak, M. (1983a) Comparison of initiation of protein synthesis in prokaryotes, eukaryotes and organelles. *Microbiol. Rev.* 47, 1–45.

Kozak, M. (1983b) Translation of insulin-related polypeptides from messenger RNAs with tandemly reiterated copies of the ribosome binding site. *Cell* 34, 971–978.

Kozak, M. (1984) Compilation and analysis of sequences upstream from the translational start site in eukaryotic mRNAs. *Nucl. Acids Res.* 12, 857–872.

Kozak, M. (1986a) Point mutations define a sequence flanking the AUG initiator codon that modulates translation by eukaryotic ribosomes. *Cell* 44, 283–292.

Kozak, M. (1986b) Influences of mRNA secondary structure on initiation by eukaryotic ribosomes. *Proc. Natl Acad. Sci. USA* 83, 2850–2854.

Kramer, R.A., De Chiara, T.M., Schaber, M.D. and Hilliker, S. (1984) Regulated expression of a human interferon gene in yeast: control by phosphate concentration or temperature *Proc. Natl Acad. Sci. USA* **81**, 367–370.

Kramer, R.A., Schaber, M.D., Skalka, A.M., Ganguly, K., Wong-Staal, F. and Reddy, E.P. (1986) HTLV-III *gag* protein is processed in yeast cells by the virus *pol*-protease. *Science* **281**, 1580–1584.

Kreus, F.A., Molendijk, L., Wullems, G.J. and Schilperoort, R.A. (1982) *In vitro* transformation of plant protoplasts with Ti plasmid DNA. *Nature* **296**, 72–74.

Krieg, P.A. and Melton, D.A. (1987) An enhancer responsible for activating trancription at the midblastula transition in *Xenopus* development. *Proc. Natl Acad. Sci. USA* **84**, 2331–2335.

Kriegler, M. and Botchan, M. (1983) Enhanced transformation by a simian virus 40 recombinant virus containing a Harvey murine sarcoma virus long terminal repeat. *Molec. Cell. Biol.* **3**, 355–339.

Kriegler, M., Perez, C.F., Hardy, C. and Botchan, M. (1984) Transformation mediated by the SV40 Tantigens: Separation of the overlapping SV40 early genes with a retroviral vector. *Cell* **38**, 483–491.

Kristiansein, B. and Lewes, C. (1986) Bacterial insecticides: recent developments. *Trends Biotechnol.* **4**, 56–58.

Kronenberg, M., Siu, G., Hood, L. and Shatri, N. (1986) The molecular genetics of the T-cell antigen receptor and T-cell antigen recognition *Ann. Rev. Immunol.* **4**, 529–591.

Krumlauf, R., Hammer, R.E., Tilghman, S.M. and Brinster, R.L. (1985) Developmental regulation of α-fetoprotein genes in transgenic mice. *Molec. Cell. Biol.* **5**, 1639–1648.

Kucherlapati, R.S., Eves, E.M., Song, K-Y., Morse, B.S. and Smithies, O. (1984) Homdogous recombination between plasmids in mammalian cells can be enhanced by treatment of input DNA. *Proc. Natl Acad. Sci. USA* **81**, 3153–3157.

Kuehn, M.R., Bradley, A., Robertson, E.J. and Evans, M.J. (1987) A potential animal model for Lesch–Nyhan syndrome through introduction of HPRT mutations into mice. *Nature* **326**, 295–298.

Kuhlemeier, C., Fluhr, R., Green, P.J. and Chua, N-H. (1987a) Sequences in the pea rbc S-3A gene have homology to constitutive mammalian enhancers but function as negative regulatory elements. *Gene Develop.* **1**, 247–258.

Kuhlemeier, C., Green, P.J. and Chua, N-H. (1987b) Regulation of gene expression in higher plants. *Ann. Rev. Plant Physiol.* **38**, 221–257.

Kühn, L.C., Barbosa, J.A., Kamark, M.E. and Ruddle, F.H. (1983) An approach to the cloning of cell surface protein genes: selection by cell sorting of mouse L cells which express HLA or 4F2 antigens after transformation with total human DNA. *Molec. Biol. Med.* **1**, 335–352.

Kühn, L.C., McClelland, A. and Ruddle, F.H. (1984) Gene transfer, expression and molecular cloning of the human transferrin receptor gene. *Cell* **37**, 95–103.

Kukuruzinska, M.A., Bergh, M.L. and Jackson, B.J. (1987) Protein glycosylation in yeast. *Ann. Rev. Biochem.* **56**, 915–944.

Kunkel, T.A. (1985) Rapid and efficient site-specific mutagenesis without phenotypic selection *Proc. Natl Acad. Sci. USA* **82**, 488–492.

Kurachi, K. and Davies, E.W. (1982) Isolation and characterisation of a cDNA coding for human factor IX. *Proc. Natl Acad. Sci. USA* **79**, 6461–6464.

Kurjan, J. and Herskowitz, I. (1982) Structure of a yeast pheromone gene (*MFα*): a putative α-factor precursor contains four tandem copies of mature α-factor. *Cell* **30**, 933–943.

Kuroda, K., Hauser, C., Rott, R., Klenk, H-D. and Doerfler, W. (1986) Expression of the influenza virus haemagglutinin in insect cells by a baculovirus vector. *EMBO J.* **5**, 1359–1365.

Lacal, J.C., Srivastava, S.K., Anderson, P.S. and Aaronson, S.A. (1986) *Ras* p21 protein with high or low GTPase activity. *Cell* **44**, 609–617.

Lacy, E., Roberts, S., Evans, E.P., Burtenshaw, M.D. and Costantni, F. (1983) A foreign β-globin gene in transgenic mice: integration at abnormal chromosome positions and expression in inappropriate tissues. *Cell* **34**, 343–358.

Laimins, L.A., Holmgren-Konig, M. and Khoury, G. (1986) Transcriptional "silencer" element in rat repetitive sequences associated with the rat insulin 1 gene locus. *Proc. Natl Acad. Sci. USA* **83**, 3151–3155.

Laimins, L.A., Khoury, G., Gorman, C., Howard, B. and Gruss, P. (1982) Host-specific activation of transcription by tandem repeats from simian viris 40 and Moloney murine sarcoma virus. *Proc. Natl Acad. Sci. USA* **79**, 6453–6457.

Lambert, P.F., Spalholz, B.A. and Howley, P.M. (1987) A transcriptional repressor encoded by BPV-1 shares a common carboxy-terminal domain with the E2 transactivator. *Cell* **50**, 69–78.

Lamppa, G., Nagy, F. and Chua, N-H. (1985) Light-regulated and organ-specific expression of a wheat Cab gene in transgenic tobacco. *Nature* **316**, 750−756.

Land, H.L., Parada, L.F. and Weinberg, R.A. (1983a) Cellular oncogenes and multistep carcinogenesis. *Science* **222**, 771−777.

Land, H.L., Parada, L. and Weinberg, R. (1983b) Tumorigenic conversion of primary embryo fibroblasts requires at least two cooperating oncogenes. *Nature* **304**, 596−602.

Langford, C.J. and Gallwitz, D. (1983) Evidence for an intron-contained sequence required for the splicing of yeast RNA polymerase II transcripts. *Cell* **33**, 519−527.

Langford, C.J., Klinz, F.J., Donath, C. and Gallwitz, D. (1984) Point mutations identify the conserved, intron-contained TACTAAC box as an essential splicing signal sequence in yeast. *Cell* **36**, 645−653.

Langone, J.J. and Vanakis, H.V. (eds) (1986) Immunoclinical techniques. Part I Hybridoma technology and monoclonal intibodies. *Methods Enzymology* Vol. 121. Academic Press, London.

Langridge, R., Eibel, H., Brown, J.W.S. and Feix, G. (1984) Transcription from maize storage protein gene promoters in yeast. *EMBO J.* **3**, 2467−2471.

Laski, F.A., Rio, D.C. and Rubin, G.M. (1986) Tissue specificity of *Drosophila* P element transposition is regulated at the level of mRNA splicing. *Cell* **44**, 7−19.

Laskey, R.A., Kearsey, S.E., Mechali, M., Dingwall, C., Mills, A.D., Dilworth, S.M. and Kleinschmidt, J. (1985) Chromosome replication in early *Xenopus* embryos. *Cold Spring Harbor Symp. Quant. Biol.*, **50**, 657−663. Cold Spring Harbor, New York.

Lau, Y.-F, Lin, C.C. and Kan, Y.W. (1983) Amplification and expression of human globin genes in chinese hamster ovary cells. *Molec. Cell. Biol.* **4**, 1469−1475.

Lauffer, L., Garcia, P.D., Harkins, R.N., Coussens, L., Ullrich, A. and Walter, P. (1985) Topology of signal recognition particle receptor in endoplasmic reticulum membrane. *Nature* **318**, 334.

Lawn, R.M. and Vehar, G.A. (1986) The molecular genetics of hemophilia. *Sci. Amer.* **254**, 40−46.

Lawrence, P.A. (1985) Molecular development: Is there a light burning in the hall? *Cell* **40**, 221.

Lawson, G.M., Knoll, B.J., March, C.J., Woo, S.L.C., Tsai, H-J. and O'Malley, B.W. (1982) Definition of 5′ and 3′ structural boundaries of the chromatin domain containing the ovalbumin multigene family. *J. Biol. Chem.* **257**, 1501−1507.

Lazo, P.A. (1985) Shuttle vectors to study somatic mutagenesis and regulation of gene expression in the immune system. *Gene* **39**, 147−153.

Learned, R.M., Learned, T.K., Haltiner, M.M. and Tjian, R.T. (1986) Human rRNA transcription is modulated by the coordinate binding of two factors to an upstream control element. *Cell* **45**, 847−857.

Leatherbarrow, R.J. and Fersht, A.R. (1986) Protein engineering. *Protein Eng.* **1**, 7−17.

Le Beau, M.M. and Rowley, J.D. (1984) Heritable fragile sites in cancer. *Nature* **308**, 607−608.

Lechler, R.I., Ronchese, F., Braunstein, N.S. and Germain, R.N. (1986) Restricted T cell antigen recognition. Analysis of the roles of Aα and Aβ using DNA-mediated gene transfer *J. Exp. Med.* **163**, 678−696.

Leder, A., Pattengale, P.K., Kuo, A., Stewart, T.A. and Leder, P. (1986) Consequences of widespread deregulation of the *c-myc* gene in transgenic mice: multiple neoplasms and normal development. *Cell* **45**, 485−495.

Leder, P., Battey, J., Lenoir, G., Moulding, C., Murphy, W., Potter, T.S and Taub, R. (1983) Translocations among antibody genes in human cancer. *Science* **222**, 766−771.

Lee, F., Mulligan, R., Berg, P. and Ringold, G. (1981) Glucocorticoids regulate expression of dihydrofolate reductase cDNA in mouse mammary tumour virus chimaeric plasmids. *Nature* **294**, 228−232.

Lee, M.G. and Nurse, P. (1987) Complementation used to clone a human homologue of the fission yeast cell cycle control gene *cdc2*. *Nature* **327**, 31−36.

Lee, W., Haslinger, A., Karin, M. and Tjian, R. (1987) Activation of transcription by two factors that bind promoter and enhancer sequences of the human metallothionein gene and SV40. *Nature* **325**, 368−372.

Lee, W., Mitchell, P. and Tjian, R. (1987) Purified transcription factor AP-1 interacts with TPA inducible enhancer elements. *Cell* **49**, 741−752.

Lee, W-H., Murphree, A.L. and Benedict, W.F. (1984) Expression and amplification of the N-myc gene in primary retinoblastoma. *Nature* **309**, 458−460.

Lee, W-H., Shew, J-Y., Hong, F.D., Sery, T.W., Donoso, L.A., Young, L.J., Bookstein, R. and Lee, E.Y-H.P. (1987) The retinoblastoma susceptibility gene encodes a nuclear phosphoprotein associated with DNA binding activity. *Nature* **329**, 642−645.

Lee, W.M.F., Schwab, M., Westaway, P. and Varmus, H.E. (1985) Augmented expression of normal c-*myc* is sufficient for cotransformation of rat embryo cells with a mutant *ras* gene. *Molec. Cell. Biol.* **5**, 3345–3356.

Leff, S.E., Evans, R.M. and Rosenfeld, M.G. (1987) Splice commitment dictates neuron-specific alternative RNA processing in calcitonin/CGRP gene expression. *Cell* **48**, 517–524.

Leff, S.E., Rosenfeld, M.G. and Evans, R.M. (1986) Complex transcriptional units: diversity in gene expression by alternative RNA processing. *Ann. Rev. Biochem.* **55**, 1091–1117.

Leffak, I.M. (1984) Conservative segregation of nucleosome core histones. *Nature* **307**, 82–85.

Lehrman, M.A., Goldstein, J.L., Brown, M.S., Russell, D.W. and Schneider, W.J. (1985a) Internalisation defective LDL receptors produced by genes with nonsense and frameshift mutations that truncate the cytoplasmic domain. *Cell* **41**, 735–743.

Lehrman, M.A., Russell, D.W., Goldstein, J.L. and Brown, M.S. (1986) Exon-Alu recombination deletes 5 kilobases from the low density lipoprotein receptor gene, producing a null phenotype in familial hypercholesterolinaemia *Proc. Natl Acad. Sci. USA* **83**, 3679.

Lehrman, M.A., Schreider, W.J., Sudhof, T.C., Brown, M.S., Goldstein, J.L and Russell, D.W. (1985b) Mutation in LDL receptor: Alu–Alu recombination deletes exons encoding transmembrane and cytoplasmic domains. *Science* **227**, 140–146.

Le Meur, M., Gerlinger, P., Benoist, C. and Mathis, D. (1985) Correcting an immune-response deficiency by creating E_α gene transgenic mice. *Nature* **316**, 38–42.

Lerner, M.R and Steitz, J.A. (1981) Snurps and Scyrps. *Cell* **25**, 298–300.

Lemischka, I., Raulet, D.H. and Mulligan, R.C. (1986) Developmental potential and dynamic behaviour of hematopoietic stem cells. *Cell* **45**, 917–927.

Levinson, A.D. (1986) Normal and activated *ras* oncogenes and their encoded products. *Trends Genet.* **2**, 81–85.

Lewin, R. (1986) Gene therapy — so near and yet so far away. *Science* **232**, 824–826.

Lichtenstein, C. (1987) Bacteria conjugate with plants. *Nature* **328**, 108–109.

Lichtenstein, C. and Draper, J. (1985) Genetic engineering of plants. In *DNA Cloning: a Practical Approach, Vol. II* (ed. D. Glover), pp. 67-127. 1RL Press, Oxford.

Lifson, J.D., Feinberg, M.B., Reyes, G.R., Rabin, L., Banapour, B., Chakrabarti, S., Moss, B., Wong-Staal, F., Steimer, K.S. and Engleman, E.C. (1986) Induction of CD4-dependent cell fusion by the HTLV III/LAV envelope glycoprotein. *Nature* **323**, 725–727.

Lindquist, S. (1986) The heat shock response. *Ann. Rev. Biochem.* **55**, 1151–1191.

Liskay, R.M. (1986) Manipulation just off target. *Nature* **324**, 13.

Littauer, U.Z. and Soreq, H. (1982) The regulatory function of poly (A) and adjacent 3' sequences in translated RNA. *Prog. Nucl. Acid Res. Molec. Biol.* **27**, 53–85.

Little, P. (1986) Restriction fragment length polymorphisms: finding the defective gene. *Nature* **321**, 558–559.

Little, P.F.R. (1982) Globin pseudogenes. *Cell* **28**, 683–684.

Littlefield, J.W. (1963) The inosinic acid pyrophosphorylase activity of mouse fibroblasts partially resistant to 8-azaguanine. *Proc. Natl Acad. Sci. USA* **50**, 568–576.

Littman, D.R., Thomas, Y., Maddon, P.J., Chess, L. and Axel, R. (1985) The isolation and sequence of the gene encoding T8: A molecule defining functional classes of T lymphocytes. *Cell* **40**, 237–246.

Loeb, D.D., Padgett, R.W., Hardies, S.C., Shehee, W.R., Comer, M.B., Edgell, M.H. and Hutchison, C.A. (1986) The sequence of a large L1 Md element reveals a tandemly repeated 5' end and several features found in retrotransposons. *Molec. Cell. Biol.* **6**, 168–182.

Logan, J. and Shenk, T. (1984) Adenovirus tripartite leader sequence enhances translation of mRNAs late after infection. *Proc. Natl Acad. Sci. USA* **81**, 3655–3659.

Löhler, J., Timpl, R. and Jaenisch, R. (1984) Embryonic lethal mutation in mouse collagen 1 gene causes rupture of blood vessels and is associated with erythropoietic and mesenchymal cell death. *Cell* **38**, 597–607.

Long, E.O. and Dawid, I.B. (1980) Repeated genes in eukaryotes, *Ann. Rev. Biochem.* **49**, 727–764.

Lopata, M.A., Cleveland, D.W. and Sollner-Webb, B., (1984) High level, transient expression of a chloramphenicol acetyl transferase gene by DEAE-dextran mediated DNA transfection coupled with a dimethyl sulfoxide or glycerolshock treatment. *Nucl. Acids Res.* **12**, 5707–5717.

Lovell-Badge, R.H. (1985) Transgenic animals: New advances in the field. *Nature* **315**, 628–629.

Loyter, A., Scangos, G.A. and Ruddle, F.H. (1982) Mechanisms of DNA uptake by mammalian cells: Fate of exogenously added DNA monitered by the use of fluorescent dyes. *Proc. Natl Acad. Sci. USA* **79**, 422–426.

Lusky, M., Berg, L., Wieher, H. and Botchan, M. (1983) Bovine papillomavirus contains an activator of gene expression at the distal end of the early transcription unit. *Molec. Cell. Biol.* **3**, 1108–1122.

Lusky, M. and Botchan, M. (1981) Inhibition of SV40 replication in simian cells by specific pBR322 DNA sequence. *Nature* **293**, 79–81.

Lusky, M. and Botchan, M. (1984) Characterisation of the bovine papilloma virus plasmid maintenance sequences. *Cell* **36**, 391–401.

Lusky, M. and Botchan, M. (1986) Transient replication of BPV-1 plasmids: *cis* and *trans* requirements. *Proc. Natl Acad. Sci. USA* **83**, 3609–3613.

Lütcke, H.A., Chow, K.C., Mickel, F.S., Moss K.A., Kern, H.F. and Scheele, G.A. (1987) Selection of AUG codons differs in plants and animals. *EMBO J.* **6**, 43–45.

Lycett, G.W., Croy R.R.D., Shirsat, A.H. and Boulter, D. (1984) The complete nucleotide sequence of a legumin gene from pea (*Pisum sativum*). *Nucl. Acids Res.* **12**, 4493–4506.

Lynch, D.R. and Snyder, S.H. (1986) Neuropeptides: multiple molecular forms, metabolise pathways and receptors. *Ann. Rev. Biochem.* **55**, 773–801.

Ma, J. and Ptashne, M. (1987) Deletion analysis of GAL4 defines two transcriptional activating segments. *Cell* **48**, 847–853.

McAleer, W.J., Buynak, E.B., Maigelter, R.Z., Wampler D.E., Miller, W.J. and Hilleman, M.R. (1984) Human hepatitis B vaccine fom recombinant yeast. *Nature* **307**, 178–179.

McBride, O.W. and Peterson, J.L. (1980) Chromosome-mediated gene transfer in mammalian cells. *Ann. Rev. Genet.* **14**, 321–345.

McClintock, B. (1948) Mutable loci in maize. *Carnegie Inst. Washington Yearbk* **47**, 155–169.

McCluskey, J., Germain, R.N. and Margulies, D.H. (1985) Cell surface expression of an *in vitro* recombinant class II/class I major histocompatibility complex gene product. *Cell* **40**, 247–257.

McDevitt, M.A., Hart, R.P., Wong, W.W. and Nevins, J.R. (1986) Sequences capable of restoring poly (A) site function define two distinct downstream elements. *EMBO J.* **5**, 2907–2913.

McDevitt, M.A., Imperiale, M.J., Ali, H. and Nevins, J.R. (1984) Requirement of a downstream sequence for generation of a poly (A) addition site. *Cell* **37**, 993-999.

McGarry, T.J. and Lindquist, S. (1985) The preferential translation of *Drosophila* hsp70 mRNA requires sequences in the untranslated leader. *Cell* **42**, 903–911.

McGrew, J., Diehil, B. and Fitzgerald-Hayes, M. (1986) Single base-pair mutations in centromere element III cause aberrant chromosome segregation of *Saccharomyces cerevisiae*. *Molec. Cell. Biol.* **6**, 530–538.

McKnight, G.S., Hammer, R.E., Kuenzel, E.A. and Brinster, R.L. (1983) Expression of the chicken transferrin gene in transgenic mice. *Cell* **34**, 335–341.

McKnight, S.L. and Kingsbury, R.L. (1982) Transcription control signals of a eukaryotic protein-coding gene. *Science* **217**, 316–324.

McKnight, S.L., Kingsbury, R.C., Spence, A. and Smith, M. (1984) The distal transcription signals of the Herpesvirus *tk* gene share a common hexanucleotide control sequence. *Cell* **37**, 253–262.

McKnight, S. and Tjian, R. (1986) Transcriptional selectivity of viral genes in mammalian cells. *Cell* **46**, 795–805.

McLauchlan, J., Gaffney, D., Whitlon, J.L. and Clements, J.B. (1985) The consensus sequence YGTGTTYY located downstream from the AATAAA signal is required for efficient formation of mRNA 3′ termini. *Nucl. Acids Res.* **13**, 1347–1368.

McLeod, M., Craft, S. and Broach, J.R. (1986) Identification of the crossover site during FLP-mediated recombination in the yeast plasmid 2 micron circle. *Molec. Cell. Biol.* **6**, 3357–3367.

McNeil, J.B. and Friesen, J. (1981) Expression of the *Herpes simplex* virus thymidine kinase gene in *Saccharomyces cerevisiae*. *Molec. Gen. Genet.* **184**, 386–393.

McTiernan, C.F. and Stambrook, P.J. (1980) Replication of DNA templates injected into frog's eggs. *J. Cell. Biol.* **87**, 45a.

McTiernan, C.F. and Stambrook, P.J. (1982) Initiation of SV40 DNA replication after microinjection into *Xenopus* eggs. *Biochem. Biophys. Acta* **782**, 295–303.

Mackett, M. and Arrand, J.R. (1985) Recombinant vaccinia virus incluces neutralizing antibodies in rabbits against Epstein-Barr virus membrane antigen gp340 *EMBO J.* **4**, 3229–3234.

Mackett, M., Smith, G.L. and Moss, B. (1982) Vaccinia virus: a selectable eukaryotic cloning and expression vector. *Proc. Natl Acad. Sci. USA* **72**, 7415–7419.

Mackett, M., Smith, G.L. and Moss, B. (1984) General method for production and selection of infectious vaccinia virus recombinants expressing foreign genes. *J. Virol.* **49**, 857–864.

Mackett, M., Yilma, T., Rose, J.K. and Moss, B. (1985) Vaccinia virus recombinants: expression of VSV genes and protective immunization of mice and cattle. *Science* **227**, 433–435.

Maeda, N. and Smithies, O. (1986) The evolution of multigene families: human haptoglobin genes. *Ann. Rev. Genet.* **20**, 81–109.

Maeda, S., Kawai, T., Obinata, M., Fujiwasa, H., Horiuchi T., Saeko, Y., Sato, Y. and Farasawa, M. (1985) Production of human α-interferon in silkworm using a baculovirus vector. *Nature* **315**, 592–594.

Magli, M-C., Dick, J.E., Huszar, D., Bernstein, A. and Phillips, R.A. (1987) Modulation of gene expression in multiple hematopoietic cell lineages following retroviral vector gene transfer. *Proc. Natl Acad. Sci. USA* **84**, 789−793.

Magram, J., Chada, K. and Costantini, F. (1985) Developmental regulation of a cloned adult β-globin gene in transgenic mice. *Nature* **315**, 338−340.

Maine, G.T., Sinha, P. and Tye, B-K. (1984) Mutants of *S. cerevisiae* defective in the maintenance of mini-chromosomes. *Genetics* **106**, 365−385.

Maniatis, T., Goodbourn, S. and Fischer, J.A. (1987) Regulation of inducible and tissue-specific gene expression. *Science* **236**, 1237−1245.

Maniatis, T., Fritsch, E.F. and Sambrook. J. (1982) *Molecular Cloning: a Laboratory Manual*. Cold Spring Harbor Laboratory, New York.

Maniatis, T., Hardison, R.C., Lacy, E., Lauer, J., O'Connell, C., Quon, D., Sim, G.K. and Efstradiatis, A. (1978) The isolation of structural genes from libraries of eukaryotic DNA. *Cell* **15**, 687−701.

Maniatis, T. and Reed, R. (1987) The role of small nuclear ribonucleoprotein particles in pre-mRNA splicing. *Nature* **325**, 673−678.

Mandal, R.K. (1984) The organisation and transcription of eukaryotic ribosomal RNA genes. *Proc. Nucl. Acid Res. Molec. Biol.* **31**, 117−160.

Mangiarotti, G., Ceccaselli, A. and Lodish, H.F. (1983) Cyclic AMP stabilizes a class of developmentally regulated *Dictyostelium discoideum* mRNAs *Nature* **301**, 616−618.

Manley, J.L. and Levine, M.S. (1985) The homeo box and mammalian development. *Cell* **43**, 1−2.

Mann, K., Mulligan, R.C. and Baltimore, D. (1983) Construction of a retrovirus packaging mutant and its use to produce helper free defective retrovirus. *Cell* **33**, 153−159.

Mardon, G. and Varmus, H.E. (1983) Frameshift and intragenic suppressor mutations in a Rous sarcoma provirus suggest *src* encodes two proteins. *Cell* **32**, 871−879.

Martinez-Arias, A. and Lawrence, P.A. (1985) Para segments and compartments in the *Drosophila* embryo. *Nature* **313**, 639−642.

Mansour, S.L., Grodzicker, T. and Tjian, R. (1985) An adenovirus vector system used to express polyoma virus tumour antigens. *Proc. Natl Acad. Sci. USA* **82**, 1359−1363.

Mantell, S.H., Matthews, J.A. and McKee, R.A. (1985) *Principles of Plant Biotechnology*. Blackwell Scientific Publications, Oxford.

Martin, G. (1987) Nomenclature for homeo-box-containing genes. *Nature* **325**, 21−22.

Martin, G.R. (1980) Teratocarcinomas and mammalian embryogenesis. *Science* **209**, 768−776.

Martin, G.R. (1982) Y-chromosome inactivation in mammals. *Cell* **29**, 721−724.

Martin, S.L., Voliva, C.F., Burton, F.H., Edgell, M.H. and Hutchinson, C.A. (1984) A large interspersed repeat found in mouse DNA contains a long open reading frame that evolves as if it encodes a protein. *Proc. Natl Acad. Sci. USA* **81**, 2308−2312.

Marzouki, N., Camier, S., Ruet, A., Moenne, A. and Sentenac, A. (1986) Selective proteolysis defines two DNA binding domains in yeast transcription factor τ. *Nature* **323**, 177.

Mason, A.J., Pitts, S.L., Nikolic, K., Szonyi, E., Wilcox, J.N., Seeburg, P.H. and Stewart, T.A. (1986) The hypogonadal mouse: reproductive functions restored by gene therapy. *Science* **234**, 1372−1378.

Mather, E.L., Nelson, K.J., Haimovich, J. and Perry, R.P. (1984) Mode of regulation of immunoglobulin μ-and δ-chain expression varies during B-lymphocyte maturation. *Cell* **36**, 329−338.

Mathis, D.J. and Chambon, P. (1981) The SV40 early region TATA box is required for accurate *in vitro* initiation of transcription. *Nature* **290**, 310−315.

Maundrell, K., Wright, A.P.H., Piper, M. and Shall, S. (1985) Evaluation of heterologous ARS activity in *S. cerevisiae* using cloned DNA from *S. Pombe*. *Nucl. Acids Res.* **13**, 3711−3722.

Mayo, K.E., Warren, R. and Palmiter, R.D. (1982) The mouse metallothionein-1 gene is transcriptionally regulated by cadmium following transfection into human or mouse cells. *Cell* **29**, 99−108.

Maxam, A.M. and Gilbert, W. (1980) Sequencing end-labelled DNA with base-specific chemical cleavages *Methods Enzymol*. Grossman, L. and Moldave, K., eds. **65**, 499−560.

Mechali, M. and Kearsey, S.E. (1984) Lack of specific sequence requirements for DNA replication in *Xenopus* eggs compared with high sequence specificity in yeast. *Cell* **38**, 55−64.

Meinkoth, J., Killary, A.M., Fournier, R.E.K. and Wahl, G.M. (1987) Unstable and stable CAD gene amplification importance of flanking sequences and nuclear environment in gene amplification. *Molec. Cell. Biol.* **7**, 1415−1424.

Mellor, A.L., Golden, L., Weiss, E., Bullman, H., Hurst, J., Simpson, E., James, R.F.L., Townsend, A.R.M., Taylor, P.M., Schmidt, W., Ferluga, J., Leben, L., Santamaria M., Atfield, G., Festenstein, H. and Flavell, R.A. (1982) Expression of murine H-2K^b histo-compatibility antigen in cells transformed with cloned *H-2* genes. *Nature* **298**, 529−534.

Mellor, J., Dobson, M.J., Kingsman, A.J. and Kingsman, S.M. (1987) A transcriptional activator is located in the coding region of the yeast *PGK* gene. *Nucl. Acids Res.* **15**, 6243–6259.

Mellor, J., Dobson, M.J., Roberts, N.A., Kingsman A.J. and Kingsman, S.M. (1985a) Factors affecting heterologous gene expression in *Saccharomyces cerevisiae*. *Gene* **33**, 215–226.

Mellor, J., Dobson, M.J., Roberts, N.A., Tuite, M.F., Emtage, J.S., White, S., Lowe, P.A, Patel, T., Kingsman, A.J. and Kingsman, S.M. (1983) Efficient synthesis of enzymatically active calf chymosin in *Saccharomyces cerevisiae*. *Gene* **24**, 1–14.

Mellor, J., Fulton, A.M., Dobson, M.J., Wilson, W., Kingsman, S.M. and Kingsman, A.J. (1985b) A retrovirus-like strategy for expression of a fusion protein encoded by yeast transposon Ty 1. *Nature* **313**, 243.

Mellor, J., Malim, M.H., Gull, K., Tuite, M.F., McCready, S., Dibbayawan, T., Kingsman, S.M. and Kingsman, A.J. (1985c) Reverse transcription activity and Ty RNA are associated with virus-like particles in yeast. *Nature* **318**, 583–586.

Melton, D., Krieg, P., Rebagliati, M., Maniatis, T., Zinn, K. and Green, M. (1984) Efficient *in vitro* synthesis of biologically active RNA and RNA hybridisation probes from plasmids containing a bacteriophage SP6 promoter. *Nucl. Acids Res.* **12**, 7035–7056.

Melton, D.A. (1985) Injected anti-sense RNAs specifically block messenger RNA translation *in vivo*. *Proc. Natl Acad. Sci. USA* **82**, 144–148.

Meneguzzi, G., Binétruy, B. Grisoni, M. and Cuzin, F. (1984) Plasmidial maintenance in rodent fibroblasts of a BPV1-pBR322 shuttle vector without immediately apparent oncogenic transformation of the recipient cells. *EMBO J.* **3**, 365–371.

Mercola, M., Goverman, J., Mirell, C. and Calame, K. (1985) Immunoglobulin heavy chain enhancer requires one or more tissue-specific factors. *Science* **227**, 266–270.

Merrill, G.F., Hauschka, S.D. and McKnight, S.L. (1984) *tk* enzyme expression in differentiating muscle cells is regulated through an internal segment of the cellular *tk* gene. *Molec. Cell. Biol.* **44**, 1777–1784.

Merrill, G., Witter, E. and Hauschka, S. (1980) Differentiation of thymidine kinase deficient mouse myoblasts in the presence of 5' bromodeoxyuridine. *Expl. Cell. Res.* **129**, 191–199.

Mertz, J.E. and Gurdon, J.B. (1977) Purified DNAs are transcribed after microinjection into *Xeropus* oocytes. *Proc. Natl Acad. Sci. USA* **74**, 1502–1506.

Messing, J. (1983) The manipulation of zein genes to improve the nutritional value of corn. *Trends Biotechnol.* **2**, 54–59.

Messing, J., Gronenborn, B., Muller-Hill, B. and Hofschneider, P.H. (1977) Filamentous coliphage M13 as a cloning vehicle: insertion of a Hind II fragment of the *lac* regulatory region in M13 replicative form *in vitro*. *Proc. Natl Acad. Sci. USA* **74**, 3642–3646.

Mechali, M. and Kearsey, S. (1984) Lack of specific sequence requirement for DNA replication in *Xenopus* eggs compared with high sequence specificity in yeast. *Cell* **38**, 55–64.

Michel, M-L., Pontisso, P., Sobczak, E., Malpiece, Y., Streeck, R.E. and Tiollais, P. (1984) Synthesis in animal cells of hepatitis B surface antigen particles carrying a receptor for polymerised human serum albumin. *Proc. Natl Acad. Sci. USA* **81**, 7700–7712.

Miesfeld, R., Godowski, P.J., Maler, B.A., and Yamamoto, K.R. (1987) Glucocorticoid receptor mutants that define a small region sufficient for enhancer activation. *Science* **236**, 423–427.

Miesfeld, R., Krystal, M. and Arnheim, N. (1981) A member of a new repeated sequence family which is conserved throughout eukaryotic evolution is found between the human α and β-globin genes. *Nucl. Acids Res.* **9**, 5931–5947.

Milanese, G., Barbanti-Brodano, G., Negrini, M., Lee, D., Corallini, A., Caputo, A, Grossi, M.P. and Ricciardi, R.P. (1984) BK virus-plasmid expression vector that persists episomally in human cells and shuttles into *Escherichia coli*. *Molec. Cell. Biol.* **4**, 1551–1560.

Milich, D.R., Thornton, G.B., Neurath, A.R., Kent, S.B., Michel, M-L., Tiollais, P. and Chisari, F.V. (1985) Enhanced immunogenicity of the pre-S region of hepatitis B surface antigen. *Science* **228**, 1195–1198.

Miller, A.D. and Buttimore, C. (1986) Redesign of retrovirus packaging cell lines to avoid recombination leading to helper virus production *Molec. Cell. Biol.* **6**, 2895–2902.

Miller, A.D., Jolly, D.J., Friedmann, T. and Verma, I.M. (1983) A transmissible retrovirus expressing human hypoxanthine phosphoribosyl transferase (HPRT): Gene transfer into cells obtained from humans deficient in HPRT. *Proc. Natl Acad. Sci. USA* **80**, 4709–4713.

Miller, A.D., Law, M-F. and Verma, I.M. (1985a) Generation of helper-free amphotropic retroviruses that transduce a dominant-acting, methotrexate resistant dihydrofolate reductase gene. *Molec. Cell. Biol.* **5**, 431–437.

Miller, A.D., Ong, E.S., Rosenfeld, M.G., Verma, I.M. and Evans, R.M. (1984b) Infectious and selectable retrovirus containing an inducible rat growth hormone minigene. *Science* **225**, 993–998.

Miller, B.L., Miller, K.Y. and Timberlake, W.E. (1985b) Direct and indirect gene replacements in *Aspergillus nidulans*. *Molec. Cell. Biol.* **5**, 1714–1721.

Miller, C.K. and Temin, H.M. (1983) High-efficiency ligation and recombination of DNA fragments by vertebrate cells. *Science* **200**, 606–609.

Miller, D.W., Safer, P. and Miller, L.K. (1987) An insect bacelovirus host-vector system for high level expression of foreign genes. In *Genetic Engineering, Vol. 8* (eds J.K. Setler & A. Hollaender), Plenum Press, New York.

Miller, J., McLachlan, A.D. and Klug, A. (1985c) Repetitive zinc-binding domains in the protein transcription factor IIIA from *Xenopus* oocytes. *EMBO J.* **4**, 1609–1614.

Miller, J.H. (1982) Carcinogens induce targeted mutations in *Escherichia coli Cell* **31**, 5–7.

Miller, J.H., Lebkowski, J.S., Griesen, K.S and Calos, M.B. (1984a) Specificity of mutations induced in transfected DNA by mammalian cells. *EMBO J.* **3**, 3117–3121.

Milner, R.J., Bloom, F.E., Lai, C., Lerner, R.A. and Sutcliffe, J.G. (1984) Brain-specific genes have identifier sequences in their introns. *Proc. Natl Acad. Sci. USA* **81**, 713–717.

Mills, J.S., Kingsman, A.J. and Kingsman, S.M. (1986) *Drosophila* ARSs contain the yeast ARS consensus sequence and a replication enhancer. *Nucl. Acids Res.* **14**, 6633–6648.

Miranda, A.F., Babiss, L.E. and Fisher, P.B. (1983) Transformation of human skeletal muscle cells by simian virus 40. *Proc. Natl Acad. Sci. USA* **80**, 6581–6585.

Mirkovitch, J., Mirault, M.E. and Laemmli, U.K. (1984) Organisation of the higher order chromatin loop: specific DNA attachment sites on the nuclear scaffold. *Cell* **39**, 223–232.

Mishina, M., Kurosaki, T., Tobimatsu, T., Morimoto, Y., Noda, M., Yamamoto, T., Terao, M., Lindstrom, J., Takahashi, T., Kuno, M. and Numa, S. (1984) Expression of functional acetyl-choline receptor from cloned cDNAs. *Nature* **307**, 604–608.

Mishina, M., Tohimatsu, T., Imoto, K., Tanaka, K-i, Fujita, Y., Fukuda, K., Kurasaki, M., Takahashi, H., Morimoto, Y., Hirose, T., Inayama, S., Takahashi, T., Kuno, M. and Numa, S. (1985) Location of functional regions of acetylcholine receptor α-subunit by site-directed mutagenesis. *Nature* **313**, 364–369.

Mishra, N.C. (1985) Gene transfer in fungi. *Adv. Genet.* **23**, 73–178.

Mitsialis, S.A., Young, J.F., Palese, P. and Guntaka, R.V. (1981) An avian tumour virus promoter directs expression of plasmid genes in *E. coli. Gene* **16**, 217–225.

Mitrani-Rosenbaum, S., Maroteaux, L., Mory, Y., Revel, M. and Howley, P. (1983) Inducible expression of human interferon β gene linked to a bovine papillomavirus DNA vector and maintained extra-chromosomally in mouse cell. *Molec. Cell. Biol.* **3**, 233–240.

Mitsuya, H. and Broder, S. (1987) Strategies for antiviral therapy in AIDS. *Nature* **325**, 773–778.

Miyajima, A., Miyajima, I., Arai, K-I. and Arai, N. (1984) Expression of plasmid R388 encoded type II dihydrofolate reductase as a dominant selective marker in *Saccharomyces cerevisiae*. *Molec. Cell. Biol.* **4**, 407–414.

Miyamoto, C., Chizzonite, R., Crowl, R., Rupprecht, K., Kramer, R., Schaber, M., Kumar, G., Poonian, M. and Ju, G. (1985) Molecular cloning and regulated expression of the human *c-myc* gene in *Escherichia coli* and *Saccharomyces cerevisiae*: comparison of the protein products. *Proc. Natl Acad. Sci. USA* **82**, 7232–7236.

Miyamoto, C., Smith, G.E., Farrell-Towt, J., Chizzonite, R., Sumovers, M.D. and Ju, G. (1985) Production of human *c-myc* protein in insect cells infected with a bacelovirus expression vector. *Molec. Cell. Biol.* **5**, 2860–2865.

Moldave, K. (1985) Eukaryotic protein synthesis. *Ann. Rev. Biochem.* **54**, 1109–1149.

Moncollin, V., Miyamoto, N.G., Zheng, X-M. and Egly, J.M. (1986) Purification of a factor specific for the upstream element of the adenovirus-2-major later promoter. *EMBO J.* **5**, 2577–2584.

Montell, C., Fisher, E.F., Caruthers, M.H. and Berk, A.J. (1983) Inhibition of RNA cleavage but not polyadenylation by a point mutation in mRNA 3' consensus sequence AAUAAA. *Nature* **305**, 600–605.

Montiel, J.F., Norbury, C.J., Tuite, M.F., Dobson, M.J., Mills, J.S., Kingsman, A.J. and Kingsman, S.M. (1984) Characterisation of human chromosomal DNA sequences which replicate autonomously in *Saccharomyces cerevisiae*. *Nucl. Acids Res.* **12**, 1049–1068.

Montencourt, B.S. (1983) Trichoderma *reesei* cellulases. *Trends Biotechnol.* **1**, 156–160.

Moore, C.L. and Sharp, P.A. (1985) Accurate cleavage and polyadenylation of exogenous RNA substrate. *Cell* **41**, 845–855.

Moore, H.P. and Kelly, R.B. (1986) Re-routing of a secretory protein by fusion with human growth hormone sequences. *Nature* **321**, 443–446.

Moore, H-P.H. and Kelly, R.B. (1985) Secretory protein targeting in a pituitary cell line: Differential transport of foreign secretory proteins to distinct secretory pathways. *J. Cell. Biol.* **101**, 1773–1781.

Moore, H-P.H., Walker, M.D., Lee, F. and Kelly, R.B. (1983) Expressing a human proinsulin cDNA in a mouse ACTH-secreting cell. Intracellular storage, proteolytic processing, and secretion on stimulation. *Cell* **35**, 531–538.

Moreland, R.B., Nam, H.G., Hereford, L.M. and Fried, H.M. (1985) Identification of a nuclear localisation signal of a yeast ribosomal protein. *Proc. Natl Acad. Sci. USA* **82**, 6561–6565.

REFERENCES

Morelli, G., Nagy, F., Fraley, R.T., Rogers, S.G. and Chua, N.H. (1985) A short conserved sequence is involved in the light inducibility of a gene encoding ribulose 1, 5-bisphosphate carboxylase small subunit of pea. *Nature* **315**, 200−204.

Morgan, W.D., Williams, G.T., Morimoto, R.I., Greene, J., Kingston, R.E. and Tjian, R. (1987) Two transcriptional activatiors, CCAAT-Box-Binding transcription factor and heat shock transcription factor interact with a human hsp 70 gene. *Molec. Cell. Biol.* **7**, 1129−1138.

Morrison, S., Johnson, M.J., Herzenberg, L.A. and Oi, V.T. (1984) Chimeric human antibody molecules: mouse antigen-binding domains with human constant region domains. *Proc. Natl Acad. Sci. USA* **81**, 6851−6855.

Moss, B. and Flexner, C. (1987) Vaccinia virus expression vectors. *Ann. Rev. Immunol.* **5**, 305−324.

Moss, B., Smith, G.L., Gerin, J.L. and Purcell, R.H. (1984) Live recombinant vaccinia virus protects chimpanzees against hepatitis B. *Nature* **311**, 67−69.

Mount, S.M. and Rubin, G.M. (1985) Complete nucleotide sequence of the *Drosophila* transposable element copia: homology between the *copia* and retroviral proteins. *Molec. Cell. Biol.* **5**, 1630−1638.

Mueller, P.P. and Hinnebusch, A.G. (1986) Multiple upstream AUG codons mediate translational control of *GCN4*. *Cell* **45**, 201−207.

Muesing, M.A., Smith, D.H. and Capon, D.J. (1987) Regulation of mRNA accumulation by a human immunodeficiency virus *trans*-activator protein. *Cell* **48**, 691−701.

Mulligan, R.C. and Berg, P. (1980) Expression of a bacterial gene in mammalian cells. *Science* **209**, 1422−1428.

Mulligan, R.C. and Berg, P. (1981) Selection for animal cells that express the *Escherichia coli* gene coding for xanthine-guanine phosphoribosyl transferase. *Proc. Natl Acad. Sci. USA* **78**, 2072−2076.

Mulligan, R.C., Howard, B.H. and Berg, P. (1979) Synthesis of rabbit β-globin in cultured monkey kidney cells following infection with SV40 β-globin recombinant genome. *Nature* **277**, 108−114.

Munro, S. and Pelham, H.R.B. (1987) A C terminal signal prevents secretion of Luminal ER proteins. *Cell* **48**, 899−907.

Munson, L., Stormo, G., Niece, R. and Reznikoff, W.S. (1984) *lac Z* translation initiation mutations *J. Mol. Biol.* **177**, 663−683.

Murray, A.W. (1985) Chromosome structure and behaviour. *TIBS*, **10**, 112−115.

Murray, A.W. and Szostak, J.W. (1983a) Pedigree analysis of plasmid segregation in yeast. *Cell* **34**, 961−970.

Murray, A.W. and Szostak, J.W. (1983b) Construction of artificial chromosomes in yeast. *Nature* **305**, 189−193.

Murray, A.W. and Szostak, J.W. (1985) Chromosome segregation in mitosis and meiosis. *Ann. Rev. Cell. Biol.* **1**, 289−315.

Murray, A.W., Schultes, N.P. and Szostak, J.W. (1986) Chromosome length controls mitotic chromosome segregation in yeast. *Cell* **45**, 529−536.

Murray, K., Bruce, S.A., Hinnen, A., Wingfield, P. van Erd, P.M.C.A., de Reus, A. and Schellekens, H. (1984) Hepatitis B virus antigens made in microbial cells immunise against viral infection. *EMBO J.* **3**, 645−650.

Murre, C., Reiss, C.S., Bernabeer, C., Chen, L.B., Barakoff, S.J. and Seidman, J.G. (1984) Construction expression and recognition of an H-2 molecule lacking its carboxyl terminus. *Nature* **307**, 432−436.

Myers, R.M., Lerman, L.S. and Maniatis, T. (1985) A general method for saturation mutagenesis of cloned DNA fragments. *Science* **229**, 242−247.

Nabel, G. and Baltimore, D. (1987) An inducible transcription factor activates expression of human immuno-deficiency virus in T cells. *Nature* **326**, 711−713.

Nabeshima, Y-i, Fujii-Kurigama, Y., Muramatsu, M. and Ogata, K. (1984) Alternative transcription and two modes of splicing result in two myosin light chains from one gene. *Nature* **308**, 333−338.

Nagai, K. and Thorenson, H.C. (1984) Generation of β-globin by sequence specific proteolysis of a hybrid protein produced in *Escherichia coli*. *Nature* **309**, 810−812.

Nagata, S., Mantei, N. and Weissmann, C. (1980) The structure of one of the eight or more distinct chromosomal genes for human interferon-α *Nature* **287**, 401−408.

Nagata, Y., Diamond, B. and Bloom, B.R. (1983) The generation of human monocyte/macrophage cell lines. *Nature* **306**, 597−599.

Nagawa, F. and Fink, G.R. (1985) The relationship between the TATA sequence and transcription initiation sites at the *HIS1* gene of *Saccharomyces cerevisiae*. *Proc. Natl Acad. Sci. USA* **82**, 8557−8561.

Nagy, F., Morelli, G., Fraley, R.T., Rogers, S.G. and Chua, N-H. (1985) Photoregulated expression of a pea rbcS gene in leaves of transgenic plants. *EMBO J.* **4**, 3063.

Nasmyth, K.A. (1982) Molecular genetics of yeast mating type. *Ann. Rev. Genet.* **16**, 439−500.

Nasmyth, K. and Shore, D. (1987) Transcriptional regulation in the yeast life cycle. *Science* **237**, 1162–1170.

Nasmyth, K., Shillman, D. and Kipling, D. (1987) Both positive and negative regulators of HO transcription are required for mother cell specific mating type switching in yeast. *Cell* **48**, 579–587.

Nietz, M. and Carbon, J. (1985) Identification and characterization of the centromere from chromosome XIV in *Saccharomyces cerevisiae Molec. Cell. Biol.* **5**, 2887–2893.

Nellen, W., Silan, C. and Firtel, R.A. (1984) DNA-mediated transformation in *Dictyostelium discoideum*: regulated expression of an actin gene fusion. *Molec. Cell. Biol.* **4**, 2890–2898.

Nester, E.W., Gordon, M.P., Amasino, R.M. and Yanofsky, M.F. (1984) Crown gall: a molecular and physiological analysis. *Ann. Rev. Plant Physiol.* **35**, 387–413.

Neuberger, M.S. (1983) Expression and regulation of immunoglobulin heavy chain gene transfected into lymphoid cells. *EMBO J.* **2**, 1373–1378.

Neuberger, M.S., Williams, G.T. and Fox, R.O. (1984) Recombinant antibodies expressing novel effector functions. *Nature* **312**, 604–608.

Neuberger, M.S., Williams, G.T., Mitchell, E.B., Jouhal, S.S., Flanagan, J.G. and Rabbitts, T.H. (1985) A hapten-specific chimaeric IgE antibody with human physiological effector function. *Nature* **314**, 268–270.

Neumann, E., Shaefer-Ridder, M., Wang, Y. and Hofschneider, P.H. (1982) Gene transfer into mouse lyoma cells by electroporation in high electric fields. *EMBO J.* **1**, 841–845.

Nevins, J.R. (1983) The pathway of eukaryotic mRNA formation *Ann. Rev. Biochem.* **52**, 441–466.

Newmark, P. (1982) Cancer genes–processed genes–jumping genes. *Nature* **296**, 393–394.

Norcross, M.A., Bentley, D.M., Margulies, D.H and Germain, R.N. (1984) Membrane Ia expression and antigen presenting accessory cell function of L cells transfected with class II major histocompatibilty complex genes. *J. Expl. Med.* **160**, 1316.

Norstedt, G. and Palmiter, R. (1984) Secretory rhythm of growth hormone regulates sexual differentiation of mouse liver. *Cell* **36**, 805–812.

Nose, M. and Wigzell, H. (1983) Biological significance of carbohydrate chains on monoclonal antibodies. *Proc. Natl Acad. Sci. USA* **80**, 6632–6636.

Nurse, P. (1985) Cell cycle control genes in yeast. *Trends Genet.* **1**, 51–55.

Nusse, R. (1986) The activation of cellular oncogenes by retroviral insertion. *Trends Genet.* **2**, 244–247.

O'Brien, R.L., Brinster, R.L. and Storb, U. (1987) Somatic hypermutation of an immunoglobulin transgene in κ transgenic mice. *Nature* **326**, 405–409.

O'Brien, S.J. (1987) *Genetic Maps 1987*, Vol. 4. Cold Spring Harbor Laboratory, New York.

Ochi, A., Hawley, R.G., Hawley, T., Shulman, M.J., Trgunecker, A., Köhler G. and Hozumi, N. (1983) Functional immunoglobulin M production after transfection of cloned immunoglobulin heavy and light chain genes into lymphoid cells. *Proc. Natl Acad. Sci. USA* **80**, 6351–6355.

Ochoa, S. (1983) Regulation of protein synthesis initiation in eukaryotes. *Arch. Biochem. Biophys.* **223**, 325–349.

Ogden, J.E., Stanway, C., Kim, S., Mellor, J., Kingsman, A.J. and Kingsman, S.M. (1986) Efficient expression of the yeast *PGK* gene depends on an upstream activation sequence but does not require 'TATA' sequences. *Molec. Cell. Biol.* **6**, 4335–4343.

O'Hare, K., Benoist, C. and Breathnach, R. (1981) Transformation of mouse fibroblasts to methotrexate resistance by a recombinant plasmid expressing a prokaryotic dihydrofolate reductase. *Proc. Natl Acad. Sci. USA* **78**, 1527–1531.

O'Hare, K. and Rubin, G.M. (1983) Structures of P transposable elements and their sites of insertion and excision in the *Drosophita melanogaster* genome. *Cell* **34**, 25–35.

Ohashi, P.S., Mak, T.W., Van den Elsen, P., Yanagi, Y., Yoshikai, Y., Calman, A.E., Terhorst, C., Stobo, J.D. and Weiss, A. (1985) Reconstitution of an active surface T3/T-cell antigen receptor by DNA transfer. *Nature* **316**, 606–609.

Ohta, Y. (1986) High efficiency genetic transformation of maize by a mixture of pollen and exogenous DNA. *Proc. Natl Acad. Sci. USA* **83**, 715–719.

Oi, V.T., Morrison, S.L., Herzenberg, L.A. and Berg, P. (1983) Immunoglobulin gene expression in transformed lymphoid cells. *Proc. Natl Acad. Sci. USA* **80**, 825–829.

Okamoto, T. and Wong-Staal, F. (1986) Demonstration of virus-specific transcriptional activators in cells infected with HTLV-III by an *in vitro* cell-free system. *Cell* **47**, 29–35.

Okayama, H. and Berg, P. (1982) High efficiency cloning of full length cDNAs. *Molec. Cell. Biol.* **2**, 161–170.

Okayama, H. and Berg, P. (1983) A cDNA cloning vector that permits expression of cDNA inserts in mammalian cells. *Molec. Cell. Biol.* **3**, 280–289.

Old, L.J. (1985) Tumor Necrosis Factor (TNF). *Science* **230**, 630–632.

Old, L.J. (1987) Tumour necrosis factor: polypeptide mediator network. *Nature* **326**, 330–331.

Old, R.W. and Primrose, S.B. (1985) *Principles of Gene Manipulation*, 3rd edn. Blackwell Scientific Publications, Oxford.

O'Malley, R.P., Mariano, T.M, Sukierka, J. and Mathews, M.B. (1986) A mechanism for the control of protein synthesis by adenovirus VA RNA$_1$. *Cell* **44**, 391–400.

Ondek, B., Shepard, A. and Herr, W. (1987) Discrete elements within the SV40 enhancer region display different cell-specific enhancer activities. *EMBO J.* **6**, 1017–1025.

Orgel, L.E. and Crick, F.H.C. (1980) Selfish DNA: The ultimate parasite. *Nature* **284**, 604–607.

Orkin, S. (1986) Reverse genetics and human disease. *Cell* **47**, 845–850.

Ornitz, D.M., Palmiter, R.D., Hammer, R.E., Brinster, R.L., Swift, G.H. and MacDonald, R.J. (1985) Specific expression of an elastase–human growth hormone fusion gene in pancreatic acinar cells of transgenic mice. *Nature* **313**, 600–602.

Orr, W., Komitopoulou, K. and Kafatos, F.C. (1984) Mutants suppressing in *trans* chorion gene amplification in *Drosophila*. *Proc. Natl Acad. Sci. USA* **81**, 3773–3777.

Orr-Weaver, T.L. and Spradling, A.C. (1986) *Drosophila* chorion gene amplification requires an upstream region regulating s18 transcription. *Molec. Cell. Biol.* **6**, 4624–4633.

Orr-Weaver, T.L. and Szostak, J.W. (1983) Yeast recombination: The association between double-strand gap repair and crossing-over. *Proc. Natl Acad. Sci. USA* **80**, 4417–4421.

Orr-Weaver, T.L., Szostak, J.W. and Rothstein, R.J. (1981) Yeast transformation: A model system for the study of recombination. *Proc. Natl Acad. Sci. USA* **78**, 6354–6358.

Orr-Weaver, T.L., Szostak, J.W. and Rothstein, R.J. (1982) Genetic applications of yeast transformations with linear and gapped plasmids. In *Methods in Enzymology* (eds R. Wu, L. Grossman and K. Moldave) Vol. 101, pp. 228–245. Academic Press, New York.

Osheim, Y.N., Miller, O.L. and Beyer, A.L. (1985) RNP particles at splice junction sequences on *Drosophila* chorion transcripts. *Cell* **43**, 143–151.

Ostrowski, M.C., Richard-Foy, H., Wolford, R.G., Berard, D.S. and Hager, G.L. (1983) Glucocorticoid regulation of transcription at an amplified episomal promoter. *Molec. Cell. Biol.* **3**, 2045–2057.

Ott, D.E., Alt, F.W. and Marcu, K.B. (1987) Immunoglobulin heavy chain switch region recombination within a retroviral vector in murine pre-B cells. *EMBO J.* **6**, 577–584.

Otten, L.A.B.M. and Schilperoort, R.A. (1978) A rapid microscale method for the detection of lysopine and nopaline dehydrogenase activities. *Biochem. Biophys. Acta* **527**, 497–500.

Ow, D.W., Wood K.V., DeLuca M., de Wet, J.R., Helinski, D.R. and Howell, S.H. (1986) Transient and stable expression of the firefly luciferase gene in plant cells and transgenic plants. *Science* **234**, 856–858.

Ozaki, L.S., Svec, P., Nussenzweig, R.S., Nussenzweig, V. and Godson, G.N. (1983) Structure of the *Plasmodium knowlesi* gene coding for the circumsporozoite protein. *Cell* **34**, 815–822.

Ozias-Akins, P. and Lörz, H. (1984) Progress and limitations in the culture of cereal protoplasts. *Trends Biotechnol.* **2**, 119–123.

Padgett, R.A., Grabowski, P.J., Konarska, M.M., Seiler, S. and Sharp, P.A. (1986) Splicing of messenger RNA precursors. *Ann. Rev. Biochem.* **55**, 1119–1150.

Paietta, J.V. and Marzluf, G.A. (1985) Gene disruption by transformation in *Neurospora crassa*. *Molec. Cell. Biol.* **5**, 1554–1559.

Palese, P. and Kingsburg, D.W. (1983) *The Genetics of Influenza Viruses*. Springer-Verlag, Vienna.

Palmer, T.D., Hock, R.A., Osborne, W.R.A. and Miller, A.D. (1987) Efficient retrovirus mediated transfer and expression of a human adenosine diaminase gene in diploid skin fibroblasts from an adenosine diaminase-deficient human. *Proc. Natl Acad. Sci. USA* **84**, 1055–1060.

Palmiter, R.D., Behringer, R.R., Quaife, C.J., Maxwell, F., Maxwell, I.H. and Brinster, R.L. (1987) Cell lineage ablation in transgenic mice by cell-specific expression of a toxin gene. *Cell* **50**, 435–443.

Palmiter, R.D. and Brinster, R.L. (1985) Transgenic mice. *Cell* **41**, 343–345.

Palmiter, R.D. and Brinster, R.L. (1986) Germ-line transformation of mice. *Ann. Rev. Genet.* **20**, 465–499.

Palmiter, R.D., Brinster, R.L., Hammer, R.E, Trumbauer, M.E., Rosenfeld, M.G., Birnberg, N.C. and Evans, R.M. (1982a) Dramatic growth of mice that develop from eggs micro-injected with metallothionein growth hormone fusion genes. *Nature* **300**, 611–615.

Palmiter, R.D., Chen, H.Y. and Brinster, R.L. (1982b) Differential regulation of metallothionein–thymidine kinase fusion genes in transgenic mice and their offspring. *Cell* **29**, 701–710.

Palmiter, R.D., Norstedt, G., Gelinas, R.E., Hammer, R.E. and Brinster, R.L. (1983) Metallothionein–human GH fusion genes stimulate growth of mice. *Science* **222**, 809–814.

Palmiter, R.D., Wilkie, T.M., Chen, H.Y. and Brinster, R.L. (1984) Transmission distortion and mosaicism in an unusual transgenic mouse pedigree. *Cell* **36**, 869–877.

Palzkill, T.G., Oliver, S.G. and Newton, C.S. (1986) DNA sequence analysis of ARS elements from chromosome III of *Saccharomyces cerevisiae*: identification of a new conserved sequence. *Nucl. Acids Res.* **14**, 6247–6264.

Panicalli, D., Davis, S.W., Weinberg, R.L. and Paoletti, E. (1983) Construction of live vaccines by using genetically engineered poxviruses: Biological activity of recombinant vaccinia virus expressing influenza virus hemagglutinin. *Proc. Natl Acad. Sci. USA* **80**, 5364–5368.

Panicalli, D. and Paoletti, E. (1982) Construction of poxviruses as cloning vectors: Insertion of the thymidine kinase gene from herpes simplex virus into the DNA of infectious vaccinia virus. *Proc. Natl Acad. Sci. USA* **79**, 4927–4931.

Panzeri, L., Landonio, L., Stotz, A. and Philippsen, P. (1985) Role of conserved sequence elements in yeast centromere DNA. *EMBO J.* **4**, 1867–1874.

Panzeri, L. and Philippsen, P. (1982) Centromeric DNA from chromosome VI in *S. cerevisiae* strains. *EMBO J.* **1**, 1605–1611.

Paolella, G., Lucero, M., Murphy, M.H. and Baralle, F.E. (1983) The Alu family repeat promoter has a tRNA-like bipartite structure. *EMBO J.* **2**, 691–696.

Parent, S.A., Fenimore, C.M. and Bostian, K.A. (1985) Vector systems for the expression, analysis and cloning of DNA sequences in *S. cerevisiae*. *Yeast* **1**, 83–139.

Parker, B. and Stark, G. (1979) Regulation of simian virus 40 transcription: sensitive analysis of the RNA species present early in infections by virus or viral DNA. *J. Virol.* **31**, 360–369.

Parker, C.S. and Topol, J. (1984) A *Drosophila* RNA polymerase II transcription factor contains a promoter-region-specific DNA-binding activity. *Cell* **36**, 357–369.

Parker, R. and Guthre, C. (1985) A point mutation in the conserved hexanucleotide at a yeast 5' splice junction uncouples recognition, cleavage and ligation. *Cell* **41**, 107–118.

Parker, R.C., Varmus, H.E. and Bishop, J.M. (1984) Expression of v-*src* and chichen c-*src* in rat cells demonstrates quatitative differences between pp60^{v-src} and pp60^{c-src}. *Cell* **37**, 131–139.

Parkhurst, S.M. and Cories, V.G. (1985) *Forked*, gypsys and suppressin *Drosophila*. *Cell* **41**, 429–437.

Parkman, R. (1986) The application of bone marrow transplantation to the treatment of genetic diseases. *Science* **232**, 1373–1378.

Paro, R., Goldberg, M.L. and Gehring, W.J. (1983) Molecular analysis of large transposable elements carrying the *white* locus of *Drosophila melanogaster* *EMBO J.* **2**, 853–860.

Parslow, T., Jones, S.D., Bond, B. and Yamamoto, K.R. (1987) The immunoglobulin octanucleotide: independent activity and selective interaction with enhancers. *Science* **235**, 1498–1501.

Parsons, K.A., Churnley, F.G. and Valent, B. (1987) Genetic transformation of the fungal pathogen responsible for rice blast disease. *Proc. Natl Acad. Sci. USA* **84**, 4161–4166.

Pasek, M., Goto, T., Gilbert, W., Zink, B., Schaller, H., MacKay, P., Leadbetter, G. and Murray, K. (1979) Hepatitis B virus genes and their expression in *E. coli*. *Nature* **282**, 575–579.

Pasleau, F., Tocci, M.J., Leung, F. and Kopchick, J.J. (1985) Growth hormone gene expression in eukaryotic cells directed by the Rous sarcoma virus long terminal repeat or cytomegalovirus immediate early promoter. *Gene* **38**, 227–232.

Passananti, C., Davies, B., Ford, M. and Fried, M. (1987) Structure of an inverted duplication formed as a first step in a gene amplification event: implications for a model of gene amplification. *EMBO J.* **6**, 1697–1703.

Paszkowski, J., Shillito, R.D., Saul, M., Mandak, V., Hohn, T., Hohn, B. and Potrykus, I. (1984) Direct gene transfer to plants. *EMBO J.* **3**, 2717–2722.

Patthy, L. (1985) Evolution of the proteases of blood coagulation and fibrinolysis by assembly from modules. *Cell* **41**, 657–663.

Pattishall, K.H., Acar, J., Burchall, J.J., Goldskin, F.W. and Harvey, R.J. (1977) Two distinct types of trimethoprim resistant dihydrofolate reductase specified by R plasmids of different compatibility groups. *J. Biol. Chem.* **252**, 2319–2323.

Patzer, E.J., Nakamura, G.R., Hershberg, R.D., Gregory, T.J., Crowley, C., Levinson, A.D. and Eichberg, J.W. (1986) Cell culture derived recombinant HBsAG is highly immunogenic and protects chimpanzees from infection with hepatitis B virus. *Bio/Technol.* **4**, 630–636.

Paucha, E., Kalderon, D., Richardson, W.D., Harvey, R.W. and Smith, A.E. (1985) The abnormal location of cytoplasmic large T is not caused by failure to bind to DNA or to p53. *EMBO J.* **34**, 3235–3240.

Pavlakis, G.N. and Hamer, D. (1983) Regulation of a metallothionein-growth hormone hybrid gene in bovine papillomavirus. *Proc. Natl Acad. Sci. USA* **80**, 397–401.

Pearse, B.M.F and Crowther, R.A. (1987) Structure and assembly of coated vesicles. *Ann. Rev. Biophys. Biophys. Chem.* **16**, 49–68.

Pederson, D.S., Thoma, F. and Simpson, R.T. (1986) Core particle, fiber and transcriptionally active chromatin structure. *Ann. Rev. Cell. Biol.* **2**, 117–147.

Pelham, H.R.B. and Brown, D.D. (1980) A specific transcription factor that can bind either the 5S RNA gene or 5S RNA. *Proc. Natl Acad. Sci. USA* **77**, 4170–4174.

Pelham, H.R.B., Wormington, W.M. and Brown, D.D. (1981) Related 5S RNA transcription factors in *Xenopus* oocytes and somatic cells. *Proc. Natl Acad. Sci.* **78**, 1760–1764.

Pennica, D., Holmes, W.E., Kohr, W.J., Harkins, R.N., Vehar, G.A., Ward, C.A., Bennett, W.F., Yelverton, E., Seeburg, P.H., Heyneker, H.L. and Goeddel, D.V. (1983) Cloning and expression of human tissue-type plasminogen activator DNA in *E. coli. Nature* **301**, 214–221.

Perkus, M.E, Piccini, A., Lipinskas, B.R. and Paoletti, E. (1985) Recombinant vaccinia virus: immunization against multiple pathogens. *Science* **229**, 981–984.

Perlman, D. and Halvorsen, H.O. (1983) A putative signal peptidase recognition site and sequence in eukaryotic and prokaryotic signal peptides. *J. Molec. Biol.* **167**, 391–409.

Perucho, M., Hanahan, D., Lipsich, L. and Wigler, M. (1980) Isolation of the chicken thymidine kinase gene by plasmid rescue. *Nature* **285**, 207–210.

Perucho, M., Hanahan, D. and Wigler, M. (1980) Genetic and physical linkage of exogenous sequences in transformed cells. *Cell* **22**, 309–317.

Pestka, S. (1986) *Interferons. Methods Enzymol.* **119**, Academic Press, London.

Pestka, S., Langer, J.A., Zoon, K.C. and Samuel, C.E. (1987) Interferons and their actions. *Ann. Rev. Biochem.* **56**, 727–779.

Peterson, J.B. and Ris, H. (1976) Electron microscopic study of the spindle and chromosome movement in the yeast *Saccharomyces cerevisiae. J. Cell. Sci.* **22**, 219–   .

Petes, T.D. (1980) Molecular genetics of yeast. *Ann. Rev. Biochem.* **49**, 845–876.

Pfeifer, K., Prezant, T. and Guarente, L. (1987) Yeast HAP1 activator binds to two upstream activation sites of different sequence. *Cell* **49**, 19–27.

Pfeiffer, P. and Hohn, T. (1983) Involvement of reverse transcription in the replication of the plant virus CaMV: a detailed model and test of some aspects. *Cell* **33**, 781–784.

Pfeiffer, S.R. and Rothman, J.E. (1987) Biosynthetic protein transport and sorting by the endoplasmic reticulum. *Ann. Rev. Biochem.* **56**, 829–853.

Piatigorsky, J. (1984) Lens crystallins and their gene families. *Cell* **38**, 620–621.

Pickett-Heaps, J.R., Tippit, D.H. and Porter, K.R. (1982) Rethinking mitosis. *Cell* **29**, 729–744.

Pieler, T., Hamm, J. and Roeder, R.G. (1987) The 5S gene internal control region is composed of three distinct sequence elements, organised as two functional domains with variable spacing. *Cell* **48**, 91–100.

Pieler, T., Oei, S.L., Hamm, J., Engelke, U. and Erdmann, V.A. (1985) Functional domains of the *Xenopus laevis* 5S gene promoter. *EMBO J.* **4**, 3751–3756.

Pierce, J.G. and Parsons, T.F. (1981) Glycoprotein hormones: structure and function. *Ann. Rev. Biochem.* **50**, 465–95.

Pikielny, C.W. and Rosbash, M. (1985) mRNA splicing efficiency in yeast and the contribution of non conserved sequences. *Cell* **41**, 119–126.

Pikielny, C.W., Teem, J.L. and Rosbash, M. (1983) Evidence for the biochemical role of an internal sequence in yeast nuclear mRNA introns: implications for U1 RNA and metazoan mRNA splicing. *Cell* **34**, 395–403.

Pinkert, C.A., Ornitz, D.M., Brinster, R.L. and Palmiter, R.D. (1987) An albumin enhancer located 10 kb upstream functions along with its promoter to direct efficient liver specific expression in transgenic mice. *Gene Develop.* **1**, 268–277.

Pinkert, C.A., Widera, G., Cowing, C., Heber-Katz-E., Palmiter, R.D., Flavell, R.A. and Brinster, R.L. (1985) Tissue-specific, inducible and functional expression of the Eα^d MHC class II gene in transgenic mice. *EMBO J.* **4**, 2225–2230.

Pious, D., Krangel, M.S., Dixon, L.L., Parham, P. and Strominger, J.L. (1982) HLA antigen structural gene mutants selected with an allospecific monoclonal antibody. *Proc. Natl Acad. Sci. USA* **79**, 7832–7836.

Pirrotta, V., Hadfield, C. and Pretorius, G.H.J. (1983) Microdissection and cloning of the *white* locus and the 3B1–3C2 region of the *Drosophila* X chromosome. *EMBO J.* **2**, 927–934.

Platt, T. (1986) Transcription termination and the regulation of gene expression. *Ann. Rev. Biochem.* **55**, 339–372.

Poole, S.J., Kauvar, L.M., Drees, B. and Kornberg, T. (1985) The *engrailed* locus of *Drosophila*: structural analysis of an embryonic transcript. *Cell* **40**, 37–43.

Popko, B., Puxett, C., Lai, E., Shine, H.D., Readhead, C., Takahashi, N., Hunt, S.W., Sidman, R.L. and Hood, L. (1987) Myelin deficient mice: expression of myelin basic protein and generation of mice with varying levels of myelin. *Cell* **48**, 713–721.

Popovic, M., Sarngadharan, M.G., Read, E. and Gallo, R.C. (1984) Detection, isolation and continuous production of cytopathic retroviruses (HTLV-111) from patients with AIDS and pre AIDS *Science* **224**, 497–500.

Porteous, D.J. and van Heyningen, V. (1986) Cystic fibrosis: from linked markers to the gene. *Trends Genet.* **2**, 149–152.

Porter, R.R. (1958) Separation and isolation of fractions of the rabbit gamma globulin containing the antibody and antigenic combining sites. *Nature* **182**, 670–671.

Potter, H., Weir, L. and Leder, P. (1984) Enhancer dependent expression of human κ immunoglobulin genes introduced into mouse pre-B lymphocytes by electroporation. *Proc. Natl Acad. Sci. USA* **81**, 7161–7165.

Potter, S. (1982) DNA sequence of a foldback transposable element in *Drosophila melanogaster*. *Nature* **297**, 201–204.

Proudfoot, N.J. and Brownlee G.G. (1976) 3' non-coding region sequences in eukaryotic mRNA. *Nature* **263**, 211–214.

Proudfoot, N.J. and Maniatis, T. (1980) The structure of a human α-globin pseudogene and its relationship to α-globin gene duplication. *Cell* **21**, 537–544.

Ptashne, M. (1986a) *A Genetic Switch*. Blackwell Scientific Publications, Oxford.

Ptashne, M. (1986b) Gene regulation by proteins acting nearby and at a distance. *Nature* **322**, 697–701.

Pyeritz, R.E. (1984) Treatment of inborn errors of metabolism by transplantion. *Nature* **312**, 405–406.

Queen, C. and Baltimore, D. (1983) Immunoglobulin gene transcription is activated by downstream sequence elements. *Cell* **33**, 741–748.

Quinn, T.C., Mann, J.M., Curran, J.W. and Piot, P. (1986) AIDS in Africa: an epidemiologic paradigm. *Science* **234**, 955–963.

Ramagopal, S. (1987) Differential mRNA transcription during salinity stress in barley. *Proc. Natl Acad. Sci. USA* **84**, 94–99.

Rassoulzadegan, M., Leopold, P., Vailly, J. and Cuzin, F. (1986) Germ line transmission of autonomous genetic elements in transgenic mouse strains. *Cell* **46**, 513–519.

Rathore, K.S. and Goldsworthy, A. (1985) Electrical control of root regeneration in plant tissue cultures. *Bio/Technol.* **3**, 1107–1109.

Ratner, L., Haseltine, W., Patarca, R., Livak, K.J., Starach, B., Josephs, S.F., Doran. E.R., Ratalski, A., Whitchorn, E.A., Baremeister, K., Ivanolt, L., Pelteway S.R., Pears on M.L., Lanteubergen II. A., Papas, T.S., Ghrayeb, J., Chang, N.T., Galio, R.C. and Wong-Staal (1985) Complete nucleotide sequence of the AIDS virus HTLV III. *Nature* **313**, 277–284.

Raulet, D.H., Garman, R.D., Saito, H. and Tonegawa S. (1985) Developmental regulation of T cell receptor gene expression. *Nature* **314**, 103–107.

Rautmann, G. and Breathnach, R. (1985) A role for branch points in splicing *in vivo*. *Nature* **315**, 430–432.

Razzaque, A., Chakrabarti, S., Joffee, S. and Seidman, M. (1984) Mutagenesis of a shuttle vector plasmid in mammalian cells. *Molec. Cell. Biol.* **4**, 435–441.

Readhead, C., Popko, B., Takahashi, N., Shine, H.D., Saavedra, R.A., Sidman, R.L. and Hood, L. (1987) Expression of a myelin basic protein gene in transgenic shiverer mice: correction of the dysmyelinating phenotype. *Cell* **48**, 703–712.

Reck, W., Collick, A., Norris, M.L., Barton, S.C. and Surani, M.A. (1987) Genomic imprinting determines methylation of parental alleles in transgenic mice. *Nature* **328**, 248–251.

Reddy, R. and Busch, H. (1983) Small nuclear RNAs and RNA processing. *Proc. Nucl. Acid Res. Molec. Biol.* **30**, 127–163.

Reddy, V.B., Beck, A.K., Garramone, A.J., Vellucci, V., Lustbader, J. and Bernstine, E.G. (1985) Expression of human choriogonadotropin in monkey cells using a single simian virus 40 vector. *Proc. Natl Acad. Sci. USA* **82**, 3644–3648.

Reddy, V.B., Gosh, P.K., Lebowitz, P., Piatak, M. and Weissman, S. (1979) Siman virus 40 early mRNAs I. Genomic localisation of 3' and 5' termini and two major splices in mRNA from transformed and lytically infected cells. *J. Virol.* **30**, 279–296.

Reddy, V.B., Thimmappaya, B., Dhar, R., Subramanian, K.N., Zain, B.S., Pan, J., Ghosh, P.K., Celma, M.L. and Weissman, S.M. (1978) The genome of SV40 virus. *Science* **200**, 494–502.

Reed, R. and Maniatis, T. (1985) Intron sequences involved in lariat formation during pre-mRNA splicing. *Cell* **41**, 95–105.

Reed, R. and Maniatis, T. (1986) A role for exon sequences and splice site proximity in splice site selection. *Cell* **46**, 681–690.

Reeder, R.H. (1984) Enhancers and ribosomal gene spacers. *Cell* **38**, 349–351.

Reeves, R., Gorman, C.M. and Howard, B. (1985) Minichromosome assembly of non-integrated plasmid DNA transfected into mammalian cells. *Nucl. Acids Res.* **13**, 3599–3615.

Regulski, M., Harding, K., Kostriken, R., Karch, F., Levine, M. and McGinnis, W. (1985) Homeo box genes of the antennapedia and bithorax complexes of *Drosophila*. *Cell* **43**, 71–80.

Reickl, P.A., Merrick, W.C., Siekiarka, J. and Mathews, M.B. (1984) Regulation of a protein synthesis initiation factor by adenovirus-associated RNA. *Nature* **313**, 196–200.

Reinert, J. and Yeoman, M.M. (1982) *Plant Cell and Tissue Culture: a Laboratory Manual.* Springer-Verlag, Berlin.

Reiss, C.S., Evans, G.A., Margulies, D.H., Seidman, J.G. and Burakoff, S.J. (1983) Allospecific and virus-specific cytolytic T lymphocytes are restricted to the N or C1 domain of H-2 antigens expressed on L cells after DNA-mediated gene transfer. *Proc. Natl Acad. Sci. USA* **80**, 2709–2712.

Remaut, E., Stanssens, P., Simons, G. and Fiers, W. (1986) Use of the phage lambda P_L promoter for high level expression of human interferons in *Escherichia coli*. *Methods Enzymol.* **119**, 366–376.

Remmers, E.F., Yang, J-Q. and Marcu, K.B. (1986) A negative transcriptional control element located upstream of the murine c-*myc* gene. *EMBO J.* 5, 899–904.

Reynolds, G.A., Basu, S.K., Osborne, T.F., Chin, D.J., Gil, G., Brown, M.S., Goldstein, J.L. and Luskey, K.L. (1984) HMG CoA reductase: a negatively regulated gene with unusual promoter and 5′ untranslated region. *Cell* **38**, 275–285.

Rhodes, D. and Klug, A. (1986) An underlying repeat in some transcriptional control sequences corresponding to half a double helical turn of DNA. *Cell* **46**, 123–132.

Rich, A., Nordheim, A. and Wang A. H-J. (1984) The chemistry and biology of left handed Z-DNA. *Ann. Rev. Biochem.* 53, 791–846.

Richardson, W.D., Roberts, B.L. and Smith, A.E. (1986) Nuclear location signals in polyomavirus large T. *Cell* **44**, 77–85.

Richmond, T.J., Finch, J.T., Rushton, B., Rhodes, D. and Klug, A. (1984) Structure of the nucleosome core particle at 7 Å resolution. *Nature* **311**, 532–537.

Riedel, H., Dull, T.J., Schlessinger, J. and Ullrich, A. (1986) A chimaeric receptor allows insulin to stimulate tyrosine kinase activity of epidermal growth factor receptor. *Nature* **324**, 68–70.

Rijsewijk, F.A.M., van Lohuizen, M. & Van Ooyen, A. and Nusse, R. (1986) Construction of a retroviral cDNA version of the int-1 mammalianoncogene and its expression *in vitro Nucl. Acids Res.* **14**, 693–702.

Rio, D.C., Clark, S.G. and Tjian, R. (1985) A mammalian host-vector system that regulates expression and amplification of transfected genes by temperature induction. *Science* **227**, 23–28.

Rio, D.C., Laski, F.A. and Rubin, G.M. (1986) Identification and immunochemical analysis of biologically active *Drosophila* P element transposase. *Cell* **44**, 21–22.

Rio, D.C. and Rubin, G.M. (1985) Transformation of cultured *Drosophila melanogaster* cells with a dominant selectable marker. *Molec. Cell. Biol.* 5, 1833–1838.

Robbins, D.M., Ripley, S., Henderson, A. and Axel, R. (1981) Transforming DNA integrates into the chromosome. *Cell* **23**, 29–39.

Roberts, J.M., Buck, L.B. and Axel, R. (1983) A structure for amplified DNA. *Cell* **33**, 53–63.

Roberts, J.M. and Weintraub, H. (1986) Negative control of DNA replication in composite SV40–bovine papillomavirus plasmids. *Cell* **40**, 741–752.

Robertson, E., Bradley, A., Kuehn, M. and Evans, M. (1986) Germ-line transmission of genes introduced into cultured pluripotential cells by retroviral vector. *Nature* **323**, 445–448.

Robertson, E.J. (1986) Pluripotential stem cell lines as a route into the mouse germ line. *Trends Genet.* 2, 9–13.

Robertson, M. (1986) Gene therapy: desperate appliances. *Nature* **320**, 213–214.

Robson, A.B. and Martin, M.A. (1985) Molecular organisation of the AIDS retrovirus. *Cell* **40**, 477–480.

Robson, B. and Garnier, J. (eds) (1986) *Introduction to Proteins and Protein Engineering.* Elsevier, Amsterdam.

Rochaix, J.D., van Dillewijn, J. and Rahire, M. (1984) Construction and characterisation of autonomously replicating plasmids in the green unicellular alga *Chlamydomonas reinhardii*. *Cell* **36**, 925–931.

Rogers, J. (1983) CACA sequences—the ends and the means? *Nature* **305**, 101–102.

Rogers, S., Klee, H., Byrne, M., Horsch, R. and Fraley, R. (1988) Improved vectors for plant transformation: expression cassette vectors and new selectable markers. *Methods in Enzymology.* Academic Press (in press).

Rogers, S.G., Bisaro, D.M., Horsch, R.B. Fraley, R.T, Hoffmann, N.L., Brand, L., Elener, J.S. and Lloyd, A.M. (1986) Tomato golden mosaic virus A component DNA replicates autonomously in transgenic plants. *Cell* **45**, 593–600.

Roggenkamp, R., Dargatz, H. and Hollenberg, C.P. (1985) Precursor of β-lactamase is enzymatically inactive: Accumulation of the preprotein in *Saccharomyces cerevisiae*. *J. Biol. Chem.* **260**, 1508–1512.

Roitt, I. (1988) *Essential Immunology*, 6th edn, Blackwell Scientific Publications, Oxford.

Rosahl, S., Schell, J. and Willmitzer, L. (1987) Expression of a tuber-specific storage protein in transgenic tobacco plants: demonstration of an esterase activity. *EMBO J.* **6**, 1155–1159.

REFERENCES

Rosen, C.A., Sodroski, J.G. and Haseltine, W. (1986) Post-transcriptional regulation accounts for the transactivation of the human T-lymphotropic virus type III. *Nature* **319**, 555–559.

Rosenberg, S., Barr, P.J., Najarian, R.C. and Hallewell, R.A. (1984a) Synthesis in yeast of a functional oxidation resistant mutant of human α1-antitrypsin. *Nature* **312**, 77–80.

Rosenberg, S.A., Grimm, E.A., McGrogan, M., Doyle, M., Kawaslci, E., Koths, K. and Mark, G.F. (1984b) Biological activity of recombinant human interleukin-2 produced in *E. coli. Science* **223**, 1412–1415.

Rosenberg, S.A. and Lotze, M.T. (1986) Cancer immunotherapy using interleukin-2 and interleukin-2 activated lymphocytes. *Ann. Rev. Immunol.* **4**, 681–711.

Rosenberg, U.B., Preiss, A., Seifert, E., Jackle, H. and Knipple, D.C. (1985) Production of phenocopies by *Kruppel* antisense RNA injection into *Drosophila* embryos. *Nature* **313**, 703–706.

Rosenberg, U.B., Schroder, C., Preiss, A., Kienlin, A., Cote, S., Riede, I. and Jackle, H. (1986) Structural homology of the product of the *Drosophila* kruppel gene with *Xenopus* transcription factor IIIA. *Nature* **319**, 336–339.

Rosenfeld, M.G., Amara, S.G. and Evans, R.M. (1984) Alternative RNA processing: Determining neuronal phenotype. *Science* **225**, 1315–1320.

Rosenfeld, P.J. and Kelly, T.J. (1986) Purification of nuclear factor I by DNA recognition site affinity chromatography. *J. Biol. Chem.* **261**, 1398–1408.

Rosensweig, B., Liao, L.W. and Hirsh, D. (1983) Sequence of the *C. elegans* transposable element Tc1 *Nucl. Acids Res.* **11**, 4201–4209.

Roth, G.E., Blanton, H.M., Hager, L.J. and Zakian, V.A. (1983) Isolation and characterisation of sequences from mouse chromosomal DNA with ARS function in yeasts. *Molec. Cell. Biol.* **3**, 1898–1908.

Rothman, J.E. (1987) Protein sorting by selective retention in the endoplasmic reticulum and Golgi stack. *Cell* **50**, 521–522.

Rothman, J.E. and Kornberg, R.D. (1986) An unfolding story of protein translocation. *Nature* **322**, 209–210.

Rothstein, R. (1985) Cloning in yeast. In *DNA Cloning, Vol. II* pp. 45–66 (ed. D.M. Glover), IRL Press, Oxford.

Rothstein, S.J., Lazarus, C.M., Smith, W.E., Baulcombe, D.C. and Gatenby, A.A. (1984) Secretion of a wheat α-amylase expressed in yeast. *Nature* **308**, 662–665.

Rubin, G.M. and Spradling, A.C. (1983a) Genetic transformation of *Drosophila* with transposable element vectors. *Science* **218**, 348–353.

Rubin, G.M. and Spradling, A.C. (1983b) Vectors for P element-mediated gene transfer in *Drosophila. Nucl. Acids Res.* **11**, 6341–6351.

Rubin, J.S., Prideaux, V.R., Willard, H.F., Dulhanty, A.M., Whitmove, G.F. and Bernstein, A. (1985) Molecular cloning and chromosomal localisation of DNA sequences associated with a human DNA repair gene. *Molec. Cell. Biol.* **5**, 398–405.

Ruddle, F.H., Hart, C.P. and McGinnis, W. (1985) Structural and functional aspects of the mammalian homeo-box sequences. *Trends Genet.* **1**, 48–51.

Rudolph, H., Koenig-Rauseo, I. and Hinnen, A. (1985) One-step gene replacement in yeast by cotransformation. *Gene* **36**, 87–95.

Ruley, H.E. (1983) Adenovirus early region 1A enables viral and cellular transforming genes to transform primary cells in culture. *Nature* **304**, 602–606.

Ruskin, B., Krainer, A.R., Maniatis, T. and Green, M.R. (1984) Excision of an intact intron as a novel lariat structure during pre-mRNA splicing *in vitro. Cell* **38**, 317–331.

Russell, I. (1986) Will a recombinant DNA yeast be able to solve whey disposal problems? *Trends Biotechnol.* **4**, 107–108.

Ryder, K., Silver, S., DeLucia, A.L., Fanning, E. and Tegtmeyer, P. (1986) An altered DNA conformation in origin region I is a determinant for the binding of SV40 large T antigen. *Cell* **44**, 719–725.

Sadofsky, M., Connelly, S., Manley, J.L. and Alwine, J.C. (1985) Identification of a sequence element on the 3' side of AAUAAA which is necessary for simian virus 40 late mRNA 3' end processing. *Molec. Cell. Biol.* **5**, 2713–2719.

Sairam, M.R. and Bhargavi, G.N. (1985) A role for the glycosylation of the α subunit in transduction of biological signal in glycoprotein hormones. *Science* **229**, 65–67.

Saito, T., Weiss, A., Miller, J., Norcross, M.A. and Germain, R.N. (1987) Specific antigen-Ia activation of transfected human T cells expressing murine Ti $\alpha\beta$–human T3 receptor complexes. *Nature* **325**, 125–130.

Sakaki, T., Oeda, K., Miyoshi, M. and Ohkawa, H. (1985) Characterisation of rat cytochrome P-450$_{MC}$ synthesised in *Saccharomyces cerevisiae. J. Biochem.* **98**, 167–175.

Sakonju, S. and Brown, D.D. (1982) Contact points between a positive transcription factor and the *Xenopus* 5S RNA gene. *Cell* **31**, 395–405.

Saloman, F., Deblaere, R., Leemans, J., Hernalsteens, J.P., Van Montagu, M. and Schell, J. (1984) Genetic identification of functions of TR-DNA transcripts in octopine crown galls. *EMBO J.* **3**, 141–146.

REFERENCES

Sambrook, J., Rodgers, L., White, J. and Gething, M.J. (1985) Lines of BPV-transformed murine cells that constitutively express influenza virus hemagglutinin. *EMBO J*. **4**, 91–103.

Sambucetti, L.C., Schaber, M., Kramer, R., Crowl, R. and Curran, T. (1986) The *fos* gene product undergoes extensive post-translational modification in eukaryotic but not in prokaryotic cells. *Gene* **43**, 69–77.

Sánchez-Herrero, E., Vernos, I., Marco, R. and Morata, G. (1985) Genetic organisation of *Drosophila* bithorax complex. *Nature* **313**, 108–113.

Sandri-Goldin, R.M., Goldin, A.L., Levine, M. and Glorioso, J.C. (1981) High frequency transfer of cloned herpes simplex virus type 1 sequences to mammalian cells by protoplast fusion. *Molec. Cell. Biol*. **1**, 743–752.

Sanger, F.S., Nicklen, S. and Coulson, A.R. (1977) DNA sequencing with chain-terminating inhibitors. *Proc. Natl. Acad. Sci. USA* **74**, 5463–5467.

Sapienza, C., Peterson, A.C., Rossant, J. and Balung, R. (1987) Degree of methylation of transgenes is dependent on gamete of origin. *Nature* **328**, 251–254.

Sapienza, C. and St-Jacques, B. (1986) 'Brain specific' transcription and the evolution of the identifier sequence. *Nature* **319**, 418–420.

Sarver, N., Byrne, J.C. and Howley, P.M. (1982) Transformation and replication in mouse cells of a bovine papillomavirus pML2 plasmid vector that can be rescued in bacteria. *Proc. Natl Acad. Sci. USA* **79**, 7147.

Sawadogo, M. and Roeder, R.G. (1985) Interaction of a gene-specific transription factor with the adenovirus major late promoter upstream of the TATA box region. *Cell* **43**, 165–175.

Sarver, N., Rabson, M., Yang, Y-C., Byrne, J. and Hawley, P. (1984) Localisation and analysis of bovine papillomavirus type 1 transforming functions. *J. Virol*. **52**, 377–388.

Scahill, S.J., Devos, R., van der Heyden, J. and Fiers, W. (1983) Expression and characterisation of the product of a human immune interferon cDNA gene in Chinese hamster ovary cells. *Proc. Natl Acad. Sci. USA* **80**, 4654–4658.

Scangos, G., Huttner, K.M., Juricek, D.K. and Ruddle, F.H. (1981) DNA-mediated gene transfer in mammalian cells: molecular analysis of unstable transformants and their progression to stability. *Molec. Cell. Biol*. **1**, 111–120.

Scangos, G. and Ruddle, F. (1981) Mechanisms and applications of DNA-mediated gene transfer in mammalian cells—a review. *Gene* **14**, 1–10.

Scarpulla, R.C. and Nye, S.H. (1986) Functional expression of rat cytochrome C in *Saccharomyces cerevisiae*. *Proc. Natl Acad. Sci. USA* **83**, 6352–6356.

Schafer, W., Gorz, A., and Kahl, G. (1987) T-DNA integration and expression in a monocot crop plant after induction of *Agrobacterium*. *Nature* **327**, 529–531.

Schaffner, W. (1980) Direct transfer of cloned genes from bacteria to mammalian cells. *Proc. Natl Acad. Sci. USA* **77**, 2163–2167.

Schatz, G. and Butow, R.A. (1983) How are proteins imported into mitochondria? *Cell* **32**, 316–318.

Schauer, I., Emr, S., Gross, C. and Schekman, R. (1985) Invertase signal and nature sequence substitutions that delay intercompartmental transport of active enzyme. *J. Cell. Biol*. **100**, 1664–1675.

Schekman, R. (1985) Protein localisation and membrane traffic in yeast. *Ann. Rev. Cell. Biol*. **1**, 115–143.

Schekman, R. and Novick, P. (1982) The secretory process and yeast cell surface assembly. In *The Molecular Biology of the Yeast* Saccharomyces cerevisiae (eds J. Strathern, E.W. Jones and J. Broach). Cold Spring Harbor Laboratory, New York. Vol. 2, pp. 361–398.

Scherer, S. and Davis, R.W. (1979) Replacement of chromosome segments with altered DNA sequences constructed *in vitro*. *Proc. Natl Acad. Sci. USA* **76**, 4951–4955.

Schimke, R.T. (1984) Gene amplification in cultured animal cells. *Cell* **37**, 705–713.

Schleyer, M. and Neupert, W. (1985) Transport of proteins into mitochondria: translocational intermediates spanning contact sites between outer and inner membranes. *Cell* **43**, 339–350.

Schlissel, M.S. and Brown, D.D. (1984) The transcriptional regulation of *Xenopus* 5S RNA genes in chromatin: the roles of active stable transcription complexes and histone H1. *Cell* **37**, 903–913.

Schneuwly, S., Klemenz, R. and Gehring, W. (1987) Redesigning the body plan of *Drosophila* by ectopic expression of the homoeotic gene *Antennapedia*. *Nature* **325**, 816–818.

Scholer, H.R. and Gruss, P. (1984) Specific interaction between enhancer-containing molecules and cellular components. *Cell* **36**, 403–411.

Scholnick, S.B., Morgan, B.A. and Hirsh, J. (1983) The cloned dopa decarboxylase gene is developmentally regulated when reintegrated into the *Drosophila* genome. *Cell* **34**, 37–45.

Schlissel, M.S. and Brown, D.D. (1984) The transcriptional regulation of *Xenopus* 5S RNA genes in chromatin: The roles of active stable transcription complexes histone H1. *Cell* **37**, 903.

Schmid, H.P., Akhayat, O., Martins de Sa, C., Puvion, F., Koehler, K. and Scherrer, K. (1984) The prosome: an ubiquitous morphologically distinct RNP particle associated with

REFERENCES

repressed mRNPs and containing specific ScRNA and a characteristic set of proteins. *EMBO J.* **3**, 29–34.

Schneider, R.J., Weinberger, C. and Shenk, T. (1984) Adenovirus VA1 RNA facilitates the initiation of translation in virus infected cells. *Cell* **37**, 291–298.

Schultz, L.D., Tanner, J., Hofmann, K.J., Emini, E.A., Condra, J.H., Jones, R.E., Kuff, E. and Ellis, R.W. (1987) Expression and secretion in yeast of a 400-kDa envelope glycoprotein derived from Epstein–Barr virus. *Gene* **54**, 113–123.

Schümperli, D., Howard, B.H. and Rosenberg, M. (1982) Efficient expression of *Escherichia coli* galactokinase gene in mammalian cells. *Proc. Natl Acad. Sci. USA* **79**, 257–261.

Schirm, S., Jirceny, J. and Schaffner, W. (1987) The SV40 enhancer can be dissected into multiple segments each with a different cell type specificity. *Genes Develop.* **1**, 65–74.

Schwartz, D.C. and Cantor, C.R. (1984) Separation of yeast chromosome-sized DNAs by pulsed field gradient gel electrophoresis. *Cell* **37**, 67–75.

Schweitzer, E.S. and Kelly, R.B. (1985) Selective packaging of human growth hormone into synaptic vesicles in a rat neuronal (PC12) cell line. *J. Cell. Biol.* **101**, 667–676.

Scott, M.G. (1985) Monoclonal antibodies–approaching adolescence in diagnostic immunoassays. *TibTech.* **3**, 170–175.

Scott, M.P. (1987) Complex loci of *Drosophila. Ann. Rev. Biochem.* **56**, 195–229.

Scott, M.P. and Weiner, A.J. (1984) Structural relationships among genes that control development: sequence homology between the Antennafedia, Ultrabithorax and fushi tarazu loci of *Drosophila. Proc. Natl Acad. USA* **81**, 4115–4119.

Scott, M.P., Weiner, A.J., Hazelrigg, T.L., Polisky, B.A., Pirotta, V., Scalenghe, F. and Kaufman, T.C. (1983) The molecular organisation of the *Antennapedia* locus of *Drosophila. Cell* **35**, 763–776.

Scott, R.W., Vogt, T.F., Croke, M.E. and Tilghman, S. (1984) Tissue specific activation of a cloned α-fetoprotein gene during differentiation of a transfected embryonal carcinoma cell line. *Nature* **310**, 562–567.

Searle, P.F., Stuart, G.W. and Palmiter, R.D. (1985) Building a metal-responsive promoter with synthetic regulatory elements. *Molec. Cell. Biol.* **5**, 1480–1489.

Searles, L.L., Jokerst, R.S., Bingham, P.M., Voelker, R.A. and Greenleaf, A.L. (1982) Molecular cloning of sequences from a *Drosophila* RNA polymerase II locus by P element transposon tagging. *Cell* **31**, 585–592.

Seeburg, P.H., Colby, W.W., Hayflick, J.S., Capon, D.J., Goeddel, D.V. and Levinson, A.D. (1984) Biological properties of human C-Ha-ras1 genes mutated at codon 12. *Nature* **312**, 71–75.

Sefton, B.M. (1986) The viral tyrosine protein kinases. *Curr. Topics Microbiol. Immunol.* **123**, 39–72.

Seguin, C. and Hamer, D.H. (1987) Regulation *in vitro* of metallothionein gene binding factors. *Science* **235**, 1382–1386.

Seidman, M.M., Dixon, K., Razzaque, A., Zagursky, R.J. and Berman, M.L. (1985) A shuttle vector plasmid for studying carcinogen induced point mutations in mammalian cells. *Gene* **38**, 233–237.

Selden, R.F., Skoskiewicz, M.J., Howie, K.B., Russell, P.S. and Goodman, H.M. (1986) Regulation of human insulin gene expression in transgenic mice. *Nature* **321**, 525–528.

Sen, R. and Baltimore, D. (1986a) Multiple nuclear factors interact with the immunoglobulin enhancer sequences. *Cell* **46**, 705–716.

Sen, R. and Baltimore, D. (1986b) Inducibility of κ immunoglobulin enhancer binding protein NF-κB by a post-translational mechanism. *Cell* **47**, 921–928.

Sengupta-Gopalan, C., Reichert, N.A., Barker, R.F., Hall, T.C. and Kemp, J.D. (1985) Developmentally regulated expression of the bean β-phaseolin gene in tobacco seed. *Proc. Natl Acad. Sci. USA* **82**, 3320–3324.

Sequira, L. (1983) Mechanisms of induced resistance in plants. *Ann. Rev. Microbiol.* **37**, 51–79.

Serfling, E., Jasin, M. and Schaffner, W. (1985) Enhancers and eukaryotic gene transcription. *Trends Genet.* **1**, 224–230.

Setzer, D.R. and Brown, D.D. (1985) Formation and stability of the 5S RNA transcription complex. *J. Biol. Chem.* **260**, 2483–2492.

Setzer, D.R., McCrogan, M., Nunberg, J. and Schimke, R.T. (1980) Size heterogeneity in the 3′ end of the dihydrofolate reductase messenger RNAs in mouse cells. *Cell* **22**, 361–370.

Shah, D.M., Horsch, R.B., Klee, H.J., Kishore, G.M., Winter, J.A., Tumer, N.E., Hironaka, C.M., Sanders, P.R., Gasser, C.S., Aykent, S., Siegel, N.R., Rogers, S.G. and Fraley, R.T. (1986) Engineering herbicide tolerance in transgenic plants. *Science* **233**, 478–481.

Shalloway, D., Coussens, P.M. and Yaciuk, P. (1984) Overexpression of the c*src* protein does not induce transformation of NIH 3T3 cells. *Proc. Natl Acad. Sci. USA* **81**, 7071–7075.

Shampay, J., Szostak, J.W. and Blackburn, E.H. (1984) DNA sequences of telomeres maintained in yeast. *Nature* **310**, 154–157.

Shani, M. (1985) Tissue specific expression of rat myosin light-chain 2 gene in transgenic mice. *Nature* **314**, 283–286.

Sharp, P.A. (1983) Conversion of RNA to DNA in mammals: Alu-like elements and pseudogenes. *Nature* **301**, 471–472.

Sharp, P.M., Tuohy, T.M.F. and Mosurski, K.R. (1986) Codon usage in yeast: cluster analysis clearly differentiates highly and lowly expressed genes. *Nucl. Acids Res.* **14**, 5125–5143.

Shatkin, A.J. (1984) mRNA cap binding proteins: essential factors for initiating translation. *Cell* **40**, 223–224.

Shaw, G. and Kamen, R. (1986) A conserved AU sequence from the 3' untranslated region of GM-CSF mRNA mediates selective mRNA degradation. *Cell* **46**, 659–667.

Shaw, P., Sordat, B. and Schibler, U. (1986) The two promoters of the mouse α amylase gene are differentially activated during parotid gland differentiation. *Cell* **40**, 907–912.

Shepherd, J.C.W., McGinnis, W., Carrasco, A.E., Robertis, E.M. and Gehring, W.J. (1984) Fly and frog homeo domains show homologies with yeast mating type regulatory proteins. *Nature* **310**, 70–71.

Sherman, F. and Stewart, J.W. (1982) Mutations altering initiation of translation of yeast iso-1-cytochrome C.: contrast between the eukaryotic and prokaryotic initiation process. In *The Molecular Biology of the Yeast Saccharomyces, Vol. 2, Metabolism and Gene Expression* (eds J. Strathern, E.W. Jones and J.R. Broach), pp. 301–355, Cold Spring Harbor Laboratory, New York.

Sherwood, J.L. and Fulton, R.W. (1982) The specific involvement of coat protein in tobacco mosaic virus cross protection. *Virology* **119**, 150–158.

Shih, C., Padhy, L.C., Murray, M. and Weinberg, R.A. (1981) Transforming genes of carcinomas and neuroblastomas introduced into mouse fibroblasts. *Nature* **290**, 261–264.

Shillito, R.D., Saul, M.W., Paszkowski, J., Muller, M. and Potrykus, I. (1985) High efficiency direct gene transfer to plants. *Bio/Technol.* **3**, 1099–1103.

Shimizu, K., *et al.* (1983) Three human transforming genes are related to the viral *ras* oncogenes. *Proc. Natl Acad. Sci. USA* **80**, 2112–2116.

Shimotohno, K. and Temin, H.M. (1981) Formation of infectious progeny virus after insertion of herpes simplex thymidine kinase into DNA of an avian retrovirus. *Cell* **26**, 67–77.

Shimotohno, K. and Temin, H.M. (1982) Loss of intervening sequences in genomic mouse α-globin DNA inserted in an infectious retrovirus vector. *Nature* **299**, 265–268.

Shine, J. and Dalgarno, L. (1975) Determinant of cistron specificity in bacterial ribosomes. *Nature* **254**, 34–38.

Short, N.J. (1987) Are some controlling factors more equal than others? *Nature* **326**, 740–741.

Shortle, D., DiMaio, D. and Nathans, D. (1981) Directed mutagenesis. *Ann. Rev. Genet.* **15**, 265–294.

Shuey, D.J. and Parker, C.S. (1986a) Bending of promoter DNA on binding of heat shock transcription factor. *Nature* **323**, 459–461.

Shuey, D.J. and Parker, C.S. (1986b) Binding of *Drosophila* heat-shock gene transcription factor to the hsp70 promoter. *J. Biol. Chem.* **261**, 7934–7940.

Simonsen, C.C. and Levinson, A.D. (1983) Isolation and expression of an altered mouse dihydrofolate reductase cDNA. *Proc. Natl Acad. Sci. USA* **80**, 2495–2499.

Simpson, J., Schell, J., van Montagu, M. and Herrera-Estrella, L. (1986) The light inducible and tissue specific expression of a pea LHCP gene involves an upstream element combining enhancer and silencer-like properties. *Nature* **323**, 551–553.

Simpson, J., Timko, M.P., Cashmore, A.R., Schell, J., van Montagu, M. and Herrera-Estrella, L. (1985) Light-inducible and tissue-specific expression of a chimaeric gene under control of the 5' flanking sequence of a pea chlorophyll a/b-binding protein gene. *EMBO J.* **4**, 2723–2729.

Singer, S.J., Maher, P.A. and Yaffe, M.P. (1987a) On the translocation of proteins across membranes. *Proc. Natl Acad. Sci. USA* **84**, 1015–1019.

Singh, H., Sen, R., Baltimore, D. and Sharp, P.A. (1986) A nuclear factor that binds a conserved sequence motif in transcriptional control elements of immunoglobulin genes. *Nature* **319**, 154–158.

Singer, S.J., Maher, P.A. and Yaffe, M.P. (1987b) On the transfer of integral proteins into membranes. *Proc. Natl Acad. Sci. USA* **84**, 1960–1964.

Sitoyama, C., Liau, G. and de Crombrugghe, B. (1985) Pleiotropic mutants of NIH 3T3 cells with altered regulation in the expression of both type I collagen and fibronectin. *Cell* **41**, 201–209.

Skipper, N., Sutherland, H., Davies, R.W., Kilburn, D., Miller, R.C., Warren, A. and Wong, R. (1985) Secretion of a bacterial cellulase by yeast. *Science* **230**, 958–961.

Skoglund, V., Andersson, K., Strandberg, B. and Daneholt, B. (1986) Three-dimensional structure of a specific pre-messenger RNP particle established by electron microscope tomography. *Nature* **319**, 560–564.

REFERENCES

Slack, J.M. W. (1983) *From egg to embryo. Developmental and Cell Biology, Vol. 13*, Cambridge University Press.

Slamon, D.J., Clark, G.M., Wong, S.G., Levin, W.J., Ullrich, A. and McGuire, W.L. (1987) Human breast cancer: correlation of relapse and survival with amplification of the HER-2/neu oncogene. *Science* **235**, 177−181.

Small, J.A., Scangos, G.A., Cork, L., Jay, G. and Khoury, G. (1986) The early region of human papovavirus JC induces dysmyelation in transgenic mice. *Cell* **46**, 13−18.

Smeekens, S., Bauerle, C., Hageman, J., Keegstra, K. and Weisbeck, P. (1986) The role of the transit peptide in the routing of precursors towards different chloroplast compartments. *Cell* **46**, 365−375.

Smith, G.E., Fraser, M.J. and Summers, M.D. (1983a) Molecular engineering of the *Autographa californica* nuclear polyhedrosis virus genome: deletion mutations within the polyhedrin gene. *J. Virol.* **46**, 584−593.

Smith, G.E., Ju, G., Ericson, B.L., Moschera, J., Lahm, H.-W., Chizzonite, R. and Summers, M.D. (1985) Modification and secretion of human interleukin 2 produced in insect cells by a baculovirus expression vector. *Proc. Natl Acad. Sci. USA* **82**, 8404−8408.

Smith, G.E. Summers, M.D. and Fraser, M.J. (1983b) Production of human beta interferon in insect cells infected with a baculovirus expression vector. *Molec. Cell. Biol.* **3**, 2156−2165.

Smith, G.L., Godson, G.N., Nussenzweig, V., Nussenzweig, R.S., Barnwell, J. and Moss, B. (1984) *Plasmodium knowlesi* sporozoite antigen: expression by infectious recombinant vaccinia virus. *Science* **224**, 397−399.

Smith, G.L. and Moss, B. (1983) Infectious poxvirus vectors have capacity for at least 25,000 base pairs of foreign DNA. *Gene* **25**, 21−28.

Smith, G.L., Murphy, B.R. and Moss, B. (1983a) Construction and characterisation of an infectious vaccinia virus recombinant that expresses the influenza haemagglutinin gene and induces resistance to influenza virus infection in hamsters. *Proc. Natl Acad. Sci. USA* **80**, 7155−7159.

Smith, M. (1985) *In vitro* mutagenesis. *Ann. Rev. Genet.* **19**, 423−462.

Smith, R.A., Duncan, M.J. and Moir, D.T. (1985) Heterologous protein secretion from yeast. *Science* **229**, 1219−1224.

Smithies, O., Gregg, R.G., Boggs S.S., Koralewski, M.A. and Kucherlapati, R. (1985) Insertion of DNA sequences into the human chromosomal β-globin locus by homologous recombination. *Nature* **317**, 230−234.

Snyder, M., Buchman, A.R. and Davis, R.W. (1986) Bent DNA at a yeast autonomously replicating sequence. *Nature* **324**, 87−89.

Snyder, M., Hunkapiller, M., Yuen, D., Silvert, D., Fristrom, J. and Davidson, N. (1982) Cuticle protein genes of *Drosophila*: structure organisation and evolution of four clustered genes. *Cell* **29**, 1027−1040.

Sodroski, J., Goh, W.C., Rosen, C., Campbell, K. and Haseltine, W. (1986a) Role of HTLV-III/LAV envelope in syncytium formation and cytoppathicity. *Nature* **322**, 470−474.

Sodroski, J., Goh, W.C., Rosen, C., Dayton, A., Terwilliger, E. and Haseltine, W. (1986b) A second post-transcriptional trans-activator gene required for HTLV III replication. *Nature* **321**, 412−417.

Sodroski, J., Goh, W.C., Rosen, C., Tartar, A., Porteletto, D., Burny, A. and Haseltine, W. (1986c) Replicative and cytopathic potential of HTLV III/LAV with *sor* deletions. *Science* **231**, 1549−1553.

Sodroski, J., Rosen, C., Wong-Staal, F., Salahudden, S.Z., Popovic, M., Arya, S., Gallo, R.C. and Haseltine, W.A. (1985) Trans-acting transcriptional regulation of human T-cell leukemia virus type III long terminal repeat. *Science* **227**, 171−173.

Sofer, G.K. (1986) Current applications of chromatography in biotechnology. *Bio/Technol.* **4**, 712−715.

Sollner-Webb, B. (1987) Surprises in polymerase III transcription. *Cell* **52**, 153−154.

Sollner-Webb, B. and Tower, J. (1986) Transcription of cloned eukaryotic ribosomal RNA genes. *Ann. Rev. Biochem.* **55**, 801−831.

Sommerville, J. (1984) RNA polymerase I promoters and transcription factors. *Nature* **310**, 189−190.

Sondahl, M.R., Evans, D.A., Prioli, L.M. and Silva, W.J. (1984) Tissue culture regeneration of plants in *Zea diploperennis*, a close relative of corn. *Biotechnol.* **2**, 455−458.

Sonigo, P., Alizon, M., Staskus, K., Klatzmann, R., Cole, S., Danos, O., Retzel, E., Tiollais, P., Haase, A. and Wain-Hobson, S. (1985) Nucleotide sequence of the visna lentivirus: Relationship to the AIDS virus. *Cell* **42**, 369−382.

Sorge, J., Kuhl, W., West, C. and Beutler, E. (1987) Complete correction of the enzymatic defect of type I Gauscher disease fibroblasts by retroviral-mediated gene transfer. *Proc. Natl Acad. Sci. USA* **84**, 906−910.

Soriano, P., Cone, R.D., Mulligan, R.C. and Jaenisch, R. (1986) Tissue specific expression of genes introduced into the germ line of mice by retroviral vectors. *Science* **234**, 1409−1413.

Soriano, P. and Jaenisch, R. (1986) Retroviruses as probes for mammalian development: allocation of cells to the somatic and germ cell lineages. *Cell* **46**, 19–29.

Sorensen, J.C. (1984) The structure and expression of nuclear genes in higher plants. *Adv. Genet.* **22**, 109–145.

Southern, E. (1975) Detection of specific sequences among DNA fragments separated by gel electrophoresis. *J. Mol. Biol.* **98**, 503.

Southern, P.J. and Berg, P. (1982) Transformation of mammalian cells to antibiotic resistance with a bacterial gene under the control of the SV40 early region promoter. *J. Molec. Appl. Genet.* **1**, 327–341.

Spalholz, B.A., Yang, Y. and Howley, P.M. (1985) Transactivation of a bovine papillomavirus transcriptional regulatory element by the E2 gene product. *Cell* **42**, 183–191.

Spandidos, P.A. and Wilkie, N.M. (1984) Malignant transformation of early passage rodent cells by a single mutated human oncogene. *Nature* **310**, 469–475.

Speck, N.A. and Baltimore, D. (1987) Six distinct nuclear factors interact with the 75 base pair repeat of the moloney murine leukaemia virus enhancer. *Molec. Cell. Biol.* **7**, 1101–1110.

Spena, A., Krause, E. and Dobberstein, B. (1985) Translation efficiency of zein mRNA is reduced by hybrid formation between the 5' and 3' untranslated region. *EMBO J.* **4**, 2153–2158.

Spiker, S. (1984) Chromatin structure and gene regulation in higher plants. *Adv. Genet.* **22**, 145–208.

Spirin, A.S. and Ajtkhozhin, M.A. (1985) Informosomes and polyribosome-associated proteins in eukaryotes. *Trends Biochem.* **10**, 162–165.

Spradling, A. (1981) The organisation and amplification of two clusters of *Drosophila* chorion genes. *Cell* **27**, 193–202.

Spradling, A.C., de Cicco, D.V., Wakinoto, B.T., Levine, J.F, Kalfayan, L.J. and Cooley, L. (1987) Amplification of the X-linked *Drosophila* chorion cluster requires a region upstream from the s38 chorion gene. *EMBO J.* **6**, 1045–1053.

Spradling, A.C. and Mahowald, A.P. (1980) Amplification of genes for chorion proteins during oogenesis in *Drosophila melanogaster*. *Proc. Natl Acad. Sci. USA* **77**, 1096–1100.

Spradling, A.C. and Rubin, G.M. (1983a) Transposition of cloned P elements into *Drosophila* germ line chromosome. *Science* **218**, 341–347.

Spradling, A.C. and Rubin, G.M. (1983b) The effect of chromosomal position on the expression of the *Drosophila* Xanthine dehydrogenase gene. *Cell* **34**, 47–57.

Sreekrishna, K. and Dickson, R.C. (1986) Construction of strains of *Saccharomyces cerevisiae* that grow on lactose *Proc. Natl Acad. Sci. USA* **82**, 7909–7913.

Srienc, F., Bailey, J.E. and Campbell, J.L. (1985) Effect of ARS/1 mutations on chromosome stability in *Saccharomyces cerevisiae*. *Molec. Cell. Biol.* **5**, 1676–1684.

Srivastava, S.K., Yuasa, Y., Reynolds, S.H. and Aaronson, S.A. (1985) Effects of two major activating lesions on the structure and conformation of human *ras* oncogene products. *Proc. Natl Acad. Sci. USA* **82**, 38–42.

Stachel, S.E. and Nester, E.W. (1986) The genetic and transcriptional organisation of the *vir* region of the A6 Ti plasmid of *Agrobacterium tumefaciens*. *EMBO J.* **5**, 1445–1454.

Stachel, S.E., Timmerman, B. and Zambryski, P. (1986) Generation of single stranded T-DNA molecules during the initial stages of T-DNA transfer from *Agrobacterium tumefaciens* to plant cells. *Nature* **322**, 706–711.

Stachel, S.E. and Zambryski, P.C. (1986) *Agrobacterium tumefaciens* and the susceptible plant cell: a novel adaptation of extracellular recognition and DNA conjugation. *Cell* **47**, 155–157.

Staneloni, R.J. and Leloir, L.F. (1982) The biosynthetic pathways of the asparagine-linked oligosaccharides of glycoproteins. *Crit. Rev. Biochem.* **12**, 298–326.

Stanway, C.A., Mellor, J., Ogden, J.E., Kingsman, A.J. and Kingsman, S.M. (1987) The USA of the yeast *PGK* gene contains functionally distinct domains. *Nucl. Acids Res.* **15**, 6855–6873.

Stark, G.R. and Wahl, G.M. (1984) Gene amplification. *Ann. Rev. Biochem.* **53**, 447–491.

Steller, H. and Pirrotta, V. (1985) A transposable P vector that confers selectable G418 resistance to *Drosophila* larvae. *EMBO J.* **4**, 167–171.

Stenlund, A., Zabielsk, J., Ahola, H. Moreno-Lopez, J. and Petterson, U. (1985) The messenger RNAs from the transforming region of bovine papillomavirus type I. *J. Molec. Biol.* **182**, 541–554.

Stent, G. (1985) Thinking in one dimension: the impact of molecular biology on development. *Cell* **40**, 1–2.

Sternberg, P.W., Stern, M.J., Clark, I. and Herskowitz, I. (1987) Activation of the yeast *HO* gene by release from multiple negative controls. *Cell* **48**, 567–577.

Stevens, C.F. (1985) Molecular tinkerings that tailor the acetylcholine receptor. *Nature* **313**, 350–351.

Stewart, C.L., Schnetze, S., Vanek, M. and Wagner, E.F. (1987) Expression of retroviral vectors in transgenic mice obtained by embryo infection. *EMBO J.* **6**, 383–388.

REFERENCES

Stewart, C.L., Vanek, M. and Wagner, E.F. (1985) Expression of foreign genes from retroviral vectors in mouse teratocarcinoma cells. *EMBO J.* **4**, 3701–3709.

Stewart, G.G., Murray, C.R., Panchai, C.J., Russell, I. and Sills, A.M. (1984a) The selection and modification of brewers yeast strains. *Food Microbiol.* **1**, 289–302.

Stewart, T.A. and Mintz, B. (1981) Successive generations of mice produced from an established culture line of euploid teratocarcinoma cells. *Proc. Natl Acad. Sci. USA* **78**, 6314–6318.

Stewart, T.A., Pattengale, P.K. and Leder, P. (1984b) Spontaneous mammary adenocarcinomas in transgenic mice that carry and express MTV/*myc* fusion genes. *Cell* **38**, 627–637.

Stinchcomb, D.T., Mann, C., Selker, E. and Davis, R.W. (1981) DNA sequences that allow the replication and segregation of yeast chromosomes. In *The Initiation of DNA Replication* (eds D.S. Ray and C.F. Fox) pp. 473–484, *ICN–UCLA Symposia on Molecular and Cellular Biology, Vol. 22*, Academic Press, London.

Stinchcomb, D.T., Mello, C. and Hirsch, D. (1985a) *Caenorhabditis elegans* DNA that directs segregation in yeast cells. *Proc. Natl Acad. Sci. USA* **82**, 4167–4171.

Stinchcomb, D.T., Shaw, J.E., Carr, S.H. and Harsch, D. (1985b) Extrachromosomal DNA transformation of *Caenorhabditis elegans*. *Molec. Cell. Biol.* **5**, 3484–3496.

Stinchcomb, D.T., Struhl, K. and Davis, R.W. (1979) Isolation and characterisation of a yeast chromosomal replicator. *Nature* **282**, 39–43.

Stinchcomb, D.T., Thomas, M., Kelly, J., Selker, E. and Davis, R.W. (1980) Eukaryotic DNA segments capable of autonomous replication in yeast. *Proc. Natl Acad. Sci. USA* **77**, 4559–4563.

St John, T. and Davis, R.W. (1981) The organisation and transcription of the *GAL* operon. *J. Molec. Biol.* **152**, 285–315.

Stohl, L.L. and Lambowitz, A.M. (1983) Construction of a shuttle vector for the filamentous fungus *Neurospora crassa*. *Proc. Natl Acad. Sci. USA* **80**, 1058–1062.

Storb, N., O'Brien, R.L., McMullen, M.D., Gollahon, K.A. and Brinster, R.L. (1984) High expression of cloned immunoglobulin κ gene in transgenic mice is restricted to B lymphocytes. *Nature* **310**, 238–241.

Stout, J.T., Chen, H.Y., Brennand, J., Caskey, C.T. and Brinster, R.L. (1985) Expression of human HPRT in the central nervous system of transgenic mice. *Nature* **317**, 250–252.

Stow, N.D. and Wilkie, N.M. (1976) An improved technique for obtaining enhanced infectivity with herpes simplex virus type I DNA. *J. Gen. Virol.* **33**, 447–458.

Strauss, F. and Varshavsky, A. (1984) A protein binds to a satellite DNA repeat at three specific sites that would be brought into mutual proximity by DNA folding in the nucleosome. *Cell* **37**, 889–901.

Struhl, K., Stinchcomb, D.T., Scherer, S. and Davis, R.W. (1979) High frequency transformation of yeast: autonomous replication of hybrid DNA molecules. *Proc. Natl Acad. Sci. USA* **76**, 1035.

Struhl, K. (1983a) The new yeast genetics. *Nature* **305**, 391–396.

Struhl, K. (1983b) Direct selection for gene replacement events in yeast. *Gene* **26**, 231–242.

Struhl, K. (1985) Naturally occurring poly (dA-dT) sequences are upstream promoter elements for constitutive transcription in yeast. *Proc. Natl Acad. Sci. USA* **82**, 8419–8423.

Struhl, K. (1987) Promoters, activator proteins, and the mechanism of transcriptional inititiation in yeast. *Cell* **49**, 295–297.

Struhl, K., Stinchcomb, D.T., Scherer, S. and Davis, R.W. (1979) High frequency transformation of yeast: Autonomous replication of hybrid DNA molecules. *Proc. Natl Acad. Sci. USA* **76**, 1035–1039.

Stumph, W.E., Baez, M., Beattie, W.G., Tsai, M-J. and O'Malley, B.W. (1983) Characterisation of deoxyribonucleic acid sequences at the 5′ and 3′ borders of the 100 kilobase pair ovalbumin gene domain. *Biochemistry* **22**, 306–315.

Stuart-Harris, C.H., Schild, G.C. and Oxford, J.S. (1985) *Influenza. The Viruses and the Disease*, 2nd edn, Edward Arnold, London.

Subramani, S., Mulligan, R. and Berg, P. (1981) Expression of the mouse dihydrofolate, reductase complementary deoxyribonucleic acid in simian virus 40 vectors. *Molec. Cell. Biol.* **1**, 854–864.

Sulston, J.E., Schierenberg, E., White, J.G. and Thomson, J.N. (1983) The embryonic cell lineage of the nematode *Caenorhabditis elegans*. *Devel. Biol.* **100**, 64–119.

Sumrada, R.A. and Cooper, T.G. (1987) Ubiquitous upstream repression sequences control activation of the inducible arginase gene in yeast. *Proc. Natl Acad. Sci. USA* **84**, 3997–4001.

Surosky, R.T., Newlon, C.S. and Tye, B.K. (1986) The mitotic stability of deletion derivatives of chromosome III in yeast. *Proc. Natl Acad. Sci. USA* **83**, 414–418.

Sussman, D.J. and Milman, G. (1984) Short-term, high-efficiency expression of transfected DNA. *Molec. Cell. Biol.* **4**, 1641–1643.

Sutton, A. and Broach, J.R. (1985) Signals for transcription initiation and termination in the *Saccharomyces cerevisiae* plasmid 2μm circle. *Molec. Cell. Biol.* **5**, 2770–2780.

Sutcliffe, J.G. (1978) Complete nucleotide sequence of the *Escherichia coli* plasmid pBR322. *Cold Spring Harb. Symp. Quant. Biol.* **43**, 77–90.

Swanson, L.W., Simmons, D.M., Arriza, J., Hammer, R., Brinster, R., Rosenfeld, M.G. and Evans, R.M. (1985) Novel developmental specificity in the nervous system of transgenic animals expressing growth hormone fusion genes. *Nature*, **317**, 363–366.

Swift, G.H., Hammer, R.E., MacDonald, R.J. and Brinster, R.L. (1984) Tissue specific expression of the rat pancreatic elastase 1 gene in transgenic mice. *Cell* **38**, 639–646.

Szostak, J.W. and Blackburn, E.H. (1982) Cloning yeast telomeres on linear plasmid vectors. *Cell* **29**, 245–255.

Szybalska, E.H. and Szybalski, W. (1962) Genetics of human cell lines, IV. DNA-mediated heritable transformation of a biochemical trait. *Proc. Natl Acad. Sci. USA* **48**, 2026–2034.

Tabe, L., Strachan, R., Jackson, D., Wallis, E. and Colman, A. (1984) Segregation of mutant ovalbumins and ovalbumin-globin fusion proteins in *Xenopus* oocytes. Identification of an ovalbumin signal sequence. *J. Molec. Biol.* **180**, 645–666.

Tabin, C.J., Hoffmann, J.W., Goff, S.P. and Weinberg, R.A. (1982) Adaptation of a retrovirus as a eukaryotic vector transmitting the herpes simplex virus thymidine kinase gene. *Molec. Cell. Biol.* **2**, 426–436.

Takahashi, K., Vigneron, M., Mathes, H., Wildeman, A., Zenke, M. and Chambon, P. (1986) Requirements of stereospecific alignments for initiation from the simian virus 40 early promoter. *Nature* **319**, 121–126.

Tao, W., Wilkinson, J., Stanbridge, E.J. and Berns, M.W. (1987) Direct gene transfer into human cultured cells facilitated by laser micropuncture of the cell membrane. *Proc. Natl Acad. Sci. USA* **84**, 4180–4185.

Tartaglia, J. and Paoletti, E. (1988) Recombinant vaccinia virus vaccines. *Trends Biotech.* **6**, 43–47.

Tatchell, K., Chalett, D.T., DeFeo-Jones, D. and Scolnick, E.M. (1984) Requirement of either of a pair of *ras*-related genes of *Saccharomyces cerevisiae* for spore viability. *Nature* **309**, 523–527.

Temin, H.M. (1986) Retrovirus vectors for gene transfer: efficient integration into and expression of exogenous DNA in vertebrate cell genomes. In *Gene Transfer* (ed. R. Kucherlapati), Plenum Press, New York.

Tepfer, D. (1984) Transformation of several species of higher plants by *Agrobacterium rhizogenes*: Sexual transmission of the transformed genotype and phenotype. *Cell* **37**, 959–967.

Theisen, M., Stief, A. and Sippel, A.E. (1986) The lysozyme enhancer: cell-specific activation of the chicken lysozyme gene by a far upstream DNA element *EMBO J.* 5, 719–724.

Thireos, B., Driscoll, P.M. and Greer, H. (1984) 5′ untranslated sequences are required for the translational control of a yeast regulatory gene. *Proc. Natl Acad. Sci. USA* **81**, 5096–5100.

Thomas, G., Herbert, E. and Hruby, D.E. (1986) Expression and cell type-specific processing of human preproenkephalin with a vaccinia recombinant. *Science* **232**, 1641–1643.

Thomas, K.R. and Capecchi, M.R. (1986) Introduction of homologous DNA sequences into mammalian cells induces mutations in the cognate gene. *Nature* **324**, 34–38.

Thomas, K.R., Folger, K.R. and Capecchi, M.R. (1986) High frequency targeting of genes to specific sites in the mammalian genome. *Cell* **44**, 419–428.

Thomas, P.S. (1980) Hybridisation of denatured RNA and small DNA fragments transferred to nitrocellulose. *Proc. Natl Acad. Sci. USA* **77**, 5201.

Tilburn, J., Scazzocchio, C., Taylor, G.G., Zabicky-Zissman, J.H., Lockington R.A. and Davies, R.W. (1983) Transformation by integration in *Aspergillus nidulans*. *Gene* **26**, 205–221.

Tobin, E.M. and Silverthorne, J. (1985) Light regulation of gene expression in higher plants. *Ann. Rev. Plant Physiol.* **36**, 569–593.

Toda, T., Uno, I., Ishikawa, T., Powers, S., Kataoka, T., Broek, D., Cameron, S., Broach, J., Matsumoto, K. and Wigler, M. (1985) In yeast, *RAS* proteins are controlling elements of adenylate cyclase. *Cell* **40**, 27–36.

Tonegawa, S. (1983) Somatic generation of antibody diversity. *Nature* **302**, 575–581.

Toneguzzo, F., Hayday, A.C. and Keating, A. (1986) Electric field-mediated DNA transfer: transient and stable gene expression in human and mouse lymphoid cells. *Molec. Cell. Biol.* **6**, 703–706.

Toneguzzo, F. and Keating, S. (1986) Stable expression of selectable genes introduced into human stem cells by electric field mediated DNA transfer *Proc. Natl Acad. Sci. USA* **83**, 3496–3499.

Toole, J.J., Knopf, J.L., Wozney, J.M. Sultzman, L.A., Buecker, J.L., Pittman, D.D., Kaufman, R.J., Brown, E., Shoemaker, C., Orr, E.C., Amphlett, G.W., Foster, W.B., Coe, M.L., Knutson, G.J., Fass, D.N. and Hewick, R.M. (1984) Molecular cloning of a cDNA encoding human antihaemophilic factor. *Nature* **312**, 342–347.

Topol, J., Ruden, D.M. and Parker, C.S. (1985) Sequences required for *in vitro* transcriptional activation of a *Drosophila hsp70* gene *Cell* **42**, 527–537.

Towbin, H., Staehelin, T. and Gordon, J. (1979) Electrophoretic transfer of proteins from polyacrylamide gels to nitrocellulose sheets: procedure and some applications. *Proc. Natl Acad. Sci. USA* **76**, 4350−4354.

Townes, T.M., Chen, H.Y., Lingrel, J.B., Palmiter, R.D. and Brinster, R.L. (1985) Expression of human β-globin genes in transgenic mice: effects of a flanking metallothionein−human growth hormone fusion gene. *Molec. Cell. Biol.* **5**, 1977−1983.

Tiollais, P., Pourcel, C. and Dejean, A. (1985) The hepatitis B virus. *Nature* **317**, 489−495.

Tschumper, G. and Carbon, J. (1980) Sequence of a yeast DNA fragment containing a chromosomal replicator and the *TRP*1 gene. *Gene* **10**, 157−166.

Tso, J.Y., Van Den Berg, D.J. and Korn, L.J. (1986) Structure of the gene for *Xenopus* transcription factor TFIIIA. *Nucl. Acids Res.* **14**, 2187−2200.

Tubb, R.S. (1986) Amylolytic yeasts for commercial applications. *Trends Biotechnol.* **4**, 98−104.

Tuite, M.F., Dobson, M.J., Roberts, N.A., King, R.M., Burke, D.C., Kingsman, S.M. and Kingsman, A.J. (1982) Regulated high efficiency expression of human interferon-alpha in *Saccharomyces cerevisiae*. *EMBO J.* **1**, 603−608.

Tumer, N.E., O'Connell, K.M., Nelson, R.S., Sanders, P.R., Beachy, R.N., Fraley, R.T. and Shah, D.M. (1987) Expression of alfalfa mosaic virus coat protein gene confers cross-protection in transgenic tobacco and tomato plants. *EMBO J.* **6**, 1181−1188.

Uchimaya, H., Fushimi, T., Hashimoto, H., Harada, H., Syono, K. and Sugawara, Y. (1986) Expression of a foreign gene in callus derived from DNA-treated protoplasts of rice (*Oryza sativa*). *Molec. Gen. Genet.* **204**, 204−207.

Ullrich, A., Bell, J.R., Chen, E.Y., Herrera, R., Petruzzelli, L.M., Dull, T.J., Gray, A., Coussens, L., Liao Y.-C., Tsubokawa, M., Mason, A., Seeburg, P.H., Grunfeld, C., Rosen, O.M. and Ramachandran, J. (1985) Human insulin receptor and its relationship to the tyrosine kinase family of oncogenes. *Nature* **313**, 756−761.

Ullu, E. and Tschudi, C. (1984) *Alu* sequences are processed 7SL RNA genes. *Nature* **312**, 171−172.

Ullu, E. and Weiner, A.M. (1985) Upstream sequences modulate the internal promoter of the human 7SL RNA gene. *Nature* **318**, 371−374.

Urlaub, G. and Chasin, L.A. (1980) Isolation of chinese hamster cell mutants deficient in dihydrofolate reductase activity. *Proc. Natl Acad. Sci. USA* **77**, 4216−4220.

Vaeck, M., Reynaerts, A., Hofte, H., Jansens, S., De Beuckeleer, M., Dean, C., Zabeau, M., Van Montagu, M. and Leemans, J. (1987) Transgenic plants protected from insect attack. *Nature* **328**, 33−37.

Valenzuela, P., Coit, D. and Kuo, G. (1985a) Synthesis and assembly in yeast of hepatitis B surface antigen particles containing the polyalbumin receptor. *Bio/Technol.* **3**, 317−320.

Valenzuela, P., Coit, D., Medina-Selby, M.A., Kuo, C.H., Van Nest, G., Burke, R.L., Bull, P., Urdea, M. and Graves, P.V. (1985b) Antigen engineering in yeast: Synthesis and assembly of hybrid hepatitis B surface antigen-herpes simplex lgD particles. *Bio/Technol.* **3**, 323−325.

Valuenzuela, P., Medina, A., Rutter, W.J., Ammerer, G. and Hall, B.D. (1982) Synthesis and assembly of hepatitis B virus surface antigen particles in yeast. *Nature* **298**, 347−350.

Van Arsdell, J.N., Kwok, S., Schweickart, V.L., Ladner, M.B., Gelfard, D.H and Innis, M.A. (1987) Cloning, characterisation and expression in *Saccharomyces cerevisiae* of endo-glucanase 1 from *Trichoderma reesei*. *Bio/Technol.* **5**, 60−64.

Van Beveren, C. and Verma, I.M. (1986) Homology among oncogenes. *Curr. Topics Microbiol. Immunol.* **123**, 73−98.

Van Brunt, J. (1986) Glycoprotein remodelling: There's nothing (quite) like the real thing. *Bio/Technol.* **4**, 838−839.

Van den Broeck, G., Timko, M.P., Kausch, A.P., Cashmore, A.R., van Montagu, M. and Herrera-Estrella, L. (1985) Targeting of a foreign protein to chloroplasts by fusion to the transit peptide from the small subunit of ribulose 1, 5-bisphosphate carboxylase. *Nature* **313**, 358−363.

Van den Elzen, P., Townsend, J., Lee, K.Y. and Bedbrook, J. (1985) A chimaeric hygromycin resistance gene as a selectable marker in plant cells. *Plant Molec. Biol.* **5**, 299−302.

Van der Ploeg, L.H.T. (1987) Control of variant surface antigen switching in Trypanosomes. *Cell* **51**, 159−161.

Van der Putten, H., Botteri, F.M., Miller, A.D., Rosenfeld, M.G., Fan, H., Evans, R.M. and Verma, I.M. (1985) Efficient insertion of genes into the mouse germ line via retroviral vectors. *Proc. Natl Acad. Sci. USA* **82**, 6148−6152.

Van Doren, K., Hanahan, D. and Gluzman, Y. (1984) Infection of eukaryotic cells by helper-independent recombinant adenoviruses: early region 1 is not obligatory for integration of viral DNA. *J. Virol.* **50**, 606−614.

Van Duin, M., de Wit, J., Odyk, H., Westerveld, A., Yasui, A., Koken, M.H.M, Hoeijmakers, J.H.J. and Bootsma, D. (1986) Molecular characterisation of the human excision repair gene ERCC-1: cDNA cloning and amino acid homology with yeast DNA repair gene RADIO. *Cell* **44**, 913−923.

504   REFERENCES

Van Haute, E., Joos, H., Maes, M., Warren, G., Van Montagu, M. and Schell, J. (1983) Intergenic transfer and exchange recombination of restriction fragments cloned in pBR322: a novel strategy for the reversed genetics of the Ti plasmids of *Agrobacterium tumefaciens*. *EMBO J.* **2**, 411–418.

Van Loon, A.P.G.M., Brandli, A.W. and Schatz, G. (1986) The presequences of two imported mitochondrial proteins contain information for intracellular and intra mitochondrial sorting. *Cell* **44**, 801–802.

Van Loon, A.P.G.M. and Young, E.T. (1986) Intracellular sorting of alcohol dehydrogenase isoenzymes in yeast: a cytosolic location reflects absence of an amino terminal targeting sequence for the mitochondrian. *EMBO J.* **5**, 161.

Van Zonneveld, A.J., Veerman, H. and Pannekoek, H. (1986) Autonomous functions of structural domains on human tissue-type plasminogen activator. *Proc. Natl Acad. Sci. USA* **83**, 4670–4674.

Varmus, H.E. (1984) The molecular genetics of cellular oncogenes. *Ann. Rev. Genet.* **18**, 553–612.

Vehar, G.A., Keyt, B., Eaton, D., Rodriguez, H., O'Brien, D.P., Rotblat, F., Oppermann, H., Keck, R., Wood, W.I., Harbins, R.N., Tuddenham, E.G.D., Lawn, R.M. and Capon, D.J. (1984) Structure of human factor VIII. *Nature* **312**, 337–342.

Veit, B.E. and Fangman, W.L. (1985) Chromation organisation of the *Saccharomyces cerevisiae* 2µm plasmid depends on plasmid encoded products. *Molec. Cell. Biol.* **5**, 2190–2196.

Velten, J., Velten, L., Hain, R. and Schell, J. (1984) Isolation of a dual plant promoter fragment from the Ti plasmid of *Agrobacterium tumefaciens*. *EMBO J.* **3**, 2723–2730.

Verma, I.M. (1986a) Proto-oncogene *fos*: a multifaceted gene. *Trends Genet.* **2**, 93–96.

Verma, R.S. (1986b) Oncogenetics: A new emerging field of cancer. *Molec. Gen. Genet.* **205**, 385–389.

Villeponteau, B., Lundell, M. and Martinson, H. (1984) Torsional stress promotes the DNase I sensitivity of active genes *Cell* **39**, 469–478.

Vincent, A. and Monod, J. (1986) TFIIIA and homologous genes. The 'finger' proteins. *Nucl. Acids Res.* **14**, 4385–4391.

Vitetta, E.S. and Uhr, J.W. (1985) Immunotoxins. *Ann. Rev. Immunol.* **3**, 197–212.

Vlasuk, G.P., Bencen, G.H., Scarborough, R.M., Tsai, P-K., Whang, J.L., Maack, T., Camargo, M.J.F., Kirsher, S.W. and Abraham, J.A. (1986) Expression and secretion of biologically active human atrial natriuretic peptide in *Saccharomyces cerevisiae*. *J. Biol. Chem.* **261**, 4789–4796.

Volkert, F.C. and Broach, J.R. (1986) Site-specific recombination promotes plasmid amplification in yeast. *Cell* **46**, 541–550.

Von Heijne, G. (1985) Signal sequences. The limits of variation. *J. Molec. Biol.* **184**, 99–105.

Von Heijne, G. (1986) Mitochondrial targeting sequences may form amphiphilic helices. *EMBO J.* **5**, 1335–1342.

Wain-Hobson, S., Sonigo, P., Danos, O., Cole, S. and Alizon, M. (1985) Nucleotide sequence of AIDS virus, LAV. *Cell* **40**, 9–17.

Wagner, E.F., Covarrubias, L., Stewart, T.A. and Mintz, B. (1983) Prenatal lethalities in mice homozygous for human growth hormone gene sequences integrated into the germ line. *Cell* **35**, 647–655.

Wahl, G.M., De Saint Vincent, B.R. and De Rose, M.L. (1984) Effect of chromosomal position on amplification of transfected genes in animal cells. *Nature* **307**, 516–521.

Walbot, V. (1985) On the life strategies of plants and animals. *Trends Genet.* **6**, 166–169.

Waldeck, W., Posi, F. and Zentgraf, H. (1984) Origin of replication in episomal bovine papillomavirus type 1 DNA isolated from transformed cells. *EMBO J.* **3**, 2173–2178.

Walker, M.D., Edlund, T., Boulet, A.M. and Rutter, W.J. (1983) Cell-specific expression controlled by the 5' flanking region of insulin and chymotrypsin genes. *Nature* **306**, 557–561.

Wall, R. and Kuehl, M. (1983) Biosynthesis and regulation of immunoglobulins. *Ann. Rev. Immunol.* **1**, 393–422.

Wallich, R., Bulbuc, N., Hämmerling, G.J., Katzav, S., Segal, S. and Feldman, M. (1985) Abrogation of metastatic properties of tumour cells by *de novo* expression of H-2K antigens following H-2 gene transfection. *Nature* **315**, 301–305.

Walmsley, R.W., Chan, C.S.M., Tye, B-K. and Petes, T.D. (1984) Unusual DNA sequences associated with the ends of yeast chromosomes. *Nature* **310**, 157–160.

Walter, P. and Blobel, G. (1982) Signal recognition particle contains a 7S RNA essential for protein translocation across the endoplasmic reticulum. *Nature* **299**, 691–698.

Walter, P., Gilmore, R. and Blobel, G. (1984) Protein translocation across the endoplasmic reticulum. *Cell* **38**, 5–8.

Walter, P. and Lingappa, V.R. (1986) Mechanism of protein translocation across the endoplasmic reticulum membrane. *Ann. Rev. Cell. Biol.* **2**, 499–516.

Walters, L. (1986) The ethics of human gene therapy. *Nature* **320**, 225–227.

Wang, J. (1985) DNA topoisomerases. *Ann. Rev. Biochem.* **54**, 665–699.

Wang, K., Herrera-Estrella, L., Van Montagu, M. and Zambryski, P. (1984) Right 25 bp terminus sequence of the nopaline T-DNA is essential for and determines direction of DNA transfer from *Agrobacterium* to the plant genome. *Cell* **38**, 455−462.

Wang, X.-F. and Calame, K. (1986) SV40 enhancer-binding factors are required at the establishment but not the maintenance step of enhancer-dependent transcription activation. *Cell* **47**, 241−247.

Wareing, P.F. (1984) Patterns of growth and differentiation in plants. In *Developmental Control in Animals and Plants*, 2nd edn (eds C.F. Graham & P.F. Wareing), pp. 33−51. Blackwell Scientific Publications, Oxford.

Warren, T.G. and Shields, D. (1984) Expression of preprosomatostatin in heterologous cells: Biosynthesis, post-translational processing and secretion of mature somatostatin. *Cell* **39**, 547−555.

Wasylyk, B. (1986) Protein coding genes of higher eukaryotes: promoter elements and *trans* acting factors. In *Maximising Gene Expression* (eds Reznikoff, W. and Gold, L.). Butterworth, London.

Waters, G.M. and Blobel, G. (1986) Secretory protein translocation in a yeast cell-free system can occur post translationally and requires ATP hydrolysis. *J. Cell. Biol.* **102**, 1543−1550.

Watson, M.E.E. (1984) Compilation of published signal sequences. *Nucl. Acids Res.* **12**, 5144−5164.

Weatherall, D.J. (1982) *The New Genetics and Clinical Practice*, 2nd edn. Oxford University Press.

Weaver, R.F. and Weissmann, C. (1979) Mapping of RNA by a modification of the Berk-Sharp procedure: the 5′ termini of 15S β-globin mRNA precursor and mature 10S β-globin mRNA have identical map coordinates. *Nucl. Acids Res.* **7**, 1175−1193.

Weber, F., de Villiers, J. and Schaffner, W. (1983) An SV40 'enhancer trap' incorporates exogenous sequences or generates enhancers from its own sequences. *Cell* **36**, 983−992.

Weinberg, R.A. (1985) The action of oncogenes in the cytoplasm and nucleus. *Science* **230**, 770−776.

Weinberger, J., Baltimore, D. and Sharp, P.A. (1986) Distinct factors bind to apparently homologous sequences in the immunoglobulin heavy-chain enhancer. *Nature* **322**, 846−848.

Weiner, A.M., Deininger, P.U. and Efstradiatis, A. (1986) Nonviral retroposons: Genes, pseudogenes and transposable elements generated by the reverse flow of genetic information. *Ann. Rev. Biochem.* **55**, 631−663.

Weiher, H., Konig, M. and Gruss, P. (1983) Multiple point mutations affecting the simian virus 40 enhancer. *Science* **219**, 626−631.

Weintraub, H. (1984) Histone-H1-dependent chromatin superstructures and the suppression of gene activity. *Cell* **38**, 17.

Weintraub, H. (1985) Assembly and propagation of repressed and derepressed chromosomal states. *Cell* **42**, 705−711.

Weintraub, H., Cheng, P.F. and Conrad, K. (1986) Expression of transfected DNA depends on DNA topology. *Cell* **46**, 115−122.

Weintraub, H., Izant, J.G. and Harland, R.M. (1985) Antisense RNA as a molecular tool for genetic analysis. *Trends Genet.* **1**, 22−25.

Weis, J.H., Seidman, J.G., Haisman, D.E. and Nelson, D.L. (1986) Eukaryotic chromosome transfer: production of a murine specific cosmid library from a neo^r-linked fragment of murine chromosome 17. *Molec. Cell. Biol.* **6**, 441−451.

Weiss, R., Teich, N., Varmus, H. and Coffin, J. (1985) *RNA Tumour Viruses*, 2nd edn, Cold Spring Harbor Laboratory, New York.

Weissbach, A. and Weissbach, H. (1986) Plant molecular biology. *Methods Enzymology*, Vol. *118*, Academic Press, London.

West, R.W., Chen, S., Putz, H., Butler, G. and Bannerjee, M. (1987) *GAL1-GAL10* divergent promoter regions of *Saccharomyces cerevisiae* contains negative control elements in addition to functionally separate and possibly overlapping upstream activating sequences. *Genes Devel.* **1**, 1118−1131.

West, R.W., Yocum, R.R. and Ptashne, M. (1984) *Saccharomyces cerevisiae GALI−GALIO* divergent promoter region: location and function of the upstream activating sequence UAS$_G$. *Molec. Cell. Biol.* **4**, 2467−2478.

White, R., Leppert, M., O'Connell, P., Nakamura, Y., Julier, C., Woodward, S., Silva, A., Wolff, R., Lathrop, M. and Lalouel, J.M. (1986) Construction of human genetic linkage maps 1. Progress and prospects. *Cold Spring Harbor Symp. Quant. Biol.*, Vol. 51, Cold Spring Harbor, New York.

Wickens, M. and Stephenson, P. (1984) Role of the conserved AAUAAA sequence: four AAUAAA point mutants prevent messenger RNA 3' end formation. *Science* **226**, 1045−1051.

Wilson, J.H., Berget, P.B. and Pipas, J. (1982) Somatic cells efficiently join unrelated DNA segments end-to-end. *Molec. Cell. Biol.* **2**, 1258−1269.

Wickner, W.T. and Lodish, H.F. (1985) Multiple mechanisms of protein insertion into and across membranes. *Science* **230**, 400−407.

Widom, J. (1984) DNA binding and kinking. *Nature* **309**, 312–313.

Wiederrecht, G., Shuey, D.J., Kibbe, W.A. and Parker, C.S. (1987) The *Saccharomyces* and *Drosophila* heat shock transcription factors are identical in size and DNA binding properties. *Cell* **48**, 507–515.

Wigler, M., Pellicer, A., Silverstein, S., Axel, R., Urlaub, G. and Chasen, L.D. (1979b) DNA-mediated transfer of the adenine phosphoribosyl transferase locus into mammalian cells. *Proc. Natl Acad. Sci. USA* **76**, 1373–1376.

Wigler, M., Perucho, M., Kurtz, D., Dana, S., Pellicer, A., Axel, R. and Silverstein, S. (1980) Transformation of mammalian cells with an amplifiable dominant acting gene. *Proc. Natl Acad. Sci. USA* **77**, 3567–3570.

Wigler, M., Sweet, R., Sim, G-K., Wold, B., Pellicer, A., Lacy, E., Maniatis, T., Silverstein, S. and Axel, R. (1979a) Transformation of mammalian cells with genes from prokaryotes and eukaryotes. *Cell* **16**, 777–785.

Wildeman, A., Zenke, M., Schatz, C., Wintzerish, M., Grundstrom, T., Takahashi, K. and Chambon, P. (1986) Specific protein binding to the simian virus 40 enhancer *in vitro*. *Molec. Cell. Biol.* **6**, 2098–2105.

Wiley, D.C. and Skehel, J. (1987) Structure and function of the haemagglutinin membrane glycoprotein of influenza virus. *Ann. Rev. Biochem.* **56**, 365–395.

Wiley, D.C., Wilson, I.A. and Skehel, J.J. (1981) Structural identification of the antibody-binding sites of Hong Kong influenza haemagglutinen and their involvement in antigenic variation. *Nature* **289**, 373–378.

Wiley, R.L., Rutledge, R.A., Dias, S., Folks, T., Theodore, T., Buckler, C.E. and Martin, M.A. (1986) Identification of conserved and divergent domains within the envelope region of the Aquired Immunodeficiency Syndrome virus. *Proc. Natl Acad. Sci. USA* **83**, 5038–5042.

Willecke, K., Klomfass, M., Mierau, R. and Döhmer, J. (1979) Interspecies transfer via total cellular DNA of the gene for hypoxanthine phosphoribosyltransferase into cultured mouse cells. *Molec. Gen. Genet.* **170**, 179–185.

Williams, D.A., Lemischka, I.R., Nathan, D.G. and Mulligan, R.C. (1984) Introduction of new genetic material into pluripotent haematopoietic stem cells of the mouse. *Nature* **310**, 476–480.

Williams, G. (1988) Novel antibody reagents: production and potential. *Trends Biotech.* **6**, 36–42.

Williamson, D.H. (1985) The yeast ARS element, six years on: A progress report. *Yeast* **1**, 1–15.

Willis, R.C., Jolly, D.J., Miller, A.D., Plent, M.M., Esty, A.C., Anderson, P.J., Chang, H.C., Jones, O.W., Seegmiller, J.E. and Friedman, T. (1984) Partial phenotypic correction of human Lesch–Nyhan (hypoxanthine–guanine phosphoribosyltransferase-deficient) lymphoblasts with a transmissible retroviral vector. *J. Biol. Chem.* **259**, 7842–7849.

Willmann, T. and Beato, M. (1986) Steroid-free glucocorticoid receptor binds specifically to mouse mammary tumour virus DNA. *Nature* **324**, 688–691.

Willmitzer, L., Dhaese, P., Schreier, P.H., Schlmalenbach, W., Van Montagu, M. and Schell, J. (1983) Size, location and polarity of transferred DNA encoded transcripts in nopaline crown gall tomours: common transcripts in octopine and nopaline tumours. *Cell* **32**, 1045–1056.

Wilson, W., Malim, M.H., Mellor, J., Kingsman, A.J. and Kingsman, S.M. (1986) Expression strategies of the yeast transposon Ty: a short sequence directs ribosomal frameshifting. *Nucl. Acids Res.* **14**, 7001–7015.

Wilson, I.A., Niman, H.L., Houghten, R.A., Cherenson, A.R., Connolly, M.L. and Lerner, R.A. (1984) The structure of an antigenic determinant in a protein. *Cell* **37**, 767–778.

Wilson, K.L. and Herskowitz, I. (1986) Sequences upstream of the *STE6* gene required for its expression and regulation by the mating type locus in *Saccharomyces cerevisiae*. *Proc. Natl Acad. Sci. USA* **83**, 2536–2540.

Winston, F., Chaleff, D.T., Valent, B. and Fink, G.R. (1984) Mutations affecting Ty-mediated expression of the *HIS4* gene of *Saccharomyces cerevisiae*. *Genetics* **107**, 179–197.

Winter, J.A., Wright, R.L. and Gurley, W.B. (1984) Map locations of five transcripts homologous to TR-DNA in tobacco and sunflower crown gall tumours. *Nucl. Acids Res.* **12**, 2391–2406.

Wolffe, A.P., Jordan, E. and Brown, D.D. (1986) A bacteriophage RNA polymerase transcribes through a *Xenopus* 5S RNA transcription complex without disrupting it. *Cell* **44**, 381–389.

Wood, C.R., Boss, M.A., Kenton, J.M., Calvert, J.E., Roberts, N.A. and Emtage, J.S. (1985) The synthesis and *in vivo* assembly of functional antibodies in yeast. *Nature* **314**, 446–449.

Woychik, R.P., Lyons, R.H., Post, L. and Rottman, F.M. (1984) Requirement for the 3' flanking region of the bovine growth hormone gene for accurate polyadenylation. *Proc. Natl Acad. Sci. USA* **81**, 3944–3948.

Woychick, R.P., Stewart, T.A., Davis, L.G., D'Eustachio, P. and Leder, P. (1985) An inherited limb deformity created by insertional mutagenesis in a transgenic mouse. *Nature* **318**, 36–40.

Wozney, J., Hanahan, D., Tate, V., Boedtker, H. and Doty, P. (1981) Structure of the pro α2(I) collagen gene. *Nature* **294**, 129–135.

Wright, C.M., Felber, B.K., Paskalis, H. and Pavlakis, G.N. (1986) Expression and characterisation of the *trans*-activator of HTLV-III/LAV virus. *Science* **234**, 988–993.

Wright, C.F. and Zitomer, R.S. (1985) Point mutations implicate repeated sequences as essential elements of the *CYC7* negative upstream site. *Molec. Cell. Biol.* **5**, 2951–2958.

Wright, S. de Boer, E., Grosveld, F.G. and Flavell, R.A. (1983) Regulated expression of the human β-globin gene family in murine erythroleukaemia cells. *Nature* **305**, 333–336.

Wright, S., Rosenthal, A., Flavell, R. and Grosveld, F. (1984) DNA sequences required for regulated expression of β-globin genes in murine erythroleukaemia cells. *Cell* **38**, 265–273.

Wu, B.J., Kingston, R.E. and Morimoto, R.I. (1986b) The human HSP70 promoter contains at least two regulatory domains. *Proc. Natl Acad. Sci. USA* **82**, 6070–6074.

Wu, C. (1980) The 5' ends of *Drosophila* heat shock genes in chromatin are hypersensitive to DNaseI. *Nature* **286**, 854–860.

Wu, H. and Crothers, D.M. (1984) The locus of sequence-directed and protein induced DNA bending. *Nature* **308**, 509–513.

Wu, L-C.C., Fisher, P. and Broach, J.R. (1986a) The REP1 protein of 2 micron circle is associated with the nuclear matrix. In *Yeast Cell Biology* (eds J. Hicks and C.F. Fox), A.R. Liss, New York.

Yamaizumi, M., Horwich, A.L. and Ruddle, F.H. (1983) Expression and stabilisation of microinjected plasmids containing the herpes simplex virus thymidine kinase gene and polyoma virus DNA in mouse cells. *Molec. Cell. Biol.* **3**, 511–522.

Yamamoto, K. (1985) Steroid receptor regulated transcription of specific genes and gene networks. *Ann. Rev. Genet.* **19**, 209–252.

Yancopoulos, G.D. and Alt, F.W. (1986) Regulation of the assembly and expression of variable region genes. *Ann. Rev. Immunol.* **4**, 339–369.

Yancopoulos, G.D., Blackwell, T.K., Suh, H., Hood, L. and Alt, F.W. (1986) Introduced T cell receptor variable region gene segments recombine in pre B cells. Evidence that B and T cells use a common recombinase. *Cell* **44**, 251–259.

Yancopoulos, G.D., Desiderio, S.V., Paskind, M., Kearney, J.F., Baltimore, D. and Alt, F.W. (1984) Preferential ultization of the most J$_H$-proximal V$_H$ gene segments in pre-B-cell lines. *Nature* **311**, 727–733.

Yancopoulos, G.D., Nisen, P.D., Tesfaye, A., Kohl, N.E., Goldfarb, M.P. and Alt, F.W. (1985) N-*myc* can cooperate with *ras* to transform normal cells in culture. *Proc. Natl Acad. Sci. USA* **82**, 5455–5459.

Yang, J.-Q., Remmers, E.F. and Marcu, K.B. (1986) The first exon of the *c-myc* proto-oncogene contains a novel positive control element. *EMBO J.* **5**, 3553–3562.

Yang, Y.C., Okayama, H. and Howley, P.H. (1985) Bovine papillomavirus contains multiple transforming genes. *Proc. Natl Acad. Sci. USA* **82**, 1030–1034.

Yang, Y-C., Spalholz, B.A., Rabson, M.S. and Howley, P.M. (1985) Dissociation of transforming and *trans* activation functions for bovine papillomavirus type 1. *Nature* **318**, 575–577.

Yanofsky, C. (1983) Prokaryotic mechanisms in eukaryotes. *Nature* **302**, 751.

Yanofsky, M.F., Porter, S.G., Young, C., Albright, L.M., Gordon, M.P. and Nester, E.W. (1986) The *vir* D operon of *Agrobacterium tumefaciens* encodes a site specific endonuclease. *Cell* **47**, 471–477.

Yates, J.L., Warren, N. and Sugden, B. (1985) Stable replication of plasmids derived from Epstein–Barr virus in various mammalian cells. *Nature* **313**, 812–815.

Yelton, M.M., Hamer, J.E., and Timberlake, W.E. (1984) Transformation of *Aspergillus nidulans* by using a *trpC* plasmid. *Proc. Natl Acad. Sci. USA* **80**, 7576–7580.

Yeh, E., Carbon, J. and Bloom, K.S. (1986) A tightly centromere-linked gene (*SPO15*) is essential for meiosis in the yeast *Saccharomyces cerevisiae*. *Molec. Cell. Biol.* **6**, 158–167.

Yunis, J.J. (1983) The chromosomal basis of human neoplasia. *Science* **221**, 227–236.

Zaller, D.M. and Eckhardt, L.A. (1985) Deletion of a B-cell-specific enhancer affects transfected but not endogenous immunoglobulin heavy-chain gene expression. *Proc. Natl Acad. Sci. USA* **82**, 5088–5092.

Zambriyski, P., Joos, H., Genetello, C., Leemans, J., Van Montagu, M. and Schell, J. (1983) Ti plasmid vector for the introduction of DNA into plant cells without alteration of their normal regeneration capacity. *EMBO J.* **2**, 2143–2150.

Zaret, K.S. and Sherman, F. (1982) DNA sequence required for efficient transcription termination in yeast. *Cell* **28**, 563–573.

Zaret, K. and Sherman, F. (1984) Mutationally altered 3' ends of yeast *CYC1* mRNA affect transcript stability and translational efficiency. *J. Molec. Biol.* **177**, 107–136.

Zehring, W.A., Wheeler, D.A., Reddy, P., Konopka, R.J., Kyriacou, C.P., Rosbash, M. and Hall, J.C. (1984) P-element transformation with *period* locus DNA restores rhythmicity to mutant arhythmic *Drosophila melanogaster*. *Cell* **39**, 369–376.

Zenke, M., Grundstrom, T., Matthes, H., Wintzerith, M., Schatz, C., Wildeman, A. and Chambon, P. (1986) Multiple sequence motifs are involved in SV40 enhancer function. *EMBO J.* **5**, 387–397.

Zheng, X-M., Moncollin, V., Egly, J-M. and Chambon, P. (1987) A general transcription factor forms a stable complex with RNA polymerase B (II). *Cell* **50**, 361–368.

Zieve, G.W. (1981) Two groups of small stable RNAs. *Cell* **25**, 296–297.

Ziff, E.B. (1980) Transcription and RNA processing by the DNA tumour viruses. *Nature* **297**, 491–499.

Zimmerman, S.B. (1982) The three dimensional structure of DNA. *Ann. Rev. Biochem.* **51**, 395–427.

Zimmermann, U. and Vienken, J. (1983) Electric field induced cell to cell fusion. *J. Membr. Biol.* **67**, 165–182.

Zinn, K., Di Maio, D. and Maniatis, T. (1983) Identification of two distinct regulatory regions adjacent to the human β-interferon gene. *Cell* **34**, 865–879.

Zinn, K. and Maniatis, T. (1986) Detection of factors that interact with the human β-interferon regulatory region *in vivo* by DNAase I footprinting. *Cell* **45**, 611–618.

Index

513   INDEX

516   INDEX